Integrative Human Biochemistry

Andrea T. Da Poian • Miguel A. R. B. Castanho

# Integrative Human Biochemistry

## A Textbook for Medical Biochemistry

Second Edition

 Springer

Andrea T. Da Poian
Instituto de Bioquímica Médica
Leopoldo de Meis
Federal University of Rio de Janeiro
Rio de Janeiro, Rio de Janeiro, Brazil

Miguel A. R. B. Castanho
Institute of Biochemistry and Institute of
Molecular Medicine, School of Medicine
University of Lisbon
Lisbon, Portugal

ISBN 978-3-030-48742-3      ISBN 978-3-030-48740-9    (eBook)
https://doi.org/10.1007/978-3-030-48740-9

This Springer imprint is published by the registered company Springer Nature Switzerland AG
The registered company address is: Gewerbestrasse 11, 6330 Cham, Switzerland

*This book is a tribute to the legacy of Leopoldo de Meis for his inspiration to younger generations.*

*Thanks, Leopoldo*

*The second edition is dedicated to Franklin Rumjanek, a biochemist with clear views out-of-the-box, where life itself challenges the scientific mind*

*Thanks, Franklin Rumjanek*

# Foreword to the First Edition

## Leopoldo De Meis' Legacy: A Biochemistry Textbook with a Difference

This is a comprehensive and concise basic biochemistry textbook for health science students. This readership is often overwhelmed by conventional textbooks, which cover many topics in great depth. Indeed, although this information is necessary for those aiming to become biochemists, it is excessively detailed for the interests of future nurses, physicians, and dentists. The authors—experienced teachers and researchers aware of the needs of health science students—have devised a book specifically for this community.

To this end, the book starts off with a description of the molecules of life and rapidly moves on to cover metabolism and related fields, such as the control of body weight. The book is therefore devoted to human metabolism. Given that its audience is health science students, only those topics considered of relevance for humans are presented. One of the hallmarks of current developments in the life sciences is the merge of classical disciplines. Consequently, the book encompasses pure biochemical information in the framework of related fields such as physiology, histology, and pharmacology. The final chapters on the regulation of metabolism during physical activity and the control of body weight clearly reflect this multidisciplinary perspective.

The presentation of metabolism is organized around the concept of the generation and management of energy. Unlike most textbooks, here the synthesis of ATP is described first in a very detailed way, after which the metabolic pathways that feed ATP synthesis are addressed. This logical approach to presenting material was advocated by Leopoldo de Meis, one of the greatest biochemistry teachers and educators of our time. In this regard, this book is a tribute to Leopoldo.

The structural aspects of macromolecules are consistently shown in the figures, and the fundamental notion that reactions are the result of molecular interactions is reiterated throughout the book. Given that in most university degrees Molecular Biology and Genetics are now taught in separate courses, the reader is provided with a description of nucleic acids, faithfully referred to as "polymers of saccharide conjugates," in the chapter dealing with the families of biological molecules. However, the reader will not find information on DNA and RNA typical of conventional textbooks.

Another interesting feature of the book is the use of "boxes," which develop singular concepts in a more informal manner. This presentation technique is highly illustrative and reader friendly. Furthermore, key experiments that have opened up new concepts are explained, thus helping students to appreciate that scientific knowledge derives from the work of researchers, some of which are depicted in caricatures. Finally, each chapter includes a set of up-to-date and well-chosen references, which will help those students wishing to delve further into specific fields.

In summary, this textbook provides a modern and integrative perspective of human biochemistry and will be a faithful companion to health science students following curricula in which this discipline is addressed. Similarly, this textbook will be a most useful tool for the teaching community.

Institute for Research in Biomedicine                                   Joan Guinovart
Barcelona, Spain
International Union of Biochemistry and Molecular Biology, IUBMB
Barcelona, Spain

# Foreword to the Second Edition

In the Preface to the First Edition of this book, the authors state "Our goal in this endeavor is not writing just another piece of literature in biochemistry. We aim at a different textbook." In that goal they succeeded impressively. The second edition of their book, though, takes this goal to another level! The main text of the second edition is largely the same as the first, except that the authors have clarified some instances that needed more attention. The authors cover the vast scope of biochemistry and illustrate important themes and principles by presenting examples, whenever possible, from human biochemistry—hence the title of the book! This edition, like the first, provides outstanding background into the biochemical processes that produce or consume energy. It is especially strong in its coverage of intermediary metabolism, which is undergoing a renaissance of interest due to the increasing prevalence of obesity and diabetes and to the emerging connections between metabolism and cancer biology. The authors are particularly adept at integrating basic and clinical aspects of metabolism, forging clear links between pathological conditions and the specific defects in biochemical pathways that underlie these conditions. In this way, they bring new life into an area that is often considered boring by medical and graduate students. Many examples of this approach to teaching underlying chemistry and physics of biological processes, while focusing on humans, arise in the case studies.

Both of us (DMJ and JPA) have strong interests in history and the challenging case studies presented in the second edition are very compelling from an historical perspective. They are rich with information related to the history of medicine as well as familiar newsworthy events. Some of the cases are literally ripped from the headlines! As such they are compelling both from a scientific viewpoint and for the perspective they provide on the progress of biochemistry and our great predecessors in science. These challenging case studies help students understand complex biochemical material as it applies to clinical situations and provide an ideal (and entertaining) mechanism for self-assessment. In our opinion, the first edition of this book was an excellent introduction to biochemistry which we felt was especially appropriate to medical students or students planning to go to medical school. But, the addition of the 30 case studies to the second edition has lifted this book into the level of pedagogical preeminence such that, in our opinion, it should be mandatory reading for all medical students.

In the Foreword to the First Edition, Joan Guinovart wrote: "In summary, this textbook provides a modern and integrative perspective of human biochemistry and will be a faithful companion to health science students...." We cannot improve upon these words. We can only add that the second edition, with its numerous and compelling case studies, will be more than a "faithful companion to health science students," and it will illuminate and inspire their paths forward to excellence in their chosen careers in a time that desperately needs more and better health scientists. This book will be a treasured resource for medical students throughout their careers, from the classroom to clinical rotations and beyond. It will also be of great value to basic scientists, particularly those with an interest in pursuing translational research. We sincerely congratulate the authors on the truly excellent treatise they have created, and we wish all of the readers an exciting and worthwhile intellectual journey.

Honolulu, HI, USA                                                           David M. Jameson
Dallas, TX, USA                                                            Joseph P. Albanesi

# Preface to the First Edition

Traditional lecture classes in biological sciences are being challenged by modern forms of communication. Modern communication tends to be more visual and less interpretative in nature. In lectures, the didactics are changing vastly and rapidly; the deductive power of mathematics is complemented by the intuitive clarity of movie simulations, even if the first is fully embedded in the scientific method and the latter are mere artistic configurations of a faintly perceived reality. It is a general trend in modern societies that the most effective communication is more condensed and focused, contextualizes the information, and is disseminated across multiple media. Textbooks do not escape this reality. A modern scientific textbook to be effective should be a means of communication that needs to address specific issues of interest, place these issues in a broader interdisciplinary context, and make use of modern visualization tools that represent reality within the state of the art available in scientific research.

We have shaped this book based on many years of biochemistry teaching and researching. We hope to stimulate other teachers to actively rethink biochemical education in health sciences and "contaminate" students with the passion for biochemical knowledge as an essential part of the indefinable but fascinating trick of nature we call life. "We're trying for something that's already found us," Jim Morrison would say.

## Presentation of Book Structure

Our goal in this endeavor is not writing just another piece of literature in biochemistry. We aim at a different textbook. Biochemistry is defined as the study of the molecular processes occurring in living organisms, which means that it comprises the network of chemical and physical transformations that allow life to exist.

However, this intrinsic integrative nature of biochemistry may be lost if it is taught as lists of molecules' types and metabolic pathways. In this book, we intend to introduce the biochemistry world in an actual integrative way. For this, our option was to focus on human biochemistry, presenting the molecular mechanisms of cellular processes in the context of human physiological situations, such as fasting, feeding, and physical exercise. We believe that this will provide to the reader not only information but knowledge (as very well represented in the cartoon from Hugh MacLeod's gapingvoid).

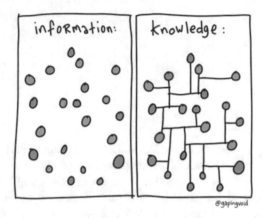

Information vs. knowledge cartoon. (Reproduced with permission from Hugh MacLeod's gapingvoid, gapingvoid.com)

The reader will find innovative approaches and deviations relative to the usual contents of classical textbooks. Part I deals with the importance of molecular-scale knowledge to reason about life, health, and disease (Chap. 1); the basic chemistry and physics of living systems (Chap. 2); and the systematization of biomolecules in chemical families, privileging molecular structure and dynamics instead of dealing with molecules as shapeless names (Chap. 3). Basic drug discovery concepts are presented to reinforce the importance of integrative biochemical reasoning. Drug discovery is a very important part of modern Medicinal Chemistry bridging Biochemistry to Pharmacology and Biotechnology. Part I prepares the student for Part II, which is devoted to metabolisms. Part II starts with the fundamentals of regulation of series of reactions in which kinetic considerations are endowed with mathematical accuracy (Chap. 4), and, by extension, the key concepts in the regulation of metabolism (Chap. 5). To introduce energy metabolism, we first explore the mechanisms of ATP synthesis (Chap. 6) to create in the reader a need to know from where cellular energy comes from. The catabolism of major biomolecules follows naturally (Chap. 7). Metabolic responses to hyperglycemia (Chap. 8), hypoglycemia (Chap. 9), and physical activity (Chap. 10) are used to introduce and contextualize several metabolic pathways and to illustrate the integrative interplay between different processes in different tissues. Finally, control of body weight and the modern metabolic diseases are explored (Chap. 11), placing biochemistry in a human health perspective, prone to be explored in later stages of health sciences students' training, when pathologies and clinical problems are addressed.

The option for the integrative view implied that sometimes complex topics have been reduced to their essence. This is the case of cholesterol synthesis, which is addressed but not described in detail, and the pentose–phosphate pathway, which is presented in the context of fatty acid synthesis, although its other functions are summarized in a box. For the synthesis of purines and pyrimidines, the reader is referred to specialized literature. Vitamins are a heterogeneous group of molecules not directly related to their structure or reactivity; vitamins seen as a family of molecules are an anachronism and are not the theme of any section of the book. Also, the reader will not find in this book matters that are typically taught in Molecular Biology programs such as the replication, transcription, and translation of informative molecules.

It is also important to mention that biochemical nomenclature is a permanent challenge for the teacher and the student. The rich history and multidisciplinary nature of biochemistry have determined that nomenclature is not always clear or coherent. Coexistence of common and systematic names is frequent, and different names have been consecrated by the use of different communities of biochemists.

The most prominent example is the case of saccharides/sugars/carbohydrates. While all designations are common, carbohydrate is probably the one preferred by most professionals in different disciplines. Yet, this name relates to a profound chemical equivocation of "carbon hydrate": Many molecules of this family have a hydrogen:oxygen atom ratio of 2:1 as in water, which makes the empirical formula $C_m(H_2O)_n$. The illusion of a hydrate is obvious but has no chemical sense. Respecting the chemical accuracy, we preferred the name saccharide in Part I, in which the chemical nature of biomolecules was presented and discussed, and reserved the name "carbohydrate" to discuss metabolic processes and dietary implications, for instance. The use of different names for different contexts and different implications is intrinsic to biochemistry.

Because biochemistry is made of biochemists and good ideas in addition to molecules, key historical experiments are used as case studies to ignite discussion and facilitate learning. Key historical experiments are excellent for classroom use, steering dynamic discussions between teachers and students. This is the perfect environment for teaching, learning, and showing that biochemistry is not only useful in shaping the future of humanity but is also fascinating and appealing.

Urucureá, Brazil                                                                                        Andrea T. Da Poian
                                                                                                             Miguel A. R. B. Castanho

# Preface to the Second Edition

Teaching needs confident biochemists, claims an article that once caught our attention [1]. Learning needs engaged and enthusiastic students, we add. When challenged to prepare a second edition of *Integrative Human Biochemistry*, we were driven by the ambition of building confidence in teaching and enthusiasm in learning. Biochemistry is not a dull bunch of molecular names connected by reaction arrows so it should not be taught—or learned, which is worse!—as such. Biochemistry is a thrilling science as lively as life itself. So, in addition to the updated contents of the first edition, we propose readers—students, teachers, or any others interested in the molecular world—a journey through 30 challenging case studies based on movies, novels, documentaries, paintings, and other creations far beyond canonic academic exercises. From the eyes of reindeers to criminal minds, from surrealistic paintings to unrealistic science, from extravagant celebrities to athletes pushing the limits of human physiology, from antique scientific book relics to modern TV shows, from sewage to gastronomy, biochemistry is everywhere teasing those daring to look out of the box. The challenging cases we selected are an invitation to ignite minds and turn classrooms into discussion arenas, where teaching and learning arise from an interactive and integrative view to the world we live in.

Each chapter of the second edition ends with 2–3 cases that have a set of questions associated. Insights into the answers are also provided to help steer teaching/learning interactions key to productive lectures, PBL (problem-based learning) or traditional tutorials, or e-learning approaches. Hopefully, the reader will find in this book the inspiration to find and develop their own cases and enrich the experience further than the limits of the book. Biochemistry provides endless opportunities.

The power of creativity in the construction of scientific knowledge, artistic outcomes, and learning process is compelling but the unification of science, art, and pedagogy has been overlooked. "Science and art are oftentimes presented as two distinct fields with very little overlap. But the anatomist and the artist, scientist and storyteller, have far more in common than meets the eye. Upon further investigation, it is clear that science and art cannot exist without the other; they are co-constitutive. Scientists must think creatively in order to form theories (the Earth is round, it circles the Sun) that often appear outlandish not only to laymen but even to their own colleagues. The theory of evolution, presented by Charles Darwin in 'On The Origin of Species', required creativity that was then supported by the evidence-revealed empirical analysis" [2]. In the second edition of the book we bring biochemistry, art, communication, entertainment, culture, and teaching/learning into a common path toward the reader.

Except for the 30 challenging cases now presented, the remaining of the book is nearly unchanged compared to first edition. Some details were corrected or improved for readability. Some specific contents were updated.

We were touched by the contribution and feedback we had from our readers, mainly students. Thank you.

Rio de Janeiro, Brazil                                               Andrea T. Da Poian
Lisbon, Portugal                                               Miguel A. R. B. Castanho

# References

1. Thomson J (2014) Teaching needs confident biochemists. Biochem (Lond) 36:28–30
2. Reagan N, Amiel J (2013) Chapter 1: "Creation": when art and science collide. In: Hollywood chemistry. ACS symposium series, vol 1139, pp 3–16, ISBN13: 9780841228245

# Acknowledgments

The authors acknowledge Rita Aroeira and Emília Alves (ULisboa, Portugal) for administrative support and Cláudio Soares (ITQBUNL, Portugal) for his critical contributions to some of the pictured molecular structures.

The authors thank Ana Coutinho, Ana Salomé Veiga, Antônio Galina, Cláudio Soares, and José Roberto Meyer-Fernandes for their critical reading of the manuscript and helpful suggestions.

# Contents

**Part II   The Interplay and Regulation of Metabolism**

# Part I
# The Molecules of Life

# Chapter 1
# Introduction: Life Is Made of Molecules!

Studying molecules is the key to understand life. A commonly accepted definition of life, known as the NASA (North American Space Agency) hypothesis, states that "Life is a self-sustained chemical system capable of undergoing Darwinian evolution" (Fig. 1.1). The link between molecules and life may be hard to explain but it is simple to illustrate.

In this introduction, we have selected three examples that are enough to show that knowledge on molecules is essential to reason about life itself, health, and disease:

1.  Searching for the origin of life is a chemical "adventure" through the molecules of primitive Earth and their reactivity.
2.  Viruses are amazing molecular machines, too simple to be considered living beings for most researchers, but with a tremendous ability to interfere with the course of life, sometimes tragically.
3.  The world of drug discovery and development consists of molecules being designed, synthesized, and interacting with other molecules in silico, in vitro, and in vivo with the end goal of interfering with vital physiologic processes.

It is all about molecules. It is all about life.

© Springer Nature Switzerland AG 2021                                                                 3
A. T. Da Poian, M. A. R. B., *Integrative Human Biochemistry*,
https://doi.org/10.1007/978-3-030-48740-9_1

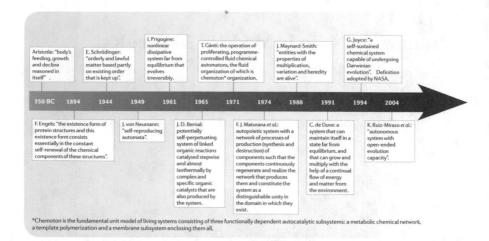

*Chemoton is the fundamental unit model of living systems consisting of three functionally dependent autocatalytic subsystems: a metabolic chemical network, a template polymerization and a membrane subsystem enclosing them all.

**Fig. 1.1** Timeline for the definitions of life or living beings. (Figure reprinted from Moreva & López-Garcia, Nat. Rev. Microbiol. 7:306–311, 2009, with permission from Macmillan Publishers Ltd.)

## 1.1 Selected Illustrative Example #1: The Molecular Origin of Life

Nothing better than trying to answer the question "what was the origin of life?" to realize that molecules are the key to life. Since the pioneering work of Aleksandr Oparin, the origin and evolution of life are decoded based on the chemistry of molecules containing carbon. By introducing this concept, Oparin truly revolutionized the way science interprets life. Nowadays, there are two main hypotheses to explain the evolution of the complexity of molecular organization into what one today calls cells, the so-called "replicator" and "metabolism" hypotheses (Fig. 1.2). These hypotheses are based on two specific characteristics common to all living beings: despite tremendous diversity among species, all life forms are organized in cells, and all cells have a replicator polymer (DNA) and a membrane with restricted permeability (a "membrane" having lipids in

Aleksandr Oparin (1894 - 1980)

its composition). Therefore, it is not surprising that the prevailing hypotheses to explain the origin of life are indeed models that elaborate on the appearance of the replicator polymer and the compartmentalization. The replicator polymer is essential to transmit the molecular inheritance from generation to generation, and a membrane forming a compartment that separates the ancestral cell from its environment is essential to ensure that the molecules in this space can react among each other in a controlled and self-regulated way (a "proto-metabolism"), with minimal impact of fluctuations in environmental conditions. These two aspects are consensual among researchers who study the origin of life, but the chronological order of events that resulted in cells as we know today is far from being established.

### 1.1.1   The Replicator Hypothesis

According to the replicator hypothesis, life started with a molecule that was randomly formed but had the ability to replicate itself. This is an extremely unlikely event, hardly possible to occur twice in the universe, but one may work on the hypothesis that it has occurred. The obvious first "choice" for a replicator molecule is DNA, the ubiquitous replicator nowadays, but this leaves us in a paradox: proteins are needed to generate DNA and DNA is needed to generate proteins. What came first then? It may be that DNA had an ancestor with self-catalytic activity. RNA is eligible as such ancestor. RNA is not as chemically stable as DNA, so it is not so well suited to store information for long periods of time, but it can still constitute genetic material (many viruses, such as HIV, SARS-CoV-2, or dengue virus, have RNA genomes). Concomitantly, its conformation dynamics enables catalytic activity. A perfect combination for the original replicator. The introduction of mutations and other errors in replication led to evolution and selection. How this process was coupled to the appearance of a metabolism is hard to conceive, but confinement of replicators into separated environments may have favored some chemical reactions that evolved in their restrained space to form metabolism (Fig. 1.2).

### 1.1.2   The Metabolism Hypothesis

An alternative model skips the Achilles heel of the replicator hypothesis. Here, the origin of life is not dependent on a starting event that is nearly impossible. The key process was the confinement of small molecules that reacted among themselves. In some cases, organized ensembles of molecules may have formed stable reaction cycles that became increasingly complex. The result was the creation of metabolism and complex polymer molecules, including replicators (Fig. 1.2). Naturally, the boundaries of the confined environment where these reactions took place had to allow for selective permeation of matter. Permeation allowed growth and replication.

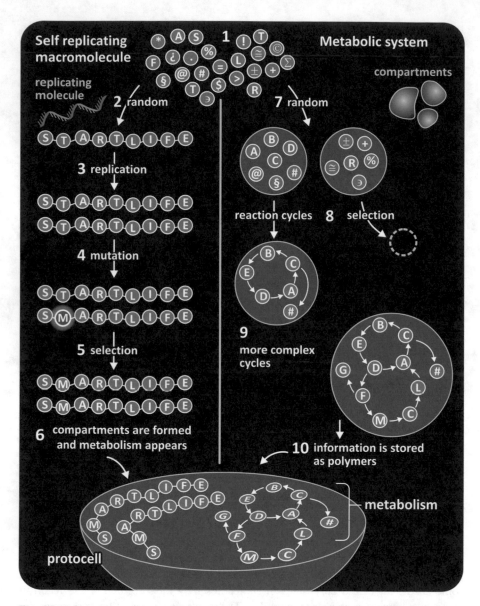

**Fig. 1.2** Schematic representation of the replicator (left, route 2-6) and metabolism (right, route 7-10) hypotheses to describe the origin of life. Both models are molecular in nature and agree on the critical roles of a replicator molecule and compartmentalization but differ on the sequence of events. (Figure reproduced with permission from Shapiro, Investigacion y Ciencia 371, 2007)

Nowadays, virtually all cell membranes are formed non-exclusively but mostly of lipids. Lipids are synthetic products of metabolism. So, what could have been the predecessors of lipid membranes in the confinement of the first "proto-metabolic" reactions? Cavities in the outer layers of rocks are a possibility. Phospholipids or other surfactant molecules may have started as coatings that, for their intrinsic dynamics and capability to expand and seal, may have evolved into membranes. Lipids and other surfactants can form three-dimensional structures other than lamellae that may have contributed to confine chemical systems (Fig. 1.3).

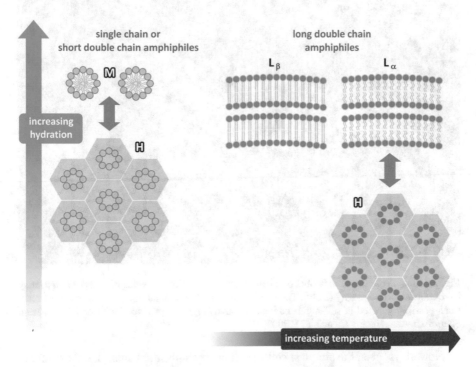

**Fig. 1.3** Types of three-dimensional assemblies of lipids. The structure of lipid assemblies depends on the degree of hydration and the molecular structure of lipids. Lipids may organize in different ways: rigid bilayers ($L_\alpha$), fluid bilayers ($L_\beta$), micelles (M), or hexagonal (H) phases

The metabolism evolved toward self-regulation creating homeostasis, a situation in which a balance exists and small to moderate perturbations trigger responses that tend to reestablish the original, equilibrated balance. The ability of certain metabolites (intermediate molecules in a complex sequence of reactions in a living system) to activate or inhibit specific reactions in the metabolism was a major contribution to homeostasis (Fig. 1.4).

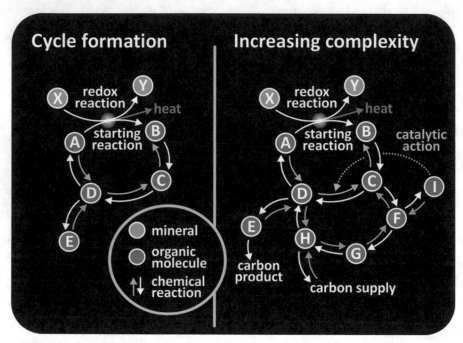

**Fig. 1.4**  The evolution of networks of chemical reactions. Simple cycles of reactions (*left*) may have evolved in complexity (*right*). The interference of certain metabolites on the course of reactions possibly resulted in self-regulated metabolisms. (Figure reproduced with permission from Shapiro, Investigacion y Ciencia 371, 2007)

Nowadays, even the simplest cells, such as mycoplasma bacteria, are extremely complex systems from the chemical/molecular point of view. Considering natural evolution, all metabolisms in all living cells are related by historical bonds, and the main metabolic sequences in living beings can be drawn as "metabolic maps" (Fig. 1.5). To understand how these complex mechanisms work and do not conflict with each other, one must bear in mind that not all reactions depicted in Fig. 1.5 operate in the same species; the ones that do may not be present in all cells; in case they do, they may not operate in the same cell compartment; in case they do, they may not be working at the same time. "Complex" does not mean "confusing." But the complexity of the metabolism as a whole is usually so high that in practice one tends to refer to "metabolisms" to designate the sectorial metabolisms in short. The word may be misleading because it may leave the impression that there are several independent metabolisms. Metabolisms are not independent of each other, and they are highly correlated, even those occurring in different organs. The need for metabolic regulation extends to the whole human body.

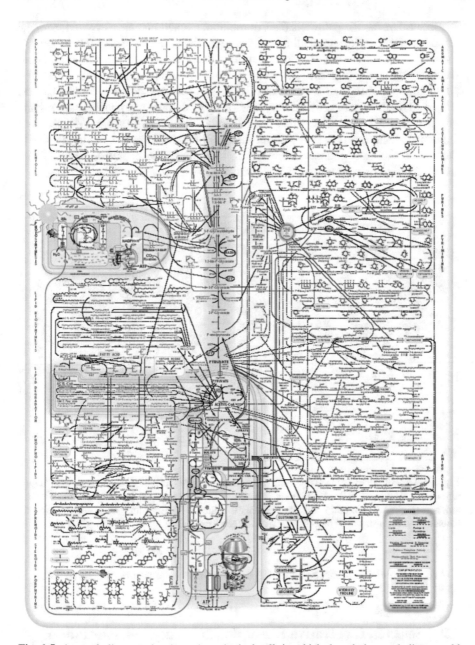

**Fig. 1.5** A metabolic map showing a hypothetical cell, in which the whole metabolism would gather many different sectorial metabolisms: amino acids, phospholipids, steroids, lipids, saccharides, etc. (note that not all cells perform all sectorial metabolisms). (Figure reproduced with permission from International Union of Biochemistry and Molecular Biology (IUBMB))

Because metabolic pathways (sets of metabolic reactions) have evolved from the same historical background, all molecules in all living cells are also related by historical links. Their common roots determined that, despite all the apparent molecular diversity, nearly all molecules in all cells can be grouped in few families. It is also intriguing at first glance that with so many chemical elements known to man (Fig. 1.6), cells rely heavily on very few of them: hydrogen, oxygen, carbon, and nitrogen are 99% of the atoms that make a cell. How can this apparent nonsense be explained? It all resorts to the common ancestor of all living cells in all living world: these were the most abundant elements in solution in the primitive ocean. These were the founding resources and life evolved from them.

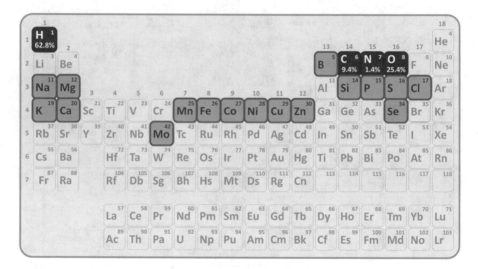

**Fig. 1.6** Periodic chart of the elements. In living beings, very few elements are needed to "build" almost the totality of cells (highlighted in *red*), and some elements are only present in trace amounts (highlighted in *pink*). Yet, the elements that are rare may be absolutely essential to life. Cobalt (Co), for instance, is part of vitamin B12 (see Box 3.8)

We shall revisit in a more detailed manner the chemical nature of cells in Chap. 3.

## 1.2   Selected Illustrative Example #2: Viruses—Molecular Machines Interfering with Life

Viruses are not considered by most researchers as living beings. They are on the edge of life, able to interfere with homeostasis. They have similar molecular constituents compared to cells (proteins, lipids, nucleic acids, etc.), but there are important differences. Above all, viruses lack a metabolism of their own. Their simplicity is not a consequence of ancestry nor does it relate to any surviving form of primitive life. Instead, might be a consequence of parasitism and regressive evolution. Alternatively, viruses may have been part of the cells. Minimal genome sizes imply faster reproduction rates for viruses and are therefore an evolutionary advantage. One may argue that viruses lack a metabolism of their own but are physical entities that are able to self-replicate and evolve, thus living beings. Even so, it is questionable whether they may be considered living because they do not replicate or evolve independently of cells. Virtually, all parasites need a host to survive and multiply, but viruses are also not able to evolve independently: they are dependent on cells to evolve because they do not have a complete own machinery of molecular synthesis.

Viruses–cell interactions are mostly physical in the early stages of the cellular infection as no chemical reactions are involved (no new covalent bonds are created nor destroyed). Let's consider as an example the influenza virus, the virus that causes flu (there are three types of influenza viruses, A, B, and C, being the influenza A virus the major cause of seasonal flu). influenza A virus is an enveloped virus, whose genome consists of eight single-stranded RNAs that encode 11 or 12 proteins (Fig. 1.7). The virus has the protein hemagglutinin A (HA) on its surface. This protein mediates virus entry into the host cells by binding to a saccharide, the sialic acid, linked to molecules (glycans) present on the cell surface, which act as the virus receptor. HA recognizes sialic acid due to a very precise combination of hydrogen bonds and ionic interactions, among others, between well-defined atoms on the protein and on the sialic acid molecule (Fig. 1.7). These atoms, both the ones in the protein and in the saccharide, are at precise distances and orientations relative to each other so that a unique combination of forces creates a strong binding between them. Influenza A virus may establish contact with many cells in the human body but will only bind to those having sialic acid-containing receptor on its surface (mainly cells of the upper respiratory tract epithelium). Consequently, these are the cells that can be preferentially infected by the virus.

Virus–cell attachment (more precisely HA–sialic acid binding) induces virus internalization through endocytosis. The virus is enclosed in a vesicle in the cytosolic space. Upon acidification of the endocytic vesicle medium, HA is cleaved and undergoes conformational changes that result in the exposure of a terminal hydrophobic segment, called fusion peptide, which binds to the endocytic vesicle membrane. Entropy balance then serves as a driving force (see Sect. 3.1) that makes the fusion peptide to insert into the endocytic vesicle membrane. Afterward, additional changes in the conformation of the protein will bring the viral envelope and the vesicle membrane together. They are both lipid bilayers, so they collapse. Ultimately, they fuse completely and the viral content is no longer separated from the cytoplasm. The viral RNA molecules proceed to the nucleus, where they are transcribed and replicated. The transcribed viral mRNAs are translated using the cellular protein synthesis machinery. The newly synthesized viral genome is packed by some of the viral proteins forming the nucleocapsid, whereas the viral surface proteins migrate to the cell surface through the cellular secretory pathway. The nucleocapsid then associates to the surface proteins at the plasma membrane, and new viruses bud from the cells ready to initiate another infection cycle.

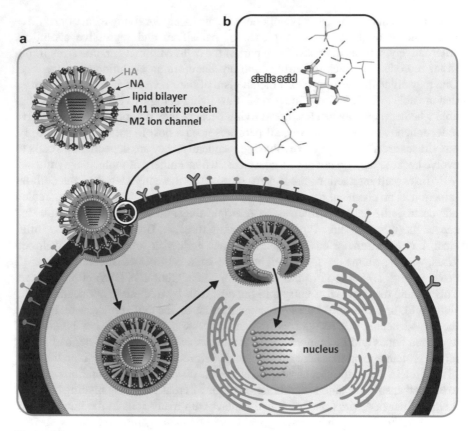

**Fig. 1.7** Role of hemagglutinin A–sialic acid interaction during influenza virus infection cycle. (**a**) Main events of influenza virus entry into a host cell. Virus binding to a sialic acid-containing receptor on cell surface is the first step of the entry process. The orientation, chemical nature, and distance of the binding amino acids of hemagglutinin A (HA) are such that sialic acid is able to engage in hydrogen bonds and other attractive forces. Upon acidification of endocytic vesicles inside the cell, HA undergoes conformational changes that bring viral and cellular membranes in contact, leading to collapse of the membranes (named fusion) from which the viral content is released inside the cell. (**b**) Zooming of segments of HA protein backbone contacting the sialic acid (protein carbon backbone in green; sialic acid carbon backbone in yellow)

When two different influenza virus strains infect the same cell, the RNA of both may coexist in the nucleus. Scrambling of RNAs originates new virus having random mixtures of the genetic material of both strains. The combined viruses are usually not functional, but occasionally a new strain of increased potency may be formed. For instance, it is possible that strains of avian or porcine flu combine with human flu to form new human flu strains. These events, combined with random mutations in viral proteins, may result in extremely lethal viruses. This was the case in 1918, when a flu strain, mistakenly named "Spanish flu," killed millions of people around the world (see Box 1.1). A mutation in a single amino acid in the HA binding site to receptors in an avian virus was enough to make it able to infect human tissues (Fig. 1.8). A small change in a molecule with a tragic impact on mankind.

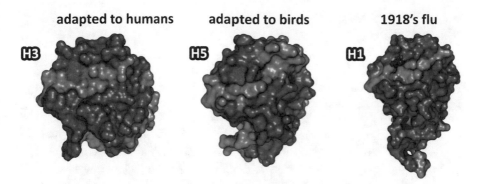

**adapted to humans**          **adapted to birds**          **1918's flu**

**Fig. 1.8** Surface structures of the globular domain of influenza virus HA variants. Hemagglutinin 3 (H3; segment 94-261, PDB 2HMG) is adapted to human cells. Hemagglutinin 5 (H5; segment 50-256, PDB 1JSM) is adapted to birds. Hemagglutinin 1 (H1; segment 56-271, PDB 1RUZ) resulted from a mutation in the amino acid residue in position 190 of bird HA, located in the region that binds cellular receptors, making it able to bind specific forms of sialic acid present in human cells. A glutamate residue (highlighted in *orange*) was replaced by an aspartate residue (highlighted in *red*). Glutamate and aspartate are very similar amino acids (see Fig. 3.33), but the slight difference between them was enough to cause a tragic pandemic of flu among humans—the "Spanish flu" or "1918 flu"

---

**Box 1.1 "Spanish Flu": Terrible and Almost Forgotten**

Between April 1918 and February 1919, the world suffered the worst pandemic of modern times. Probably, it was the worst pandemics since the Black Death plague in the fourteenth century. Influenza, the virus causative of flu, infected hundreds of million people and killed, directly or indirectly, 50,000,000–100,000,000, figures so high that are hard to estimate. Europe was also being devastated by World War I. The mobility of the armies and the precarious medical assistance conditions helped spread the disease. Moreover, the horrors of war and the censorship of the news from the fronts distracted the attention of mankind to the real dimension of the pandemic, which still remains largely ignored.

Despite the common name "Spanish flu," the disease did not start in Spain. Having a less tight censorship on the news because of its neutrality, Spain became a privileged source of information about the disease, which may have led to the impression that the disease was somehow related to Spain. In reality, the pandemic is believed to have started in the Kansas State region, in the USA, in March 1918. The new virus strain caused sudden effects, killing in a few days. In the worst cases, the patients suffered headaches, pain all over, fever, cyanosis, cough with blood, and nasal bleeding. Most deaths were associated with pneumonia, which was a consequence of opportunistic infection of lungs by bacteria. The histological properties of lungs were transformed, and there was accumulation of fluids that literally suffocated the victims, like a drowning.

(continued)

**Box 1.1** (continued)

The electron microscope was invented in the 1940s. Before this technical breakthrough, it was very difficult to study viruses. Other breakthroughs have followed, such as the development of super-resolution optical microscopes and PCR (polymerase chain reaction) technique, but the molecular singularities of the 1918 virus were still a challenge. The quest for the sequence of amino acids of the proteins of the 1918 strain is a story of persistence and devotion. In 1940, Johan Hultin, a medical student, spent the summer in Alaska. He heard about Teller Mission, a small missionary settlement that literally disappeared in November 1918. Seventy-two victims of flu were buried in a common grave. Later, Hultin matured the idea of recovering the 1918 flu virus from the bodies of the Teller Mission victims, presumably conserved in the Alaska permafrost. In the summer of 1951, he joined efforts with two colleagues from Iowa University, a virologist and a pathologist, and returned to Alaska to visit the former Teller Mission, meanwhile renamed Brevig Mission. With previous consent from the local tribe, Hultin obtained samples from lung tissue of some of the 1918 victims. The team tried to isolate and cultivate the virus using the most advanced techniques available but they did not succeed. It was an extreme disappointment. Hultin quit his PhD studies and specialized in pathology. Forty-six years later, in 1997, he was retired in San Francisco (California, USA) and read a scientific paper on a study of the genes of the 1918 flu strain obtained from 1918 to 1919 autopsies using PCR. Enthusiastically, Hultin resumed the intention of studying the samples from Teller/Brevig Mission. One of Hultin's colleagues from Iowa had kept the samples since 1951 until 1996! The samples had been disposed the year before! Hultin did not quit and asked permission to repeat the 1951 sample collection. This time he found the body of an obese young woman, whose lungs had been protected by the low temperatures and the layer of fat around them. The complete genome from the 1918 flu strain was obtained from these samples.

The hemagglutinin sequence of the 1918 strain (H1) was reconstructed from the genome of the virus. The sialic acid binding site underwent mutations in the amino acid residues relative to avian flu (H5) that enlarged the binding site, enabling the mutated viruses to bind and infect human cells. The modern studies on the phylogenetic tree of the flu viruses, which now include data from samples from South Carolina, New York, and Brevig, all from 1918, relate the origin of the virus to an avian strain found in a goose (Alaska 1917) (see below figure). Although this hypothesis is not totally consensual among researchers, the fear that new unusually deadly flu strains adapted to humans evolve from avian flu strains persists and is a matter of thorough surveillance of health authorities around the world. In 2020, another respiratory virus, SARS-CoV-2, causative of the COVID-19, crossed the inter-species barrier, presumably from bats to humans, and a world wide pandemic started.

(continued)

**Box 1.1** (continued)

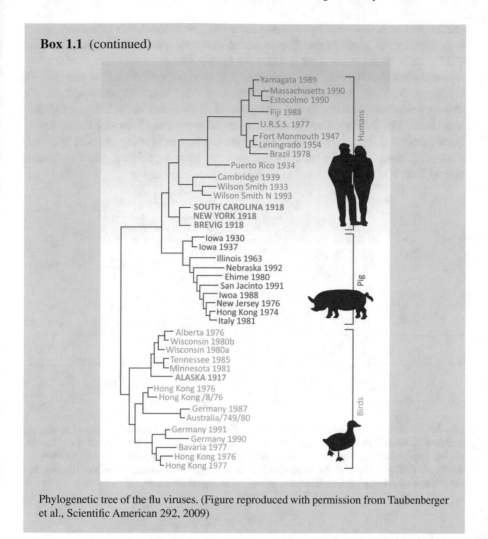

Phylogenetic tree of the flu viruses. (Figure reproduced with permission from Taubenberger et al., Scientific American 292, 2009)

## 1.3   Selected Illustrative Example #3: Molecules as Tools— Drug Discovery and Development

Designing new drugs that can be developed into new medicines demands knowledge on the role of different molecules in different pathologies. A molecular-level target is needed for the drug candidate, and the researcher needs to have an idea on how they are going to interact so that the target can be inhibited or activated. A drug candidate that is targeted to a protein, such as an enzyme or a membrane receptor, for instance, needs a binding site where it can react or attach both strongly and selectively. "Selectively" means it will discriminate this site from all others in the same target or in any other molecule of the body. The uniqueness of the binding site is granted by the precise arrangement of the atoms in space. Ideally, only that site has the right atoms at the right distance, in the right orientation to maximize inter-molecular attraction forces (see the example of HA–sialic acid binding in Fig. 1.8). Hydrogen bonding, ionic/electrostatic forces, van der Waals interactions, etc., all these are dependent on the spatial arrangement of elements of both drug candidate and target. The Beckett–Casy model for opioid receptors illustrates the basics of these principles (Fig. 1.9). In addition to the efficacy in binding to its target, drug molecules need not to be exceedingly toxic or have significant other undesired effects, which is directly related to its reactivity and selectivity.

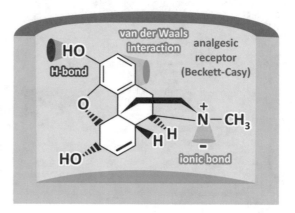

**Fig. 1.9** Beckett–Casy hypothesis for the binding of an analgesic molecule (such as morphine, which is illustrated) to an opioid receptor. While the exact structure of the receptor is not known, the key interaction/attraction forces were identified: electrostatic attraction, H bonds, and van der Waals interactions. The receptor is so specific for the ligand that chiral molecules (bearing the same chemical groups but with different orientation in space) do not fit

The same reasoning applies to complex therapeutic molecules such as antibod-ies. Let us take as an example one of the antibodies that targets the protein gp120 on the surface of the HIV (Fig. 1.10). This protein is responsible for binding to the receptors and co-receptors for the virus on the T-lymphocytes, this being the first step of infection. When an antibody binds to gp120, it may block the access of gp120 to the receptors and/or co-receptors, thus preventing infection. Anti-HIV antibodies are hopes for future therapeutics although the rate of mutations in gp120 and the presence of glycosylated groups on gp120 surface pose problems that are difficult to overcome.

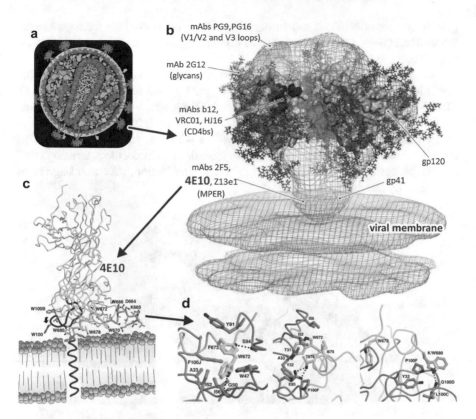

**Fig. 1.10** Therapeutic antibodies against HIV. (**a**) Illustration showing HIV particle, highlighting the envelope proteins (pink), and the capsid (purple), which surrounds viral nucleic acid. Besides the capsid, the virion is loaded with several other proteins with different functions in the replication cycle (Figure reproduced from Goodsell, The Machinery of Life, 2009). (**b**) Broadly neutralizing monoclonal antibodies target specific motifs (epitopes) at the surface of the envelope proteins, gp120 and gp41. The model depicted was generated gathering scientific data from different sources. The contour of the envelope proteins and viral membrane is shown in *gray*; what is known from the structure of gp120 is shown in *colors*. The glycosidic (saccharidic) part of the protein is shown in green and blue. MPER stands for membrane proximal external region and refers to the proteic part of gp41 most close to the viral membrane (Figure reproduced from Burton & Weiss, Science 239:770–772, 2010, with permission from The American Association for the Advancement of Science). (**c**, **d**) Some antibodies, such as 4E10, target MPER. At the contact points between the amino acids of antibody and gp41, attractive forces such as ionic, H bonds, and van der Waals contribute to a strong binding. The chemical nature and the spatial arrangement of the amino acids confer selectivity to antibody–epitope interaction. (Figure reproduced from Cardoso et al., Immunity 22:163–173, 2005, with permission from Elsevier)

Some researchers devote their work to antibody engineering, i.e., the manipulation of antibodies for a specific purpose. Some try to find the smallest portion of an antibody that is still active, so that antibody therapy can be made simpler and more cost-effective. Manipulating antibodies demands knowledge on the molecular-level interactions they perform with their antigens. At this level, the forces that are responsible for selectivity and strength of binding are not different from those that small molecules (such as the opioids in Fig. 1.9) establish with their molecular targets, but the number of bonds (hydrogen bonding, electrostatic interactions, van der Waals interactions, etc.) involved may be higher, resulting in extreme selectivity and very strong binding.

The whole process of devising and studying drugs (frequently termed "pipeline") has three main stages: research, development, and registration (Fig. 1.11). The research stage is typically, but not exclusively, carried out at universities and academic research centers. During this phase, relevant targets for selected pathologies are identified, and a molecule to interfere with that target is selected. This molecule is a drug candidate that can be improved. Such molecule is termed "lead" and the process of improvement is termed lead optimization. Research stage takes several years (rarely less than 5).

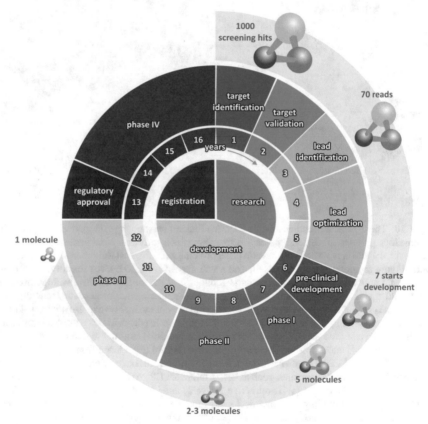

**Fig. 1.11** The drug discovery and development process, generally termed "pipeline" in pharmaceutical industries. There are three main stages: research, development, and registration (*center*), distributed over several years (*numerical timeline*). Each stage is divided in sub-phases. During the research stage, relevant targets for determined pathologies are identified, and molecules to interfere with those targets ("leads") are selected. The development stage may proceed for the next 7 years, during which the drug leads are tested for safety and efficacy in carefully designed animal and human clinical trials. At the ending of each phase, the results are evaluated; safety and/or efficacy issues may prevent further tests. For each nearly 1000 molecules starting the process, only one ends the last stage successfully. This failure ratio, generally named "attrition rate," is incredibly high. Moreover, not all molecules are granted approval to enter the market for regulatory reasons, and those that enter the market are still monitored afterward (phase IV)

The pre-clinical development is the first step is the development stage and the last step before clinical trials. Pre-clinical studies consist in as many experiments in vivo and in vitro (both in cells and artificial systems) as needed to ensure that a certain optimized lead is safe at a certain dosage range when prepared as a specific selected formulation using a defined mode of administration. The goal is to minimize risks to the lowest possible level when administrating the optimized lead to humans. Efficacy comes after safety in the priority list. This is the reason why the first tests in human (phase I clinical trials) are performed in few healthy volunteers, not patients. At this phase, safety is tested using conservative doses of the compounds under study. Tolerability, absorption, and distribution in the body and excretion are followed. Phase II clinical trials include patients, and efficacy, besides safety, is also tested. The drugs are administered to up to several hundred individuals for several weeks or a few months, typically. The dose range of the drug is improved during the trials. It should be stressed that all trials are scientifically controlled for the statistical significance of the results. The placebo effect (Box 1.2) is also accounted for in the trials. The process of designing clinical trials, data collection, and meaningful data analysis for reliable conclusion is, by itself, a complex discipline.

**Box 1.2 The Placebo Effect: The Power of Nothing**
*(Based on "The power of nothing" by Michael Specter in The New Yorker December 12, 2011, Issue)*
A placebo is a simulated or otherwise medically ineffectual treatment for a disease or other medical condition intended to deceive the recipient. Sometimes patients given a placebo treatment will have a perceived or actual improvement in a medical condition, a phenomenon commonly called the placebo effect.

For most of human history, placebos were a fundamental tool in any physician's armamentarium—sometimes the only tool. When there was nothing else to offer, placebos were a salve. The word itself comes from the Latin for "I will please." In medieval times, hired mourners participating in Vespers for the Dead often chanted the ninth line of Psalm 116: "I shall please the dead in the land of the living." Because the mourners were hired, their emotions were considered insincere. People called them "placebos." The word has always carried mixed connotations. Placebos are often regarded as a "pious fraud" because bread pills, drops of colored water, and powders of hickory ashes, for instance, may sometimes lead to a perceived improvement in patients.

The first publicly acknowledged placebo-controlled trial—and still among the most remarkable—took place at the request of King Louis XVI, in 1784. The German physician Franz Anton Mesmer had become famous in Vienna for a new treatment he called "animal magnetism," and he claimed to have discovered a healing fluid that could "cure" many ailments. Mesmer became highly

(continued)

**Box 1.2** (continued)

sought after in Paris, where he would routinely "mesmerize" his followers—one of whom was Marie Antoinette. The King asked a commission of the French Academy of Sciences to look into the claims. (The members included the chemist Antoine Lavoisier and Joseph Guillotin—who invented the device that would eventually separate the King's head from his body.) The commission replicated some of Mesmer's sessions and, in one case, asked a young boy to hug magnetized trees that were presumed to contain the healing powers invoked by Mesmer. He did as directed and responded as expected: he shook, convulsed, and swooned. The trees, though, were not magnetic, and Mesmer was denounced as a fraud. Placebos and lies were intertwined in the public mind.

It was another 150 years before scientists began to focus on the role that emotions can play in healing. During the World War II, Lieutenant Colonel Henry Beecher met with more than 200 soldiers, gravely wounded but still coherent enough to talk; he asked each man if he wanted morphine. Seventy-five percent declined. Beecher was astounded. He knew from his experience before the war that civilians with similar injuries would have begged for morphine, and he had seen healthy soldiers complain loudly about the pain associated with minor inconveniences, like receiving vaccinations. He concluded that the difference had to do with expectations; a soldier who survived a terrible attack often had a positive outlook simply because he was still alive. Beecher made a simple but powerful observation: our expectations can have a profound impact on how we heal.

There is also a "nocebo effect." Expecting a placebo to do harm or cause pain makes people sicker, not better. When subjects in one notable study were told that headaches are a side effect of lumbar puncture, the number of headaches they reported after the study was finished increased sharply.

For years, researchers could do little but guess at the complex biology of the placebo response. A meaningful picture began to emerge only in the 1970s, with the discovery of endorphins, endogenous analgesics produced in the brain.

Phase III clinical trials are a replica of the phase II trials, but several thousands of individuals are enrolled and treatment may be extended in time. Phase III is thus a refinement of phase II both in terms of efficacy and safety. Rare events such as unlikely undesired off-target effects that may have not been detected in phase II are now more likely to be detected. Concern for rare undesired off-target effects that may jeopardize the safety of drugs, even to small and very specific subpopulations of patients, is always present, even after the drug has been approved as a medicine for clinical use. This is sometimes termed "phase IV" and consists in screening how the drugs perform in the "real world," outside a tightly controlled environment.

Regulatory approval follows phase III and precedes phase IV and initiates the registration stage. The results of the development of the drug, from molecule to man, from bench to bedside, are submitted to regulatory agencies, which assess the results and conclusions of the whole clinical development based on the evaluation carried out by independent experts. The need for that specific new drug and how innovative it is when compared with existing drugs for the same purpose are also taken into consideration. The decision on allowing a specific molecule to be part of a new medicine or not belongs to these agencies.

The numbers associated with the difficulty in developing a successful drug, which is later incorporated in a new medicine, are absolutely impressive. For each 1–5 million "new chemical entities" (molecules tested for their pharmacological interest), only 1000 have positive results in in vitro tests, from which only 70 are selected as leads, which are then optimized to form seven drug candidates that enter clinical trials. Out of these seven, only 2–3 reach phase III clinical trials and only one is approved by regulatory entities. The whole process takes around 15 years to complete (Fig. 1.11) and has an estimated total cost of several millions of USD for each approved new drug, on average. It is important to point out that these are very crude numbers that vary a lot for different areas of medicine, but they serve to illustrate the efforts needed to continuously fight against disease progress. Reducing the attrition rate (fraction of new chemical entities that fail during the drug development process), accelerating the whole process, and making it more cost-effective are hugely demanding but urgently needed tasks.

# Selected Bibliography

Akst J (2011) From simple to complex. The Scientist (January):38–43

Garwood J (2009) The chemical origins of life on Earth. Soup, spring, vent or what? Lab Times:14–19

Kawaguchi Y, Shibuya M, Kinoshita I, Yatabe J, Narumi I, Shibata H, Hayashi R, Fujiwara D, Murano Y, Hashimoto H, Imai E, Kodaira S, Uchihori Y, Nakagawa K, Mita H, Yokobori S, Yamagishi A (2020) DNA Damage and Survival Time Course of Deinococcal Cell Pellets During 3 Years of Exposure to Outer Space. Front. Microbiol. 11:2050

Lombard J, López-García P, Moreira D (2012) The early evolution of lipid membranes and the three domains of life. Nat Rev Microbiol 10:507–515

Moran U, Phillips R, Milo R (2010) SnapShot: key numbers in biology. Cell 141:1262

Moreira D, López-García P (2009) Ten reasons to exclude viruses from the tree of life. Nat Rev Microbiol 7:306–311

Raoult D (2014) Viruses reconsidered. The discovery of more and more viruses of record-breaking size calls for a reclassification of life on Earth. The Scientist (March):41–45

Raoult D, Forterre P (2008) Redefining viruses: lessons from mimivirus. Nat Rev Microbiol 6:315–319 [see also comment by Wolkowicz R, Schaechter M (2008). What makes a virus a virus? Nat Rev Microbiol 6:643]

## Challenging Case 1.1: The Origin of Life and Panspermia

### *Source*

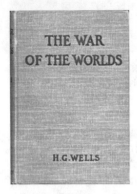

This case is based on the classic novel by H.G. Wells, *The War of the Worlds*,[1] that although written more than a century ago, remains contemporary, having been adapted many times to different media, from a 1938 radio play to a Hollywood film directed by Steven Spielberg in 2005. It is also a source of inspiration for other productions. It brilliantly highlights the collective wishful thinking expressed by science fiction writers that not only are we not alone in the universe but also that life's origin may follow similar trajectories, a point that is still hotly debated today.

First edition cover of the novel *The War of the Worlds*

### *Case Description*

In real life, humans do not seem to display much love or respect for their planet as judged by the overwhelming neglect that has led to widespread degradation of air, land, and sea. This unfortunate human trait is not justifiable in terms of ignorance—everyone knows what's happening—and yet the visible global damage now approaching the point of no return does not scare anyone into action. Curiously, as far as fiction goes, all it takes to provoke an about-turn is a threat, or actual invasion by aliens. Then, humans join forces and zealously guard the Earth against all odds and sophisticated weaponry. Although the record shows numerous tales of attempted alien invasion, perhaps *The War of the Worlds* translates faithfully our angst regarding the unknown.

Although *The War of the Worlds* was written in 1898, it still holds its own against contemporary productions, such as Carl Sagan's "Contact" (1985); "Independence Day" (1996), by Roland Emmerich; and "Arrival" (2016), by Denis Villeneuve. Briefly, *The War of the Worlds* begins when British astronomers detect strange explosions on the surface of Mars. This is followed by the arrival of a meteorite in Woking, Surrey. The narrator then witnesses the emergence of the Martians from artificial cylinders. They were described as big and grayish and with oily brown skin. The Martians did not agree with the local atmosphere and Earth's gravity, and so they retreat back into the cylinders from where they go on masterminding the

---

[1] Wells HG (1898) The war of the worlds, 1st edn. Harper & Bros (US) Publishers.; 287 p, Heinemann.

invasion. From the start, they reject human attempts to establish contact by promptly dispatching the ambassadors to be with powerful heat rays. After the peace talks fail, war begins and there is much destruction and pain. The Martian heat rays and a black smoke go on producing havoc, and gradually it dawns on the main characters that along with destruction the Martians are also feeding on the humans that get caught by their tripod fighting machines. They also try some mars forming (as opposed to terraforming) by spreading some weed that grows luxuriantly wherever there is abundant water available. The humans are taking a beating, but suddenly when everything seems to be lost and the Earth is finally going to be relinquished, the Martians are found dead. Every one of them. It turns out they were killed by earthy pathogens against which they had no immunity. And so the story ends.

*The War of the Worlds* has been amply discussed by scholars and many interpretations were put forward. Surely, having witnessed wars in Europe, Wells had many reasons to be ambivalent about technology, as well as pessimistic about the human's ability to cope with the environment. Thomas Henry Huxley, Darwin's bulldog, also asserted that *The War of the Worlds* was an example of the struggle for survival. Wells himself betrays his xenophobic views vis-à-vis his description of the Martians. The list goes on. On the other hand, as a biologist, Wells must have given some thought to the question of the origin of life when writing *The War of the Worlds*. This is clearly expressed as the plot unravels.

In *The War of the Worlds*, human armies are inoffensive, powerless to stop the advance of Martians over our territory. Yet, the powerful aliens soon fall dead from the contagion by bacteria. Our ancestor enemies were merciless: Martians had no resistance to Earth bacteria. The importance and need for antibiotics are certainly felt differently after listening to the adaptation of the visionary novel of H.G. Wells to radio or Spielberg's movie.

## Questions

1. Select from the text above a passage that indicates that the appearance of life in the universe depends on the same scenario as that on Earth.
2. What does the fact that the Martians were feeding on humans tell about the physiology of the aliens?
3. After less than 80 years from the beginning of the mass use of antibiotics, these miracle drugs are failing. Resistant infections kill hundreds of thousands of people around the world each year, and there are now dozens of so-called superbugs.
   (a) Relate the excessive use of antibiotics with the emergence of bacterial resistance.
   (b) Is excessive use a consequence of excessive prescription by doctors?

## Biochemical Insights

1. H.G. Wells' text mentions that the Martian weed had a very strict requirement
   for water, in the same way as the terrestrial plants, or for that matter, all other life
   forms on Earth. Indeed, whenever astrobiologists—scientists seeking life in the
   universe (see more about astrobiology in the Challenging Case 1.3)—investigate
   a planet in our solar system or outside (exoplanets), the first question they ask is:
   "Is there water there?" Why water? Water has many physical–chemical properties that according to our knowledge of life and its origins make it a very versatile solvent (see Chap. 2). For example, water is a dipole (the same molecule has
   two opposite charges), that is, water molecules are able to surround various types
   of polar solutes and, in doing so, dissolve them. This is only possible because the
   interactions between water and several solutes are energetically more favorable
   than the interactions between water molecules. Also, the polar nature of water
   makes it immiscible with hydrophobic compounds such as lipids. This kind of
   interaction favors the spontaneous formation of membranes and also helps to
   organize the structures typical of different intracellular cellular organelles. In
   other words, water is able to organize structures in which hydrophobic compounds prevail. Water is also a reagent that participates in many biochemical
   reactions of the living cell, such as enzymatic reactions. And so on and so forth.
   In contrast, other types of solvents are more limited in terms of their physical–
   chemical properties. In addition, water is ubiquitous in the universe. It is thought
   that in our solar system, some planets and their moons contain frozen water or
   liquid water underground. So, it is reasonable to propose that, if conditions were
   right, life in other planets could have a similar origin as far as water is concerned.
   Naturally, other building blocks would have to be involved (see below). Similar
   arguments could be put forward for the element carbon. We know for a fact that
   life, as we know it, is based on the chemistry of carbon. Why carbon? Why was
   this element selected as a major constituent of living beings? Certainly, that was
   not a capricious choice. Carbon is able to form bonds with itself and with many
   different elements, such as nitrogen, oxygen, and sulfur. These in turn can form
   many radicals that, by reacting with each other, can form various compounds.
   The ability to form bonds that are at the same time not too stable or unstable also
   counts favorably within the context of life. In general, stable compounds tend to
   be unreactive, and very unstable compounds do not remain intact for a significant length of time. Another relevant property of carbon is associated with rotation of bonds that permit the formation of tri-dimensional structures that add to
   the variety of carbon compounds. Rigid structures, on the other hand, limit the
   way in which compounds can interact with other molecules. In addition to reactivity, there seems to be no limit to the size and variety of carbon compounds. It
   is estimated that today approximately nine million carbon compounds exist in
   nature. So, it is not surprising that among all these possibilities, some actually
   did contribute to establish the intricate networks of interacting compounds that
   are the basis of the biochemical reactions (chemical reactions of the living cell).

Carbon is also abundant in the universe. It is formed in the stars and is one of the main constituents of nebulae. Therefore, an exoplanet that has water and carbon could in principle contain the main ingredients that eventually would give origin to life. Even though we could consider the elements to be universal, this does not mean that living beings on other planets besides Earth would be exactly the same as the ones we observe here. They may have similar biochemistry, however. As we know, the diversity of living beings is conditioned by evolution. By examining the fossils of animals that existed millions of years ago, we can appreciate that those living forms are as different from many extant ones as alien life would be from the terrestrial ones.

2. The feeding habits of the Martian invaders are similar to the earthlings. Despite their exotic properties and advanced technology, they too cannot escape from the universal dependency on energy. Hence, the Martians had to harness energy through the ingestion of other living beings just as we do on Earth. Besides, just like us, depending on what we eat, there might be nefarious consequences. If one ingests contaminated food, containing either bacteria or parasites, we may become sick. Sometimes fatally, like the Martians in the story. Their organisms did not agree with some of the components of humans, presumably the billions of bacteria that live in our gut and that normally are harmless to us. Again, H.G. Wells' description of the aliens allow us to perceive that he was a true evolutionist and as such believed that depending on the environment and the chemistry that would occur with the existing elements, life elsewhere could generate a myriad of creatures.

3. Misuse of antibiotics spreads across many different private and professional activities and cannot be assigned exclusively to excessive prescription. "The increasing global availability of these drugs over time since the 1950s, their uncontrolled sale resulting in broad-spectrum antibiotics being prescribed when not indicated, as well as antibiotic use in livestock feed at low doses for growth promotion and the releasing of large quantities of these medicines into the environment during pharmaceutical manufacturing"[2] are the main causes of excessive environmental presence of antibiotics in a broad sense. This has driven a massive process of killing of antibiotic-sensitive bacteria. Concomitantly, antibiotic-resistant bacteria thrive and multiply, colonizing areas that would naturally be occupied by antibiotic-sensitive bacteria. The ubiquitous presence of antibiotics has thus led to a fast and prosperous selection process for antibiotic-resistant bacteria.

---

[2] Riva E (2015) The war of the worlds and antibiotic resistance: a case study for science teaching. Biochemist 37:26–28.

## Final Discussion

Although we don't really know how life appeared on Earth, there are several hypotheses based exclusively on physical–chemical explanations, as addressed in section 1.1. The general idea is that, during the prebiotic era (before life), small molecules such as amino acids, monosaccharides, and purine and pyrimidine bases were synthesized from existing elements (carbon, nitrogen, oxygen, hydrogen) using thermal, electric, or UV energy. This has been experimentally confirmed. In time, small molecules reacted with each other and eventually generated polymers. It is also known that membranes and, therefore, compartments can form spontaneously when amphipathic hydrocarbons are exposed to aqueous solutions. Compartment formation as such was an important step because by increasing the local concentrations of the reagents, several reactions would be favored. The next stages would comprise the organization of the molecules into coherent pathways that were the precursors of metabolism.

The "appearance" of life could thus be regarded as resulting from the gradual increase in complexity that ultimately manifested itself by displaying the recognized features of a living cell. As life evolved on Earth, multicellular organisms were formed, lived for a while, and went extinct. This brief description of the events from prebiotic chemistry to living beings sums up the scientific thought, which is broadly consensual among scientists, giving or taking this or that pathway. By accepting the materialistic view on the origin of life on Earth, it is plausible to generalize it to the rest of the universe. When astrobiologists ponder about extraterrestrial life, they invariably question whether the planet in question lies within the comfort zone determined by its location in relation to that of the nearest star (the planet can't be too cold or too hot), whether it has water and an atmosphere. By doing so, they agree that life elsewhere followed the same trajectory as on Earth. Challenging Case 1.3 will revisit this theme.

*This case is a contribution of Prof. Franklin D. Rumjanek (Instituto de Bioquímica Médica Leopoldo de Meis, Universidade Federal do Rio de Janeiro, Brazil). In memoriam (1945-2020).*

# Challenging Case 1.2: Science in Fiction—Mighty Weakness vs. Micropower

## Source

In 2015, the Biochemical Society, UK, dedicated a special theme of its magazine *The Biochemist* to "Biochemistry on screen." One of the articles, by Nick Loman and Jennifer Gardy, titled "Contagion: a worthy entrant in the outbreak film genre," pinpoints the realistic and unrealistic aspects of the scientific setting in the fictional Steven Soderbergh's movie "Contagion" (2011). The exploration of the overlap between science and fiction in an intense drama paved with fear creates an opportunity to address the molecular aspects of viral epidemiology.

Cover of magazine *The Biochemist* vol. 37, No. 6, December 2015

## Case Description

The clash between powerful rulers and unprotected victims of abuse is a common theme in art, sometimes to convey alert messages for a better society, other times just to fuel an intense storyboard. Occasionally, these dramas happen in scientific scenarios, and unusual characters such as aliens or microbia star in the main roles of the script. This was certainly the case of *The War of the Worlds* (see Challenging Case 1.1), as well as the Steven Soderbergh's film "Contagion" (2011). Both examples deserve attention as the search for the boundaries that divide science from fiction is certainly more challenging and enticing than meets the eye at first glance. H.G. Wells' original novel is completely fictional and tells the story of a Martian invasion of Earth. "Contagion" is a film that mirrors recent scares of viral outbreaks such as severe acute respiratory syndrome (SARS), COVID-19, or Ebola. In both cases, it is evident how the discovery, development, mass production, and distribution of antibiotic and vaccines (since 1940s) revolutionized medicine and changed human societies in every single aspect from agriculture to war and … art. Importantly, it is also implied how the emergence of antibiotic resistance or anti-vaccine attitudes may revert past achievements and changes the tide in favor of microbes.

"Contagion" is focused on routine aspects of epidemiology work, as Nick Loman and Jennifer Gardy put it: *"the paper notebooks and whiteboard sketches trying to link cases to each other, the frustration that comes with trying to keep a case count going across hundreds, if not thousands, of jurisdictions, and the constant querying around precisely whose budget things are to come out of. Contagion gets a great deal right. The fictional virus, MEV-1, is borrowed from the Nipah virus (a newly emerged zoonosis that causes severe disease in both animals and humans) including the bat-to-pig-to-human "spillover" event that launches the outbreak, and the SARS* *coronavirus, which, like its celluloid cousin, used Hong Kong and its crowded apartment complexes and hotels as a stepping stone to global spread. The uncertainty in the early days of the outbreak is also clear, from the medical doctor who can't offer a diagnosis to the epidemiologist trying to calculate $R_0$, the virus' reproductive ratio to estimate the potential scale of the problem."*

Uncertainty spreads in the population, causing fear, misplaced belief in sham remedies, and manipulation, a set as dangerous as viruses themselves. We have all witnessed these events in real life during recent viral outbreaks or pandemics, from influenza to Ebola, SARS-CoV-2, dengue, or Zika. On the other hand, fiction has its own limitations and even scientific scenarios must be adapted for storytelling. Realism would cost a lot of repetitive testing, incomplete models, iterative thinking, collective data sharing, and other sorts of action difficult to fit in a linear, thrilling, clear-cut narrative. Surely, a pandemic striking one in every 12 people in our planet would take more than two epidemiologists to handle the case, and buzzwords, like sequencing or receptor entry, have more precise meanings for biochemists than lay audiences can perceive. Interestingly, in sequencing, reality has overcome fiction as modern portable sequencing technologies can read all nucleic acids from pathogen and host cells present in a sample rapidly. Phylogenetic trees of outbreaks can then be used to identify epidemiological patterns, related hotspot communities, and cross-border spread, for instance. Overall, "Contagion" helps in realizing that scientific knowledge is the key to manage and ultimately eradicate viral deadly menaces. *"The question when it comes to the next great pandemics is not if, but when".*[3] In 2020, SARS-CoV-2 reminded the world of this reality.

## *Questions*

1. $R_0$, the "basic reproduction number" of an infection, is a tool that epidemiologists use to understand an infectious disease's potential to spread among a population. It describes the average number of people that one sick person will ultimately infect during an outbreak—a disease with an $R_0$ of 2, for example (something like SARS-CoV-2 or Ebola virus $R_0$), is one in which every one sick person will go on to infect two more people. The $R_0$ of a disease isn't set in stone—it will vary over the course of an outbreak or epidemic and depends on characteristics of the host, pathogen, and environment—but the general rule is that when $R_0$ is less than 1, the outbreak will eventually die out.
   (a) For Ebola, calculate the progression in number of people affected as the virus spreads.
   (b) How do you translate its progression into a mathematical formulation?
   (c) Search the literature and find which virus has the highest $R_0$. Compare disease progression for this case and Ebola ($R_0 = 2$).

2. Why viral genome sequencing and knowledge on the receptors used by the viruses to entry the cells are crucial to devise strategies to develop drugs against viruses?

---

[3] Loman N, Gardy J (2015) Contagion: a worthy entrant in the outbreak film genre. Biochemist 37:22–25.

3. Why antibacterial and antifungal drugs (antibiotics) are not effective as antivirals?
4. RNA viruses have a higher mutation rate when compared to DNA viruses. Why is this so and why is this a challenge in drug and vaccine development?
5. Why so many pandemics, both in movies and real life, start in Southeast Asia, such is the case in "Contagion?"
6. In "Contagion" a vaccine is developed in 6 months. Is this realistic?

## Biochemical Insights

1. First infection: 1
   Secondary infection: 2 + 1
   Tertiary infection: 4 + 2 + 1
   (...)                ... + 8 + 4 + 2 + 1

$$N = R_0^n + R_0^{n-1} + R_0^{n-2} + \cdots = \sum_{i=0}^{n} R_0^{n-i}$$

$$t = n \times \Delta t$$

($N$, total number of infected people; $n$, order of infection; $\Delta t$, the average interval between consecutive infections)

   The highest $R_0$ values are found for measles virus and rotavirus infections, which have $R_0 \sim 15$ (sometimes even higher), AS illustrated in the following figure:

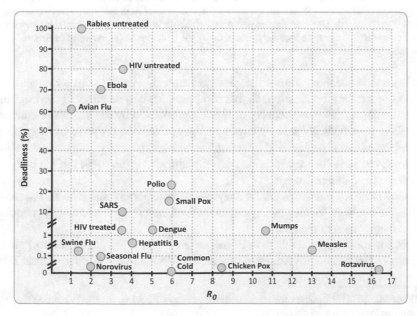

Deadliness–$R_0$ correlation for several infections. (Figure adapted from https://www.theguardian.com/news/datablog/ng-interactive/2014/oct/15/visualised-how-ebola-compares-to-other-infectious-diseases)

The number of affected people is expected to progress as depicted in the following figure. $\Delta t$ were chosen so that the model data could come close to the experimental data of the Ebola outbreak in Liberia in 2014.[4] Other outbreaks cannot be described using such a simple model and demand more sophisticated calculations using, e.g., fraction of susceptible and exposed individuals, who can either recover and survive or die, and duration of incubation and infectiousness and fatality rate.

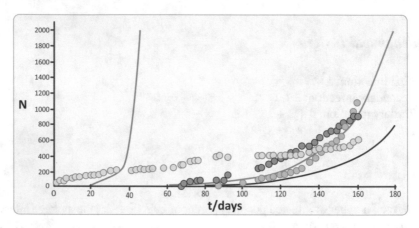

Infection progression for viruses with different $R_0$. Expected theoretical progression (*solid* lines) of the number of people infected with a virus with a $R_0 = 15$ (experimental value for measles virus) or $R_0 = 2$ (experimental value for Ebola) for $\Delta t = 18$ (*orange* and *green* lines, respectively). $\Delta t = 18$ was selected so the theoretical expectation could fit the data from an Ebola outbreak in Liberia in 2014[4] (*green* circles). For the sake of comparison, $R_0 = 2$ for $\Delta t = 21$ is also represented (i.e., slower rate of transmission). The data for the outbreaks of Ebola in Guinea (*yellow* circles) and Sierra Leone (*blue* circles) are not adequately described by the simple model developed in Answer 2 and need more sophisticated models[4]

2. Genome sequencing allows the determination of viral protein sequences and their posterior identification, so that viral proteins can be targeted by drugs and antibodies. Likewise, discovery of the receptors used by the viruses to enter the cells is important because they may be druggable, i.e., prone to be targeted by molecules that abrogate viral infection through receptor blocking or inactivation. Moreover, receptor identification may explain why some tissues in the body are infected and others are not.
3. Many antibacterial drugs inhibit enzymes or transporters that take part in bacterial metabolism. Viruses lack a metabolism of their own. Instead, they modulate the metabolism of the host.
4. DNA is much more stable than RNA (see Fig. 3.29). Frequent chemical alterations in RNA cause high rates of mutation in RNA viruses as compared to DNA

---

[4]Althaus CL (2014) Estimating the reproduction number of Ebola virus (EBOV) during the 2014 outbreak in West Africa. PLoS Curr 6.

ones. Additionally, RNA viruses use their own RNA-dependent RNA polymerases to replicate, which lack the proofreading activities present in DNA polymerases. Mutations affect the structure of proteins, which may alter the binding sites of drugs and antibodies (epitopes).

5. Places like big cities in Asia are ideal settings for new virus to emerge and propagate because street markets with live animals are abundant, population density is high, and as cosmopolitan urban communities, lots of people are entering and leaving every day. SARS started in China and spread to the world from Hong kong; COVID-19 pandemic started in Wuhan, China. New influenza strains spread throughout the world every year starting from this area, albeit not exclusively.

6. Developing a new vaccine is very laborious and time consuming. Besides creating the vaccine prototype, testing itself demands years of thorough data retrieval and processing. On top, production is usually not trivial and is thus time demanding. In this regard, "Contagion" is not realistic.

## Final Discussion

Science fiction seems to have evolved into the new era of science in fiction. Movies such "E.T., The extraterrestrial" (1982) and "The Rock" (1996), in which biochemistry is central for script, made the transition. The successes of TV series like "The Big Bang Theory" and "Breaking Bad" consolidated the awareness of the asset science setting and scientists represent to Hollywood and entertainment industry. The National Academy of Sciences of the USA created the program "The Science & Entertainment Exchange" ("The Exchange," http://scienceandentertainmentexchange.org/), which connects entertainment industry professionals with top scientists and engineers to create a synergy between accurate science and engaging storylines in both film and TV programming. The goal of "The Exchange" is to use the vehicle of popular entertainment media to deliver sometimes subtle, but nevertheless powerful, messages about science.

The frontier where science meets art, communication, and entertainment is certainly fertile for teaching and learning biochemistry inasmuch the public is educated in science matters, so facts are distinguished from fiction and the best of both worlds can be appreciated for the benefit of all.

## Challenging Case 1.3: Mars—NASA's Laboratory for the Origin of Life … On Earth!

### Source

This case is based on real data published in the scientific literature. NASA, the National Aeronautics and Space Administration of the USA, has a Program focused on Mars. Among the aims of the Mars Exploration Program, two are very relevant to biochemistry: to determine if Mars ever supported life and to determine if chemical

events taking place on Mars can be extrapolated to what is thought as being the early stages of life on Earth. The end goal of NASA is to prepare for human exploration of Mars, which also demands knowledge on human biochemistry in hypobaric conditions. Find out more at https://mars.nasa.gov/.

NASA's Curiosity rover exploring Mars surface found ancient organic molecules embedded within sedimentary rocks that are billions of years old. (https://www.jpl.nasa.gov/news/news.php?feature=7154; courtesy of NASA/JPL-Caltech)

### Case Description

The mission statement of the Mars Exploration Program is clear: *"The goal (…) is to explore Mars and provide a continuous flow of scientific information and discovery through a carefully selected series of robotic orbiters, landers and mobile laboratories interconnected by a high-bandwidth Mars/Earth communications network."* In fact, the Program is a science-driven, technology-enabled study of Mars as a planetary system in order to understand, among other issues, the potential for Mars to have hosted life and how Mars compares to and contrasts with Earth in present or past times.

The quest for signs of life on Mars is the most interesting endeavor for a biochemist spectator of the Mars Exploration Program. To begin with, identification of the signs of life is not trivial as living beings are made of non-living matter. The molecular "building blocks" of life are not alive themselves. Life is a complex organization of matter, and the frontier between living and non-living is not sharp for a biochemist. So, what to look for when looking for life? To help them in the complex task of finding the answer to this question, NASA scientists created what they called the "Life Detection Ladder," from the basic requirements essential for molecules to organize and originate life to other more sophisticated forms of very specific molecular interactions. In creating the ladder, researchers started with the NASA definition of life, *"Life is a self-sustaining chemical system capable of Darwinian evolution,"* and considered the specific features of the one life known, which is terrestrial life.[5]

Water, presence of molecular building blocks for use, an energy source, and gradients (i.e., confinement of sets of molecules) are the basic essential conditions to originate life. The presence of moderately complex organic molecules deviating from background bulk concentration and having chiral centers would be a subsequent stage in the signs of life path. Functional macromolecules, metabolism, and highly adaptive systems are higher-order signs, but they are not so relevant because no signs of such kind have been detected yet in any circumstance outside Earth. In practice, the quest for signs of life so far has only been a matter of detecting water and simple organic molecules and coping with elusive hints of eventual bacteria-like encapsulation structures that may have allowed molecular confinement/compartmentalization.

Life on the surface of Mars is unlikely given the presence of very reactive oxygen species that "attack" organic molecules, degrading them. However, inside rocks and deep below the surface, using geothermal energy, life is, in principle, possible … Inasmuch there is water! This is the reason why NASA researchers are searching for evidence of liquid water that exists or has existed on Mars. In January 2018,[6] NASA reported that researchers have found eight sites where thick deposits of ice beneath Mars surface are exposed in faces of eroding slopes. The ice was likely deposited as snow long ago, providing clues about climate history. The sites are both in northern and southern hemispheres of Mars, at latitudes from about 55° to 58°, equivalent on Earth to Scotland or the tip of South America. Six months later, in July 2018, Italian scientists from the European Space Agency reported indirect evidence that there is a frozen water lake in Planum Australe, at the Martian south pole, having liquid water underneath.[7] This lake extends through 20 km and reaches 1.5 km deep. Liquid state is putatively assigned to high contents in Mg, Na, and Ca ions, from the rocky soil, and pressure caused by the thick ice layer on the surface.

Ten years earlier, the Phoenix mission had already detected and analyzed buried water ice at 68° north latitude. Ice was directly exposed on the surface and sublimated, from solid into water vapor. Recent studies reached similar findings.[8] It is known that that most of the current Martian atmosphere consists of carbon dioxide. If water and gases in the atmosphere reacted and formed the carbonate minerals on the Martian surface by chemical reactions, the presence of these minerals is a clue that water had been present for long enough for life to have developed. If this was the case, sedimentary rocks would certainly keep traces of microscopic life forms, like protobacteria: Martian fossils! The claim that fossils of microbes had been

---

[5] Life detection ladder. Astrobiology newsletters. NASA. https://astrobiology.nasa.gov/research/life-detection/ladder/.

[6] Dundas CM et al (2018) Exposed subsurface ice sheets in the Martian mid-latitudes. Science 359:199–201.

[7] Orosei R et al (2018) Radar evidence of subglacial liquid water on Mars. Science 361:490–493.

[8] Piqueux S et al (2019) Widespread shallow water ice on mars at high and mid latitudes. Geophys Res Lett 46(24).

found in Martian meteorite ALH84001[9] originated one of the most intense and passionate scientific controversies in modern science together with cloning of the sheep Dolly, both occurring in 1996. The hypothesis was finally considered unreasonable because it was not totally supported by evidence.[10]

In 1976, NASA's two Viking landers made first the attempt to retrieve organic matter on Mars but did not succeed. In June 2018, new data was disclosed by NASA on the existence of organic matter on Mars, namely, in 3-billion-year-old mudstones in the Gale crater region.[11] Three years before, in 2015, the Curiosity rover drilled into the mudstone called "Mojane," and the analysis of the cuttings yielded organic molecules. The analyses were carried out automatically at a miniaturized automatic lab; the results were not trivial to interpret but, in the end, genuine Martian organics were identified. Interestingly, the organic matter was found in rocks formed about 3.5 billion years ago when Mars was drying out, although water persisted in Gale crater for thousands to millions of years more. The new finding shows that organic molecules located near the surface resist solar radiation. There is no compelling evidence for a biological origin of these molecules, but it persists as a legitimate working hypothesis. In case this hypothesis is true, one cannot tell if life had existed sometime in the past or if it is still there, somewhere.

A second study,[12] published in the same month and in the same journal as the study on Gale crater organic matter, showed that methane, the simplest organic molecule ($CH_4$), in the Martian atmosphere has large seasonal variations, which can only be accounted for if small localized sources of methane are released from the Martian surface or subsurface reservoirs. In other words, there must be a "methane cycle." Cycles of organic matter are amenable to life. Methane can be created by geological interaction between rocks, water, and heat, or it could be a product of microbes (or protomicrobes) that release waste methane. Again, there is no compelling evidence for life on Mars, past or present, but the opposite cannot be granted either, so more definite conclusions demand additional information from missions to Mars. At present, two new rovers are planned to be launched, one from NASA and one from the European Space Laboratory. The European will drill much deeper than Curiosity did. The next NASA rover will collect rocks that will be brought back to Earth on a later mission, so a more detailed analytic study will be possible.

## Questions

1. Why is so much importance given to finding water?
2. Since there is methane, why couldn't methane replace water in originating life?

---

[9] McKay DS et al (1996) Search for past life on Mars: possible relic biogenic activity in Martian meteorite ALH84001. Science 273:924–930.

[10] Treiman A. Traces of ancient Martian life in meteorite ALH84001: An outline of states in late 2003. https://planetaryprotection.nasa.gov/summary/alh84001.

[11] Anders E et al (1996) Evaluating the evidence of part life on Mars. Science 274:2119–2125.

[12] Eigenbrode JL et al (2018) Organic matter preserved in 3-billion-year-old mudstones at Gale crater, Mars. Science 360:1096–1101.

3. Silicon is abundant on Mars. Why is it not reasonable to assume it may replace carbon in the process of originating life?

## Biochemical Insight

1. The origin of life on Earth is unconceivable without liquid water. The elemental composition of Mars is not radically different from that of Earth. Silicon, oxygen, iron, calcium, and potassium are abundant in the crust. The red color of the planet is due to iron oxides. Hydrogen is present as water ice, whereas carbon occurs mainly as carbon dioxide in the atmosphere and, depending on climate seasons, as dry ice (solid carbon dioxide) at the poles. The atmosphere also has molecular nitrogen. The chemistry of these elements demands a medium (solvent) that enables and facilitates reactions; otherwise, chemical evolution cannot take place. The present climate of Mars is too cold for liquid water, but the occurrence of liquid water in the past is a hypothesis not yet excluded.
2. At Earth's atmospheric pressure, methane is liquid at temperatures below $-162$ °C, which is too extreme for the origin of life as chemical reactions are slowed down at low temperatures. Moreover, methane is not polar and cannot participate in H-bonding sharing. See Chap. 2 for more details on this subject.
3. Silicon has an atomic number of 14 and has chemical properties similar to carbon. Silicon however has high affinity for oxygen. C–O bond energy is $-360$ kJ/mol, while Si–O is 452 kJ/mol. Also, Si–Si bond energy is 226 kJ/mol, in contrast with 356 kJ/mol for the C–C bond. Overall, these figures show that compounds having Si–Si multiple bonds are unstable compared to compounds based on C–C scaffolds, and silicates (Si–O rich compounds) are very stable, which is why Si is deposited in minerals in the crust of Earth and Mars and life based on C–C scaffolds cannot translate to Si–Si scaffolds.

## Final Discussion

### Astrobiology

Controversy and speculation based on misinterpretation or overinterpretation of data, such as occurred in the public debate of possible life signs in ALH84001, has led to the need to carry out robust, unbiased, interdisciplinary work to achieve evidence-based irrefutable conclusions in the field of the potential for life to exist beyond the Earth. Astrobiology emerged and consolidated in this context, gathering contributions from astronomers, physicists, chemists, biochemists, geologists, climatologists, etc. On Mars, astrobiological activities have evolved from "follow the water" approach to a "seek the signs of life" strategy, in which evidence of cells preserved in rocks is researched for. However, in hundreds of millions of years, these eventual fossils may have been destroyed. For this reason, astrobiology is now focused in evidence that may come from shielded regions well beneath the surface of the Red Planet.

**Extraterrestrial Exporting of Life**

Microbe resistance to extreme environments is sometimes amazing. Even in extreme conditions of pH, temperature, and pressure, some species of microbes thrive and develop. Surprisingly, some forms of life have been found in the stratosphere, in spite of being a dry, cold, and low pressure (hypobaric) environment severely irradiated with UV, such as is on … Mars. So, it is reasonable to assume Terrestrial microbial life may persist on Mars, leading to colonization in the event cells that are transported by human devices to Mars. It is not likely that microbes on the surface of spacecrafts can survive to exposure to cosmic rays, but the competence of some microbes to repair radiation-caused damages is surprising (see Box 2.2), and the existence of pockets of microbes shielded in niche locations during traveling cannot be excluded[13]. Naturally, survival during the trip does not guarantee conditions to grow and multiply in other planets, but the risk of colonization does exist to some extent. Above all, this issue raises the need for awareness of the biochemistry of extremophiles on Earth and the need to account for co-colonization of other planets with Earth microbes in case of human presence,[14] which is the reverse of the novel by H.G. Wells, *The War of the Worlds* (see Challenging Case 1.1): Reality is not about microbes protecting the Earth, but about microbes being potential biological assault weapons on other planets.

Close to our planet, the "pristine environment" of the moon may have already been broken with a multicellular complex form of life: an Israeli spacecraft that crash-landed on the moon in April 2019 was carrying Tardigrades—often called water bears or moss piglets—which are creatures under a millimeter long that can survive being heated to 150 °C, frozen at very low temperatures, or being dehydrated for long periods. When dried out, they retract their heads and their eight legs, and shed almost all of the water in their body, which makes their metabolism slow tremendously. If reintroduced to water decades later, they are able to reanimate. The survival of Tardigrades is questionable, and the biological impact of the crash of the spacecraft, if any, is yet to be seen but, as BBC reporters wrote in good humor: *"And alternatively, there is definitely some great source material for a sci-fi/horror movie. Attack of the Moss Piglets from the Moon? We'd watch it."*[15]

---

[13] Kawaguchi Y, Shibuya M, Kinoshita I, Yatabe J, Narumi I, Shibata H, Hayashi R, Fujiwara D, Murano Y, Hashimoto H, Imai E, Kodaira S, Uchihori Y, Nakagawa K, Mita H, Yokobori S, Yamagishi A (2020) DNA Damage and Survival Time Course of Deinococcal Cell Pellets During 3 Years of Exposure to Outer Space. Front. Microbiol., 26 August, https://doi.org/10.3389/fmicb.2020.02050.

[14] Webster CR et al (2018) Background levels of methane in Mars' atmosphere show strong seasonal variations. Science 360:1093–1096.

[15] DasSarma P, DasSarma S (2018) Survival of microbes in Earth's stratosphere. Curr Opin Microbiol 43:24–30.

# Chapter 2
# The Chemistry and Physics of Life

Our idea of the interior of a cell at the molecular scale is often rather naive. If one could see the interior of a cell with molecular resolution, an aqueous solution of molecules with the cellular organelles suspended would not be seen. The molecular crowding, particularly the macromolecular crowding, inside a cell is such that the interior of a cell is more like a gel than a solution. Molecular packing is so dense that it is hard for macromolecules to diffuse freely. The ubiquitous presence of the cytoskeleton and macromolecular assemblies in a space that is highly restricted due to cellular organelles makes the interior of cells tightly packed (Fig. 2.1). Nevertheless, it is a highly hydrated environment, where solvation is made by water molecules and voids are filled by water that solubilizes ions and small molecules. Thus, virtually all exposed molecules in a cell are under the chemical and physical influence of water. The interior of a cell is not an aqueous solution, but the chemical reactions of the living cells are typical chemical reactions of aqueous solutions.

Life started in water and, chemically speaking, it is still dominated by water. Even the elemental composition of the molecules in a cell was determined by water. With few exceptions, the abundance of elements in a cell reflects the abundance of the same elements in the oceans (Fig. 2.2). Iron, phosphorus, and nitrogen are among the exceptions: they are more abundant in cells on average than in seawater due to their so-called chemical utility. The electronic structure of nitrogen makes it appropriate as an electron donor to establish dative covalent bonds, also known as dipolar or coordinate bonds. Phosphorus is an amazing element for its capabilities in coordination chemistry. The chemistry of phosphate, $PO_4^{3-}$, is so useful to cells nowadays that phosphorylation/dephosphorylation is an ubiquitous mechanism to activate or inhibit enzymes or determine the direction a molecule will take in a metabolic sequence, for instance. Iron may have once been abundant as soluble $Fe^{2+}$, which was afterward oxidized to $Fe^{3+}$ upon the appearance of molecular oxygen ($O_2$) in the Earth's atmosphere. $Fe^{3+}$ formed oxides and hydroxides that precipitated, making iron less abundant in seawater. Nonetheless, iron was already being used by cells, and its "chemical utility" determined that cells resisted environmental

© Springer Nature Switzerland AG 2021
A. T. Da Poian, M. A. R. B., *Integrative Human Biochemistry*,
https://doi.org/10.1007/978-3-030-48740-9_2

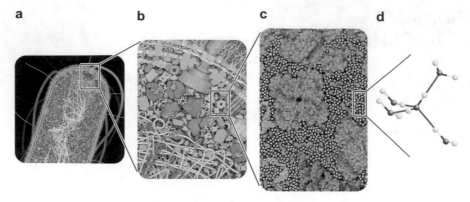

**Fig. 2.1** Crowding in the molecular organization of cells. (**a**) A simple cell such as bacteria (e.g., *Escherichia coli*) may be used as an example. (**b**) Cellular interior is a dense tight packing of macromolecules (for instance, proteins and nucleic acids) and smaller molecules such as nucleotides and amino acids. (**c**) In cytoplasm, the external surface of macromolecules, small molecules, and ions are in direct contact with water (amino acids, saccharides, ATP, and many other small organic molecules are shown in *pink*; metal ions are represented in *red*; phosphate ions are *yellow* and *orange*; and chloride ions are in *green*; water molecules are colored in *pale blue*). (**d**) Water molecules are polar and prone to establish hydrogen bonds. Although the cell interior is not a solution, its chemistry and physics are dominated by the properties of water. (Panels **a–c** are reproduced from Goodsell, The Machinery of Life, 2009)

pressure and kept this element at higher levels in their composition. Iron is able to participate in coordination chemistry and organometallic complexes, hemoglobin being an example.

Elements that are very abundant in the crust of the Earth but not in the oceans, such as aluminum (8.2% of the atoms) or silicium (28% of the atoms!), are also not frequent in cells because they are mostly involved in insoluble oxides. Only 1% of the total atoms in cells are not hydrogen (62.8%), oxygen (25.4%), carbon (9.4%), nor nitrogen (1.4%). Yet, many elements that are only present in trace amounts may be part of molecules or processes essential to life, such as boron (B), cobalt (Co), cupper (Cu), manganese (Mn), or molybdenum (Mo).

So, water imposed severe constraints on the molecular evolution of cells. And still does! Water is a small but an amazing molecule. In spite of its simplicity and abundance, water still fascinates chemists. In particular, its polarity and ability to establish hydrogen bonds are determinant to influence chemical reactions in regions where water serves as a dominant solvent, which means almost the whole cell (lipid membranes are the main exception). If it wasn't for hydrogen bonding, for instance, water would not be a liquid at temperature and pressure ranges humans are adapted to. Methane, with a molecular structure that can be compared to water, is not as nearly important as water in the history of life: methane's boiling point is −162 °C (Fig. 2.3).

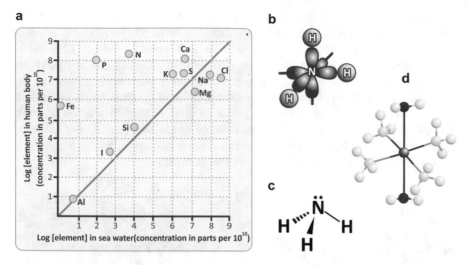

**Fig. 2.2** Chemical elements of life other than C, O, and H. (**a**) The abundance of the chemical elements in the human body matches the abundance of such elements in seawater, with few exceptions. Elements such as carbon, hydrogen, and oxygen are not represented in the graph because they are extremely abundant (figure reproduced with permission from Dobson et al. Foundations of Chemical Biology, 2001). (**b**) In ammonia ($NH_3$), nitrogen binds three hydrogens, and the covalent bonds are formed with the *s* orbitals of H (gray) and three *p* orbitals of N (*red*). (**c**) A fourth *p* orbital has electrons that are available to participate in dative (coordination) bonds. A simplified representation of the molecule symbolizes these electrons as •• over N. (**d**) Oxygen atoms in water molecules or dioxygen ($O_2$) molecules also have electrons available to participate in dative bonds, for instance, with metal ions. This is the case of the tetraamminediaquacopper(II) cation, $[Cu(NH_3)_4(H_2O)_2]^{2+}$, in which a central $Cu^{2+}$ ion accepts electrons from two water and four ammonia molecules. Binding of dioxygen to iron in hemoglobin follows the same principle

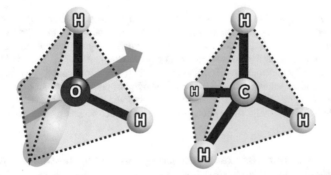

**Fig. 2.3** Water vs. methane: similar molecular geometry but different polarity (charge asymmetry axis represented in blue from the slightly higher electronic density in the non-bonding O orbitals—green—to the lower electronic density in H) and different hydrogen bonding capabilities, which determines very different boiling points: 100 °C to water, −162 °C to methane

Molecules that are polar, mainly those that can establish hydrogen bonding with water, can be distributed in the human body by simple diffusion, such as in blood and/or cerebrospinal fluids, while non-polar molecules have low solubility and need special carrier systems, such as lipoproteins (see Sect. 3.1.2). Glucose (Fig. 2.4), for instance, does not need a carrier to be distributed in the human body, and its concentration in blood can reach very high values with no solubility problems. Cholesterol is the opposite as its solubility in aqueous environment is very low, so it would form crystals and precipitate if it was not kept in non-polar environments, such as the hydrocarbon core of lipid membranes or the interior of lipoproteins. Some pathological situations, such as gallstones, relate to the low solubility of cholesterol-related molecules, which leads to formation of aggregates (due to the so-called entropic effect; see Sect. 3.1) that in turn nucleate small crystalline structures that grow into "stones" that can measure up to few centimeters.

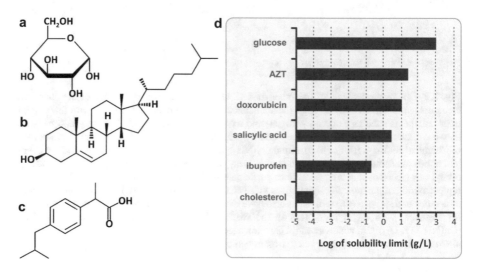

**Fig. 2.4** Differential solubility of organic molecules in aqueous medium. (**a**) Very hydrophilic molecules are polar, such as glucose, which, like water, can establish several hydrogen bonds per molecule. (**b**) In contrast, cholesterol, having only a single polar group, the alcohol (-OH), is very hydrophobic (non-polar) and is not soluble in aqueous media (right panel), such as blood plasma. (**c**) Some biological molecules and drugs, such as ibuprofen, for instance, may be intermediate cases and have limited solubility due to the presence of hydrophobic and hydrophilic groups in the same molecule (ibuprofen has a phenyl ring and a carboxylic group), which is very important for what is called their ADME: absorption, distribution, metabolization, and excretion by the human body

Solubility is as important for pharmacology as it is for physiology. A drug that precipitates and crystallizes in the blood, or even in the stomach after being swollen, is hardly effective as it cannot be absorbed and/or distributed in the body. Non-polar drugs are usually mixed with other molecules in formulations that prevent drug aggregation in aqueous environment. Solubility is one of the key parameters considered to devise drug development strategies.

The tendency that drugs have to locate in aqueous or in hydrophobic regions of tissues, such as lipoproteins, lipid bilayers, or adipose depots, is studied in pharmacology measuring the partition of the drug between octanol and water, two immiscible solvents. Octanol is an organic solvent, largely apolar. Depending on their own polarity, solutes will distribute more extensively to octanol or water. The ratio of the

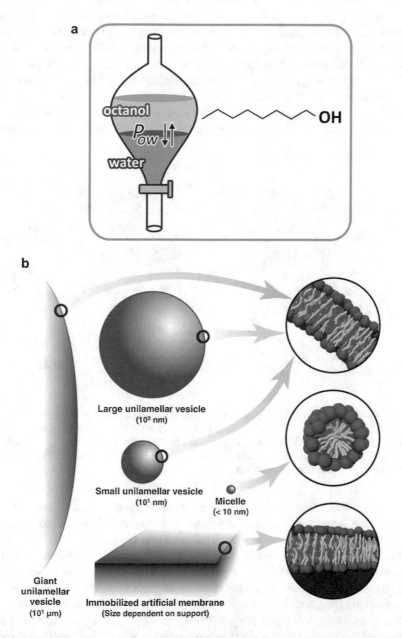

**Fig. 2.5** Methods used to evaluate the hydrophobicity of drugs. (**a**) Octanol is an amphiphilic molecule with an acyl chain of eight carbons and a polar alcohol group (–OH). Its hydrophobicity prevents its miscibility with water. (**b**) More recent techniques use different approaches (artificial lipid bilayers in suspension). (Figure reproduced from Ribeiro et al. Trends Pharmacol Sci 31:449–454, 2010, with permission of Elsevier)

equilibrium concentrations of both phases is constant, regardless of the total amount of solute or the volume of each phase, which is the reason why this ratio is referred to as partition constant, $P_{ow}$. This parameter is used to estimate the tendency molecules under study (e.g., a drug candidate for a future medicine) have to interact with membranes and other lipid structures, such as lipoproteins or lipid droplets. Despite the ubiquity and simplicity of the octanol/water approach, octanol is a poor replica of lipids, and more modern alternatives exist, like working with aqueous suspensions of lipid vesicles (Fig. 2.5). Lipid vesicles in suspension are lipid bilayers of very well-defined composition. Although they lack many characteristics of biological membranes (receptors, transporters, cytoskeleton anchorage, clustering of specific lipids, etc.), they are much more realistic as biological membrane mimetics than octanol.

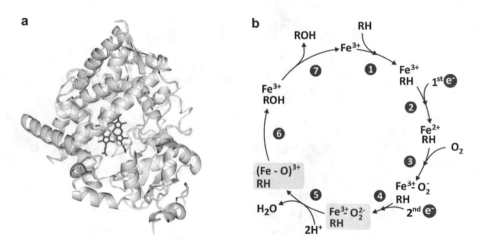

**Fig. 2.6** Structure and function of cytochrome P450, an important detoxifying mechanism in the human body. (**a**) The structure of cytochrome P450 complex (PDB 1W0E) with a heme group in its core (colored in *red*), in which an iron ion resides in the center. (**b**) The iron ion binds oxygen to add an OH group to organic molecules (generically represented by RH), rendering them more soluble, prone to be excreted through urine

The polarity of molecules is also determinant for their excretion. Polar molecules are easier to excrete for their solubility in blood and urine. The main strategy of the human body to eliminate xenobiotics (molecules that are not natural constituents of human tissues) consists in grafting hydroxyl (OH) groups, so that they become more polar and, therefore, more soluble in aqueous fluids. This is an efficient method that can be applied to a wide diversity of molecules, serving the purpose of low specificity for a broad protection of the human body against multiple toxic molecules. The molecular complex responsible for polyhydroxylation of different compounds is cytochrome P450, a big proteic complex that contains iron (Fig. 2.6).

Hydrophobicity (more accurately one would say the "entropic effect") is important not only for the absorption, distribution, and excretion of molecules, as discussed in this section, but also for holding together the most ubiquitous of all supramolecular non-covalent structures of cells: the lipid bilayer membranes. The importance of this issue is such that it will be kept for detailed explanation in a later section (Sect. 3.1).

## 2.1   The Basics of Chemistry in Cells and Tissues

The boundaries between chemistry and physics at the molecular scale are hard to establish. The interface between both is a rich scientific field referred to as physical chemistry or chemical physics. These disciplines deal typically with molecular structure and the way reactivity is affected by it. A functional distinction that is very practical for those working with molecules is to consider chemical reactions are all transformations of matter that involve formation or breakdown (or both) of covalent bonds. Transformations that do not involve alteration of covalent bonds are considered physical processes. Thus, light being absorbed by the molecules on the skin surface by a protective "sunscreen" cream constitutes a physical process, while UV radiation reaching the skin cells and damaging DNA duc to covalent bond cleavage is chemistry. On a macroscopic scale, this functional frontier between chemistry and physics may lose intuitive sense: a plumber cutting a metallic or PVC tube would then be doing chemistry as he is actually destroying chemical bonds in doing so.

In practice, a clear definition of what is chemistry by opposition to what is physics is not needed or useful. Many professionals use both and do not really mind or think about naming what they are doing in terms of chemistry vs. physics classification.

The "chemical life" of cells is very rich and diverse. Many different kinds of reactions may occur. Probably, all kind of reactions occurring in aqueous medium that are listed in the most complete organic chemistry textbooks can be found in cells. In this chapter, we will focus only on those most important to understand human metabolic regulation, the current core of human integrative biochemistry (i.e., biochemistry in relation to other disciplines such as histology, physiology, pharmacology, and even anatomy, so that a global perception of human body homeostasis is achieved). It should be stressed that such reactions benefit from the presence of water as solvent. As interaction with solvent molecules affects the electronic distribution of molecules, their reactivity is affected by the solvent. If natural evolution was based at molecular level on a different solvent (e.g., octanol), the "portfolio" of the chemical reactions of life would be different.

Oxidation–reduction reactions are among the most important reactions of the living world. As the name implies, oxidation–reduction reactions are those where electrons are donated (oxidation) or received (reduction). Because electrons in cells do not remain isolated, individually, as they would in vacuum in outer space, they are transferred between molecules or ions, and therefore oxidation and reduction

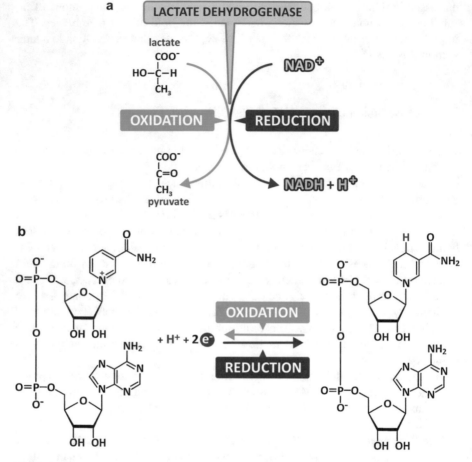

**Fig. 2.7** An example of an oxidation–reduction reaction. (**a**) Oxidation of lactate to form pyruvate, catalyzed by the enzyme lactate dehydrogenase. Lactate OH group is transformed in a C=O (carbonyl) group and electrons were transferred to NAD⁺ in the process. (**b**) Transformation of NAD⁺ in NADH, in which the non-bonding orbitals in an N atom enables "electron storage"

coexist. They are thus oxidation–reduction reactions. Many different molecules may be oxidized or reduced. Some are particularly well adapted as reducing agents, such as NADH (Fig. 2.7); others are particularly well adapted to be part of a chain of successive electron transfers, such as some metalloproteins. Metalloproteins have metallic elements in their composition that facilitate reception and donation of electrons. The electron transport system in mitochondria, for instance, has several of these proteins (see Sect. 6.2.3).

Acid–base reactions constitute another class of extremely important and ubiquitous reactions in the "living world." These are reactions in which a proton (in abstract, H⁺) is either donated (by an acid) or received (by a base). In aqueous environment, such as in the almost totality of the cell, H⁺ does not exist as such because water molecules capture the proton forming $H_3O^+$, or donate a proton, forming $OH^-$, the hydroxide anion. These chemical species ($OH^-$, $H_3O^+$, and $H_2O$) are all related, and equilibrium among them may be reached as the following:

$$2H_2O \leftrightarrows H_3O^+ + OH^-$$

The equilibrium constant for this reaction is:

$$K_{eq} = \frac{\left[H_3O^+\right]\left[OH^-\right]}{\left[H_2O\right]^2}$$

Because [$H_2O$] (molar concentration of water) is constant at any given temperature and pressure, the so-called ionic product of water, $K_w$, is used instead for its simplicity:

$$K_w = \left[H_3O^+\right]\left[OH^-\right]$$

Nearly at 25 °C, $K_w = 1 \times 10^{-14}$ mol² dm⁻⁶ (i.e., $1 \times 10^{-14}$ M²). This may seem rather pointless at a first glance, but it is from here that one can conclude that in pure water, pH value is 7. In pure water, one ion of $H_3O^+$ is formed for each $OH^-$:

$$\left[H_3O^+\right]_{eq}\left[OH^-\right]_{eq} = 10^{-14} M^2 \Leftrightarrow$$

$$\left[H_3O^+\right]_{eq}^2 = 10^{-14} M^2 \Leftrightarrow$$

$$\left[H_3O^+\right]_{eq} = 10^{-7} M^2$$

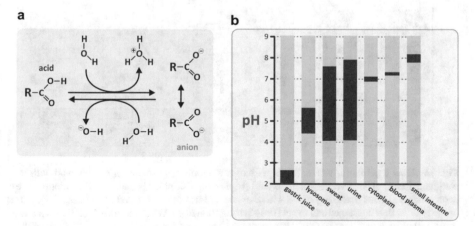

**Fig. 2.8** Acid–base reactions and pH in the human body. (**a**) Protonation (right to left) and deprotonation (left to right) of a carboxylic acid. (**b**) Ranges of pH variation in different environments of the human body

Therefore:

$$-\log\left[H_3O^+\right]_{eq} = 7 \Leftrightarrow pH = 7$$

So, in pure water, at temperature nearly 25 °C, the pH is 7.

The pH of the cytoplasm is approximately 7, but it is manipulated in certain cellular organelles. In the human body as a whole, the pH values of different environments are very diverse, from the extremely acidic gastric juice to the basic intestinal lumen medium (Fig. 2.8).

Other kinds of reactions, such as addition–elimination or nucleophilic substitutions, are also very frequent, but acid–base reactions will continue to be the focus of our attention due to the importance the control of pH has in homeostasis. Variations in pH cause variations in the protonation/deprotonation of proteins and other biological molecules, which in turn affect their solubility and function. Take the example of enzymes: protonation or deprotonation of chemical groups on the structure of the protein causes variation in charge, leading to new sets of attractions and repulsions between different parts of the molecule, which may cause that some segments of the protein contract and others become looser, impairing or facilitating the optimal function of the enzymatic catalysis (this will become more clear in Sect. 3.3 of Chap. 3, where we address the structure of proteins in detail). The same impact on molecular structure and function applies for other biological molecules, such as

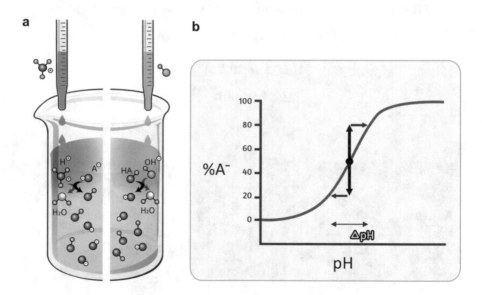

**Fig. 2.9** How a pH buffer works. (**a**) A pH buffer solution is a mixture of a weak acid with its conjugated base or vice versa. When an acid (represented in pink) is added to the solution, the deprotonated species (base, $A^-$) reacts with the acid to form the protonated form (acid, HA). Part of the $H_3O^+$ is thus consumed and the pH drop is thus attenuated. When the added solution is a base (represented in green), part of the $OH^-$ is consumed by reaction with HA, the acidic form of the buffer, and the rise in pH is thus attenuated. (**b**) Additions of significant amounts of acids or bases to buffered solutions result in modest variation in pH as long as protonated (HA) and deprotonated ($A^-$) buffer species coexist

polysaccharides. Therefore, stabilizing pH in order to guarantee proper structure and function of biological molecules is very important. This is not saying that the pH should be the same in all tissues, or in all cells of the same tissue or in all organelles of the same cell. pH is actively controlled in different anatomical, histological, and cellular environments, but it is not the same in all cases.

Blood plasma pH, for instance, is very strictly controlled and only allowed to vary in a very restricted range around 7.4. This is not surprising because the efficiency with which hemoglobin transports $O_2$ is very much dependent on pH. Nevertheless, $CO_2$, which is a molecule that has the potential to largely impact on pH, diffuses freely in the blood:

$$CO_2 + H_2O \leftrightarrows H_2CO_3 \leftrightarrows HCO_3^- + H^+; \quad H^+ + H_2O \rightarrow H_3O^+$$

Then, how is it possible for our body to cope with $CO_2$ diffusion and still maintain a blood plasma pH tightly controlled, centered at pH 7.4? The answer resides on a deceivingly simple mechanism of pH control named "pH buffering" (Fig. 2.9).

pH buffers are no more than mixtures of weak acids or weak bases with their conjugated bases or acids, respectively. In practice, an aqueous solution of a weak acid or a weak base in equilibrium is a pH buffer because these chemical species dissociate in moderate extent, forming mixtures that are the pH buffers themselves. The reason why these mixtures work in a way to conserve pH is related to the basic principles of chemical equilibrium. Consider a generic weak acid, represented by HA for the sake of simplicity, in aqueous solution:

---

**Box 2.1 pH Buffers and the Origin of the Henderson–Hasselbalch Equation**

The initial situation of a weak acid dissociation, before strong acid addition, is described by:

$$HA + H_2O \leftrightarrows A^- + H_3O^+$$

$$K_{eq} = \frac{\left[A^-\right]\left[H_3O^+\right]}{[HA]}$$

After strong acid (HA′) addition, [$H_3O^+$] raises:

$$HA' + H_2O \rightarrow A'^- + H_3O^+$$

But HA dissociation equilibrium constant is the same in both cases, so:

$$K_{eq,i} = K_{eq,f} \Leftrightarrow \frac{\left[A^-\right]_i\left[H_3O^+\right]_i}{[HA]_i} = \frac{\left[A^-\right]_f\left[H_3O^+\right]_f}{[HA]_f} \Leftrightarrow \frac{\left[H_3O^+\right]_f}{\left[H_3O^+\right]_i} = \frac{\left[A^-\right]_i[HA]_f}{\left[A^-\right]_f[HA]_i}$$

(continued)

**Box 2.1** (continued)

(i and f stand for the initial and final equilibria, respectively)

This equation shows that as $\dfrac{\left[H_3O^+\right]_f}{\left[H_3O^+\right]_i} > 1$, the weak acid equilibrium is

perturbed in a way to favor formation of HA, thus consuming $H_3O^+$. All HA′ is converted to $H_3O^+$, and, in the absence of the weak acid, the concentration of $H_3O^+$ would increase correspondingly. In the presence of the weak acid, part of the $H_3O^+$ is consumed by $A^-$ to form HA, thus attenuating the drop in pH caused by the strong acid. This is the molecular mechanism of pH buffering (see also Fig. 2.9 for a pictured explanation).

If a strong base is used, the same principle applies, this time with the "intermediation" of the ionic product of water:

$$BOH \rightarrow B^+ + OH^-; \quad OH^- + H_3O^+ \leftrightarrows 2H_2O$$

Also, if a weak base is used instead of a weak acid, the same buffering capacity exists.

The Henderson–Hasselbalch equation is a very simple and robust way to show how pH is expected to evolve when $[A^-]$ and $[HA]$ are perturbed and change when new equilibria are formed upon the appearance of new acids or bases in solution. The Henderson–Hasselbalch equation:

$$pH = pK_a + \log \frac{\left[A^-\right]}{\left[HA\right]}$$

is no more than a simple rewriting of the equation that gives the equilibrium constant (here named $K_a$ to stress it refers to an acid):

$$K_a = \frac{\left[A^-\right]\left[H_3O^+\right]}{\left[HA\right]} \Leftrightarrow \left[H_3O^+\right] = K_a \frac{\left[HA\right]}{\left[HA^-\right]} \Leftrightarrow -\log\left[H_3O^+\right]$$

$$= -\log K_a - \log \frac{\left[HA\right]}{\left[A^-\right]} \Leftrightarrow pH = pK_a + \log \frac{\left[A^-\right]}{\left[HA\right]}$$

A graphical schematic representation of this equation (see below figure) shows that pH varies very little when strong acids or bases are added to buffers, mainly when pH is within the $pK_a \pm 1$ range. This range is centered around the point where $[A^-] = [HA]$, as implied by the Henderson–Hasselbalch equation when pH = $pK_a$. Naturally, the efficiency of the buffer increases with $[HA]$

(continued)

**Box 2.1** (continued)

because higher [HA] imply that more $H_3O^+$ of $OH^-$ can be added to the solution before [$A^-$] reaches zero, a point where the buffer mechanism becomes exhausted.

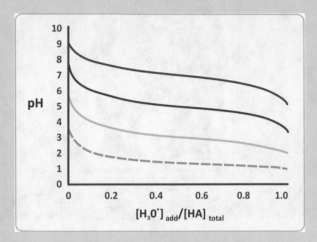

Changes in pH in a buffered solution are much less severe than they would be in the unbuffered solution. Examples of pH variation in a solution initially at pH 9 upon addition of a strong acid, with a total concentration of HA of 0.1 M, when the p$K_a$ of HA is 7 (blue), 5 (red), or 3 (yellow). The dashed green curve indicates the pH drop if HA was absent (unbuffered solution, $0 < [H_3O^+]_{ad} < 0.1$ M). The concentration of the $H_3O^+$ added to the solution is $[H_3O^+]_{add}$ and is represented relative to the total concentration of HA used in the buffer ($[HA]_{total}$). The buffering zone (nearly horizontal pH variation lines) is mainly in the pH range that lies within p$K_a \pm 1$, and the buffering capacity (i.e., the limits of $[H_3O^+]_{add}$ in the buffer zone) depends on $[HA]_{total}$. This example illustrates how big amounts in added $H_3O^+$ cause modest changes in pH in the range p$K_a \pm 1$. The case in which a base is added to an acidic solution is similar. $OH^-$ concentration raises, which is concomitant with a decrease in $H_3O^+$, and the situation is symmetrical to the one obtained here (raise in pH instead of drop)

Although the Henderson–Hasselbalch equation is central in understanding fundamental pH buffer chemistry, its historical root is in medicine. Lawrence Henderson was a doctor interested in pH plasma alterations in pathology. He created the concept of acid–base equilibrium. Karl Hasselbalch studied the effect of $CO_2$ in hemoglobin with Christian Bohr (whom the Bohr effect was named after—see Sect. 3.3.3). The works of both doctors open the way for the scientific study of respiratory and metabolic perturbation of blood plasma pH equilibrium, a significant breakthrough in early twentieth-century physiology.

(continued)

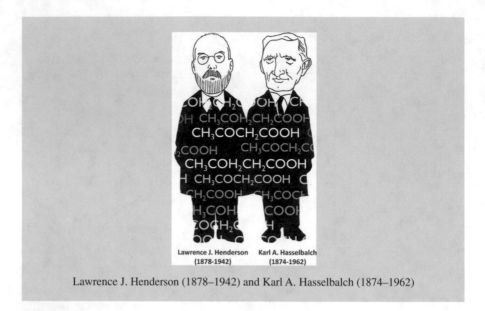

Lawrence J. Henderson (1878–1942) and Karl A. Hasselbalch (1874–1962)

$$HA + H_2O \leftrightarrows A^- + H_3O^+$$

$$K_{eq} = \frac{\left[A^-\right]\left[H_3O^+\right]}{\left[HA\right]}$$

Upon the addition of a strong acid, HA′ (by definition, strong acids have nearly complete dissociation), the concentration of $H_3O^+$ raises:

$$HA' + H_2O \rightarrow A'^- + H_3O^+$$

The newly formed $H_3O^+$ have an impact on the equilibrium of the weak acid, which will progress in the reverse order (formation of HA) in order to consume part of the $H_3O^+$. The extent of this consumption of $H_3O^+$ can be calculated based on the equilibrium constant, which remains unaltered: the concentrations of $H_3O^+$ and HA increase and the concentration of $A^-$ decrease up to the point where $K_{eq}$ is kept (see Box 2.1).

## 2.1.1   Principal Biological Buffers

Generalization of the previously mentioned concepts and equations should be made cautiously for several reasons:

1. From the point of view of scientific accuracy, activity coefficients should be used in addition to concentrations in all the abovementioned equations. Activity coefficients are used in thermodynamics to account for deviations from ideal behavior in a mixture of chemical substances. For practical reasons, it is assumed that activity coefficients do not alter equilibrium constants in the experimental conditions addressed.
2. Water dissociates to form $H_3O^+$; therefore $H_3O^+$ ions are always present in aqueous solutions. In the abovementioned equations and reasoning, the ionic product of water was never considered, which is valid in circumstances where the acidic species are present in concentration much higher than those involved in the ionic product of water: ($[H_3O^+] = 10^{-7}$ M)
3. We have explored the concept of buffers in situations in which chemical equilibrium exists. This is not always the case in biological situations.

Despite of all these limitations, the general concept of buffer still applies to cells and organs. Equations should be applied judiciously to chemical problems in living systems, but the concept of pH buffer in vivo is still valid. Wherever weak acids or bases are present in cells or body fluids, they contribute to form a buffer. Ions, such as $H_2PO_4^-/HPO_4^{2-}$ ($pK_a = 6.8$), are very important to buffer the cytoplasmatic pH (notice that the pH of cytoplasm is about 7.0, well within the buffering range of $H_2PO_4^-/HPO_4^{2-} = pK_a \pm 1 = 6.8 \pm 1$). On the other hand, $H_2CO_3/HCO_3^-$ is much more important to buffer plasma pH. The $-COO^-$ and $-NH_3^+$ groups present in plasma proteins also contribute to pH buffering, but not as much as the hydrogenocarbonates ($H_2CO_3/HCO_3^-$): while proteins contribute to 24% of the buffering capacity in plasma, $H_2CO_3/HCO_3^-$ contribute to 75% (the remaining 1% is a modest contribution of $HPO_4^{2-}/H_2PO_4^-$).

The dominance of $H_2CO_3/HCO_3^-$ in buffering plasma is not surprising because $CO_2$ diffuses freely in plasma, regardless of a certain ability hemoglobin has to bind $CO_2$. In erythrocytes, $CO_2$ is extensively converted to di-hydrogencarbonate by the enzyme carbonate-dehydratase:

$$CO_2 + H_2O \rightarrow H_2CO_3$$

$H_2CO_3$ is then present in plasma, where it is in equilibrium with hydrogencarbonate (also known under the popular name of "bicarbonate"), $HCO_3^-$:

$$H_2O + H_2CO_3 \leftrightharpoons HCO_3^- + H_3O^+ \quad pK_a = 6.1 \,(\text{in plasma})$$

$HCO_3^-$ exists in plasma with typical concentrations in the 22–26 mM range.

It should be stressed that the $H_2CO_3/HCO_3^-$ equilibrium in vitro has $pK_a = 3.8$, quite different from the $pK_a$ in plasma (6.1), which is said to be an apparent $pK_a$ because it is under the influence of the enzymatic production of $H_2CO_3$. To account for the change in pH due to fluctuation in the plasma levels of $CO_2$ one can apply the Henderson–Hasselbalch equation to the multiple equilibria:

$$CO_2 + H_2O \leftrightarrows H_2CO_3$$
$$\underline{H_2CO_3 + H_2O \leftrightarrows HCO_3^- + H_3O^+}$$
$$CO_2 + 2H_2O \leftrightarrows HCO_3^- + H_3O^+$$

$$K_{a,apparent} = \frac{\left[HCO_3^-\right]\left[H_3O^+\right]}{\left[CO_2\right]}$$

The water concentration is constant and is thus incorporated in $K_{a,apparent}$; the variables are on the right-hand side of the equation and the left-hand side of the equation is constant. It would be an unnecessary complication to have [$H_2O$], which is constant, in the right-hand side of the equation.

Now the question arises on what is the most appropriate way to express [$CO_2$] because, at the pressure of 1 atm and usual temperatures, $CO_2$ is a gas and molar units is not well suited for gases. [$CO_2$] should then be converted to partial pressure, $pCO_2$:

$$\left[CO_2\right] = 0.03 \times pCO_2$$

for [$CO_2$] in millimolar and $pCO_2$ in millimeter of mercury. Thus:

$$K_{eq,apparent} = \frac{\left[HCO_3^-\right]\left[H_3O^+\right]}{0.03\,pCO_2}$$

$$pH = 6.1 + \log\frac{\left[HCO_3^-\right]}{0.03\,pCO_2}$$

This equation helps in anticipating the drop in pH when $pCO_2$ increases, but it should be emphasized that an increase in $pCO_2$ also causes an increase in [$HCO_3^-$], mitigating the impact of the fluctuation of plasma levels of $CO_2$ in pH.

## 2.2  More than Only Chemistry: There Is Physics Too

Given the importance of chemical reactions in cells and in the organisms as a whole, one tends to forget the importance of physical processes in life. They are also important to understand life at the molecular level. Apart from all the issues related to the solubility of molecules in aqueous media, which were discussed in the beginning of this chapter, the interaction of light and biomolecules, for instance, is of utmost importance. This is clear when studying the vision of animals, the photosynthesis in plants, or the bioluminescence of microorganisms, just to name a few examples. Regarding human biochemistry in the health sciences field, the topics selected are the early events of vision and the effects of ultraviolet (UV) light in tissues such as the skin and hair. Other topics, such as the effects of radioactive and other non-optical radiation on humans, are left out of this book, although they are interesting and very relevant in particular situations.

When radiation interacts with matter, different kinds of phenomena may occur, such as absorption, scattering, or diffraction. Absorption implies that the energy of radiation matches the energy needed to change molecular states, this meaning that the energy of radiation is used in the change of the molecular configuration of nuclei and/or electrons so that the radiation extinguishes in the process (it is "absorbed"). This happens when a photon enters our eyes and triggers a radical change in conformation of retinol in the retina, thereby initiating the physiological process that ends

**Fig. 2.10** Dorothy Mary Crowfoot Hodgkin in her study at home portrayed by Maggi Hambling (oil on canvas, 1985). The four-handed scientist representation symbolizes her unusual working and entrepreneurial capacity. Dorothy Hodgkin made extremely valuable contributions to the advancement of biochemistry. She revealed the structure of cholesterol, penicillin, insulin, and vitamin B12, among other molecules. She also contributed to the refinement of X-ray diffraction spectroscopy to unravel the structure of proteins. A structural model of insulin is on the left. (Figure reproduced with permission of the National Portrait Gallery, London, UK)

with visual perception. Scattering relates to the radiation that interacts with molecules and is not absorbed, but causes the molecules to emit secondary radiation,
having the same energy or not. This is what happens when light impinges on matter
and is not absorbed: the radiation causes oscillations on the electronic cloud of molecules that in turn causes simultaneous emission of radiation, the scattered light.
Scattering is responsible for the white color of milk, blue color of the sky, or reddish
color of the sunset. Excessive light scattering in human eyes causes blurring in
vision. Diffraction is a particular kind of scattering that occurs when radiation
impinges on matter that has voids of size comparable to wavelength of the radiation.
X-rays, for instance, are diffracted by molecular crystals because the distances
between the nuclei, the chemical bonds' length, are similar to its wavelength, i.e., in
the order of tenth of nanometer. The spatial pattern of diffracted X-rays can be analyzed to reveal the 3D structure of crystallized molecules, even if they are quite big
and complex, such as proteins, which lead to historical landmark discoveries in
Biochemistry (Fig. 2.10).

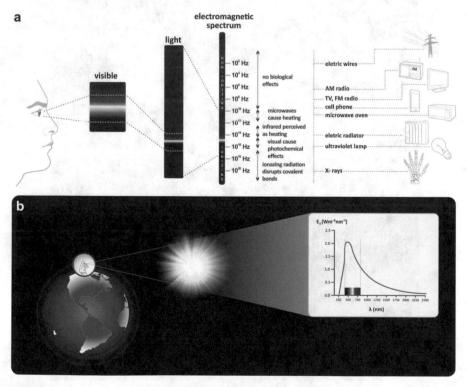

**Fig. 2.11** The electromagnetic spectrum in biochemistry and technology. (**a**) The electromagnetic
spectrum with the typical frequencies of radiation, the devices that use them, and their typical
biochemical/physiological effects. The narrow range of optical radiation (in practical terms, the
radiation that can be conducted using lenses and mirrors), to which human eyes are sensitive, is
highlighted. (**b**) Spectrum of total solar radiation power that reaches the Earth's surface. The maximum power range is coincident with optical radiation one calls "visible light," the radiation many
animal eyes, including human, have adapted to

**Box 2.2  The Toughest Cell on Earth**
In 1956, Arthur Anderson, working in Oregon (USA), was studying if canned food could be sterilized using high doses of gamma radiation. Even though he was using doses enough to destroy all forms of life known to man to sterilize meat, the food subsequently spoiled. A bacterium, *Deinococcus radiodurans*, was isolated and found to be responsible for this result. Later studies revealed that it is extremely resistant to ionizing radiation of different frequency, desiccation, and oxidizing agents. While a bacterium like *E. coli* can stand up to 200–800 joules per kg, i.e., gray, Gy, *D. radiodurans* can stand up to 5000 Gy. Human cells have much lower tolerance levels: below 5 Gy.

From the chemical point of view, the DNA composition of *D. radiodurans* is not different from that of other organisms. The key to resistance is not the chemical nature of DNA but the efficacy of the repair mechanisms. Radiation interacts with the chemical bonds of the DNA similarly in all cells, causing occasional breakings in these bonds when its energy is high enough. It is impossible for the vast majority of cells to cope with very frequent and simultaneous breaking of the DNA covalent bonds. *D. radiodurans*, however, has its DNA packed in toroids and multiple copies of the genome, usually 4–10. The whole repair machinery enables the reconstruction of the complete genome from shattered bits in hours. Bacterium-to-bacterium transfer of DNA may also play a role in increased resistance.

One interesting question that arises from the amazing properties of *D. radiodurans* is "Could a bacterium like this have developed in a planet where there are no environments with high doses of ionizing radiation?" It has been suggested that the origin of *D. radiodurans* is extraterrestrial because it has acquired resistance to a set of very harsh physical–chemical environmental conditions, like the ones expected in Mars. However, the bacterium is genetically, biochemically, and microbiologically very similar to other bacteria, and there is no other evidence for Martian forms of life. As dehydration and radiation cause very similar types of DNA damage, it is possible that resistance to radiation is a side effect from selective pressure toward resistance to dehydration. *D. radiodurans* is extremely well adapted to dryness.

The remarkable biochemical skills of *D. radiodurans* may make it a powerful ally to clean contaminated radioactive areas, in which the bacterium can survive and operate. This kind of approach is named bioremediation. Radioactive toxic waste would be processed by bacteria that would naturally colonize risk areas.

The extreme resistance of *D. radiodurans* inspired researchers and science communication professional, who frequently call it "Conan bacterium." In 1998, the *Guinness Book of World Records* listed *D. radiodurans* the "most radiation-resistant life form."

(continued)

**Box 2.2** (continued)

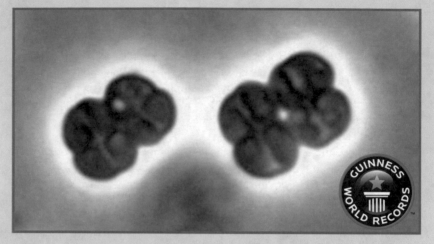

*Deinococcus radiodurans* **micrograph. (Courtesy of Sandra P. Santos and Célia Romão, ITQB-UNL, Portugal)**

The all range of frequencies (i.e., energies) of radiation known so far constitutes the so-called electromagnetic spectrum (Fig. 2.11a). It includes low-energy (low frequency, long wavelength) radiation, such as radio and television radiation, as well as high-energy (high frequency, short wavelength) radiation, such as X- and gamma-rays, capable of ionizing molecules and disrupting covalent bonds. Very low-energy radiation hardly causes changes in molecules and therefore is not prone to produce effects in living organisms. High-energy radiation (starting from the far, high-frequency, UV radiation) is able to disrupt molecules by destroying chemical bonds and is therefore a potential hazard to the chemistry of life (an amazing exception is presented in Box 2.2). It is not surprising, therefore, that natural phenomena in living beings involving radiation occur in a limited range of energies, from microwaves to near (lower frequencies) UV. Even so, one has to wonder why many animal eyes, including human, use a much smaller fraction of radiation, from about 400 nm (blue light) to about 800 nm (red light). The answer is simple and can be found in molecular evolution toward optimized use of environmental resources: the 400–800 nm range accounts for the most abundant sun radiation reaching the surface of planet Earth (Fig. 2.11b). Naturally, this dictated the course of evolution of vision. The same happened with photosynthetic organisms.

In the human eye, rhodopsin, a membrane protein (opsin) that exists in retina rod cells, has a covalently bound retinal molecule, which adopts two stable isomeric

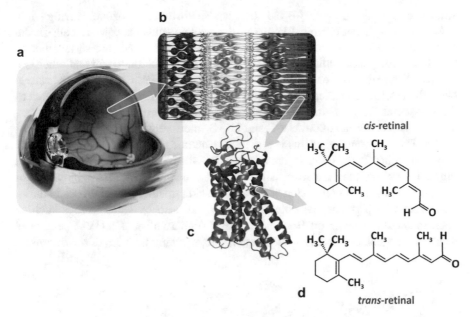

**Fig. 2.12** Eye and vision. (**a**) Anatomical cross section of the eye. (**b**) Retina, where rod cells are located. (**c**) Sensory rhodopsin (PDB 2I35), which binds retinal. (**d**) Retinal isomeric forms, which interconvert due to light absorption

forms: cis and trans (Fig. 2.12). When the protein-attached cis-retinal absorbs light, it converts to trans-retinal, this being the triggering event of vision. Not only retinal changes in shape, from a bent structure to a nearly linear arrangement, but it also detaches from opsin. All trans-retinal (i.e., the "linear" retinal, with all double bonds in the trans-configuration) does not fit the pocket formed by the opsin transmembrane helices as the *cis* isomer does, causing tensions in the protein structure, which has to change conformation (shape) to adapt. The rhodopsin conformational changes trigger a series of intracellular reactions that ultimately cause an electric current along the cell. The intensity of this current is proportional to the intensity of light that reached the retina. So, an electric impulse is created that is detected by the ganglion cell and then the optic nerve. Nerve fibers reach the back of the brain (occipital lobe), where images are created. This region is called the primary visual cortex. Some of the visual fibers go to other parts of the brain to help control eye movements, the responses of the pupils and iris, and behavior.

The effects of UV light on the skin and hair constitute other examples on how the radiation–matter interaction is important in natural processes. In practical terms, UV are divided in three categories, A to C, according to their biological effects. UV-A has the lowest energy range (320–400 nm). UV-C has the highest energy range (200–290 nm) and is, to a great extent, filtered by the ozone layer in the atmosphere. A part of the UV-B radiation is also absorbed by ozone. Having sufficient energy to break chemical bonds in the nucleic acids of exposed cells in the skin layers near the surface, UV-C and, to a less extent, UV-B are serious health threats regarding skin tumors, mainly in prolonged exposition to sun and other UV sources. Whenever the ozone layer is threatened, the risk of skin cancer raises. Nevertheless, the immediate most noticeable effects of UV-B are erythemas, such as those associated with "sunburns" in short-term solar overexposure. Being moderate- to high-energy radiation, UV-C and UV-B interfere with many molecular processes. The consequence of this is that they penetrate shallow in the skin (they are totally absorbed in the outer layers of the skin—the epidermis and nearby dermis; Fig. 2.13), where they are responsible for some naturally occurring biochemical processes, such as the conversion of the amino acid tyrosine in melanin (a sun radiation–protection pigment that colors the skin), but also by some lesions such as detachment of the epidermis from the dermis ("sunburn blisters"). UV-A penetrates deeper in the skin because it is not so energetic, reaching the core of the dermis. It

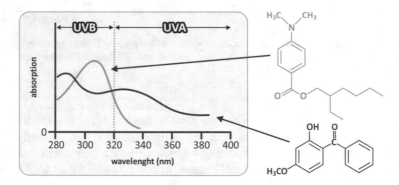

**Fig. 2.13** Two UV-absorbing molecules commonly found in sunscreen formulations: octyldimethylaminobenzoate (OD-PABA), in blue, and 2-hydroxy-4-methoxybenzophenone (HM-BZP) in red. The chemical group responsible for the absorption of radiation ("chromophore") is the aromatic (benzenic) ring, which is present in both molecules and many other components of sunscreens as well

also stimulates the formation of melanin ("suntan"), which provides some protection of the skin against the effects of solar radiation. Nevertheless, contrary to popular belief, the extent of this protection is very limited (sun protection factor, SPF, of about 2–4, far below the minimum recommended SPF of 15 for direct skin exposition to sun radiation).

The SPF is determined from the UV exposition time that is needed for the appearance of a minimal erythema (i.e., erythema after 24 h). Specifically, it is the ratio of the time needed for a minimal erythema in protected skin over the time needed in unprotected skin. It is determined indoors with a light source that is meant to reproduce the noontime sun. The SPF is mainly useful to have a quantitative scale of UV-B protection when sunscreens are used. A sunscreen with a SPF of 10 filters 90% of the UV-B light, for instance (in other words, a person that would have sunburn after a 20-min exposition to sun, using a sunscreen of SPF 10, has the same

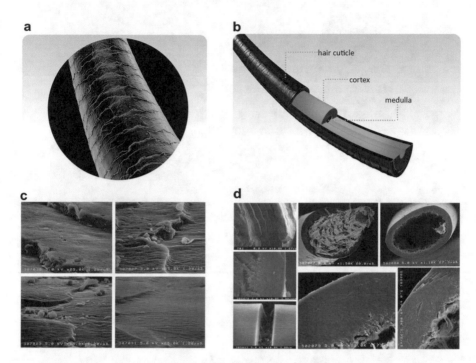

**Fig. 2.14** The microscopic structure of hair and how it is affected by UV radiation. (**a**) Hair surface as seen under the electronic microscope reveals scales. (**b**) The external layer of the hair shaft, named cuticle, is formed of proteic plaques that slide over each other due to the lubrication of lipids, which makes hair flexible. (**c**, **d**) Intense UV exposure leads to "fusion" of the plaques in the whole cuticle. "Fused" cuticles lose flexibility and break easily (**d**). (Panels **c** and **d** were reproduced from Ruetsch et al. in Comprehensive Series in Photosciences 3:175–205, 2001, with permission from Elsevier)

effect after 200 min). Sunscreen creams, lotion, sprays, gels, or other topical formulations have organic molecules in their composition that absorb UV light. Notice that the most common of these compounds have aromatic rings in their structure (Fig. 2.13), which are groups that absorb UV light, and/or inorganic metallic compounds (for instance, zinc oxide) that scatter UV light (the scattering efficacy is higher for UV than for visible light).

Although the effects of UV radiation on the skin usually monopolize one's attention, the biochemistry involved on the damage caused by UV radiation on other tissues, such as hair and eyes, is worthy of consideration. Hair is an extremely well-adapted cellular organization: it conjugates flexibility and strength. Hair has an external proteic cuticle, an inner core named medulla, and an intermediate layer

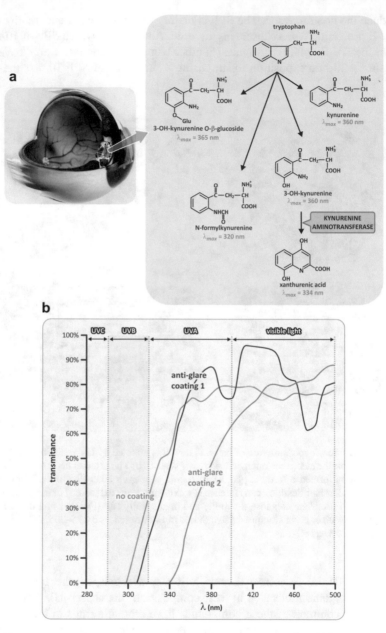

**Fig. 2.15** Natural eye preventive protection against UV. (**a**) A family of tryptophan derivatives in the crystalline protects the eye against UV radiation. Each of these molecules absorbs UV light at a slightly different wavelength range. Acting together, they are able to filter a broad range of UV light entering the eye. (**b**) Light filtering by vitreous material having two different possible coatings (transmittance is the fraction of light intensity that crosses the material). Wearing glasses may be a form to reinforce the protection of eyes against UV radiation. Depending on the materials used in their fabrication, lenses may absorb UV and/or visible light. Transparent lenses do not interfere significantly with visible light but may filter, at least partially, UV light. Sunglasses having good quality lenses usually filter a substantial fraction of visible light (therefore the term "dark glasses" also used for sunglasses) and most UV light

called cortex (Fig. 2.14). The cuticle is formed of proteic plaques that are lubricated by the presence of lipids. The plaques are rigid, but lubrication allows flexibility. Exposure to UV light, mainly UV-B, causes "fusion" of these plaques, turning the cuticle in a rigid not-flexible structure as a whole. The hair then becomes prone to break. Hair with broken terminals loses its natural appearance, and UV protection of hair is one of the main topics in cosmetic research. Loss of coloration due to oxidation of melanin pigments caused by the UV radiation is another factor that contributes to the loss of natural appearance.

While the solar UV-B radiation can be blocked and prevented from entering and damaging our eyes using glasses (not necessarily using dark lenses because UV rays are colorless, so transparent materials may be efficient in blocking UV light), one should ask ourselves how natural molecular evolution coped with the unavoidable exposure of eyes to UV radiation. The answer lays in the same concept as chemical sunscreens one uses to protect the skin. All light enters the eye through the crystalline (also named "lenses," Fig. 2.15). There are natural molecules that absorb UV light that are present in the crystalline. To ensure an efficient filtering of the whole range of UV light, human eyes' molecules evolved to create a family of compounds derived from tryptophan, an amino acid that enters the constitution of proteins. All these molecules absorb UV light at slightly different wavelength ranges. The effect of the sum of all the spectra is a generalized blocking of UV radiation at the entrance of the eye. This is the natural barrier to UV radiation that protects the retina.

It should be stressed that cells have biochemical processes that repair the damage caused by the UV radiation on nucleic acids and other molecules. As this section focuses on light–matter interaction, this subject will not be further addressed, but it is important to keep in mind that even when molecular and cellular lesions occur due to the action of radiation, repair mechanisms exist. Pathological situations arise when both preventive and repair mechanisms fail to account for all the aggressions radiation exposure imposes.

## Selected Bibliography

Dobson CM, Gerard JA, Pratt AJ (2001) Foundations of chemical biology, Oxford chemistry primers. Oxford University Press, Oxford
Gensler WJ (1970) Physical versus chemical change. J Chem Educ 47:154–155
Hubel DH (1995) Eye, brain, and vision. Scientific American Library Series, New York, NY

## Challenging Case 2.1: Going Deep into Acidosis

### Source

This case is inspired by the experience of freediving, one of the daring forms humans use to push the limits imposed by nature. Umberto Pelizzari, an Italian freediver born in 1965, established world records in all the existing disciplines of freediving, being considered among the best of all time. In the profile section of his website, he describes in a very personal perspective the experience of freediving, summarized as the following: *"The scuba diver dives to look around, the freediver dives to look inside."*[1]

Umberto Pelizzari. (Picture reproduced with permission from of his website)

### Case Description

Athletes who practice freediving aim to achieve the longest or the deepest breath-hold dive without the use of any breathing apparatus such as a scuba gear. Currently, there are 11 recognized disciplines of competitive apnea defined by AIDA (International Association for the Development of Apnea) and CMAS (Confédération Mondiale des Activités Subaquatiques). Depending on the discipline, the competitors attempt to attain great depths, times, or distances on a single breath, assisted or not by weight to descend or inflatable bag to ascend. As examples of more recent AIDA recognized world records, we have, for static apnea, 11 min and 35 s, by Stéphane Mifsud (men), in 2009, and 9 min and 2 s, by Natalia Molchanova (women), in 2013; and for constant weight apnea, 130 m in 3 min and 55 s, by Alexey Molchanov (men), and 107 m in 3 min and 44 s, by Alessia Zecchini (women), both in 2018 (see more information in AIDA website[2]).

Freedivers face as one of their major challenges the interruption of both pulmonary $O_2$ uptake and $CO_2$ excretion, and the primary consequence of this situation is an inadequate $O_2$ delivery to tissues, a condition known as hypoxia. Although hypoxia is experienced by many vertebrates, especially those living in aquatic environments, it is

---

[1] http://www.umbertopelizzari.com/en/profilo/.

[2] https://www.aidainternational.org/worldrecords.

a condition to which human beings, in principle, are not well adapted. Hypoxia is particularly dangerous to brain functioning. The brain of an adult human corresponds to about 2% of its body weight, but accounts for more than 20% of the organism's energy consumption, as it is the most sensitive organ to oxygen deficit. Brains cannot work anaerobically; very short periods of oxygen deficiency cause unconsciousness as a protective response against brain damage. After few minutes without oxygen, neuron damage may become irreversible, drastically affecting the nervous system functions.

So, under such threatening condition, why are humans attracted to freediving? Probably inspired by aquatic mammals and birds, such as seals, sea lions, whales, and penguins, humans used freediving as a means to gather food: it was a matter of survival for some ancient cultures. Diving was later used in war defenses, such as underwater barricades, and to harvest resources, such as sponges, used for bathing in ancient Greek, as well as pearls, retrieved by divers more than 2000 years ago, mainly in the Indian Ocean. But the development of freediving as a recreative activity has to be explained by other human necessities. In his website, Umberto Pelizzari makes a reflection about his personal perspective of the experience of freediving: *"From the depth of 100 m and more, headlong in the abyss, the heartbeat gets slower, the body disappears, and all the feelings take a new form. The only thing that remains in us is the soul. We take a long jump into the soul, which seems to absorb the universe. Every time I ascend, I am making a choice: it is me who is re-discovering myself in my human dimensions, metre by metre, to come up to see the light again. It often happens that I am asked what is there to see deep down in the sea? Maybe the only possible answer is that one does not descend this way to look around, but to look into himself. In the deep I look for myself. This is a mystical experience bordering on the divine. So deep down, I am immensely alone but inside it feels as if all humanity is with me. It is by being human that I surpass the limits we set for ourselves and diving makes us one with the sea and its surroundings. It is here that I become one with the sea and discover my true self. The scuba diver dives to look around, the freediver dives to look inside."*

Being a matter of survival or a whim of human beings, breath-holding and, consequently, $O_2$ deprivation cause a drastic increase in the anaerobic metabolism, which produces lactate and $H_3O^+$. This may significantly decrease blood pH, resulting in a condition known as metabolic, or non-respiratory, acidosis. Additionally, the retention of the $CO_2$ produced in cellular metabolism causes an increase in the partial pressure of $CO_2$ ($pCO_2$) in the blood, which may overcome the buffering capacity of the $H_2CO_3/HCO_3^-$ system, contributing to the establishment of the acidosis condition, in this case, a respiratory acidosis.

A number of physiological adaptations allow the organism to deal with the hypoxic stress as well as with the acid–base challenge imposed by diving.[3] This includes a group of cardiovascular responses collectively known as the "diving response," first described in the classic works of the physiologist Per Fredrik Scholander. The diving response is caused by simultaneous activation of the sympathetic and parasympathetic components of the nervous system and is characterized by (1) bradycardia, i.e., the deceleration of the heartbeat; (2) a selective peripheral

---

[3] Ostrowski A et al (2012) The role of training in the development of adaptive mechanisms in freedivers. J Hum Kinet 32:197–210.

vasoconstriction; (3) an adrenal hormonal reaction; and (4) hypometabolism, a remarkable reduction of body metabolic activities.

Regarding acidosis specifically, peripheral vasoconstriction results in a retention of lactic acid and $CO_2$ in the tissues where they are produced, delaying their release into the circulation and, thus, minimizing the metabolic and respiratory components of acidosis.

## Questions

1. The central chemoreceptors responsible for the increase in the rate and depth of breathing are primarily sensitive to changes in the local concentration of hydrogen ions ($H^+$; protons). Based on this fact, explain why one can affirm that the respiratory rate is mainly controlled by the partial pressure of $CO_2$ (p$CO_2$) in the blood (i.e., blood $CO_2$ concentration).

2. The impressive records reported for all disciplines of competitive apnea imply that training improves the adaptative mechanisms of the cardiorespiratory system that minimize the hypoxia-induced tissue damage. Discuss the adaptations that allow elite freedivers or artistic swimmers, for instance, to increase their tolerance to long periods without breathing.

3. The next figure shows the results of an experiment performed with caimans.[4] Blood lactate concentration, p$CO_2$, and pH were measured during a cycle of diving and surface breathing. Analyze the results and discuss what happens with the metabolic and respiratory components of acidosis during the diving and when the animal returns to the surface and breathes.

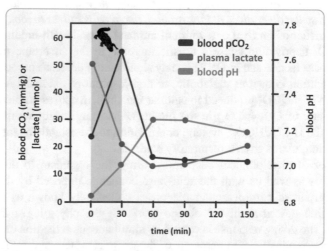

Blood pH variation after a cycle of diving (blue bar—between 0 and 30 min) and surface breathing in caiman (*Caiman latirostris*). (Figure adapted from the article cited in the footnote 4)

---

[4] Jackson DC (2004) Acid-base balance during hypoxic hypometabolism: selected vertebrate strategies. Respir Physiol Neurobiol 141:273–283.

4. Blood pH should be tightly controlled, meaning that the $pCO_2$ vs. $HCO_3^-$ levels have to be maintained steady. The pH of blood should be around 7.4. According to the American Association for Clinical Chemistry (AACC), acidosis is characterized by a pH of 7.35 or lower. Making use of the Henderson–Hasselbalch equation, evaluate whether the $HCO_3^-/H_2CO_3^-$ buffer suffices to account for pH regulation within physiological limits.

## *Biochemical Insights*

1. $CO_2$ dissolved in the blood is converted to di-hydrogencarbonate, or carbonic acid ($H_2CO_3$), by carbonase-dehydratase (or carbonic anhydrase), an enzyme present in the erythrocytes. $H_2CO_3$ dissociates to bicarbonate (hydrogen carbonate, $HCO_3^-$) and hydrogen ions ($H^+$). Therefore, an increase in $CO_2$ concentration in the blood causes an increase in $H^+$ concentration, which, by definition, decreases the pH of the blood. However, it is important to have in mind that the blood–brain barrier (BBB) is not permeable to $H^+$. On the other hand, $CO_2$ easily diffuses across the BBB. Therefore, although changes in plasma pH are not able to stimulate central chemoreceptors, an increase in blood $pCO_2$ directly affects the cerebrospinal fluid (CSF) pH, and is much more effective in triggering respiratory center activity (see below figure).

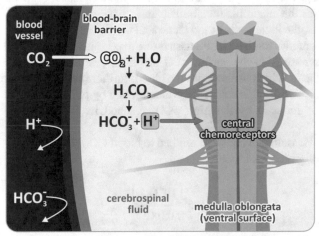

Central chemoreceptors' stimulation by $H^+$. Since BBB is not permeable to $H^+$, the $H^+$-induced activation of central chemoreceptors is indeed affected by changes in blood $pCO_2$. $CO_2$ easily diffuses across BBB reaching the cerebrospinal fluid (CSF), where it is converted in $H_2CO_3$ by carbonic anhydrase, which then dissociates to $HCO_3^-$ and $H^+$

2. Elite freedivers develop the ability to shift the respiratory stimulus to higher $pCO_2$. However, while this adaptation allows them to hold their breath for longer periods, it may lead to a dangerous decrease in $O_2$ blood saturation during the diving. Thus, freedivers must develop adaptations to increase their tolerance to anoxia as well as to decelerate $O_2$ consumption in the body (see detailed discussion about these adaptations in reference 3).

    Different components of the diving response contribute to the adaptation to long-term breath-holding, including bradycardia, hypometabolism, and selective peripheral vasoconstriction. Bradycardia is a diving reflex that generally occurs in sea mammals, being induced by the breath-holding itself but also as a response to the submersion of the face in water, especially when it is cold. The heartbeat may decrease to even 20 beats/min, and this, together with the reduction of body metabolic activities as a whole (hypometabolism), decreases the general cellular demands on $O_2$. Additionally, the selective peripheral vasoconstriction directs the circulating blood to the heart and central nervous system, maintaining $O_2$ delivery to these organs for longer periods without breathing and slowing the rate of $O_2$ depletion by limiting the use of $O_2$ by other tissues. This ensures that the cerebral oxygenation is reduced only slightly until the end of the breath-hold. Another important adaptation is related to the $O_2$ reserve in elite divers' lungs. It is about threefold higher than in non-divers, allowing that arterial blood $O_2$ saturation goes close to 100% in the well-trained athletes, even after a breath-hold for some minutes.

3. During the diving, the low $O_2$ delivery to the peripheral tissues, especially the muscles, increases the anaerobic metabolism and, consequently, lactic acid production. However, the increase in blood lactic acid concentration is not so prominent because this acid is retained in the tissues due to peripheral vasoconstriction. When the diver returns to the surface and breathes, peripheral circulation is recovered, and lactic acid is flushed into the blood, further increasing acidosis in the first 30 min. However, this condition is ameliorated by pulmonary hyperventilation, which is induced by the high $pCO_2$ reached in the end of the diving period.

4. The Henderson–Hasselbalch equation applied to blood $CO_2/H_2CO_3/HCO_3^-$ buffering is quite straightforward:

$$pH = 6.1 + \log \frac{\left[ HCO_3^- \right]}{0.03 pCO_2}$$

This is valid for 1 atm pressure at 37 °C.
The conversion factor has units 0.03 mEq $L^{-1}$ $mmHg^{-1}$.
When the ratio $[HCO_3^-]/pCO_2$ changes, the difference in pH, $\Delta pH$, is:

$$\Delta pH = \log \left[ \left( \frac{pCO_2}{\left[ HCO_3^- \right]} \right)_{initial} \middle/ \left( \frac{pCO_2}{\left[ HCO_3^- \right]} \right)_{final} \right]$$

For the sake of simplicity, $\left(\dfrac{pCO_2}{\left[HCO_3^-\right]}\right)$ may by represented by $x$ and so:

$$\Delta pH = \log\left[x_{initial} / x_{final}\right] = \log\left[x_{initial} / \left(x_{initial} + \Delta x\right)\right]$$

$\Delta x$ is the variation in $x$: $\Delta x = x_{final} - x_{initial}$.
This equation is equivalent to:

$$\Delta pH = \log\left[\dfrac{1}{1+\dfrac{\Delta x}{x_{initial}}}\right] = \log\left[\dfrac{1}{1+f_x}\right]$$

$f_x$ is the increase in $x$ relative to the initial value, $x_{initial}$.

This other form of the Henderson–Hasselbalch equation allows for calculating the variation in pH when $x$ varies by a certain percentage (see next figure) in the simplistic scenario of having the buffering system as the only regulator of pH. The onset of acidosis corresponds to a drop in pH of circa 0.05, which is expected for a moderate increase of $x$ of only 12%.

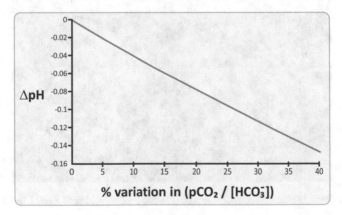

pH dependence on the % variation in $pCO_2/[HCO_3^-]$. A moderate increase of 12% in $pCO_2/[HCO_3^-]$ would lead to a pH drop of 0.05, i.e., onset of acidosis, if the $HCO_3^-/H_2CO_3^-$ buffer was the only protective mechanism against respiratory acidosis

Realistically, pH homeostasis in blood depends on three main mechanisms: buffering, respiratory regulation of $pCO_2$, and renal regulation of $[H_3O^+]$ and $[HCO_3^-]$. These factors concur to maintain steady values of $\dfrac{pCO_2}{\left[HCO_3^-\right]}$. Buffering is not very powerful, but it is certainly very fast as it does not depend on metabolic or physiological processes. The acid–base buffering of the blood is rapidly responsive and therefore is the forefront mechanism against respiratory acidosis.

## Final Discussion

The metabolic adaptations during breath-hold diving also include hormonal responses. As a result of stress caused by breath-holding, an intense production of catecholamines by the adrenal glands occurs. This induces another important adaptation related to the diving response: "the spleen effect," which consists of spleen contractions resulting in the injection of an additional supply of oxygenated erythrocytes into the bloodstream (spleen stores erythrocyte-rich blood). This increases the blood oxygen transport capacity enhancing the resistance to the breathhold diving.

The spleen effect was firstly described for the *Ama* (means "sea woman"), a traditional group of Japanese pearl divers. They are mainly women and are known to wear white clothes (see below picture).

*Ama* female divers in Mie Prefecture, in Japan. (Picture credit: Mikimoto Pearl Island)

More recently, a very interesting study published in the scientific periodical *Cell*[5] revealed a genetic basis for the spleen effect. A research group from the University of Copenhagen performed genomic analyses of the Bajau, a population living in Southeast Asia (see localization in the below map) also known as "Sea Nomads," whose subsistence lifestyle is entirely based on breath-hold diving. These people are known by their extraordinary breath-holding abilities, spending most of their

---

[5] Ilardo MA et al (2018) Physiological and genetic adaptations to diving in Sea Nomads. Cell 173:569–580.

time underwater and diving over 70 m using only a set of weights and a pair of wooden goggles.

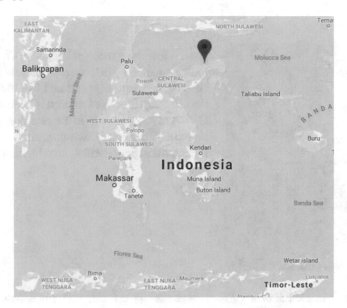

The researchers first compared Bajau's spleen size with that of another population, the Saluan people, who, although living very close to the Bajau, are not genetically related to them and have a completely different lifestyle, mainly based on agriculture. They found that the Bajau spleen was on average about 50% bigger than the Saluan spleen.

After the comparative genomic analyses, it was possible to find a genetic basis for this difference (see also a comment of the result published in the *National Geographic* magazine[6]). So, the natural selection of a genetic variant allowed the Bajau to be adapted to their means of survival, explaining, at least in part, their impressive diving abilities.

---

[6] https://news.nationalgeographic.com/2018/04/bajau-sea-nomads-free-diving-spleen-science/

## Challenging Case 2.2: Reindeer Know the Color of UV Light

### Source

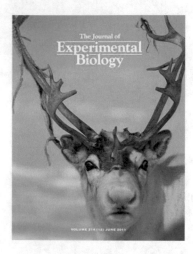

Glen Jeffery and colleagues (University College London) discovered that reindeers can see UV (ultraviolet) light and use this perception to find food and stay safe. Sensitivity to UV light was known in insects but is rare in mammals. The study was published in 2011, in the scientific periodical *Journal of Experimental Biology*.[7]

Cover of the periodical *Journal of Experimental Biology*, vol. 214, 2011. (Reproduced with permission; image credit: Kia Krarup Hansen)

### *Case Description*

Living in the Arctic in the winter is challenging for vision: the sun barely rises in the middle of the day, and blue light, the visible light most efficiently scattered, is not abundant. UV light is more efficiently scattered but not perceptible to the vision of almost all animals. However, an amazing exception was found: reindeers can see UV light! Considering snow reflects most UV light that falls on it, this is an enormous advantage. Color patterns not perceived by other species are recognized by reindeers. This is the case of urine waste in the soil, meaning the presence of predators or competitors. Likewise, food such as lichens contrast more with the snow background, making it easier for reindeers to find them. The furs of wolves, for instance, also contrasts more pronouncedly with snow, making camouflage and stealth attacks on reindeers much more difficult. Interestingly, it was known that rodents and some species of bats perceive UV light in other environments, but the advantage of this capability remains elusive in these species.

Jeffery's team first shone LED light of different wavelengths, including UV, into the eyes of 18 anaesthetized reindeers while recording with an electrode whether nerves in the eye fired, indicating that the light had been seen. The UV light

---

[7] Hogg C et al (2011) Arctic reindeer extend their visual range into the ultraviolet. J Exp Biol 214:2014–2019.

triggered a response in the eyes of all the reindeer. Then they used a UV camera in the Artic to analyze patterns in the environment, as if the camera was an artificial reindeer eye to give a "black and white" idea of the presence of UV stimulus. That was when they noticed the effect caused by urine, lichens, and fur.

Later, Jeffery proposed that sensitivity to UV may be the reason why reindeers, like some other mammals and some birds, avoid power lines.[8] They see UV discharges as "corona" of light along power cables and also as intermittent bright flashes from cable insulators. Although this is a plausible explanation for reindeers to avoid power lines during the Arctic dark winter, it is puzzling that the effect during the day, when UV light is everywhere, is the same.[9]

In addition to the interesting facts revealed about reindeer biology, the research raises important perspectives on eye health. In human eyes, UV light causes damages, sometimes irreversible, to the human retina. Reindeers, in contrast, handle UV light without negative consequences. The molecular basis for UV sensitivity and protection are not known. It may be that reindeers have a unique way of protecting themselves. If so, how can we learn from them? *"We can learn a lot from studying the fundamental biology of animals and other organisms that live in extreme environments. Understanding their cell and molecular biology, neuroscience, and other aspects of how they work can uncover the biological mechanism that meant they can cope with severe conditions. This knowledge can have an impact on animal welfare and has the potential to be taken forward to new developments that underpin human health and wellbeing,"* said rightly by Douglas Kell, Chief Executive of the agency that funded the study of Glen Jeffery.[10]

## Questions

1. Why can humans not see UV light?
2. When using a UV camera to image a person, what differences do you expect to see before and after the person applies a sunscreen cream on his/her face?
3. Plan a simple lab experiment to test the efficacy of sunscreen creams and glass lenses in blocking UV light transmission.
4. Are humans sensitive to IR radiation?

---

[8] Tyler N et al (2014) Ultraviolet vision and avoidance of power lines in birds and mammals. Conserv Biol 28:630–631

[9] Cressey D (2014) Why reindeer steer clear of power lines, Nature News.

[10] EarthSky team (2012) Reindeer see a twilight world in UV light. EarthSky News. http://earthsky.org/earth/reindeer-see-a-twilight-world-in-uv-light.

## *Biochemical Insight*

1. In addition to the wavelength of absorption of visual pigments, human lenses do not let UV radiation through into the eye due to the presence of tryptophan derivatives (see Fig. 2.15).

2. This is the situation presented by the video titled "How the sun sees you" (https:// youtu.be/o9BqrSAHbTc), which readers are encouraged to watch. Sun light, UV light included, is reflected by the person in front of the camera. UV light that is reflected or scattered by the skin (or cloths, lenses of glasses, or creams applied on the skin) and impinges on the camera lens is collected by the camera and appears white in the image. On the contrary, UV that is absorbed does not reflect or scatter into the camera and the image is thus black. Therefore, areas of the skin reflecting UV appear white, while areas absorbing UV appear black. When applying sunscreen cream, the areas covered by the cream appear black, as seen in the following figure. Likewise, areas covered by lenses of glasses appear black. This holds even if the glasses are not sunglasses as glass transparent to visible light may absorb UV light.

Sunscreen cream covering part of the cheek contrasts with the unprotected skin. The cream absorbs the UV light with negligible absorption of visible light, so it appears black in the UV camera. (Frame from the video "How sun sees you," https://youtu.be/o9BqrSAHbTc. Image credit: Thomas Leveritt)

3. The following experiment is very simple, fast, and informative:

   (a) Use a quartz glass surface (like the quartz glass used in cuvettes for UV–Vis absorption spectroscopy). Quartz glass is transparent (does not absorb) to UV light. Alternatively, you can use special plastic, transparent to UV light.

   (b) Spread a thin layer of sunscreen cream of different protection factors or a regular hydration cream on the cuvette surface.

   (c) Place the surface covered with cream layer in the sample compartment of a UV–Vis spectrophotometer, perpendicular to the light beam that impinges on the sample compartment.

4. Yes, but not as visible radiation. Humans perceive IR radiation as heat.

(d) Register the UV–Vis spectrum covering the UV and visible region on the spectrum (e.g., 250–500 µm). Do not forget to account for the blank.
(e) Check the cutoff wavelength of light absorption.

Comparing the absorption spectra of the sunscreen creams and the hydration cream allows for studying the reason behind different protective powers. The same experiment can be carried out with glass lenses. In this case, place the lens perpendicular to the light beam directly in the sample compartment. Compare lenses from sunglasses and regular transparent lenses. See typical results for these experiments in the following figure.

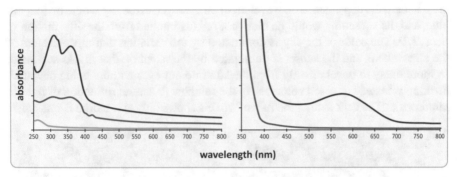

Sunscreen cream UV-Vis absorption spectra. (**a**) UV–Vis absorption spectra showing the wavelength dependence of the absorbance of sunscreens with factors 50 (red) and 30 (blue), compared to a regular hand cream (green). (**b**) UV–Vis absorption spectra of lenses of sunglasses (red) and regular transparent glasses (blue). Cutoff absorption wavelengths of 400 and 375 nm are detected for the glasses, and 400 nm for the sunscreen creams, meaning the sunscreen and glasses protect for UV light having wavelength lower than these values. The difference between sunglasses ("dark" lenses) and regular glasses with transparent lenses is the absorption of light in the visible region of the radiation spectrum

## *Final Discussion: Examples of Missing and Misplaced Colors in Literature and Music*

Colors no one has seen cannot be named or described. An arbitrary name can be given but arbitrary perceptions are impossible to describe. Unperceived realities are non-existing. They cannot be described, so cannot be communicated, so are no part of the tangible nature. *Flicts* is a book for children written by a famous Brazilian writer and artist, Ziraldo, about a different color that does not fit in the rainbow, on the flags of the world or anywhere else. The unknown color itself, named Flicts, is the main character of the book. It feels misplaced and sad for not having the strength of color red or the greatness of yellow or the peace that resides in blue. [Bear in mind that blue is the color of the poets, interestingly.] Flicts finally finds its place being the color of the dark side of the moon. The side very few human have seen. Neil Armstrong, the astronaut who first stepped on the moon, confirmed to Ziraldo that the dark side of the moon is flicts after reading the book. *Flicts* was first published in 1969, the same year Neil Armstrong landed on the moon. The book was a tremendous success, with more than 400,000 copies sold in Brazil and translation to several other languages.

Curiously, "The dark side of the moon" is the title of one of the bestselling rock albums of modern history, by the band Pink Floyd. It has a song titled "Any colour you like." It's instrumental … No lyrics. Further enticing our correlations, the iconic album covers pictures a glass prism causing refraction of light, which divides into different colors as the refraction angle depends on wavelength (violet, having lower wavelength, has the higher refraction angle—see the album cover in the next figure). Should the dependence on wavelength be the reverse, the sky would be red, not blue, and the sunsetting would be blue, not red (as implied from the information in Sect. 2.2). The color of the sky is dominated by the radiation that most scatters in the atmosphere, and the sunset is dominated by the radiation that is less scattered, so more likely to travel directly from the sun into our eyes without being deviated. Strikingly, instead of seven colors, as in the rainbow (a natural refraction of light in raindrops), "The dark side of the moon" covers shows only six: indigo is missing.[11]

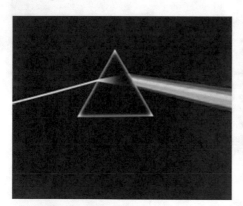

Cover and back cover of "The dark side of the moon." Indigo color is missing. Another curiosity: Imagine many covers and back covers placed side by side and realize they would form a continuum in which a light beam is endlessly decomposed and recombined. Image by Storm Thorgerson/ Hipgnosis from cover and back cover of Pink Floyd's album "The dark side of the moon," 1973. (Reproduced with permission of Paul Loasby)

---

[11] Harris J (2006) The dark side of the moon: the making of the Pink Floyd masterpiece. Harper Perennial, London. ISBN: 9780007232291.

## Challenging Case 2.3: *Malacocephalus*—The Strange Case of the Radioactive Fish from the Deep Sea

### Source

The centennial library of the School of Sciences of the University of Lisbon (Portugal) hides some surprising relics. Among them, a scientific report from 1911, presenting a presumed radioactive fish captured offshore from Sesimbra, Lisbon, in Portugal, caught our attention.[12]

Cover of the fascicle 1 of the "Memórias do Museu Bocage," 1909. (https://archive.org/details/memoriasdomuseub01muse/page/n2)

### Case Description

Balthazar Osório (1855–1926, Portugal) was a surgeon and a zoologist. He devoted his life to medicine, teaching zoology, and researching on sea fish and crustaceans. Balthazar Osório published a preliminary note in 1911 about a finding in a fish, *Malacocephalus laevis*, which seems bizarre in present days: a radioactive pocket present in the gut.

Balthazar was puzzled by the ancient practice of the fishermen of Sesimbra in baiting named "candil." They used to fish "rat fish" (*Malacocephalus laevis*; see

---

[12] Osorio B (1911) Phenomenos de phosphorescencia manifestados n'um liquido extrahido d'um peixe da profundidade do Oceano. In: Memorias do Museu Bocage, vol III, pp 1–10. [In Portuguese; available online at https://archive.org/stream/memoriasdomuseub01muse#page/n147 ]. Translated title: Phosphorescence phenomena exhibited by a liquid extracted from a fish from deep sea. A communication of this work was published at the *Comptes rendus des scéances de la Société de Biologie de Paris*, vol LXXII, Pg 432, 16th March 1912, Une propriété singulière d'une bactérie phosphorescente (premiére note de B Osorio présentée par H. Coutière).

next figure) at 400 m deep, squeeze its belly, and scrub the glowing substance coming out of the belly on the inner side of skin previously removed from shark species such as *Scyllium canicula* and *Pristiurus artedi* (*Galeus melastomus*). The inner part of the skin would still be covered with a muscle tissue layer and would preserve the liquid glowing for more than 20 h. Even if the skin was not hydrated for some days and had lost its brightness, glowing was recovered if the skin was immersed in seawater. The skin would then be cut in pieces and the glowing pieces put above the hooks in the fishing line to attract fish. Candil could also be used as bait.

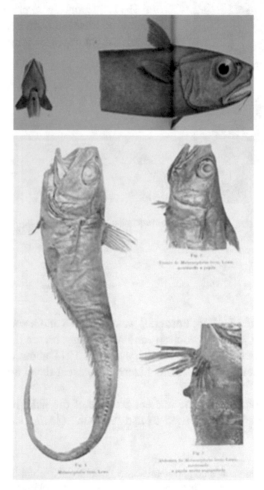

(*Upper panel*) *Malacocephalus laevis* as drawn in the report of the Challenger (UK) expedition (1873–1876) by Albert Günther. The apertures in the belly were noticed and highlighted in the illustration. (*Lower panel*) Illustration of Balthazar's paper: The protrusion in the belly is also highlighted. The common name in Sesimbra, Portugal, is "peixe rato" ("rat fish"), but in other parts, it is named softhead grenadier, armed grenadier, or smooth-headed rattail, for instance

According to Balthazar's description, the substance was collected from a protuberance near the anal orifice of the "rat fish" (see previous figure), was dense/jelly and yellowish in color, and glowed blue very intensively. The biological role of the liquid was elusive and the early twentieth century technology did not allow to observe the fish in its own environment, 400 m deep. He hypothesized that the fish would release the illuminating substance in order to get a better vision around to hunt or attract preys but could not test this hypothesis. Balthazar also noticed that the protuberance formed by the organoid that contained the glowing matter could eventually serve as a lamp. He devoted his efforts to unravel the chemical nature of the glowing substance and its photophysical properties. Two main reasons convinced Balthazar that the glow would emanate from bacteria: (1) the muscle tissue in the inner side of the shark skin is an excellent growing medium for bacteria, and (2) the recovery of glow after immersion of the dried skin in seawater seemed "reviviscence," an old uncommon name for "life recovery" typical in bacterial populations. In a follow-up work, in 1915, he reported the confirmation of this hypothesis.[13] The jelly substance indeed contained bacteria. He cultivated and observed the bacteria on a microscope. The bacteria looked like bacillus and would have 1.6–2.6 μm length and half this distance in width.

Another naturalist, named Fisher, had described a smaller (1.15–1.75 μm) glowing bacterium found in the waters of the "Indies" (Caribean): *Bacillus phosphorescens*. Nevertheless, Balthazar was in the presence of bacteria living enclosed inside fish, probably in symbiosis, while Fisher described free-floating bacteria. More importantly, Fisher's bacillus would glow white, slightly bluish, while Balthazar's bacillus would glow pure blue. "Blue is the light from burning carbon fuels", he described. Based on the observations at the microscope, on the very specific environment of the species (inside a specific closed anatomic structure of a fish), and the unique characteristics of the color of the glowing light, Balthazar was convinced he discovered a new species and named it *Bacillus malacocephali* to stress the intimate relationship between the hosting fish (*Malacocephalus*) and the bacterial species he was classifying.

The light emitted by the bacteria was in itself a mystery Balthazar was committed to solve. For this, he used the technology available in the 1910s:

1. Spectral analysis of the light using a glass prism to find what basic colors (wavelengths in modern analysis) it was made of. He was convinced the light was due to "phosphorescence" of chemical elements in bacterial composition and was determined to identify the element(s) through the identity "fingerprint" mixture of lights emitted.
2. Photographic analysis of the glow to study the ability to be registered in different kinds of photographic paper. In the early days of photography technology, it was not obvious that all visible light would interact with light-sensitive doped surfaces the same way.

---

[13] Osorio B (1915) Uma propriedade singular de uma bactéria luminosa. In: Arquivos da Universidade de Lisboa, vol II, pp 67–76. [In Portuguese; available online at https://archive.org/details/arquivodaunivers2191univ] Translated title: A singular property of a glowing bacterium.

3. The electroscope assay to check if light emission is associated with radioactivity. Radioactivity was known but the biological effects of radioactive radiation were not. The hypothesis that the bacteria could contain radioactive elements in enough quantity to glow strong seemed plausible at that time. The laws of physics of electricity and electrostatics were not as developed and widely disseminated as today.

Balthazar found that the spectrum of light was made of a continuum of colors (wavelengths) between the "B and F lines of Fraunhofer." Fraunhofer's set of "lines" were wavelength/color markers typical of atmospheric gases that were used to establish a standard scale. B and F "lines" are marks that correspond to 686.7 nm and 486.1 nm, respectively, perceived as colors red and blue, respectively (see next figure).

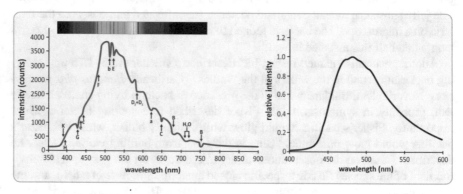

(*Left panel*) Fraunhofer lines in the visible spectrum of light. When sunlight traverses the atmosphere, its gases absorb light of well-defined wavelengths, so that the sunlight spectrum at the surface of the Earth has decreased intensity at those wavelengths. These sharp wavelengths at which intensity is sharply decreased are named Fraunhofer lines (indicated by the *red arrows* in the intensity spectrum and the *black lines* in the visible spectrum in the *top*). Lines B and F correspond to atmospheric H-containing gases and $O_2$ absorption of sunlight. (*Right panel*) A normalized bioluminescence spectrum of emission of *Photobacterium phosphoreum* ("*Bacillus malacocephali*") using a modern spectrofluorometer

A continuum spectrum was puzzling as chemical elements and simple diatomic molecules have very well-defined wavelengths of emission. Balthazar was hoping to find a new chemical element. "*If the rocks and the atmosphere have revealed most metals and metalloids known, it is reasonable to assume that the sea, given its immense extension, or the land covered by the sea, will also reveal new elements,*" he wrote. Nature goes much further than human expectation and Balthazar got a puzzling result. A new element could not be inferred … And it was not a known element either. The answer had to wait for developments in the area of metabolism. Chemistry and physics were developing scientific disciplines in the 1910s but biochemistry was still incipient.

The analyses with the photographic paper were less puzzling. Using paper with silver bromide from the still existing manufacturer Kodak, he found that the light from *Bacillus malacocephali* would "impress" the paper quickly: 30-s exposure in standard conditions was enough, which was considered relevant given "the small quantity" of bacteria used. Balthazar then used a filter with a photo imprinted between the source of light (glowing tubes) and the photographic paper and obtained a reproduction of the photographs used as filters. Moreover, he used the glow to illuminate a silver coin and photograph it. A clear photo was obtained. With proper exposition time, the glowing light was as competent to generate photos as daylight. Importantly, bacterial suspension and extracts of bacteria had the same result.

The experiment with the electroscope is straightforward but led Balthazar to a misleading conclusion. Strangely, the description of the procedure is detailed and seems to be carried out thoroughly, duly controlled but, even so, not reliable. He "loaded" the electroscope (see below figure), meaning static electricity potential was applied, and the two metallic foils made of aluminum repealed each other. When the fish organoid was placed near the pole, the two foils came together. To make sure the result was due to the light and not the organic tissues of the fish, he emptied the organoid of glowing matter and repeated the procedure. The two foils did not change position. He then repeated the experiment with seawater containing the glowing liquid. The two foils moved closer to each other albeit not as extensively as with the fish organoid. The conclusion was obvious: the glowing extract taken from *Malacocephalus laevis* was radioactive!

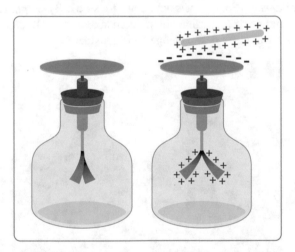

Basic scheme of an electroscope. Two metallic foils are suspended from a metallic wire connected to a metallic surface on top. When the electroscope is loaded, electrons are provided or removed from these metallic systems through electrostatics. The two foils then acquire the same charge and repel each other. When neutrality of the metal parts is recovered, the two foils come together

## A Modern Interpretation

In the 1910s, scientific knowledge did not allow Balthazar Osorio to understand that the conclusion on a radioactive anatomic structure in a living being was not plausible. The low availability of radioactive elements in nature and the extreme ionizing effects of this class of radiation make Balthazar's conclusion barely possible. Failure in finding a chemical element to which to assign light emission and the weak knowledge yet available on radiochemistry pushed him into a wrong conclusion, which was also one that seemed possible at that time. Today's explanation relays on knowledge in photochemistry, radiochemistry, and biochemistry of metabolism that became available much later than the 1910s. It is now known that more than half of the approximately 300 species of macrourids (the taxonomic family that includes the *Malacocephalus* genera) have bulbous ventral light organs (named photophores) that are situated anterior to the anus and from which a duct opens into the perianal groove of the rectum. Lenses and reflector systems (see next figure) may be present, and the light may be diffused throughout specialized regions of the ventral musculature. The bulbs harbor symbiont bacteria that can readily be cultured in vitro. Isolates from species from three genera other than *Malacocephalus* have been identified as *Photobacterium phosphoreum*, but it is likely that other macrourids harbor the same species. This bacterium is adapted to low temperature and high pressure in contrast to other related species such as *P. fischeri* or *Baneckea harveyi*. It is not known for sure if *P. phosphoreum* is *Bacillus malacocephali* after being reclassified by modern taxonomy or if, on the contrary, the work of Balthazar was never acknowledged and considered.

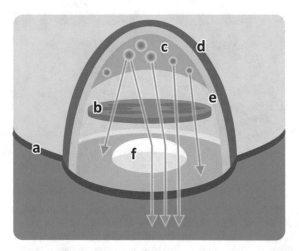

Basic structure of a photophore, located in the dermis with an aperture in the epidermis (a). The aperture consists of a lens made of chitin or protein (f). The bioluminescence-producing cells, such as *Photobacterium phosphoreum*, are generally referred to as photocytes (c). Light emitted by photocytes is reflected in a mirror surface (e) made of guanine crystals, proteins, collagen, or chitin, supported in a pigmented layer of chromatophores (d). Light wavelength/color is then refined by an absorption or interference filter film (b) before being refracted by the lens (f)

In fact, the light emitted by the symbiotic bacteria is not caused by the presence of a chemical element. If this was the case, Balthazar would have observed discrete wavelengths/colors of emission. Moreover, light emission would only be observed after absorption of light, which would make light emission at 400 m deep in the oceans impossible as sunlight reaching this depth is very faint. Instead, light emission is owed to organic molecules, which explains the continuum of wavelengths observed by Balthazar, and is part of a process that depends on the metabolism of bacteria. The process exists in many bacteria and is named bioluminescence (see next figure). An oxidation of an electron donor molecule is catalyzed by the enzyme luciferase, the electrons being transferred to molecular oxygen, $O_2$, while complexed to luciferase. This complex is unstable and highly energetic. "Excess" energy is released as a photon, a "particle" of light (or "quantum"). In some cases, this light is the observed bioluminescence; in other cases, the energy is transferred to fluorescent proteins that will emit light of well-defined wavelength/color. Different species may have their own kinds of proteins, therefore emitting light of species-specific colors. The process of conversion of energy from a chemical source into light in this case is incredibly efficient: nearly one photon emitted per molecule oxidized, i.e., 100% conversion efficacy.

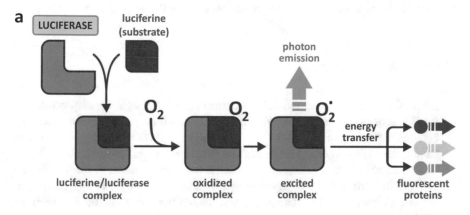

(a) Outline of bioluminescence reactions. (b) Chemical structure of the most common electron donor molecules, generally named luciferins. One species-dependent luciferin combines with an enzyme, luciferase, forming a complex that reacts with oxygen. The redox process has a "highly energetic" intermediate that releases "excess energy" as a photon, a "particle" of light, or resonates with a fluorescent protein that will capture the energy and fluorescent light of a given species-specific color

## Questions

1. Propose advantages that bioluminescence might confer to *Malacocephalus laevis*.
2. Propose hypotheses to explain why the color of bioluminescence associated with most deep-sea fishes is blue.
3. Bioluminescence-generating metabolism uses molecular oxygen ($O_2$). This means its metabolism is relatively recent and may have started for reasons other than bioluminescence. Propose an explanation.
4. Why is hosting radioactive bacteria not plausible in light of current scientific knowledge?
5. Propose an explanation for the misinterpretation of the electroscope experiment.

## Biochemical Insight

1. Bioluminescence enables animals to communicate through signals encoded in the frequency and duration of light flashes, hunt by luring preys, evade predators creating diversion, or camouflage. Fish living at depths reached by sunlight and having the ability to emit blue light from their ventral surface typically use bioluminescence for camouflage. When seen from below against the water surface background, the fish are easily detected by the shadow they cause. Blue light emission counteracts the shadowing and allows the fish to better escape from predators.
2. Water is not totally transparent. Because of its molecular structure, water absorbs energy that contributes to several vibrating modes. This energy may be light of well-defined wavelengths in the infrared region of the radiated spectrum. A very weak absorption of radiation occurs in the red-infrared transition region. Small volumes of water, such as in a test tube or a domestic drinking glass, do not absorb significant quantities of the sun radiation and appear transparent, yet higher volumes absorb red radiation and the water appears to have the complementary color: green. This is why by the seaside, at shallow depths, the clear water is transparent but, as the depth increases, the perceived water color changes to green. At even higher depths, the volume of water that absorbs light is such that red and green light are absorbed. In other words, only blue light is not yet absorbed. This is the reason why divers at depths 20–500 m see the aquatic envi-

ronments in shades of blue, the dominant color (see below figure). In fact, blue light is, in practice, the only illumination from sun available below 100 m. Therefore, camouflage demands blue light. At depths below 500 m, even blue light is absorbed due to the cumulative effect of the mass of water gradually absorbing light from the surface to the bottom of the ocean. The end result is total darkness. If water was totally transparent, sunlight would shine even at the deepest abyss trenches, kilometers away from the surface.

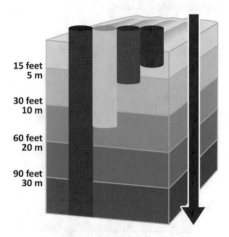

Sunlight penetration in water. Up to circa 3 m depth, all light wavelengths are not significantly absorbed; water is transparent and the sunlight remains "white." At higher depths, blue light is increasingly dominant because red and yellow radiation are absorbed. At 30 m deep, only blue light illuminates the environment. This situation persists down to 500 m, where illumination is so dim it vanishes

3. The series of chemical reactions that cause bioluminescence differ among different organisms but they all follow common principles (see the last figure in the case description section): a redox reaction brings a molecule to a higher energy state and the "excess energy" is released by means of light emission. Molecular oxygen ($O_2$) is the species that accepts the electrons, water being formed. This rudimentary form of extracting energy from oxidation of organic molecules and having $O_2$ as final electron acceptor encompasses the complex form of oxidative phosphorylation by which many animals obtain energy from nutrients. The main difference being that instead of the energy being released by light emission, energy is stored as ATP (as will be detailed in Sect. 6.2).

   $O_2$ released into the atmosphere by primitive photosynthetic processes was extremely oxidant and therefore toxic. $O_2$ can be reduced to form water, which is a form of protection by $O_2$ consumption. Cells able to reduce $O_2$ were the best fitted to survive in the new oxidative atmosphere being formed by algae ancestors (see next figure). Bioluminescence, which is a form of $O_2$ disposal, may have been a form of protection in the early times of massive atmospheric $O_2$ poisoning. In this sense, it may have been the reverse of photosynthesis: photosynthesis consumed light to synthetize organic molecules, $O_2$ being a by-product, whereas bioluminescence used organic molecules to produce light, making use of $O_2$. Later, some cells evolved into not wasting energy as light but rather to synthetize other molecules that could be used in their own chemical reactions. Some cells remained bioluminescent and took advantage of light emission in different situations. In bacteria and fungi, for instance, light emission is not essential to survival. Bioluminescent bacteria and fungi develop even in condition in which bioluminescence is halted.

Sedimentary fossil of cyanobacteria stromatolites (2000 million years) from Pilbara, Western Australia. Stromatolites produced the oxygen that accumulated in the sea and the atmosphere. Many organisms that had lived without oxygen went extinct in what was probably a huge extinction event. (Collection of the Museu Nacional de História Natural e da Ciência, Universidade de Lisboa, Portugal, Inventory reference MUHNAC-MNHN/UL.II.495)

4. Radioactive elements are present in Earth biomes in trace amounts, not enough to accumulate extensively in living cells and emit the light intensity observed in bioluminescent species. Moreover, even if this was the case, the energy of radioactive radiation is such it can break covalent bonds in organic molecules, i.e., it is highly ionizing radiation. Exposure to radioactivity, depending on duration and radiation intensity, is extremely aggressive to cells. Even extremely resistant cells, such as *Deinococcus radiodurans* (see Box 2.2), cannot survive for long in these conditions.

5. Radioactive materials discharge the electroscope because ionization caused by radioactivity neutralizes the metallic foils, therefore ending repulsion between them. Other materials able to exchange ions with the metal in the electroscope may have the same effect. Likewise, light able to remove electrons from the metal by photoelectric effect (electron release by impinging light) may have the same effect inasmuch the electroscope is charged negatively. In principle, only ultraviolet light can remove electrons from non-alkali metals, but longer wave-

lengths suffice if they are present. It may be that Balthazar observed the discharge of the electroscope due to redox and other electric phenomena in the sample where bioluminescent processes were active and/or due to the effect of blue light impinging on the impure alkali-oxidized surface of the metal of the electroscope. The first cause seems more plausible as Balthazar made a through and carefully planned set of experiments, with all proper controls; therefore, extensive oxidation of the metal was not likely.

## Final Discussion

### Scientific Knowledge Through Time

World War I in Europe did not create a proper environment for sharing knowledge among nations, and the work of Balthazar may have been overlooked. Before him, a famous naturalist, Albert Günther (1830–1914), keeper of the Department of Zoology in the British Museum, who participated in the scientific expedition of the ship "Challenger," in 1873–1876, had already described one individual specimen of *Malacocephalus laevis* caught near Pernambuco, in Brazil (shown in the first figure in this case). Before that, the only specimen known was the one classified originally by Richard Thomas Lowe (1802–1874), caught at the Madeira Island (Portugal). The British Museum conserved in its collection another specimen caught also in Madeira Island before 1915.

The glowing bacterial species was classified by Balthazar as *Bacillus malacocephali*, but it probably is *Photobacterium phosphoreum*. Because microbial taxonomy was not so well developed and exchange of knowledge and information dissemination was slow, this bacterium was classified as a new species differently by many researchers in several countries: *Micrococcus phosphoreus* Cohn 1878, *Bacillus phosphorescens* II Baumgarten 1888, *Photobacterium phosphorescens* Beijerinck 1889, *Bacillus hermesi* Trevisan 1889, *Bacillus phosphoreus* (Cohn) Mace 1901, *Photobacter phosphorescens* Beijerinck 1901, *Bacterium phosphoreum* (Cohn) Molisch 1912, *Photobacter phosphoreum* (Cohn) Beijerinck 1916, *Photobacterium phosphoreum* (Cohn) Ford 1927, *Micrococcus physiculus* Kishitani 1930, *Coccobacillus acropoma* Yasaki and Haneda 1936, and *Photobacterium profundum* Weisglass and Gavrilovic 1963, for instance.

### Luminescent Art

Luminescent bacteria are easy to cultivate in vitro and may glow bright enough to produce a reasonable "bio-lamp." In the World Exhibition of 1900, in Paris, Dubois had impressed the visitors by illuminating a pavilion, "The optics palace," with bioluminescent bacteria in suspension in 25 L containers. For the 1900s standards, the light produced was comparable to light from artificial sources in other pavilions.

In present days, bioluminescence cannot compete with artificial illumination systems, but bacterial glow has been translated into the world of artistic creation. Bioluminescent art exhibitions are a reality and the works presented are impressive. There are even festivals of bioluminescent art such as the Bioluminescence Festivals in Auckland, New Zealand.

## Bioluminescent Bays

There are many species of aquatic microorganisms that respond to mechanic stimuli with light emission. They are the reason behind the luminescence sometimes observed in the revolved water when boats move (named wake). In extreme cases, the density of luminescent microorganism in water may be such that a night swim is a dazzling spectacle. Vieques Bay in Puerto Rico, for instance, is world famous for the glowing water where waves break around moving swimmers. Elucidative photos are available in https://vieques.com/island-bioluminescent-bay/.

## Alternative Case Study: Marie Curie and the Radium Girls

Unaware of the biological effects of radioactivity, Marie Curie (Nobel Prizes in Physics and Chemistry in 1903 and 1911, respectively, for the studies on radioactivity) was exposed for decades to materials containing radium and polonium, elements she contributed to discover. Radium's bluish glow dazzled a lot of people, Marie Curie included, for being fancy. The hidden danger was only realized later. In the 1920s, colleagues of Marie Curie and radiation workers died from leukemia. Marie Curie's decades of exposure left her chronically ill and nearly blind from cataracts, and ultimately caused her death at 67 years old, from either severe anemia or leukemia. But she never fully acknowledged that her work had ruined her health. Her daughter, Irene Joliot-Curie, and son-in-law, Frederic Joliot-Curie—also Nobel Prize winners—continued her work with radioactive material. Eventually, both also died of diseases induced by radiation. What makes radium so dangerous is that it forms chemical bonds in the same way as calcium, and the human body can absorb it into the bones. Then, it eradiate cells at close range, which may cause bone tumors or bone marrow damage that can give rise to anemia or leukemia.

Marie Curie died in 1934. She was buried at the Sceaux Cemetery, in Paris. However, in 1995, her remains were transferred to the Pantheon. When she was exhumed from the original resting place, she was so radioactive she was interred in the Pantheon in an inch-thick lead coffin that prevent the radiation from crossing. Actually, since the unfortunate Curies were contaminated with radium 226, the most stable isotope of radium that has a half-life of approximately 1500 years, their remains will stay dangerously radioactive for at least that long.[14] Furthermore, since Marie Curie was not aware that her experiments were harmful, she contaminated her entire household along with many of her personal items. She frequently carried samples of radium and polonium in the pockets of her lab coat and brought them home to analyze in her spare time. She unwittingly contaminated all of her clothes; her books, notebooks, and cookbooks; and her jewelry, furniture around her home, and various other personal items.

In 1938, radioactive consumer products were banned in the USA by an act issued by the Food and Drug Administration. By that time, it was known that radioactive elements can be used as a powerful and efficient energy source, but also that their effects on health can be devastating.

Another case in which fascination for the glow of radioactive materials combined with ignorance of its biological hazards cost human lives was the case of the so-called Radium Girls, factory workers exposed to radium poisoning in the wristwatch industry. In the 1910s, a special paint containing radium was used to illuminate numbers and dials in watches so they could be seen glowing in the dark. The girls working in the factories were even encouraged to lick the tip of their radium-contaminated brushes that were used to paint the numbers and dials, so that the brush would gain a finer point. Completely unaware of the dangers concerning radium, around 4000 women hired by various radium-dial companies were poisoning themselves on a daily basis. The glow-in-the-dark paint was perceived as so harmless that the women would goof around, painting their lips and nails with it, until the first symptoms of poisoning began to surface.[15] The dial painters' first health problems turned up in the 1920s, when some of the women began suffering from fatigue, anemia, and trouble with their teeth. When dentists tried to extract the bad teeth, they were horrified to find jawbones so diseased that chunks of bone came out as well.[16] The extraction sites didn't heal, and infections set in. In many cases, the women's bodies were actually radioactive, because radium had been absorbed by their bones. Government researchers studied live and dead dial painters and used the data to calculate safe exposure levels for future generations of workers. Radium Girls have become a notorious, but tragic, chapter in the history of occupational medicine.

---

[14] Valjak D (2018) More than 80 years after she passed away, Marie Curie's remains and personal items are still dangerously radioactive. The Vintage News. www.thevintagenews.com/2018/02/27/marie-curie.

[15] Budanovic N (2018) Radium Girls female factory workers exposed to radium poisoning without their knowledge, were even encouraged to lick radium paintbrushes. The Vintage News. www.thevintagenews.com/2018/01/01/radium-girls-2.

[16] Grady D (1998) A glow in the dark, and a lesson in scientific peril. The New York Times. www.nytimes.com/1998/10/06/science/a-glow-in-the-dark-and-a-lesson-in-scientific-peril.html.

# Chapter 3
# The Families of Biological Molecules

Diversity is essential to the sustainability of living systems. This is true for species in ecosystems as it is for molecules in cells, tissues, and organisms. Yet, the same way different species are linked by common ancestors and may be grouped in taxonomic classes according to common characteristics they share, molecules may be grouped in classes and classified according to common chemical and physical characteristics. One of such characteristics is solubility in water (in other words: how polar atoms are distributed in the 3D structure of molecules). One class of biological molecules, the lipids, includes only low water solubility ("hydrophobic") molecules, this being the characteristic that defines this class. Other classes include molecules that are mostly moderately or highly soluble in water and can be recognized for the dominant presence of specific chemical groups: OH in saccharides (also referred to as "carbohydrates") and a combination of amino and carboxyl groups in amino acids. Lipids, saccharides, and amino acids may combine with molecules of its own class to form either polymers (molecules formed by successively covalently attaching smaller molecules), such as polysaccharides and proteins, or supramolecular assemblies (organized arrangements of molecules that are in contact but are not covalently attached), such as the lipid bilayer of cell membranes. It is common to find molecules and supramolecular assemblies that combine elements from different classes, such as nucleotides, which contain saccharides. Proteins are extremely versatile in this regard because proteins' interactions with saccharides, lipids, and nucleic acids (nucleotide polymers) are ubiquitous in virtually all cells. Figure 3.1 depicts the basic principles that support the organization of biological molecules in different classes.

© Springer Nature Switzerland AG 2021

A. T. Da Poian, M. A. R. B., *Integrative Human Biochemistry*,
https://doi.org/10.1007/978-3-030-48740-9_3

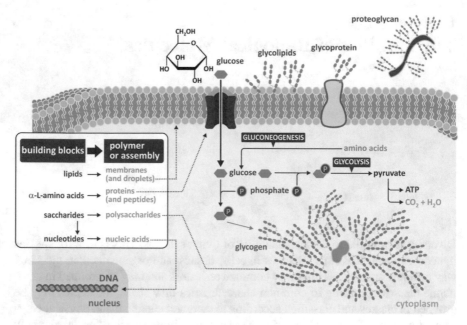

**Fig. 3.1** Families of biological macromolecules and large supramolecular assemblies. Three classes of fundamental molecules ("building blocks") are enough to order in different families most biological macromolecules (polymers) and large supramolecular assemblies (such as lipid bilayers and lipid droplets). Nucleic acids are polymers of nucleotides, which have saccharide residues in their composition. Proteins are polymers that are commonly combined with molecular residues of other classes such as saccharides (glycoproteins or proteoglycans when the saccharidic content is high). Glycolipids (glycosilated lipids—lipids with saccharide groups attached) are also common in membranes. Both building blocks and polymers/assemblies are important in the structure and functioning of cells, constituting the vast majority of matter in a cell (excluding water)

## 3.1   Lipids and the Organization of Their Supramolecular Assemblies

Lipids are highly hydrophobic molecules that nonetheless may have polar chemical groups in their composition. If part of the 3D structure of the molecule is very polar and the other is non-polar, the molecule is referred to as amphiphilic, which stresses its dual nature: while the polar part will tend to interact with water and other polar molecules, the other will tend to minimize its interaction with water and other polar molecules. Nevertheless, it is important to bear in mind that lipids are molecules in which hydrophobicity dominates, even if they are amphiphilic. This is a qualitative definition with no clear boundaries in terms of molecular structure, which is nonetheless a useful working definition because hydrophobicity grants lipids the ability to organize in supramolecular assemblies that are very distinctive from polar molecules. Take lipid bilayers as example: they are extensively organized supramolecular assemblies that are very stable and yet do not involve covalent bonds between lipid molecules. Lipids spontaneously self-associate in aqueous environments, and amphiphilic lipids in particular may self-associate in a very organized way. This results from the so-called hydrophobic effect, although the most appropriate term would be "entropic effect."

The entropic effect results from the second law of thermodynamics, which in one of its possible statements implies that all physical and chemical events tend to evolve in a way so that total entropy ("disorder") increases. Consider Fig. 3.2: strongly amphiphilic molecules of generic cylindrical or rectangular cuboid shape will spontaneously form a bilayer to minimize the contact of non-polar regions with water. The driving force for this event may be counterintuitive at first glance: the bilayer is the arrangement that corresponds to the most disordered system. This may seem absurd because we tend to focus our attention in the solute (the lipids, in this case) and forget the solvent (water); yet, the gain in entropy refers to both. The lipids become ordered relative to each other, but the contact of hydrophobic groups with water molecules imposes restrictions to the orientational freedom of water, which is very costly in terms of entropy. When two lipid molecules associate, less water molecules are forced to order and, although the lipid molecules become more ordered relative to each other, the whole molecular systems (water included) becomes more disordered, in agreement with the entropic formulation of the second law of thermodynamics. Thus, this is named the entropic effect (occasionally imprecisely referred to as "hydrophobic effect"). The same principle applies when 3, 4, 5, … $n$ molecules are considered forming large assemblies of lipids, such as lipid bilayers (see Sect. 3.1.1).

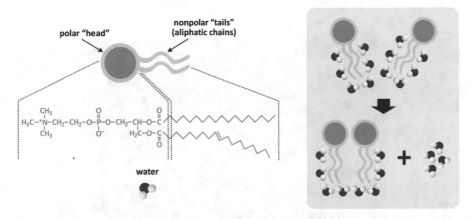

**Fig. 3.2** Simplified explanation of the entropic effect. Most membrane lipids have two aliphatic (hydrocarbon) chains and a polar "head." Polar heads usually contain phosphate (phospholipids) and other polar groups (*left* panel). Non-polar regions of lipids tend to associate with each other because fewer molecules of water get exposed to the aliphatic chains (*right* panel). Exposed aliphatic chains force molecules to orient their oxygen atom away from the hydrocarbon constituents, i.e., they force a certain degree of order to the solvent

Broadly speaking, diacyl lipids (i.e., lipids containing two acyl—aliphatic—chains) with polar "heads," such as most of the lipids found in cell membranes, have a generic shape with the characteristics depicted in Fig. 3.2, and the entropic effect compels these molecules to an ordered and parallel self-association. This is the physical process that sustains the stability of cell membranes as we shall discuss in Sect. 3.1.1. Now we will look closer into the chemical structure of diacyl lipids: although these lipids tend to assemble as bilayers, the differences in their chemical nature determine which chemical processes they are able to participate in (e.g., cell signaling) and their tendency to associate with membrane proteins and/or form specific domains in the membrane. Most triacyl lipids occurring in nature, such as triacylglycerols (often referred to as triglycerides), lack a bulk polar "head" group and therefore lack the propensity to form ordered supramolecular structures such as membranes (Fig. 3.3). Instead, they self-associate disorderly, forming aggregates, such as the lipid droplets found in cells where these lipids are synthesized (hepatocytes) or stored (adipocytes). Figure 3.3 shows histological images of the adipose tissue in which the presence of lipid aggregates occupying almost all cellular volume is detected.

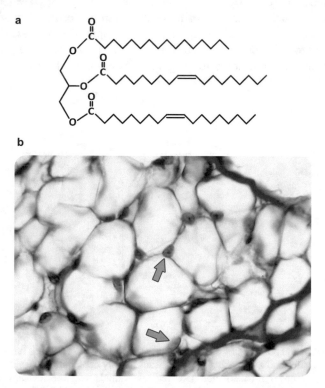

**Fig. 3.3** Lipids storage in the adipose tissue cells (adipocytes). (**a**) Triacylglycerols lack bulk polar "head" groups, which makes them not prone to self-assemble into bilayers. (**b**) Triacylglycerols aggregate in disordered agglomerated structures that can be stored adipocytes. In Panel (**b**), the lipids were extracted from the histological preparations of adipocytes, leaving white empty areas; arrows point nucleus. (Figure reprinted with the permission of Instituto de Histologia e Biologia do Desenvolvimento, Faculdade de Medicina, Universidade de Lisboa, FMUL)

Fatty acids can be considered the basic unit common to most lipids in human cells. They consist of a linear unbranched aliphatic chain ("tail") with a carboxyl group (Fig. 3.4a). The aliphatic chain can be saturated (i.e., having all the carbon–carbon bonds as simple bonds, without double or triple bonds) or unsaturated. Most fatty acids in humans have an even number of carbons, ranging from 10 to 28. Fatty acids are important molecules in the energetic metabolism of the organism, and they can be found either in "free" form (unattached to other molecules) or attached to other molecular structures through ester bonds (Fig. 3.4d, e). Free fatty acids differ from each other in the number and position of double bonds they have and in the length (in practice, the number of carbons) of the chains. The diversity attainable by changing these characteristics is virtually infinite, but, in practice, not all combinations are detected in nature (Table 3.1 displays examples of the most frequent fatty acids found in the human cells). Naturally, the nomenclature of fatty acids highlights their characteristic differences in terms of number and position of double bonds and chain length (Fig. 3.4b). The more recent nomenclature systems are very descriptive of these characteristics, but although older, traditional nomenclature systems are widely used. The recent nomenclature identifies the carbon atoms in a chain by number, starting consecutively from the carbon of the carboxyl (see Fig. 3.4b). Double bonds, if existent, are identified by the number of the carbon of lowest order forming the bond. The $\omega$, or $n$, nomenclature numbers the carbons in reverse order (see Fig. 3.4b). Another nomenclature, still popular among many biochemists, identifies carbons by Greek letters in alphabetical order ($\alpha$, $\beta$, $\gamma$, $\delta$, etc.) starting from the carbon adjacent to the carboxyl group (see Fig. 3.4b). A process of metabolic degradation of lipids is titled "$\beta$-oxidation" because of the importance of carbon $\beta$ in the reactions of this metabolic pathway (see Sect. 7.4.4).

Glycerolipids such as diacylglycerols and triacylglycerols result from the chemical conjugation of a glycerol molecule with fatty acids through esterification (Fig. 3.4c). Yet, glycerolipids may also conjugate other groups besides fatty acids. Usually, a phosphate group is used to bridge glycerol to a polar group, forming glycerol phospholipids (Fig. 3.4c). The phosphate binds both the glycerol and the polar group through phosphodiester bonds, which are ubiquitous in biochemical processes for their importance in molecular structure and energetics. Glycerophospholipids are named according to the fatty acids attached to glycerol and the chemical nature of the polar group (Fig. 3.4e).

Sphingolipids constitute another class of lipids present in many human cells, being responsible for rigid domains in the membranes. Figure 3.5 shows that although these lipids are similar in structure to glycerophospholipids, they are chemically different. These differences are responsible for the formation of rigid bilayers when conjugated to phospholipids and cholesterol.

Sterols, of which cholesterol in an example (Fig. 3.6), are extremely hydrophobic alcohols. These molecules are characterized by a flat rigid system of carbon rings, cyclopentaphenanthrene, and have some remarkable properties:

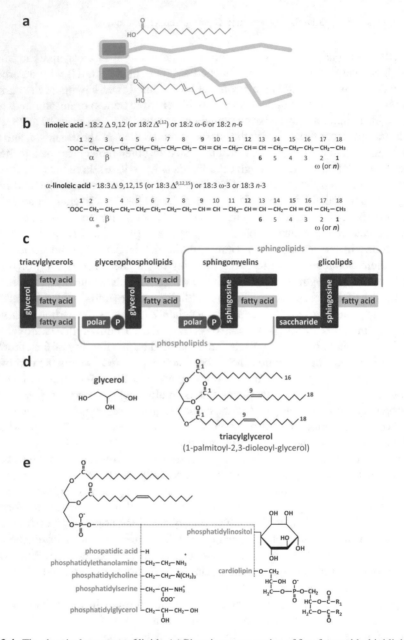

**Fig. 3.4** The chemical structure of lipids. (**a**) Pictoric representation of free fatty acids, highlighting the polar carboxylic group in blue and the non-polar aliphatic chain in light blue; double bonds impose constraints in the conformation of the chain. (**b**) Illustrative examples of the application of different nomenclature systems; recent systems (numbering carbons from the carboxylic group) are more precise and descriptive, but older systems (Greek lettering and $\omega$-numbering from the non-carboxylic ending carbon) are still widely used. Trivial (non-systematic) names (Table 3.1) were never abandoned for their "user-friendly" nature. (**c**) Fatty acids form ester bonds with other molecular structures to form more complex lipids such as triacylglycerols (detailed example in Panel **d**), glycerophospholipids, or sphingolipids. Fatty acid residues are present in all. In sphingomyelins, the polar group attached to phosphate is either choline or ethanolamine (Panel **e**). The example of specific triacylglycerol in (**d**) is 1-palmitoyl-2,3-dioleoyl-glycerol. (**e**) Phosphatidylglycerol lipids form different classes depending on the chemical nature of the "polar head" (fatty acid chains in this example are illustrative)

**Table 3.1**  Nomenclature of the most common fatty acids found in humans, both saturated (only simple carbon–carbon bonds) and unsaturated

| Number of C (chain length) | Saturated | Unsaturated | Number of double bonds | Nomenclature system | | |
|---|---|---|---|---|---|---|
| | | | | $\Delta$ | $\omega$ | $n$ |
| 14 | Myristic | | | | | |
| 16 | Palmitic | Palmitoleic | 1 | 16:1 $\Delta$9 | 16:1 $\omega - 7$ | 16:1 $n - 7$ |
| 18 | Stearic | Oleic | 1 | 18:1 $\Delta$9 | 18:1 $\omega - 9$ | 18:1 $n - 9$ |
| | | Linoleic | 2 | 18:2 $\Delta$9,12 | 18:2 $\omega - 6$ | 18:2 $n - 6$ |
| | | $\alpha$-Linoleic | 3 | 18:3 $\Delta$9,12,15 | 18:3 $\omega - 3$ | 18:3 $n - 3$ |
| | | $\gamma$-Linoleic | 3 | 18:3 $\Delta$6,9,12 | 18:3 $\omega - 6$ | 18:3 $n - 6$ |
| 20 | Arachidic | Arachidonic | 4 | 18:4 $\Delta$5,8,11,14 | 18:4 $\omega - 6$ | 18:4 $n - 6$ |
| 24 | Lignoceric | | | | | |

There are trivial names and descriptive abbreviations. Trivial names usually refer to the neutral (protonated) form of the acid, but one should bear in mind that at most physiological pHs (including plasma pH, 7.2–7.4), the acidic groups are unprotonated (pH above 4–5). Abbreviated nomenclatures highlight the number of carbon atoms in the aliphatic chain and number and position of unsaturated carbon–carbon bonds ($\Delta$). Carbon numbering varies depending on adopting modern or older ($\omega$, $n$) systems. Older systems are still widely used in some fields such as nutrition and ocasionally used in biochemistry

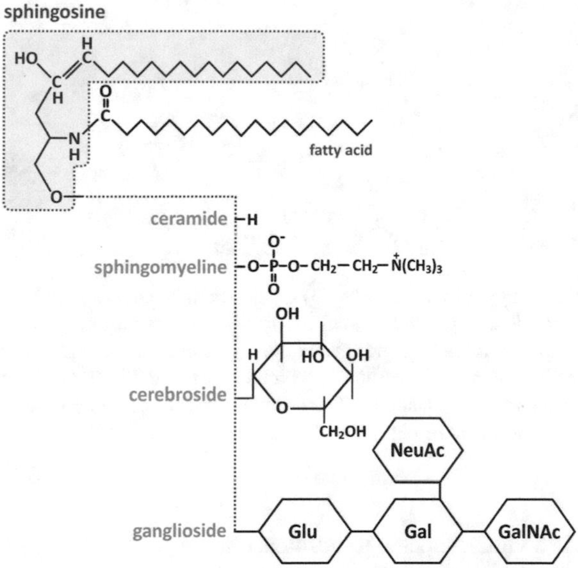

**Fig. 3.5**  Molecular structure of sphingolipids. As with phosphatidylglycerol lipids, sphingolipids form different classes depending on the chemical nature of the "polar head" (the fatty acid chain in this example is illustrative). Glu, glucose; Gal, galactose; GalNAc, $N$-acetyl-galactosamine; NeuAc, $N$-acetylneuraminidate

1. Only one small polar group present (OH in cholesterol; C=O in the equivalent ketones)
2. A flat molecule formed of four almost coplanar rings
3. When placed in lipid bilayers, it acts as a fluidity regulator (not too rigid to compromise dynamics and not too fluid to compromise barrier integrity)

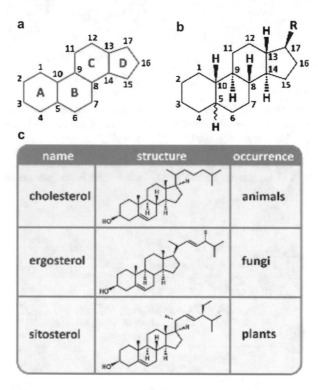

**Fig. 3.6** Chemical architecture of steroids. (**a**) Cyclopentaphenanthrene is the basic structure of (**b**) steroids, in which R is a generic radical group. The numbering system of carbons is highlighted. Cyclopentaphenanthrene is rigid, planar, and hydrophobic. (**c**) Sterols are a sub-family of steroids having an alcohol group in carbon 3, which can be found in the lipid membranes of animals, fungi, and plants but not bacteria. They are the targets of some anti-infectious drugs in humans (this issue is addressed in Fig. 3.9)

Cholesterol is a vital molecule to humans (Box 3.1). In addition to its properties in membranes, it takes part in other biochemical processes such as the synthesis of some hormones. Nevertheless, the seriousness of problems associated with its ingestion in excessive amounts in unhealthy diets, associated with its very low solubility in aqueous media, turns this hero molecule into a bad reputation killer.

It is curious to note that although cholesterol is essential to many species in nature, including humans (Fig. 3.6c), its synthesis is relatively recent in evolutionary terms as it uses molecular oxygen, which is only present in atmosphere since the advent of photosynthesis.

**Box 3.1 Cholesterol: A Hero with Bad Reputation**

High concentration ("level") of cholesterol circulating in the blood is a major risk factor for cardiovascular diseases, which affect a significant fraction of the whole population in many countries. The campaigns to prevent cardiovascular disease are often centered in reducing dietary cholesterol, which gives the impression that cholesterol is some kind of poison or toxin that should be banned. The persistent unhealthy doses of cholesterol in diet are indeed harmful, but this must not create the illusion that cholesterol is by itself malicious to cells. Virtually all molecules in excessive doses are harmful and cholesterol is not an exception. Moreover, the human body is so dependent on cholesterol that it synthesizes its own cholesterol and has homeostatic mechanisms to regulate its production in connection with several metabolic processes.

The way the human body works at biochemical and physiological levels is crucially dependent on cholesterol. This extremely hydrophobic molecule intervenes in three major processes in humans: (a) it contributes to the balance of physical–chemical properties of lipid membranes dynamics in cells; this includes the plasma membrane and the intracellular membranes; (b) it is a precursor for the bile acid synthesis; bile acids are important to absorb the lipids existing in food after ingestion; and (c) it is also a part of the synthesis of vitamin D and hormones as important as estrogen in women and testosterone in men. Because of its hydrophobicity, the solubility in aqueous media, such as blood, is very low. There are specialized structures, lipoproteins, which incorporate cholesterol and form stable emulsions in aqueous medium (see Sect. 3.1.2). When cholesterol is not contained in lipoproteins and/or when these structures deteriorate due to oxidation, cholesterol depots may be formed in the endothelial tissue of the blood vessels. These depots are typical of atherosclerosis. In the most extreme cases, the blood vessels may be interrupted and neighboring tissues are not irrigated. Lack of nutrients and oxygen (ischemia) may cause severe lesions in tissues. Vulnerable atherosclerotic plaques may also detach and clog vessels. Any of these conditions is called infarction. Myocardial infarction, commonly known as a heart attack, is one example. Cerebral infarction is another example, commonly known as stroke.

A demonstration of how cholesterol is indispensable for the human body is the Smith–Lemli–Opitz syndrome. This is a rare disease characterized by failure to thrive; mental retardation; visual problems; physical defects in hands, feet, and/or internal organs; increased susceptibility to infection; and digestive problems, among others. Smith–Lemli–Opitz syndrome is a genetic disease caused by a defect in cholesterol synthesis, namely, deficiency of the enzyme 3β-hydroxysterol-Δ7-reductase, the final enzyme in the sterol synthetic pathway that converts 7-dehydrocholesterol (7DHC) to cholesterol (the complete pathway of cholesterol synthesis is described in Box 8.8 in Chap. 8). This results in low plasma cholesterol levels and elevated levels of cholesterol precursors, including 7DHC.

### 3.1.1   The Structure of Biological Membranes

The concept illustrated in Fig. 3.2 can be extended to the association of many lipid molecules. The entropic effect will cause the molecules to self-associate orderly, ultimately forming bilayers of lipids with a hydrophobic core (Fig. 3.7). If these bilayers are extensive enough for the overall structure to bend, then the bilayer curves allow the sealing of the hydrophobic borders. All hydrophobic aliphatic chains became protected from water because the only areas in direct contact with the surrounding aqueous environment are the external and internal ("luminal") surfaces formed by polar headgroups of the lipids. This is depicted in Fig. 3.7.

Lipid vesicles form spontaneously, being the simplest models of biological membranes. Amazingly, there is not a single cell known in nature that does not have their membranes formed with lipid bilayers to some extent. Diversity arises from the different kinds of lipids used and from their combination with other molecules, but the structural arrangement of the membrane itself is configured by lipid bilayers.

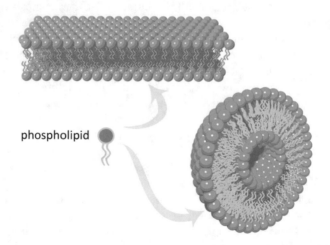

phospholipid

**Fig. 3.7**  Assembly of phospholipids into bilayers. Lipid molecules with a nearly cylindrical shape as a whole, such as phospholipids, tend to self-associate orderly due to the entropic effect (see Fig. 3.2). Extensive self-association forms lipid bilayers (*top*) that ultimately bend and curve to self-seal into vesicles (*bottom right*). Vesicles are the simplest prototype of cell membranes

Some cells (e.g., bacteria, fungi, and plants) may also have a cell wall in addition to the cellular membrane, which serves for structural and/or protection purposes.

In fact, lipid membranes are relatively malleable and fragile. Yet, such malleability and apparent fragility are very important characteristics from a dynamic point of view: membrane division and fusion, for instance, are favored as these processes do not imply covalent bonds to be broken or formed among lipids and lipid bilayers are flexible. Membrane fusion is advantageous in many biological circumstances (Fig. 3.8). In biotechnology, mainly in pharma and cosmetics, lipid vesicles are valuable tools due to their properties (Box 3.2).

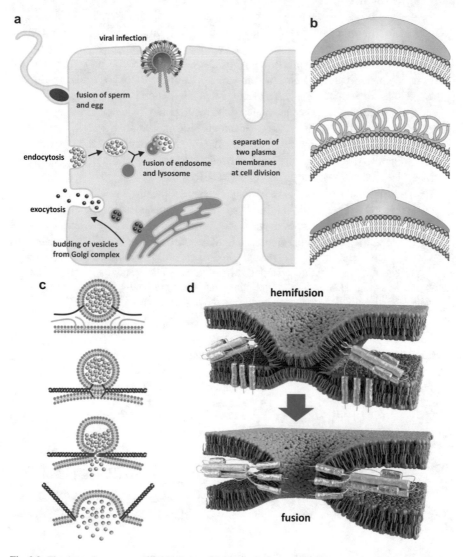

**Fig. 3.8** The dynamic structure of lipid bilayers. (**a**) Lipid membrane fusion occurs in nature associated with a plethora of cellular events. Some of these events such as endocytosis or budding of vesicles from the Golgi complex require inducing a curvature in membranes so that vesicles are formed. (**b**) This induction is accomplished by the action of specialized proteins that are curved and adhere to the surface of lipid bilayers (**b**, *top*), stabilize spontaneous curvatures (**b**, *middle*), or insert into one leaflet of the bilayer only, thus forcing the membrane to bend (**b**, *bottom*). (**c**, **d**) Membrane fusion implies that two different bilayers come together, which is due to the action of proteins that insert the two bilayers and enroll conformational changes that lead to the contact of the lipid bilayers. First, a common bilayer is formed from the mix of the two membranes (hemifusion state), and then total fusion merges the two entities that were initially enclosed in their own membranes. Panel (**c**) shows the example of membrane fusion during neurotransmission, namely, a neurotransmitter being released at a synapse. The proteins responsible for fusion are named soluble NSF attachment receptors, SNAREs (represented by green, pink, and blue filaments). Panel (**d**) shows the details of hemifusion and fusion of an envelope virus such as influenza virus, HIV, SARS-CoV-2, or dengue virus with the target cell. In this case, viral proteins at the surface of the virus insert the membrane of the target cell and undergo conformational changes ending in fusion and consequent release of viral contents in the cytosol

## Box 3.2 Lipid Vesicles in Pharma and Cosmetics

For many decades, lipids were considered relatively inert biologically, with functions of storage of energy in the adipose tissue or constitution of a matrix for cell membranes. Thus, in general, there was little interest in research to discover the properties, structures, biosynthetic pathways, biological utilization, and other functions of lipids. In the present, the situation is completely opposite. Lipids are regarded as important biological molecules that in addition to being energy stores and membrane components, participate in the regulation of many biochemical processes in cells and endocrine physiological regulation in the human body. Moreover, lipids are now important tools in pharma and cosmetic industries because they can be used in formulations that distribute and deliver drugs or other biologically active molecules in the human body. In most of these formulations, the lipids self-assemble in bilayers that form extensive vesicular systems, liposomes (see below figure), able to encapsulate molecules having desirable functions. It is a very versatile system as hydrophobic molecules may be accumulated in the lipid areas and hydrophilic molecules may remain solubilized in the aqueous spaces inside the vesicles.

In cosmetic applications, the liposomes may be part of formulations to be applied topically in the skin. The lipids help in diffusing the whole formulation through the outer layers of the skin. Moreover, the simple fact that lipids and water are forming an emulsion will help the formulation hydrating the desired areas of the skin.

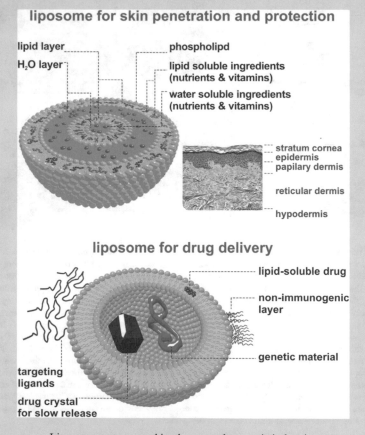

Liposome systems used in pharma and cosmetic industries

(continued)

**Box 3.2** (continued)

Liposomal encapsulation can substantially improve the action of a drug, such as the decreased toxicity observed with amphotericin B (see Fig. 3.9). Conventional amphotericin B has been generally considered the drug of choice for many types of systemic fungal infections. These infections are a major threat to those whose immune systems are compromised, such as patients undergoing chemotherapy for cancer, bone marrow transplant recipients, and AIDS patients. However, amphotericin B is very toxic, thus limiting its utility. For these patients, who have a high rate of morbidity and mortality, there is a dosage form distinct from conventional amphotericin B, which consists of amphotericin B complexed with two phospholipids in approximately a 1:1 drug-to-lipid molar ratio: 1,2-Dimyristoyl-glycero-3-phosphocholine (DMPC) and 1,2-Dimyristoyl-glycero-3-phosphorylglycerol (DMPG), present in a 7:3 molar ratio. Doxorubicin is another example. Liposomal doxorubicin is designed to target to tumor cells and spare healthy tissue, maintaining efficacy while reducing toxicity. Conventional doxorubicin, drug commonly used to treat cancer, is limited by its potential for causing a variety of severe side effects.

Researchers are developing innovative liposomes with refined drug delivery properties to be part of future medicines. Some have their surface modified with proteins and other selected polymers to target meaningful cells. Synthetic phospholipids are suitable for specific applications in liposome targeting and gene therapy. Gene therapy is based on the efficient delivery of genes to their intended targets. Researchers have successfully put DNA into liposomes and have achieved fusion of these liposomes to cells. Scientists have also succeeded in protecting these liposomes from degradation and are able to modulate their circulation time.

Because lipid bilayers are the basic structure of cell membranes, several drugs directed to bacteria and fungi target the organization of lipid bilayers of these pathogens. Several antimicrobial peptides, for instance, are cationic so they bind and disrupt the lipid membranes of bacteria, which are highly anionic. Polyene antibiotics such as nystatin B and amphotericin B bind specifically to ergosterol, causing the selective membrane permeability and lysis of fungi (Fig. 3.9) because ergosterol only exist in the membrane of fungi.

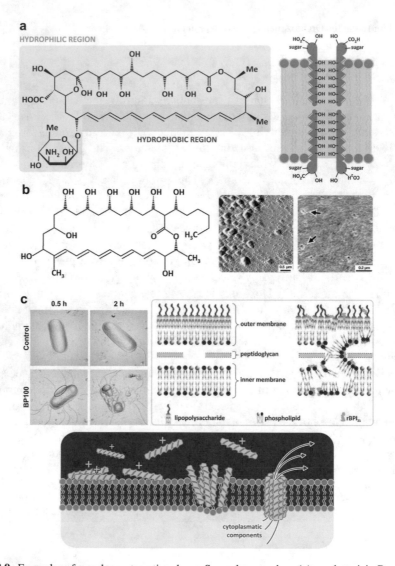

**Fig. 3.9** Examples of membrane-targeting drugs. Some drugs such as (**a**) amphotericin B, a fungicide used in the treatment of infections with *Candida* sp. among others, or (**b**) filipin, a fungicide also toxic to human cells and therefore not used in therapies, and (**c**) cationic amphipathic peptides with antibacterial properties target cell membranes. (**a**, *right*) Amphotericin B binds to ergosterol forming ordered complexes in which the hydrophobic sides of the polyene rings face the lipids and Fig. 3.9 (continued) the polar sides face each other forming a hydrophilic pore. Filipin is not so selective to ergosterol when compared to cholesterol as amphotericin B; therefore, it is more toxic to human cells. Panel (**b**) (*right*) shows atomic force microscopy images of pores created by filipin in cholesterol-containing bilayers (*arrows*). (**b**, *center*) Large structures in the surface of the lipid bilayer are filipin aggregates. Panel (**c**) (*left*) also shows atomic force microscopy images but of an individual bacterium (*E. coli*). Exposure of the bacterium to the cationic amphipathic peptide BP100 caused the collapse of the bacterial membrane. The bacterial membrane is anionic, thus attracting electrostatically the peptides, which then aggregate on the lipid bilayer causing perturbation and increasing permeability. (**c**, *bottom*) This perturbation may be caused by formation of pores or by unspecific destruction of the lipid organization (**c**, right shows the action of rBPI$_{21}$, a peptide derived from bactericidal/permeability-increasing protein potentially useful against meningitis). (Figures in Panel **b** (*right*) are reprinted with permission from Santos et al., Biophys J. 75:1869–1873, 1998. Figures in Panel **c** (*left*) are reprinted with permission from Alves et al., J. Biol. Chem. 285:27536–27544, 2010)

While lipid vesicles depicted in Fig. 3.7 are formed only by lipids (a single pure lipid or a mixture of lipids), biological membranes are often composed of lipids, proteins, and saccharides. How these components organize in the membrane has been the subject of intensive scientific research over the years (Fig. 3.10). Nowadays, a biological membrane of a human cell is regarded as a lipid bilayer having a heterogeneous distribution of lipids both in each layer and among layers. This heterogeneity leads to the formation of specific domains of lipids having defined functions. Rigid platforms, for instance, may serve to anchor proteins on the membrane (Fig. 3.11). The cell membrane is directly connected to the cytoskeleton through an array of proteins, the so-called cytoskeleton anchors. The outer surface of cell membranes may have lipids and proteins that are glycosylated (i.e., covalently attached to saccharides) contributing to a rich chemical diversity on the surface of cells (Fig. 3.10).

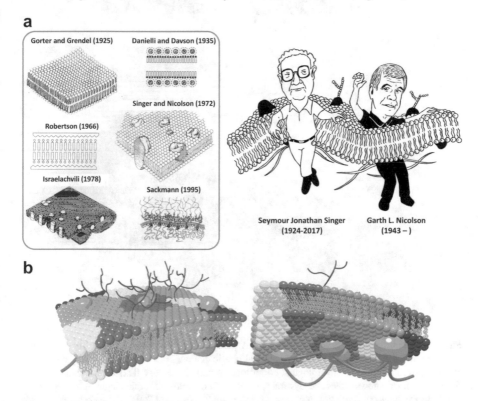

**Fig. 3.10** The structure and organization of cell membranes. (**a**) Historical evolution of the concept of cell membranes (according to Ole Moritsen; figure reprinted with permission from Biol. Skr. Vid. Selsk. 49:7–12, 1998). The lipid bilayer was described after the discovery that proteins interact with lipids. The concept that there are integral proteins embedded in the lipid bilayer forming a dynamic structure came with the fluid mosaic model by Singer and Nicolson, in 1972, which is still accepted as the basic framework of membranes. Nevertheless, from then on, the organization of cell membranes has been continuously unraveled. Lipids are now known to self-associate in lateral domains of different composition, and some membrane proteins are bound ("anchored") to the cytoskeleton (models proposed by Israelachvili and Sackmann). (**b**) The modern view on cell membranes, in which lipid colors represent heterogeneous lipid compositions. In the outer surface of cell membranes, glycosylated (i.e., saccharide-containing, *black*) lipids and proteins are present with different functionalities (see Sect. 3.2). Cytoskeleton and cytoskeleton-binding proteins are represented in brown

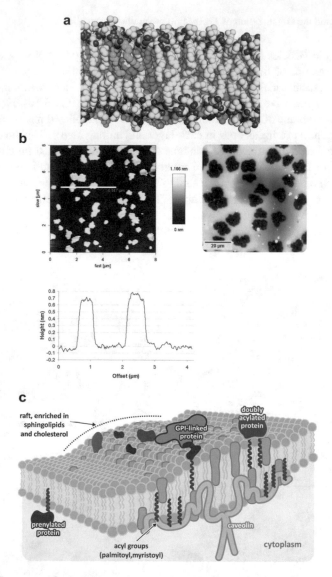

**Fig. 3.11** Heterogeneities in lipid membranes. Although phospholipids are frequently depicted in an oversimplified form with ordered stretched aliphatic chains, as in Fig. 3.2, in reality most molecules in lipid bilayers in physiological conditions in cells have very flexible and dynamic acyl chains. Panel (**a**) shows molecular dynamics simulation of lipids with one single lipid in green to highlight the bent conformation (courtesy of Dr. Claudio Soares, ITQB-UNL, Portugal). Lipids with longer and saturated chains adopt stiffer and linear conformations as they interact more tightly with each other. Mixing stiff and fluid lipids results in partial segregation of the lipids. The atomic force microscopy image of a mixture of a fluid unsaturated lipid (palmitoyl-oleoyl-phosphatidylcholine, POPC—50% molar) with a saturated rigid lipid (dipalmitoylphosphatidylcholine, DPPC—50% molar) is shown in Panel (**b**) (*upper left*); segregated areas are clear and the height profile along the line seen in the top image (*bottom graph*) shows that the segregated areas are higher, therefore corresponding to more rigid areas of the membrane. (**b**, *right*) Epifluorescence images of lipid bilayers having one of the lipids tagged with a fluorescent dye confirm segregation of both lipids (figures reprinted with permission from Franquelim et al., J. Am. Chem. Soc. 130:6215–6223, 2008; and Franquelim et al., Biochim. Biophys. Acta. 1828:1777–1785, 2013). (**c**) These more rigid areas of the membrane are more adapted to anchor proteins and membrane receptors. Ceramides and cholesterol, for instance, further enhance these characteristics. Proteins covalently attached to lipids are typically found in these rigid platforms

The lipid composition of cell membranes varies a lot from species to species, from organelle to organelle in the same cell, and from the inner leaflet to the outer leaflet in the same membrane (Fig. 3.12). This variety of compositions grants the necessary diversity to membranes so that they are specific for certain functions, in spite of the common feature to all membranes of all cells: in the end, they are all constructions based on lipid bilayers.

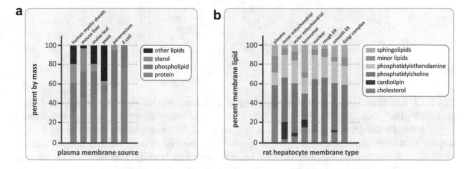

**Fig. 3.12** Although all membranes in cells in nature are constructions based on lipid bilayers, the membrane composition varies a lot among species (**a**), among organelles of the same cell, (**b**) and even between the two leaflets of the same membrane

Because lipid bilayers are hydrophobic barriers, hydrophilic molecules, such as glucose, cannot freely transverse them, which poses challenges to cells. To overcome such challenges, the cells have channels and transporters in their membranes. These are proteins specific for certain molecules or ions that facilitate or enable the translocation of such molecules or ions across the membranes. This subject will be revisited in Sect. 5.3.1, after careful consideration of protein structure.

## 3.1.2  The Structure of Lipoproteins

Lipoproteins are organized assemblies of lipids and proteins covering a wide range of sizes and densities. They circulate in the blood and are responsible for the transport of lipids among different tissues (Fig. 3.13). These assemblies have a lipidic core formed mainly by triacylglycerols and cholesterol esters, surrounded by a monolayer of phospholipids and cholesterol. The global arrangement is largely determined by the entropic effect as the monolayer of phospholipids minimizes the contact between apolar components of the core and water molecules in blood (Box 3.3).

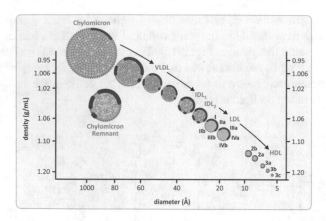

**Fig. 3.13** Lipoprotein structure, size, and density. Lipoproteins consist of a monolayer of phospholipids (lilac) and cholesterol covering a lipid droplet of triacylglycerols and cholesterol esters (represented by the orange and blue circles, respectively). There are also proteins at the surface (represented in magenta), the apolipoproteins, which are specific of each class of lipoproteins and serve for cell recognition, i.e., interact with specific receptors in cells. Lipoproteins are grouped according to their densities and sizes, although the most common nomenclature refers to density (*HDL* high-density lipoprotein, *LDL* low-density lipoprotein, *VLDL* very-low-density lipoprotein). Chylomicra are the structures formed with dietary lipids in the enterocyte (intestines) and released in plasma

---

**Box 3.3  Lipoproteins: The Burden of Lipid Transport**

Lipids have extremely low solubility in aqueous media. Therefore, as a consequence of the entropic effect (see Sect. 3.1), when placed in aqueous medium, they tend to self-associate. Most phospholipids, having a hydrophilic "head" and two hydrophobic acyl chains, tend to pack side-by-side and form bilayers. Cholesterol is not prone to form very organized supramolecular assemblies itself but is able to insert in the lipid bilayers and contribute to its stability. Triacylglycerols ("triglycerides") and esters of cholesterol (cholesteryl esters; see next figure) do not have the amphipathic properties and structural requirements to form lipid bilayers. Instead, triacylglycerols and cholesteryl esters amalgamate in an aggregate having no polar surface. These aggregates tend to be spherical, the geometry that minimizes the surface area exposed to the solvent. The lipid aggregates tend to grow until free lipids are nearly absent, unless a phospholipid monolayer covers the surface of these aggregates, forming an entropically favorable interface. The phospholipid monolayers stabilize the lipid aggregates and an emulsion is formed. Emulsion means the lipids are heterogeneously distributed in microscopic scale because the lipids are clustered in aggregates, but homogeneously distributed in macroscopic scale since the aggregates are evenly disseminated in the solvent.

(continued)

**Box 3.3** (continued)

Example of a cholesteryl ester: cholesteryl nonanonate. The molecule is composed of a cholesterol moiety and a fatty acid moiety (nonanoic acid in this case)

In the human body, very large lipid aggregates are found in the cells of the adipose tissue (adipocytes), occupying almost all cytoplasmatic space. It is a storage place (Fig. 3.3 in the main text). Smaller aggregates are found emulsified in blood, in association with specific proteins. These smaller aggregates are dragged by blood and serve as lipid transporters. In both cases, the aggregates are covered by monolayers of phospholipids having the polar groups exposed to aqueous environment and the acyl "tails" in contact with the lipids. The ensemble formed by the emulsified lipid aggregate covered with a phospholipid monolayer and associated with specific proteins is named lipoprotein (see Fig. 3.14 in the main text). The proteins themselves are named apolipoproteins. The lipoprotein, as a whole, is the lipid carrier entity; apolipoproteins' main function is binding to specific receptors so that lipids are delivered to target cells only.

Lipoproteins vary among them in the proportion of triacylglycerols, cholesteryl esthers, and proteins, which directly impacts in their compactness and density. Early studies on the properties of lipoproteins achieved separation of several classes of lipoproteins based on their different densities, so the density-based nomenclature was naturally adopted, from high-density lipoproteins (HDL) to very-low-density lipoproteins (VLDL) and chylomicra (see below table). There is a concomitant change in volume, but it is not given importance regarding lipoprotein classification. It is also worth highlighting that apolipoproteins are also divided in classes (A, B, C, etc.), and HDL are the only lipoproteins not bearing apolipoprotein B, which is a distinctive feature.

Properties of plasma lipoproteins

| Plasma lipoproteins | Density (g ml-1) | Diameter (nm) | Apolipoprotein | Physiological role |
|---|---|---|---|---|
| Chylomicron | <0.95 | 75–1200 | B48, C, E | Dietary fat transport |
| Very-low-density lipoprotein | 0.95–1.006 | 30–80 | B100, C, E | Endogenous fat transport |
| Intermediate-density lipoprotein | 1.006–1.019 | 15–35 | B100, E | LDL precursor |
| Low-density lipoprotein | 1.019–1.063 | 18–25 | B100 | Cholesterol transport |
| High-density lipoprotein | 1.063–1.21 | 7.5–20 | A | Reverse cholesterol transport |

(continued)

**Box 3.3** (continued)

During digestion, the lipids are partially degraded and emulsified in the intestinal lumen by bile acids, molecules similar to cholesterol but having several polar groups. Lipids and other nutrients are uptaken by the intestinal cells, enterocytes (see next figure). Chylomicra are formed in these cells and released in the blood. 80% to 90% of the lipids in chylomicra are triacylglycerols, which account for their low density. The remaining cargo is free cholesterol (1–3%), cholesteryl esters (3–6%), and phospholipids (7–9%). Chylomicra circulate in the blood, where degradation of their triacylglycerols into free fatty acids occurs. Part of these fatty acids is delivered to adipose tissue and peripheral tissues. The chylomicra remnants bind to liver cells that have specific receptors that recognize their proteins. The excess of nutrient uptaken after digestion is converted to lipids in the liver, where they form VLDLs in a process similar to the assembly of chylomicra in the intestine. VLDLs are released from the liver cells. In the bloodstream, they are depleted of free fatty acids meanwhile formed by the hydrolysis of triacylglycerols, resulting in the intermediary-density lipoproteins (IDL) and low-density lipoproteins (LDL), which transfer lipids to peripheral tissues having LDL receptors. Liver cells having LDL receptors also bind LDL. In contrast, HDLs transport cholesterol from peripheral tissues to the liver, where there are cells having specific receptors for HDL. The cholesterol is then used to synthesize bile acids. HDLs are the only lipoproteins that dispose of cholesterol. This characteristic renders the name "good cholesterol" to HDL-associated cholesterol in public health campaigns for lay audiences. This name makes no sense on biochemical grounds but helps to spread the message that in cardiovascular risk evaluation, it is important to differentiate between cholesterol that is being removed and cholesterol that is being incorporated. Interestingly, from the biochemical point of view, cholesterol to be disposed is associated with lipoproteins having no apolipoprotein B, and cholesterol to be incorporated is associated with apolipoprotein B-containing lipoproteins.

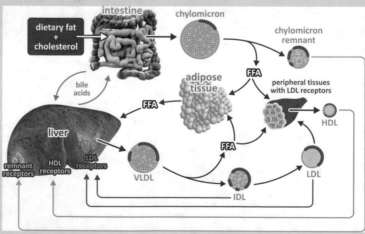

Origin and fates of plasma lipoproteins

(continued)

**Box 3.3** (continued)

LDLs are degraded inside the cells after being uptaken by endocytosis. The LDL receptor is segregated in the endocytic vesicle, which then divides in two: one empty vesicle having the receptors in the membrane returns to the surface of the cell, and the other vesicle has the proteins and lipids of the lipoproteins and joins the lysosome. LDL-derived cholesterol may then either be used in cell membranes, or to synthesize steroid hormones or bile acids, or simply be stored as cholesteryl esters. The exact destination of cholesterol in the cell depends on the type of cell and its metabolic state. Dietary cholesterol suppresses the synthesis of cholesterol by the body, and high free cholesterol levels inhibit the synthesis of LDL receptors. Cellular uptake is thus inhibited in the presence of excess cholesterol and the level of LDL in the blood increases. Moreover, the LDLs take more time to be uptaken and circulate in blood for longer periods. This increases the chances of having the LDL exposed to oxidative agents such as NO, hydrogen peroxide, or the superoxide ion. Oxidized LDLs are then removed from circulation by macrophages, but the macrophages get their properties severely altered after incorporating oxidized LDL, becoming the so-called foam cells. Foam cells accumulate in the walls of endothelia, releasing growth factors and cytokines that stimulate the migration of smooth muscle cells that proliferate in the site of accumulation of foam cells and form collagen matrices. This consists in the deposition of atherosclerotic plaques, which pose severe cardiovascular risk.

Interestingly, the proteins responsible for triacylglycerols conversion into fatty acids in the heart (the enzymes named "heart lipoprotein lipases") have much higher affinity for triacylglycerols than the corresponding proteins in the adipose tissue. The affinity parameter, $1/K_M$, which will be addressed in Sect. 4.2.1, is about tenfold higher in the heart. During starvation, the levels of plasmatic triacylglycerol drop, but delivery of fatty acids from triacylglycerols is kept in the heart even when suppressed to the adipose tissue.

Different classes of lipoproteins differ in density (due to differences in the relative amounts of proteins, phospholipids, cholesterol, triacylglycerols, and cholesterol esters in their composition), size, and specific proteins associated (see Table in Box 3.3). Nevertheless, these classes are named after the differences in density only, which relates to the most practical property that can be used for their separation in different fractions (see Fig. 3.13). Lipoproteins formed in the intestine with dietary lipids are known as chylomicra, and the remaining classes range from very-low-density lipoproteins (VLDLs) to high-density lipoproteins (HDLs). Intermediary (IDLs)- and low-density lipoproteins (LDLs) are in between.

The different classes of lipoproteins have different functions and different target tissues for their action (see Box 3.3). Target recognition depends on the specificity of the proteins present on the lipoprotein surface (referred to as apolipoproteins to highlight that only the proteic part is being addressed) for well-defined receptors.

## 3.2    Saccharides and Their Polymers and Derivatives

Saccharides, at variance with lipids, are extremely polar, therefore hydrophilic, molecules. They are linear aldehydes or ketones with hydroxyl groups bound to the carbons that do not form the carbonyls (C=O). Many of these molecules have only C, H, and O in their composition and fit the formula $(CH_2O)_n$. This spurious characteristic consecrated the designation "carbon hydrate," which is still widely used to identify saccharides despite its total inadequacy in chemical terms: not all saccharides obey to $(CH_2O)_n$, and this does not reflect a hydration of carbon, only a specific molar proportion between C, H, and O atoms. Referring to saccharides as "sugars" is equally inadequate and misleading. "Sugar" is related to a property, sweetness, which not all saccharides possess and extends to molecules other than saccharides, such as peptide sweeteners (see Box 3.4). Saccharides or oses are therefore the preferred nomenclatures for biochemists although "carbon hydrates" and "sugars" are also commonly used.

Saccharides are the most abundant biomolecules and owe this ubiquity to their reactivity and structural plasticity, which enable a great variety of functionalities, including energetic storage, cell communication, and cell protection against mechanical aggressions and dehydration. In order to understand such structural plasticity and the functionalities arising therefrom, one has to start with the basic chemistry and reactivity of saccharides. Although this is a wide and complex world in the realm of biochemistry, we will devote ourselves to the understanding of the most important saccharides in human biochemistry only. One will stick to the basics of this fascinating world for the sake of clarity and focus on processes that are foundational for other medical disciplines such as histology and physiology.

---

**Box 3.4  Sweeteners and Sugar Substitutes**
The problem of popularization of high caloric diets stimulated the search for sugar substitutes. Sucrose, the most commonly used sugar in cooking, is a natural sweet molecule from which a certain amount of energy can be used by the human body after metabolization. However, there are molecules known as high-intensity sweeteners that have many-fold the sweetening power of sucrose. Saccharin, for instance, is approximately 300-fold sweeter than sucrose when equal quantities are compared. Aspartame and acesulfame K are approximately 200-fold sweeter than sucrose. For sucralose, the ratio raises to an impressive 600-fold. A specific chemical modification in aspartame, advantame, grants an impressive 20,000-fold increase in sweetness relative to sucrose. Therefore, much less mass of sweetener is needed to achieve the sweetness of a food or beverage. Even though the "caloric content" of a unit mass of the molecule may be equivalent to sucrose in some cases, the total amount used is several orders of magnitude less and the total calories in the diet drops drastically.

(continued)

**Box 3.4**  (continued)

The chemical structure of most popular sweeteners, including sucrose and the artificial ones, is very different (see next figure). Sucrose is a disaccharide composed of the residues of the monosaccharides glucose and fructose. Sucralose is prepared from sucrose via the substitution of three hydroxyl groups for chlorides. Saccharin and acesulfame K have much different structures. Aspartame is the methyl ester of the dipeptide L-aspartyl-L-phenylalanine.

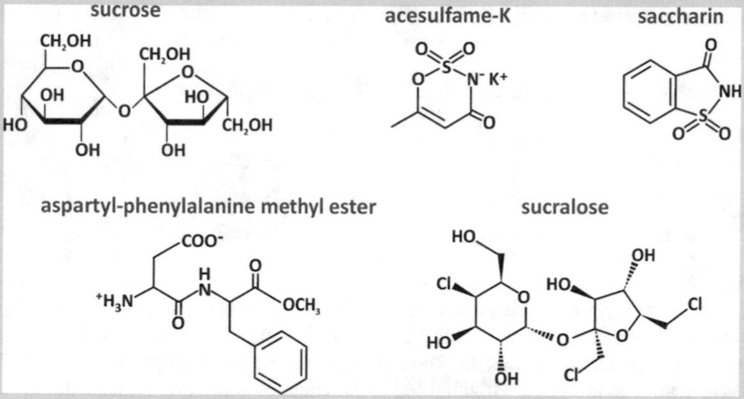

Chemical structures of most common sweeteners

The molecular structure of sweeteners must be such that they bind to a specific receptor molecule at the surface of the tongue. The receptor is coupled to a G-protein (see Sect. 5.4), which dissociates when the sweetener binds to the receptor. This dissociation leads to the activation of an enzyme that triggers a sequence of events resulting in signals that are transmitted to and interpreted by the brain. The sweetness perception depends on fine details of the interaction between the sweetener and its receptor. The importance of fine details in molecular shape to sweetness is illustrated by the case of aspartame, as its stereo isomer, L-aspartyl-D-phenylalanine methyl ester, has a bitter, not a sweet, taste.

There has been a long and continuous controversy on the impact of artificial sweeteners on health, which has driven a lot of research about the possible toxicity of their metabolic products. Saccharin has been very controversial and banned in some countries. In the body, aspartame is broken down into/absorbed as products that include aspartate, phenylalanine, and methanol, which is toxic. Phenylalanine is toxic to individuals with phenylketonuria, a genetic disease wherein individuals cannot process phenylalanine. Products containing aspartame must therefore be labeled for phenylalanine. Regardless of the controversies and limitations in their use, artificial sweeteners have an important role in the improvement of the quality of life of diabetics, who are limited in the consumption of sucrose and other saccharides.

The simplest conceivable saccharides have three-carbon chains, i.e., they are trioses ("tri" for three carbons, "ose" for saccharide). Depending on the position of the carbonyl group, C=O, which may be terminal (aldehyde) or not (ketone), the saccharide is an aldose or a ketose. In the specific case of trioses, only glyceraldehyde and dihydroxyketone are possible (Fig. 3.14). But even in these cases, two kinds of common chemical reactions in nature are possible: esterification and reduction (Fig. 3.15).

The chemical structure of glyceraldehyde deserves close attention as its central carbon (carbon number 2; C2) has four different substituents (i.e., it is bound to four different atoms or groups of atoms), being referred to as a chiral carbon or chiral center. Imagine the permutation of the H and OH substituents, for instance. A different molecule results from this switch (Fig. 3.16).

| glycerol | glyceraldehyde | dihydroxyketone | dimethyl ketone |
|----------|----------------|-----------------|-----------------|
| | O≋C−H | | |
| −C−OH | C | −C−OH | −C− |
| −C−OH | −C−OH | C=O | C=O |
| −C−OH | −C−OH | −C−OH | −C− |

**Fig. 3.14** Important glycerol-related structures. Glycerol is related to an aldehyde; glyceraldehyde, which has a ketone isomer; and dihydroxyketone, which in turn is related to acetone, also named dimethyl ketone, or propanone. The aldehyde and ketone groups of glyceraldehyde and dihydroxyketone, respectively, are highlighted in the chemical structures by a shadowed box

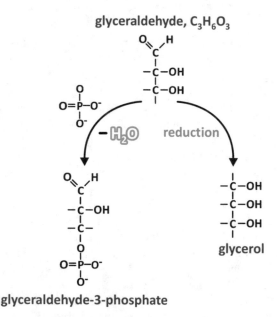

glyceraldehyde, $C_3H_6O_3$

−$H_2O$        reduction

glyceraldehyde-3-phosphate

**Fig. 3.15** Examples of very frequent reactions involving saccharides. Glyceraldehyde may be reduced to glycerol upon chemical reduction of the C=O group in carbon 1 (carbons are numbered starting with the one from the carbonyl group, similarly to fatty acids, in which carbons are numbered starting in the carboxyl group). Phosphoric acid ($HPO_4^{2-}$) may react with carbon 3, for instance, to form an ester, glyceraldehyde-3-phosphate

Although being isomers, both molecules cannot be overlapped because the orientation in space of the H and OH substituents is different. The difference is clear if the molecule is represented in a three-dimensional (3D) perspective. A closer look reveals that both molecules are mirror images of each other, i.e., they are enantiomers. To distinguish between both enantiomers of glyceraldehyde, one is named "L," and the other is named "D." These labels were arbitrarily assigned by Emil Fischer but are used to name saccharides and amino acids by extrapolation from glyceraldehyde (Fig. 3.16): D stands for right (dextro in Latin) and refers to the structure having the OH group in the chiral carbon to the right when it is pro-

Hermann Emil Fischer
(1852–1919)

jected toward the observer; L stands for left (levo in Latin) and refers to the structure having the OH group in the chiral carbon to the left when it is projected toward the observer. The chemical structure of glyceraldehyde deserves close attention as its central carbon (C2) has four different substituents (i.e., it is bound to four different atoms or groups of atoms), being referred to as a chiral carbon or chiral center. Imagine the permutation of the H and OH substituents, for instance: a different molecule results therefrom (Fig. 3.16).

| D - glyceraldehyde | L - glyceraldehyde | L - amino acid | D - amino acid |
|---|---|---|---|

**Fig. 3.16** The chiral carbon in glyceraldehyde leads to enantiomers. L- and D-glyceraldehyde are mirror images. The same nomenclature was extrapolated to amino acids (alanine is shown as example)

In chemistry research, there are two other naming conventions for enantiomers independent from each other: the R- and S-system, which is based on a classification of the substituent group based on the atomic number of atoms bound to the central atom (chiral center), or the + and − system, based on optical activity, i.e., on direction of rotation of incident plane-polarized light. Symbols + and − are sometimes replaced by d- (dextrorotatory) and l- (levorotatory), but d- and l- are easy to confuse with D- and L- and prone to misunderstanding. Both systems are more robust than Fischer's D and L because they are not dependent on the comparison with glyceraldehyde. Chemists tend to use R/S or +/−, but biochemists are still "attached" to the D/L system for a simple reason: chiral diversity among biological saccharides and amino acids is very restricted. By far the most abundant saccharides in human biochemistry are D. Interestingly L is the preferred form in amino acids. Natural evolution favored one form specifically probably because it is simpler to have only one form as the building block for saccharide polymers (polysaccharides) and amino acid polymers (proteins). A small protein with 100 amino acid residues that could be D or L would have 2100 different possible isomeric structures. Because only L-amino acids are used, only one isomeric form is allowed. Why specifically L-amino acids and D-saccharides and not the other forms? It is not clear; probably it originated from ancient primordial random processes that later propagated and converged by evolution into the specific enantiomers found in nature nowadays.

When longer carbon chains are considered, more complex saccharides are possible depending on the:

1. Length of carbon chains
2. Position of the carbonyl group in the carbon chain
3. Number and location of chiral centers

Although many different saccharides can be found in the human body, pentoses and hexoses are the most frequent in metabolic processes, so we will now focus on these molecules, namely, ribose and glucose (Fig. 3.17).

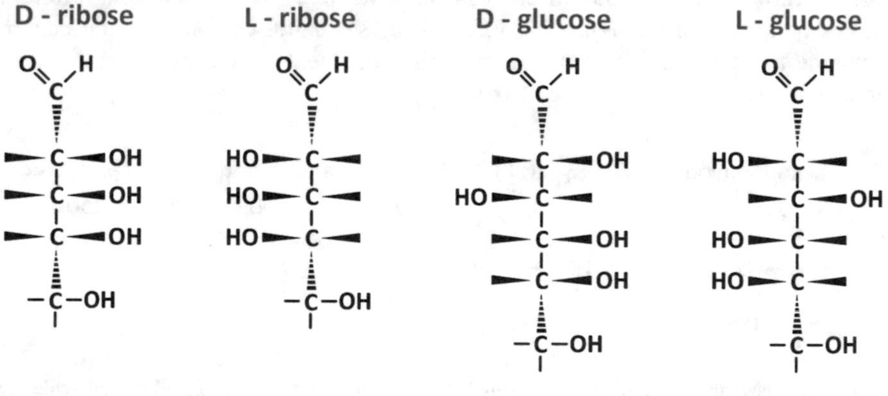

**Fig. 3.17** D- and L-isomers of ribose (pentose) and glucose (hexose). D-forms are the more relevant forms in nature

Aldoses such as glyceraldehyde, ribose, and glucose react with water, forming a hydrate. This happens at the C=O group because the oxygen atom attracts the electrons leaving the C deficient in electrons, therefore prone to interact with the electrons of water oxygen (Fig. 3.18). The reaction is reversible, so aqueous solution of aldoses contains mixtures of their aldehyde and hydrate forms. Nevertheless, pentoses and hexoses may react intramolecularly in a way that is similar to hydration. Because the carbon chain is able to bend and is dynamic (similarly to saturated aliphatic chains in lipids), the carbonyl group may contact alcohol groups (OH—hydroxyl) in the same molecule and react with it. The result is the formation of a cyclic molecule by the conversion of the carbonyl group in a hemiacetal group (Fig. 3.18). The cyclization is reversible, and in cells, the cyclic forms of ribose and glucose coexist with the linear forms, although the cyclic forms are dominant. Upon cyclization, two enantiomers are formed because the hydroxyl group in C1 may be linked to any of the two sides of the plane of the ring: in the α-anomer, the OH group in C1 is in the opposite side of the plane of the ring relative to the terminal carbon, C6 (–CH₂OH), and in the β-anomer, they are both at the same side. Glucose adopts a "chair" conformation at variance with the strict planar ring of ribose (Fig. 3.18), but α- and β-anomers exist the same way. The existence of the enantiomers has drastic implications in the polymerization of hexoses.

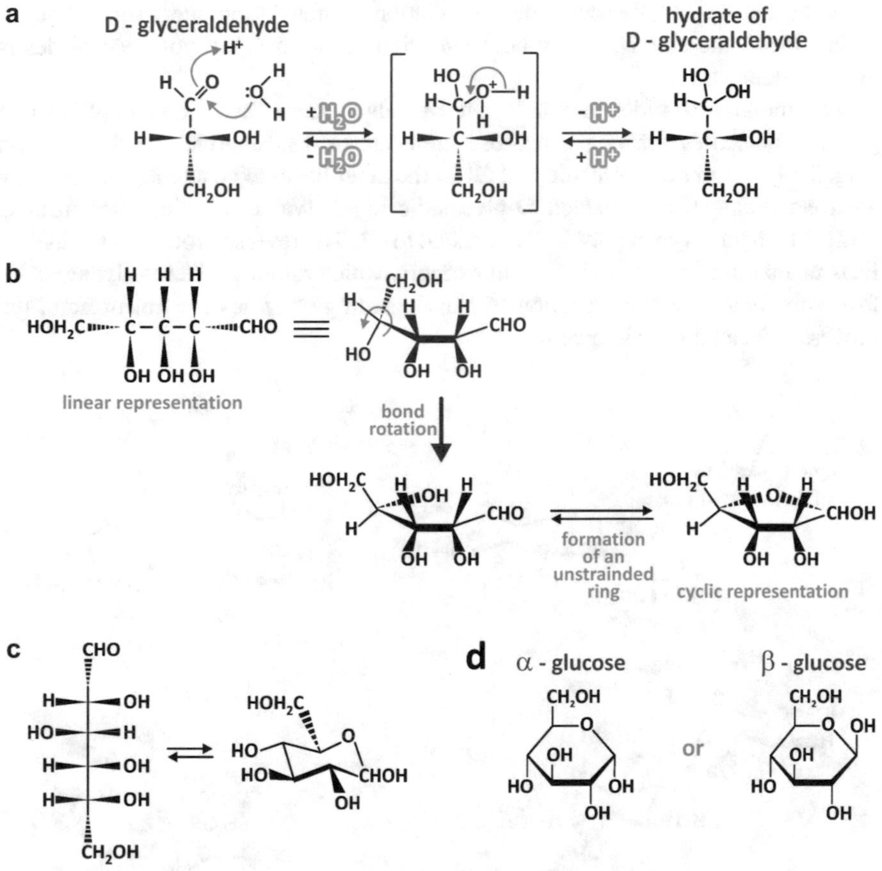

**Fig. 3.18** Hydration of saccharides. (**a**) Hydration of glyceraldehyde forms a hydrate. This reaction is reversible, so glyceraldehyde coexists with its hydrate. The carbon originally present as a carbonyl group is the only carbon with two bonds to oxygen in the hydrate. A similar reaction may occur intramolecularly in (**b**) pentoses and (**c**) hexoses. Panel (**b**) shows in detail the reaction of C=O (C1, in CHO) with the OH group in C4, analogous to hydration. A cyclic pentose is thus formed. As with hydration, the reaction is reversible and both forms coexist, although the cyclic form is more abundant. (**c**) The cyclic form of the hexose glucose is not planar as ribose is, as the molecular hexagon is flexible and adopts other conformations, such as the "chair" conformation. (**d**) Cyclization results in the formation of two anomers because the OH group formed at C1 may be placed on two different sides of the newly formed molecular ring: in the α-anomer, the OH group in C1 is in the opposite side of the plane of the ring relative to the terminal carbon, C6 (– CH₂OH), and in the β-anomer, they are both at the same side. Both the "chair" (**c**) and the more simplistic planar (**d**) representation of hexoses are used in this book

## 3.2.1  From Monomers to Polymers: Polysaccharides

Saccharides such as hexoses and pentoses may react with each other forming chains that may reach considerable size. Molecules built from the association of smaller molecules of a kind are generally named polymers, and polymers made of unit saccharides such as ribose or glucose are named polysaccharides. The units forming polysaccharides are referred to as monosaccharides. A covalent association of two monosaccharides is a disaccharide. Association of "few" monomers forms "oligosaccharides"; the size boundary between oligosaccharides and polysaccharides is not well defined.

Two monosaccharides may associate by dehydration. Take the example of two glucose molecules forming a maltose molecule (a disaccharide) by dehydration (Fig. 3.19). C4 in one molecule and C1 in the other become covalently attached by an acetal linkage, also named O-glycosidic bond. Water resulting therefrom is formed with the oxygen previously attached to C1. The reverse process is the hydrolysis of maltose into two glucose monomers, which although thermodynamically favorable, is a very slow reaction. When degrading enzymes are not present, the process is meaningless in practice.

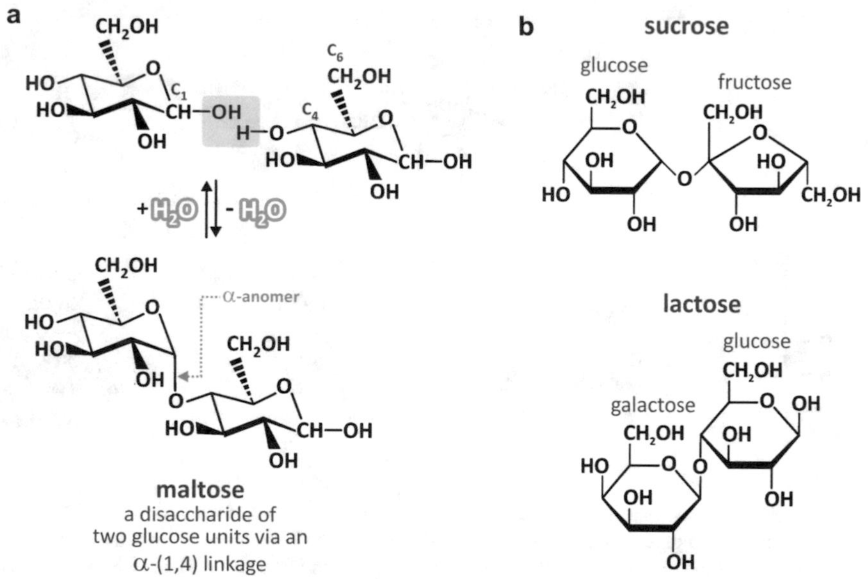

Fig. 3.19  Formation of disaccharides. (a) Two glucose molecules may associate by dehydration. When C1 in α-glucose (α anomer) reacts with C4 in another glucose, maltose is formed. Maltose is thus a disaccharide formed by linking two glucose molecules through an acetal or O-glycosidic bond. This bond is named "α-(1,4)" to stress that C1 in the α-anomer binds to C4 in the other molecule. (b) Sucrose and lactose are other examples of disaccharides. Both are formed from the conjugation of different constituent monosaccharides: glucose and fructose in the case of sucrose and galactose and glucose in the case of lactose. Sucrose involves an α-(1,2) bond, whereas lactose involves a β-(1,4) linkage

The stereochemistry (i.e., the spatial orientation of the chemical groups in the molecule) is very important because the covalent linkage of two molecules imposes restrictions on the way molecules can move in space. Depending on whether monomers are α- or β-anomers, different degrees of restriction arise. The flexibility of the conjugate is very much dependent on the enantiomers because the interaction between molecular groups in the disaccharide is very different (Fig. 3.20a). This effect is amplified in large polymers; polysaccharides may have a wide range of flexibilities, from extremely stiff and straight to coiled and deformable depending on the enantiomers used. Cellulose, for instance, is a glucose polymer formed with β-(1,4) bonds that is extremely mechanically resistant, whereas amylose is an example of a flexible polymer formed by a α-(1,4) backbone (Fig. 3.20b, c). Cellulose properties determined its evolutionary selection toward structural functions in plants, forming cell walls, which impacts in the macroscopic properties of wood, for instance. Amylose is a component of starch, a molecule that curls forming helices and is stored in plants for use in the energetic metabolism. Starch is the most common polysaccharide in human diet. Cellulose and amylose are striking examples of how apparently small details may actually determine profound differences in molecular properties and structures and therefore also in function.

Carbon 6 is also available for reaction, so linear polymers formed of C1–C4 chains may branch when C1–C6 bonds are also formed (Fig. 3.20d). Amylose turns into amylopectin when α-(1,6) links are formed. Glycogen is the human storage polysaccharide and is very similar to plant amylopectin (Fig. 3.20e). They differ only in the frequency of branching and average size of α-(1,4) segments. The advantage of having glycogen as energy storage relative to a linear (unbranched) polysaccharide relates to the fact that glycogen is enzymatically degraded by saccharide hydrolysis of the terminal units. A branched molecule has several termini which can all be degraded at the same time, making glucose readily available at high rate.

Depending on their chemical composition and stereochemistry, polysaccharides found in nature have one of three functions: (1) structural/mechanical protection, (2) energetic storage, and (3) water-binding (protection against dehydration). Several examples are in Table 3.2. Hyaluronic acid, a polysaccharide with sulfate groups ($SO_4^{2-}$ has similar properties to $PO_4^{2-}$), forms an extracellular mesh with collagen in the connective tissue, forming a flexible but resistant hydrated histological structure (Fig. 3.21).

### 3.2.2   Molecular Conjugates of Mono- and Oligosaccharides

We have seen in previous sections that saccharide monomers offer diverse possibilities of reaction, and so they are molecules that form many conjugates in nature. The most important derivatives are phosphate esters. Phosphoric acid is able to form up to three ester bonds (Fig. 3.22), although the triesters are not commonly found in nature. Yet, diesters are important and enable saccharide phosphates to form polymers (e.g., nucleic acids) or bridge saccharides with other organic molecules.

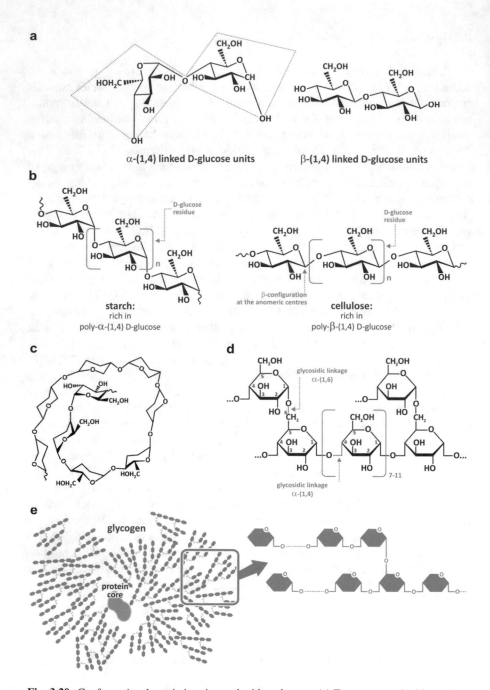

**Fig. 3.20** Conformational restrictions in saccharide polymers. (**a**) Two monosaccharides such as D-glucose forming a disaccharide have very different restrictions to articulate and move depending on whether the glycosidic bond is α-(1,4) or β-(1,4). (**b**) When several monomers bind to form a polymer, successive α-(1,4) or β-(1,4) bonds confer distinct properties to the polymer: α-(1,4) bonds enable bending between monomers, which results in curled polymers such as (**c**) amylose, and the stiffer β-(1,4) links between monomer favor linear straight polymers, such as cellulose. Therefore, cellulose is found in structural elements of plants, while amylose is used by plants as energy storage. Humans also use a poly-α-(1,4) saccharide as energy storage. (**d**) Periodic α-(1,6) branching further enables a globular organization of this polysaccharide. The final result is a regularly branched polymer of D-glucose named (**e**) glycogen. Glycogen synthesis is initiated by a protein and elongation requires several enzymes (see Sect. 8.2)

**Table 3.2** Examples of the function of polysaccharides found in nature: structural (Str), energy storage (Sto), and water-binding hydration (Wat)

| Polysaccharide | Monosaccharide 1 | Monosaccharide 2 | Bond | Branching | Location | Main function |
|---|---|---|---|---|---|---|
| *Bacteria* | | | | | | |
| Peptidoglycan | D-GlcNAc | D-MurNAc | β-(1,4) | – | Bacterial wall | Str |
| Dextran | D-Glc | – | α-(1,6) | α-(1,3) | Capsule | Wat[a] |
| *Animals* | | | | | | |
| Chitin | D-GlcNAc | – | β-(1,4) | – | Insects, crabs | Str |
| Glycogen | D-Glc | – | α-(1,4) | α-(1,6) | Liver, muscles | Sto |
| Hyaluronic acid | D-GlcUA | D-GlcUA | β-(1,4) β-(1,3) | – | Connective tissue | Str, Wat |

D-Glc: D-glucose; D-GlcNAc: *N*-acetyl-D-glucosamine; D-GlcUA: D-glucuronic acid; D-MurNAc: D-*N*-acetylmuramic acid

[a]Capsular materials like dextrans may be overproduced when bacteria are fed with saccharides to become reserves for subsequent metabolism

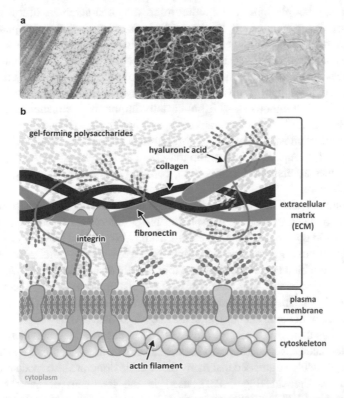

**Fig. 3.21** The extracellular matrix has hyaluronic acid in its composition. (**a**) Histological preparations of cock crest highlighting the hyaluronic acid matrix (*left*, conventional electron microscopy; *center*, platinum–carbon replica; *right*, preserved blue-dyed hyaluronic and extracellular heavily glycosylated proteins, the "proteoglycan matrix") (figures reprinted with the permission of Instituto de Histologia e Biologia do Desenvolvimento, Faculdade de Medicina, Universidade de Lisboa, FMUL). (**b**) Schematic representation of the molecular organization of extracellular matrix, which is composed of gel-forming saccharides attached to a backbone of hyaluronic acid that is intertwined among collagen fibrils. Polysaccharides form gels due to the high density of H bonds. These gels confer structure and mechanical protection to cells and retain water, which prevents desiccation of the tissues

Phosphate groups forming esters are anionic in aqueous environment in the most common biological pH ranges. This means that neutral molecules, such as glucose, become charged when esterified with a phosphate. The consequence is an increase in solubility in water and a decrease in the ability to cross lipid bilayers, for instance. This is deemed important as glucose metabolism starts by forming glucose phosphate (Fig. 3.22a).

Phosphate ester hydrolysis is a spontaneous but very slow process, which makes it under enzymatic control in cells. In addition, many chemical processes occurring in cells, such as condensation of polymers with formation of water, are unfavorable processes (there is "excess" water in most cell environments); enzymes speed the reaction but do not shift the equilibrium toward condensation. The use of phosphate derivatives of the monomers in the process of condensation facilitates the reaction as phosphates are so-called good leaving groups: they alter the reactivity of transient chemical species in the course of the mechanism of reaction.

ATP (adenosine triphosphate, Fig. 3.22b) is among the biological molecules that are saccharide derivatives and involves a phosphate ester. A phosphate diester bond bridging two other phosphates is another interesting characteristic of this molecule. The energy balance involved in the hydrolysis of phosphate–phosphate bonds makes this molecule pivotal in energetic metabolism. Divalent cations such as $Mg^{2+}$ are usually associated with ATP and other molecules having diphosphate groups. This reduces electrostatic repulsion between the oxygen atom of water and the negative charge of phosphate groups, facilitating the hydrolysis of phosphate derivatives.

Probably not so famous as ATP, but equally important in biochemistry, is coenzyme A (Fig. 3.23c). This is a relatively small but complex molecule. Amazingly, coenzyme A has phosphate and saccharide groups but owes its reactivity to a terminal thiol (–SH) group. This thiol group may bind an acetyl residue through a thioester bond, but may also bind a fatty acid, forming acyl-CoA, which is involved in lipid metabolism.

Nicotinamide adenine dinucleotide ($NAD^+$) is another interesting case of saccharide derivative that also contains phosphates. $NAD^+$ intervenes in redox reactions as it may accept and donate electrons, changing from $NAD^+$ to $NADH + H^+$ or vice versa. One extra phosphate group turns $NAD^+$ into $NADP^+$, which has similar redox properties but can only bind to specific enzymes that usually do not bind $NAD^+$. This implies that there are specific metabolic roles for $NADP^+$, distinct from $NAD^+$. The phosphates are involved in enzyme recognition but not in the redox activity itself (Fig. 3.22d). The same happens with flavin adenine dinucleotide ($FAD^+$ and $FADH_2$; Fig. 3.22e).

Nucleotides themselves deserve closer attention because they polymerize to form the so-called nucleic acids. They will be left for further discussion in the next section. To finalize, it should be stressed that many therapeutic drugs are also saccharide derivatives, such as digoxin (Fig. 3.22f), used in the treatment of heart conditions. Azidothymidine (AZT) is another example. It is an analog of thymidine that may inhibit the action of reverse transcriptase of HIV. It was the first drug used in the treatment of AIDS. Cellular enzymes convert AZT into the effective 5-triphosphate form (Fig. 3.22g). Once bound to reverse transcriptase, the azide group, N3, is responsible for chemical inhibition. Inspired by the success of AZT (Fig. 3.23), many nucleosides are now under development to create new inhibitors of HIV reverse transcriptase to fight AIDS.

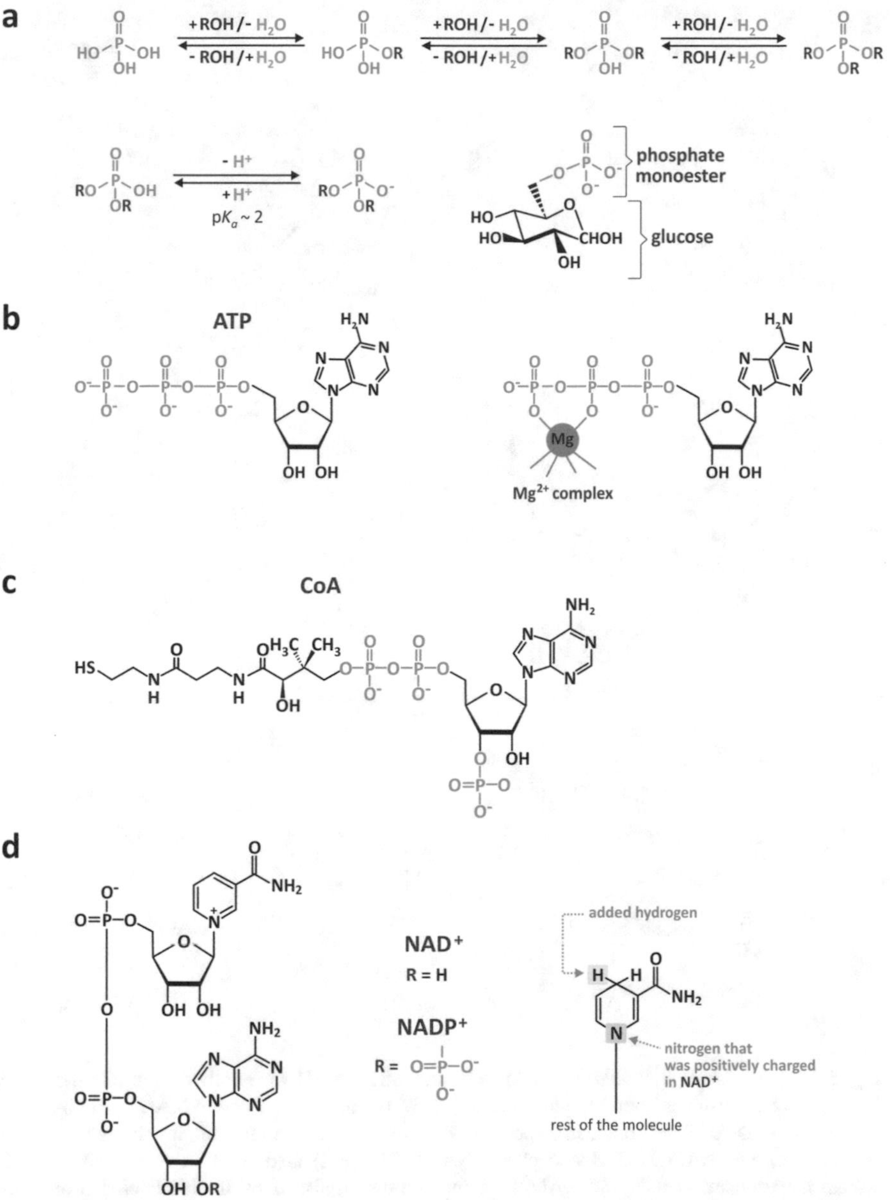

**Fig. 3.22** The importance and ubiquity of phosphates. (**a**) Phosphates form esters or diesters bridging two organic molecules. Phosphate confers an anionic charge to the newly formed chemical entity because the ionization of the phosphate group occurs at pH > 2, increasing its solubility in aqueous medium and decreasing its ability to translocate lipid membranes. This is the case for glucose-6-phosphate, which is "trapped" in the cytosol of cells, where it will be processed in different metabolic pathways. (**b**) Adenosine triphosphate, ATP, and (**c**) coenzyme A, CoA, are important biological molecules with a saccharide residue bound to a phosphate group. ATP also contains a phosphodiester bond, very important for its reactivity in cells. CoA has a couple of phosphate groups bound to each other, but its reactivity in cells is dictated by the sulfhydryl group, also named thiol (-SH).

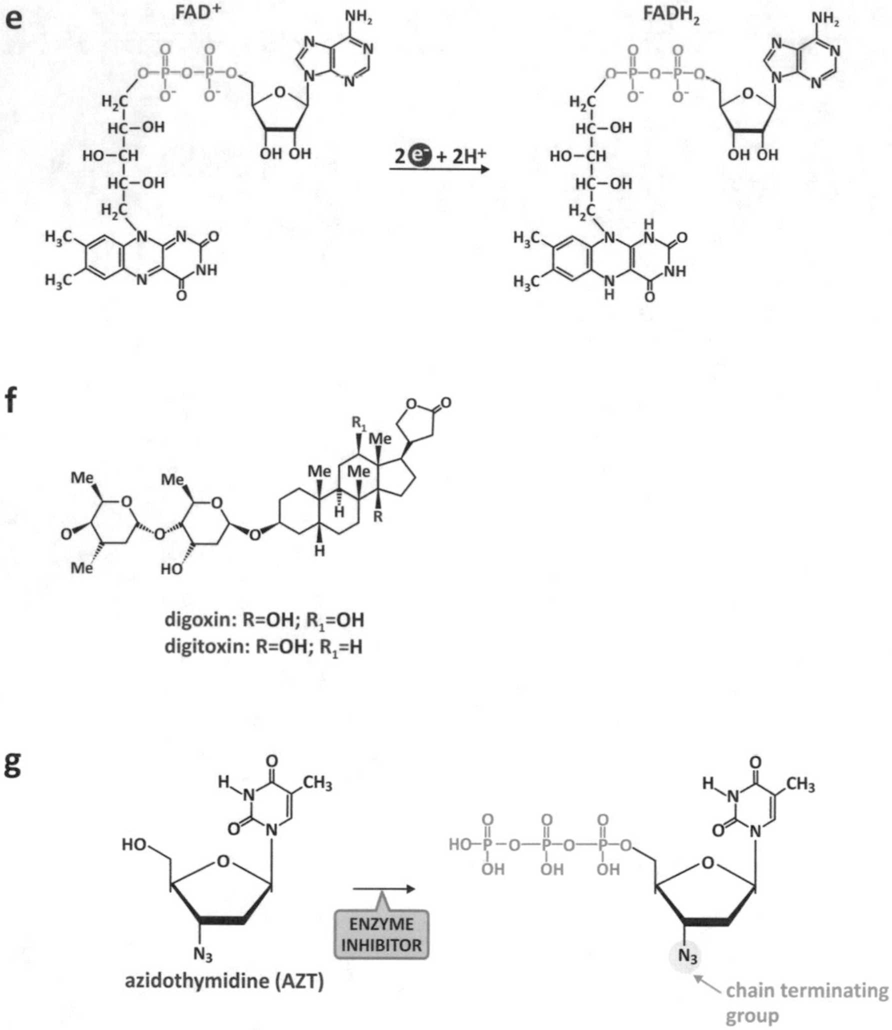

**e**

$$2\,\ominus + 2H^+$$

**f**

digoxin: R=OH; $R_1$=OH
digitoxin: R=OH; $R_1$=H

**g**

azidothymidine (AZT)

ENZYME INHIBITOR

chain terminating group

**Fig. 3.22** (continued) Nicotinamide adenine dinucleotide (NAD$^+$) is another important molecule with saccharide residues bound to phosphates. (**d**) NAD$^+$ may be reduced to NADH. Redox reactions of NAD$^+$/NADH take place in a specific cyclic residue of the molecule, involving a nitrogen atom (*right*). (**d**) NAD$^+$/NADH phosphate (NADP$^+$/NADPH) also enter in redox reactions in human metabolism. NADH and NADPH cannot be distinguished by their reducing properties because the phosphate group that distinguishes them is not related to the nitrogen atom that grants the redox properties. Yet enzymes use specifically NADH or NADPH and so there is no redundancy between these molecules. (**e**) FAD$^+$ and FADH$_2$ are molecules similar to NAD$^+$ and NADH in the adenine nucleotide and their role in metabolic redox reactions. (**f**) Digoxin and digitoxin are examples of drugs with monosaccharides in their structure; more specifically, three residues are specifically combined as part of a unique structure. (**g**) Another example of drug that is a saccharide derivative is azidothymidine (AZT), which is converted to a triphosphate in cells and is able to insert in the active center of the reverse transcriptase of HIV because it is similar to the natural substrate. However, the natural substrate does not have the N$_3$ group. The presence of this group blocks the conversion of the viral RNA into DNA

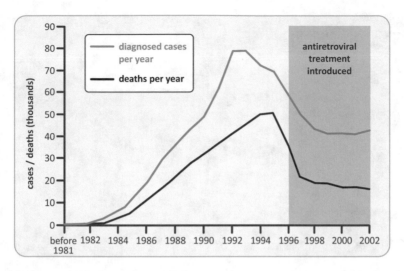

**Fig. 3.23** The use of AZT and other drugs had a very positive effect in the reduction of AIDS-caused mortality, which significantly decrease compared to other causes of death. Preventive campaigns highlighting the need to change risk behaviors had a strong impact in the spreading of AIDS in the USA, with a marked decrease in the number of diagnosed cases and deaths per year after 1993–95

### 3.2.3 Molecular Conjugates of Oligosaccharides

It is worth stressing that combination of saccharide monomers may generate a big diversity of products when compared to amino acids, for instance (Fig. 3.24). Two glucoses, for instance, can bind via six carbons in each monomer, thus being able to form 36 different molecules. Considering the anomers, the diversity increases. It is not surprising that oligosaccharides are present in the surface of cells as receptors of unique structure (see an example in Box 3.5), while amino acids form polymers (proteins) having domains with few restricted and well-defined structures. Moreover, monosaccharides or oligosaccharides are frequently formed in nature attached to proteins.

**Box 3.5 The ABO Blood Groups**
There are different blood groups according to different immunogenic molecules present in erythrocytes. The most important classification of blood groups is based on three antigens, A, B, and O, that form four groups: A, B, O, and AB—the ABO blood groups. The ABO blood group antigens are oligosaccharide chains attached to proteins and lipids located in the outer surface of erythrocytes. One single residue of a small oligosaccharide determines whether the antigen is A, B, or O (see below figure).

(continued)

**Box 3.5** (continued)

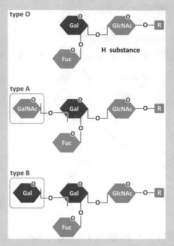

ABO group antigens: Fuc represents the monosaccharide fucose; Gal, galactose; GalNAc, *N*-acetylgalactosamine; and GlcNAc, *N*-acetylglucosamine

The immune system of an individual produces antibodies against the ABO antigens not present in his own erythrocytes. Individuals in A group will produce antibodies against B and vice versa. Type O, the most common, does not contain the last residue, which is the antigen, in its structure (in fact the original nomenclature was 0—zero—but became the letter O). So, individuals in blood group O will produce both anti-A and anti-B. Individuals in blood group AB are rare and, naturally, have no anti-A and no anti-B antibodies. This has tremendous implications in blood transfusions as a patient cannot receive erythrocytes against which he/she has antibodies. AB individuals can, in principle, receive blood from any donor; O individuals can donate blood to any individual; A and B individuals can only donate and receive blood to/from individuals belonging to the same blood group.

It is believed that ABO antibody production is stimulated when the immune system contacts in foods or in microorganisms with the saccharide antigens that are absent in the erythrocytes. The functions of the ABO blood group antigens are not known. Individuals who lack the A and B antigens are healthy, suggesting that any function the antigens have is not important, at least not in modern times.

Hemolytic disease of the newborn (HDN) is a serious medical problem that occurs almost exclusively in infants of blood group A or B who are born to group O mothers. This is because the anti-A and anti-B formed in group O individuals tend to be of the IgG type, which can cross the placenta. HDN tends to be relatively mild mainly because fetal erythrocytes do not express adult levels of A and B antigens. However, the precise severity of HDN cannot be predicted.

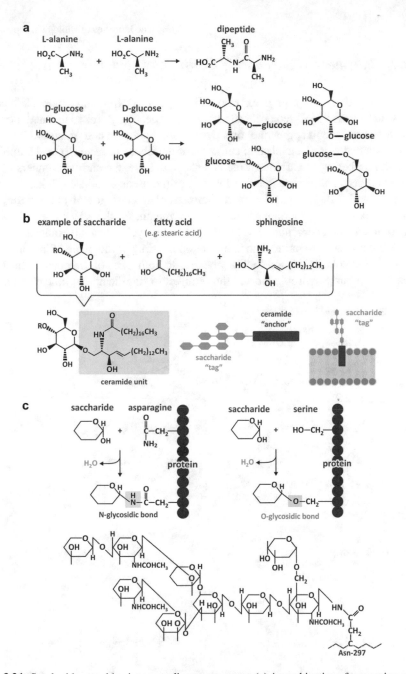

**Fig. 3.24** Saccharides combine into very diverse structures. (**a**) A combination of two amino acids generates one single dimer, but there are several ways that two monosaccharides can combine to form a disaccharide. Saccharides are better suited to form highly specific structures at the surface of (**b**) cell membranes or (**c**) proteins. (**b**) Saccharide tags are covalently bound to lipids, usually rigid lipids such as ceramide for a better anchoring to the membrane. Glycolipids (i.e., associations of saccharides and lipids) determine blood groups, for instance (see Box 3.5). (**c**) The same principle applies to oligosaccharides attached to proteins, i.e., glycoproteins. The side chain of the amino acid asparagine may react with a saccharide by dehydration forming an N-glycosidic bond (analogous to an O-glycosidic bond but involving N instead of O). Likewise, the side chain of the amino acid serine may react with a saccharide forming an O-glycosidic bond. An oligomeric sequence of saccharides attached to an hypothetical protein IgG are shown in (**c**) (bottom) as an example

### 3.2.4   Polymers of Saccharide Conjugates: Nucleic Acids

Nucleotides that compose deoxyribonucleic acid (DNA) and ribonucleic acid (RNA) are formed by 2-deoxyribose or ribose, respectively, linked to a heterocyclic base, a purine (adenine, guanine) or a pyrimidine (cytosine and uracil or thymine), and a phosphate group attached to carbon 5 of the ribose residue. To avoid ambiguity with numbering of carbons of the heterocyclic base, the carbon numbers of the ribose and deoxyribose are identified with a prime: phosphate ester linkage occurs at C5′ (Fig. 3.25). The physical and chemical characteristics of the heterocyclic bases are extremely important as they are determinant for the way nucleotide polymers (nucleic acids) organize. The bases are planar, cyclic, aromatic molecules with N and O atoms able to participate in hydrogen bonding in the plane of the ring. The bases are low polarity groups poorly solvated, so both faces of the plane of the base rings will be fairly hydrophobic and thus subject to significant entropic effects.

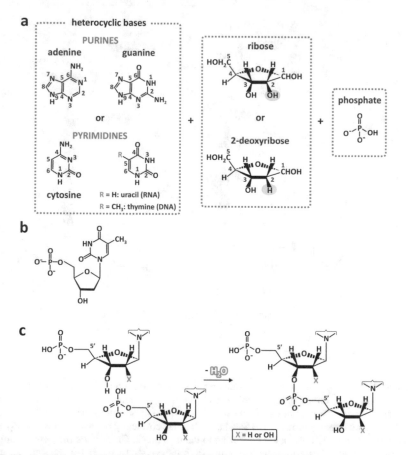

**Fig. 3.25** Conjugation of building blocks in nucleic acids. (**a**) Nucleotides are formed with a heterocyclic base, purine (adenine or guanine) or pyrimidine (cytosine, uracil, or thymine), a ribose or 2-deoxyribose, and a phosphate. The phosphate group forms a phosphodiester bond in C5 and the heterocyclic base binds to C1. (**b**) The nucleotide deoxythymidine phosphate is shown as example. (**c**) A dimer of nucleotides may be formed by dehydration, which creates a phosphodiester linkage between the monomers via C5′ and C3′. In RNA X=OH, and in DNA X=H

There are four different possible nucleotides in RNA and DNA. RNA is formed by adenosine-5′-monophosphate (AMP), guanosine-5′-monophosphate (GMP), cytidine-5′-monophosphate (CMP), and uridine-5′-monophosphate (UMP). DNA is formed by deoxyribose, which is denoted by a prefix: dAMP, dGMP, dCMP, and dTMP. dTMP stands for thymine-5′-monophosphate using deoxyribose; DNA does not contain dUMP.

Because nucleotides are phosphate monoesters, they can form additional phosphoester links to other alcohols, such as the OH groups in other nucleotides. In order words, they can polymerize by dehydration reactions. Nucleic acids are formed by phosphodiester bonds between C5′ of one nucleotide and C3′ of another nucleotide (Fig. 3.25c). The result is a linear polymer having the heterocyclic bases and the phosphate groups in opposing sides, the phosphate groups being anionic (Fig. 3.26a). Some simplified representations of nucleic acids pinpoint this characteristic (e.g., Fig. 3.26b), which remains elusive when the nucleic acid is simply represented by a sequence of letters identifying the nucleotides (T, thymine; C, cytosine; G, guanine; A, adenine; U, uracil) (Fig. 3.26c, d). By convention, nucleic acid sequence is written from the C5′ to the C3′ endings, 5′ → 3′.

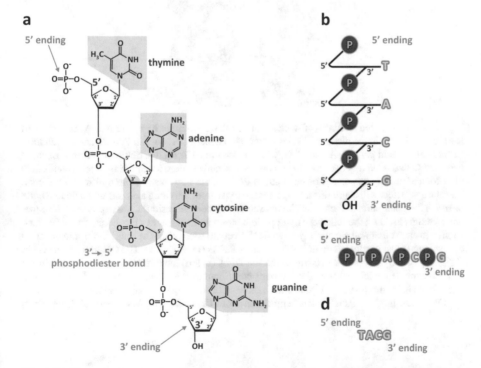

**Fig. 3.26** Natural polymers of nucleotides—nucleic acids. (**a**) Deoxyribonucleic acid (DNA) has 2-deoxyribose residues and uses thymine but not uracil. Ribonucleic acid (RNA) has ribose residues and uses uracil but not thymine. Both polymers are formed by C3′–C5′ phosphodiester bonds. (**b–d**) For the sake of simplicity, the chemical structure of the monomers is usually omitted, and other forms of presenting the nucleotide residues sequence are preferred. The simplest and more common form represents the nucleotides by a one-letter code (the first letter of the base name: T, A, C, or G). Which ending is the free, C5′ or C3′, is not explicitly mentioned, but it is established by convention that the sequences are presented in the sense 5′ to 3′

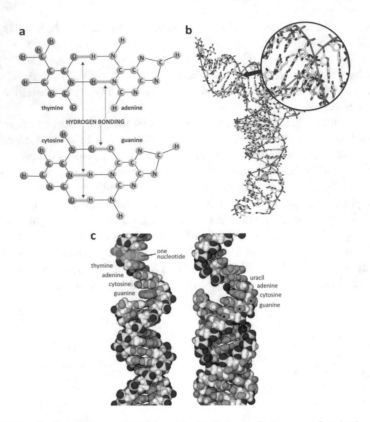

**Fig. 3.27** The detailed structure of nucleic acids. (**a**) Heterocyclic bases are flat. At the edges of heterocyclic bases, in the plane of the rings, hydrogen bonding may occur. Purines and pyrimidines fit each other, as in pairs T–A and C–G, which is known as Watson–Crick base pairing. Because bases are so flat, relatively hydrophobic on both sides, and undergo base pairing, nucleic acids may bind complementary sequences of nucleotides in the same polymer or from a different polymer. The entropic effect will cause this arrangement to twist around its long axis forming a double helix in which the polar parts of the molecule, phosphate and pentose residues, are exposed to the aqueous medium shielding the relatively hydrophobic bases. In the center of this helix, the bases stack parallel to each other and are slightly rotated relatively to each other. (**b**) This kind of organization can be found even in some domains of the transfer RNAs (PDB 2TRA). The OH groups present in C2′ groups of RNA (asterisk in **c**, right) but not DNA (**c**, left) have structural implications in the conformation of nucleic acids as these groups contribute to shield the core of the double strands from aqueous environment. (**c**) DNA forms a more stable and regular double helix because it lacks the OH group in C2′. (Panel **c** was reproduced from Goodsell, The Machinery of Life, 2009)

Unlike polysaccharides, nucleic acids are amphiphilic molecules (Fig. 3.26), so the entropic effect will be a significant driving force for folding in aqueous environment. The heterocyclic bases will tend to nucleate to minimize the contact with water molecules. The crystal structure of transfer RNA (tRNA) shows that the bases stack parallel to each other, which is favored by their strictly planar structure. In addition, the nucleic acid tends to twist along its major axis forming a helix that exposes the phosphates to the aqueous medium and has the base stacks in its core. In addition, most of the helical regions in tRNA consist of two sequences of the RNA chain running in opposite directions with bases in opposite sequences contacting each other close enough and with the adequate stereochemical arrangement to establish hydrogen bonding between them. This adequate arrangement only occurs if purines pair with pyrimidines, as in pairs A–U and G–C (Fig. 3.27). This is known as Watson–Crick base pairing. Hydrogen bonding occurs at the edges of heterocyclic bases, in the plane of the rings.

In regions in which the two opposing antiparallel sequences of tRNA have a considerable array of complementary base pairs, both RNA sequences fold into a helical structure to keep the parallel stacked base pairs in the core surrounded by the pentose and phosphate ester backbone. This is the double-helix structure frequently associated with DNA but equally present in RNA. Ribosomal RNA structure is similar to that of tRNA (Fig. 3.27). It is also worth stressing that DNA polymers may have complementary RNA polymers, which may associate and fold into helices. However, the alcohol group, OH, at C2′ makes the structure of RNA less compact due to its volume and polarity.

The structure of RNA is not only less compact, but it is also less chemically stable. The OH group at C2′ is close to the phosphate diester bond with which it can react to hydrolyze RNA (Fig. 3.28). Because RNA molecules have transient functionalities and are not stored for very long periods in the cells, this limitation of RNA is not a problem. DNA is less prone to hydrolysis because it lacks the OH group in C2′, being the molecule that natural evolution selected to store genetic information for longer periods.

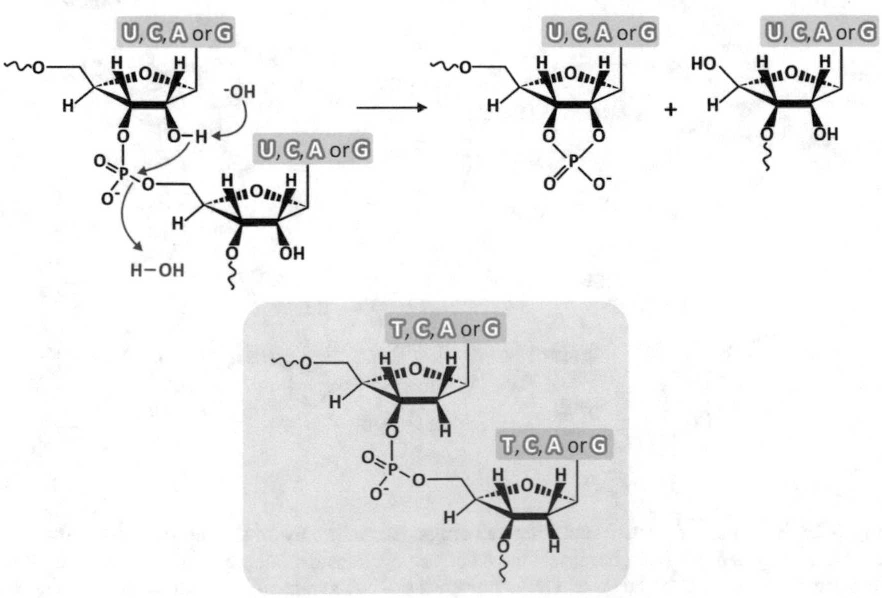

Fig. 3.28 The chemical stability of nucleic acids. Intramolecular hydrolysis of the phosphodiester bonds of RNA caused by a base (*top*). The absence of a hydroxyl group at C2′ increases the hydrolytic stability of DNA relative to RNA (*bottom shaded structures*)

Some drugs target DNA taking advantage from the parallel stacking of heterocyclic bases. Notably, most of these molecules are composed of hydrophobic planar heterocyclic groups able to intercalate the base pairs of DNA (Fig. 3.29). Some of these molecules are used in oncology because they prevent cell multiplication.

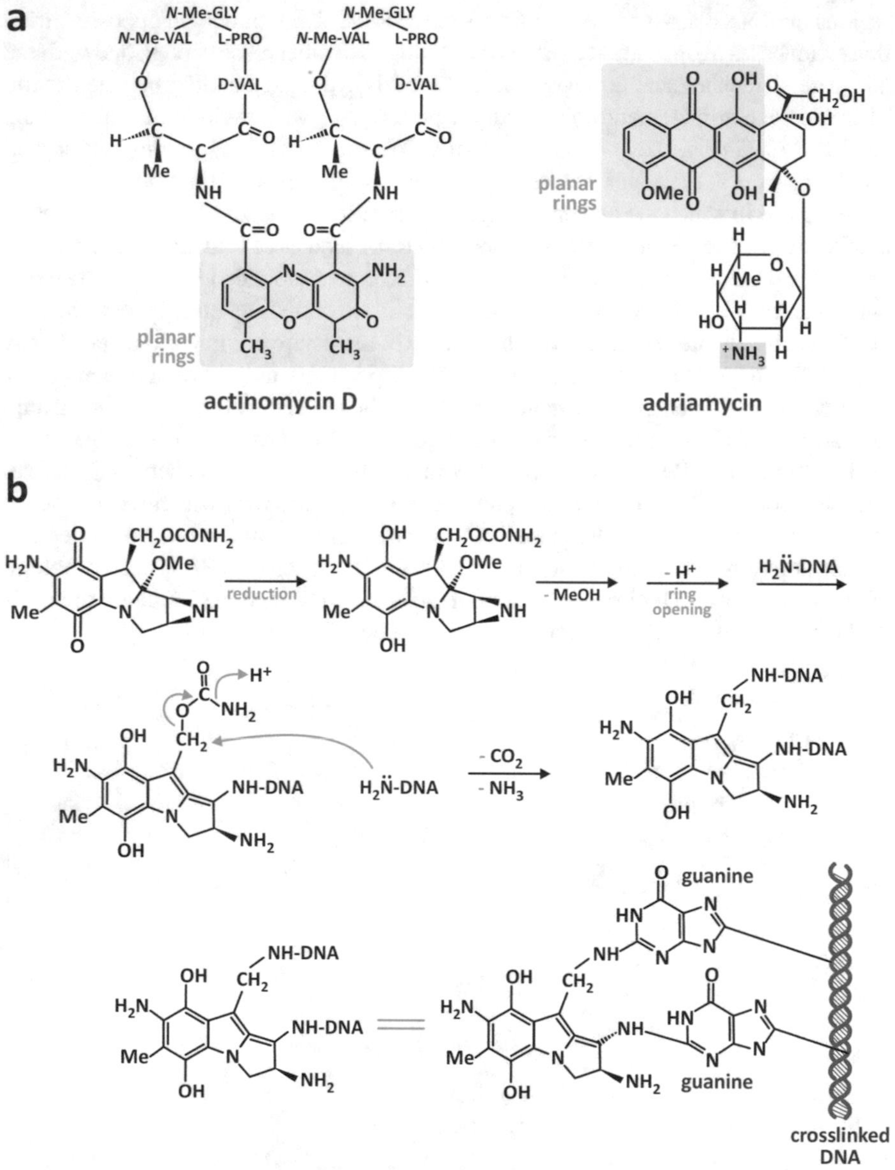

**Fig. 3.29** Examples of nucleic acids-targeted drugs. (**a**) Actinomycin D is an antibiotic with anticancer activity. It binds DNA because it has a flat polycyclic and relatively hydrophobic group able to intercalate the stacked bases of DNA, preventing RNA synthesis. (**a**) Adriamycin is also an anticancer drug that operates with the same mechanism of action: intercalation of a polycyclic flat hydrophobic group between the base pairs of DNA, preventing cell proliferation. (**b**) Mitomycin C has a different mechanism of action: it is a DNA cross-linker by covalently linking two guanines. The direct contact with these nucleotide residues is possible because this drug is a polycyclic flat and relatively hydrophobic molecule

## 3.3    Amino Acids and Their Polymers: Peptides and Proteins

Chemically speaking, amino acids are molecules that simultaneously have carboxyl (-COOH) and amine ($-NH_2$) groups. Biochemists focus on α-amino acids, in which these groups are bound to the same terminal carbon (so-called α-carbon in older organic chemistry nomenclatures), because naturally occurring proteins are polymers of α-amino acids. These amino acids have the structure depicted in Fig. 3.30. Besides the amino and carboxylic acid groups, the α-carbon (also named central carbon) attaches to a hydrogen atom and another group, so-called lateral chain and represented by R. In nature, R is one of the 20 possible groups with few exceptions that are usually derivatives of these groups.

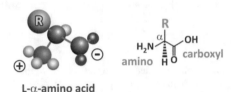

L-α-amino acid

**Fig. 3.30** The structure of α-amino acids. Depending on pH, in aqueous solution, the amino group may be protonated and the carboxylic acid deprotonated, which makes amino acids potential zwitterions, i.e., globally neutral molecules with equal number of oppositely charged groups. Being weak bases and acids, amino acids may constitute buffers themselves (see Sect. 2.1.1)

Another interesting peculiarity of naturally occurring amino acids besides being α-amino acids is that they are almost exclusively L-enantiomers as the α-carbons are chiral centers. The other enantiomer is named D. The L and D nomenclature for the stereochemistry of the amino acids was established by Emil Fisher in analogy with glyceraldehyde, which also has a single chiral center with two possible enantiomers (see Sect. 3.2). Another nomenclature, more complex and following modern rules, exists to describe the stereochemistry of amino acids, but the predominance of L-amino acids and the simplicity of the L vs. D system resulted in the long-term longevity and universality of this system.

One simple empirical rule to distinguish L- from D-enantiomers is to adopt the perspective of the chemical structure of the amino acid along the H-αC axis (Fig. 3.31). The groups COOH, NH_2, and H appear projected as the vertices of a triangle. You can now recognize "CORN" written clockwise, in L-enantiomers, or counterclockwise, in D-enantiomers. This is the CORN rule of thumb.

The chemical nature of the lateral chain, R, is determinant for biochemical processes in which amino acids participate and for the structure that proteins adopt when such amino acids are present. Broadly speaking, amino acids can be grouped in four different categories based on polarity and acidic/basic nature of R (Fig. 3.32): acidic, basic, neutral polar, and neutral non-polar. Other classification systems are based on the chemical nature of R: hydrocarbons, carboxylic acids, amides (–CONH_2), acyclic nitrogen-containing, hydroxyl, sulfur-containing, and nitrogen heterocycles. Figure 3.32 includes grouping of the amino acids according to polar-

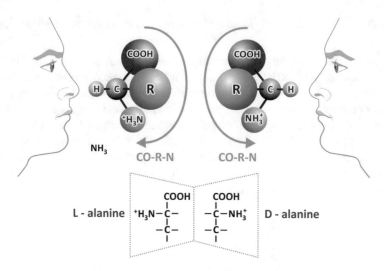

**Fig. 3.31** The L vs. D nomenclature revealed by the CORN rule of thumb. When R = H (this happens in glycine, the simplest amino acid), chirality does not exist as two equal substituents (H, in this case) are attached to the central carbon. The example of L- and D-alanine is presented, highlighting they are mirror images

ity and charge and shows the chemical nature of the side chains. Table 3.3 clarifies the relationship between the names of amino acids in extent and the three-letter and one-letter code abbreviated nomenclature. It also summarizes the most relevant properties of amino acids.

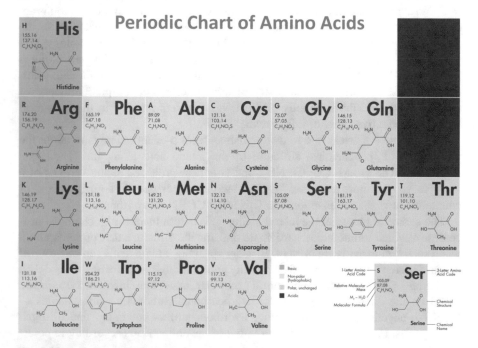

**Fig. 3.32** Periodic chart-like arrangement of the natural amino acids. (Figure reprinted with the permission of Bachem, Bubendorf, Switzerland)

**Table 3.3** Natural amino acid nomenclature (three-letter and one-letter code) and main properties

| Nomenclature rationale for 1-letter code | Amino acid | 3-letter code | 1-letter code | Main properties |
|---|---|---|---|---|
| First letter is unique | Cysteine | Cys | C | Thiol side chain susceptible to oxidization to form disulfides. |
| | Histidine | His | H | Essential amino acid with imidazole side chain. The imidazole side chain has a *pKa* of approximately 6.0, which implies that relatively small shifts in most frequent physiologically relevant pH values will change its average charge, which in turn may impact significantly on protein structure. |
| | Isoleucine | Ile | I | Essential amino acid isomer of leucine. Chiral side chain. |
| | Methionine | Met | M | Side chain possesses a S-methyl thioether, which may be a source of sulphur for cartilage healing. It has been suggested that Met is able to strengthen the structure of hair and nails because its side chains may cross react. |
| | Serine | Ser | S | Residues of Ser are found in some phospholipids (besides proteins). |
| | Valine | Val | V | Essential amino acid. Like Leu and Ile, Val is a branched-chain amino acid. |
| First letter not unique. Most frequent amino acids have priority. | Alanine | Ala | A | D-Ala occurs in bacterial cell walls and in some peptide antibiotics. Side chain is very small (methyl group). |
| | Glycine | Gly | G | Side chain consists in H, making Gly the only achiral and the smallest possible amino acid. |
| | Leucine | Leu | L | Essential branched-chain amino acid. |

(continued)

**Table 3.3** (Continued)

| | | | | |
|---|---|---|---|---|
| | Proline | Pro | P | The amine nitrogen is bound to two alkyl groups forming a cyclic side chain, which gives Pro an exceptional conformational rigidity compared to other amino acids. When Pro is involved in a peptide bonding, its nitrogen is not bound to any hydrogen, meaning it cannot act as a hydrogen bond donor, causing a disruption of α-helices and β-sheets. |
| First letter not unique and less frequent: letter with phonetic similarity or side chain chemical nature | Threonine | Thr | T | Essential amino acid. Chiral side chain. The hydroxyl group in the side chain is prone to glycosylation and phosphorylation. |
| | Arginine | Arg | R | The guanidium group in the side chain is positively charged at physiological pH ranges therefore prone to binding negatively charged groups. This group has also the ability to form multiple H-bonds. |
| | Asparagine (side chain contains N) | Asn | N | Its side chain is curiously an amide (like in peptide bonds). Owes its name to asparagus because it was first detected in asparagus juice. |
| | Aspartate | Asp | D | Essential amino acid with a benzyl side chain, which makes it fluorescent and neutral. |
| | Glutamate | Glu | E | In addition to its role in proteins and amino acid metabolism, in neurosciences Glu is a very relevant neurotransmitter. |
| | Glutamine | Gln | Q | Its side-chain is curiously an amide (like in peptide bonds) formed by replacing the side-chain hydroxyl of Glu with an amine functional group. |
| | Phenylalanine | Phe | F | Essential amino acid with a benzyl side chain, which makes it fluorescent and neutral in proteins. |

| | | | |
|---|---|---|---|
| Tyrosine | Tyr | Y | Tyr has a phenol group in the side chain, which makes it fluorescent. More importantly, the phenol group functions as a receiver of phosphate mediated by protein kinases (so-called tyrosine kinases) resulting in alterations on the activity of the target protein. |
| Tryptophan (side chain with double ring) | Trp | W | Essential amino acid having a fluorescent indole functional group in the side chain. The indole group is bulky and hydrophobic, so Trp is commonly found in lipid-contacting domains of proteins, such transmembranar regions of membrane proteins or fusion domains of viral proteins. |
| Lysine | Lys | K | Essential amino acid. Like in Arg, Lys side chain participates in hydrogen bonding and is cationic at physiological pH range, therefore prone to binding negatively charged groups. |
| (unknown amino acid) | -- | X | Undefined amino acids in peptides or proteins structure are generically represented by X. |

**Nearest first letter**

**Unkown**

It should be kept in mind that the ionization states of amino acids vary with pH, so depending on pH, amino acids may have different global charges. The example of His is presented in Fig. 3.33. His is peculiar as the side chain changes ionization ($pK_a \sim 6$) not far from the range of plasmatic and cytoplasmatic pHs. The intermediate value of the neutrality range (from pH 6 to 9.2), the so-called isoelectric point, pI, is 7.6, within the range of plasmatic and cytoplasmatic pH range, which happens only for His.

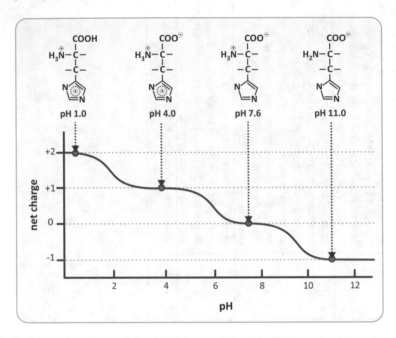

**Fig. 3.33** Schematic variation of the global charge of His with pH. The molecule has three ioniz-able groups, one acidic and two basic. Therefore, allowed global charges range from −1 to +2. However, in most common physiological pHs, the global charge is nearly nil. The $pK_a$ correspond-ing to the three deprotonations with increasing pH are 1.8, 6.0 and 9.2

### 3.3.1 From Monomers to Polymers: Peptides and Proteins

Amine and carboxylic groups may react by dehydration, forming amide bonds (Fig. 3.34a). Amide bonds connecting several amino acids form a peptide. Many amino acids connected through amide bonds form a protein. There is no precise limit to separate the number of amino acid monomers in peptides and proteins although 30 is usually taken as a reference value.

Among biochemists, amide bonds forming peptides or proteins are generally referred to as peptide bonds. Because peptide bonds are very planar (C=O, C–N, and N–H bonds are coplanar) due to electron distribution limitations imposed by specific molecular orbitals, and have the R groups in close vicinity, the chain of peptide bonds forms a polymer backbone that is not freely flexible. It articulates with spatial constraints, which means that the polymeric chain tends to adopt fixed angles between its amide groups; these angles are the ones that allow accommodat-ing the side chains of the amino acids and adapting the orientation of the amide bonds to each other (Fig. 3.34b). As shown in Fig. 3.32, there is a wide diversity of side chains in charge, polarity, and size. All these parameters influence the way a protein folds to cope with the electrostatics, hydrogen bonding, entropic effects (hydrophobicity), and occupation of 3D space. In the end, altogether these factors determine that amino acid polymers have two different preferred kinds of conformations: α-helices and β-sheets. Many other folds exist but are not as common because these two are the ones that better suit stabilization of amino acid sequences.

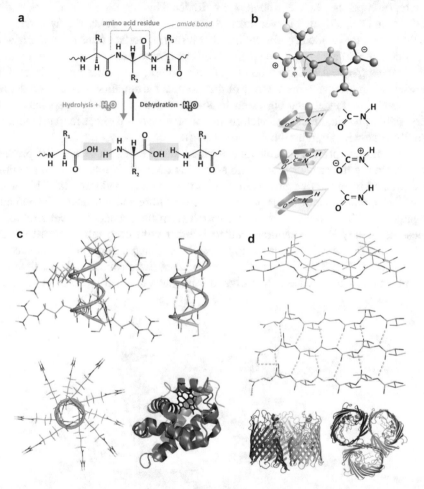

**Fig. 3.34** The conformational organization of proteins. (**a**) Dehydration reactions among amino acids lead to polymerization through amide (also known as "peptide") bonds. (**b**) The amide bonds CO, CN, and NH bonds are coplanar because of the electronic distribution among the connected OCN set of atoms. This implies that when the polymeric chain folds, flexibility is limited, and specific arrangements tend to be adopted, which include (**c**) α-helices and (**d**) β-sheets. Whether a certain sequence of amino acids adopts the conformation of α-helix, β-sheet, or any other, depends largely on the amino acids involved, their order, and environmental factors such as solvent polarity, pH, and temperature. (**c, d**) Both α-helices and β-sheets are conformations that enable the occurrence of frequent intramolecular hydrogen bonding and externalize the location of side chains. Proteins may be formed almost exclusively of α-helices, such as (**c**, bottom right) myoglobin, or (**d**, bottom) β-sheets, such as porin, or be a mixture of both. (Images of porins is a courtesy of Dr. Claudio Soares, ITQB-UNL, Portugal)

In a α-helix, the peptide bond sequence (i.e., the peptide or protein "backbone") adopts a helical structure projecting the side chains, R, to the exterior of the helix. It is a very stable structure because there are almost no constraints to spatially accommodate R and because the vast array of C=O and N–H groups in the backbone interacts strongly through frequent hydrogen bonds. Many proteins, such as myoglobin (Fig. 3.34c), are composed of several helical segments in their amino acid sequence. To facilitate protein representation and reading, helical segments are usually represented as a helical ribbon or a cylinder. This highlights the conformation of the segments, although it overlooks what specific amino acids are involved.

β-sheets are extended conformations that turn in specific points resulting in several linear amino acid residue sequences antiparallel to each other. Like in helices, this enables frequent hydrogen bonding in the protein backbone and projection on the side chain groups to the exterior of this compact arrangement. A certain degree of bending is allowed, and big extensions of β-sheets are usually associated with very stable proteic structures, such as membrane pores. β-sheet representation is usually done with straight ribbons (Fig. 3.34d).

The complete protein structure is described in three or four levels. The primary level is simply the sequence of amino acids that compose the protein, conventionally counted from the free amine terminal to the free carboxyl terminal. This elucidates the chemical nature of the protein but tells us little about what are the domains engaging α-helices, β-sheets, or none, which form the secondary-level structures. These secondary-level structures tend to interact with each other toward mutual stabilization by means of electrostatic forces, hydrogen bonding, and entropic effect contributions (Fig. 3.35). The tertiary level arises therefrom: α-helices, β-sheets, and other local arrangements that organize in space to form the protein structure

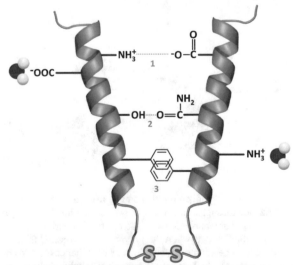

**Fig. 3.35** Intramolecular forces that contribute to stabilize protein structures. Secondary-level structures such as helices interact intramolecularly or intermolecularly through electrostatic forces (1), hydrogen bonding (2) or entropic effect, which determines exposure of polar groups such as –COO⁻ and –NH₃⁺ to aqueous solvent and association of hydrophobic groups with minimal exposure to the aqueous environment (3). Two Cys residues in contact may react through the thiol groups (-SH) in the side chains forming disulfide bonds (S–S) that strongly contribute to the structure of proteins (see as an example the structure of insulin in Fig. 3.36)

itself. Occasionally, there are different parts of the global geometry of the protein that form fairly independent and separable parts, frequently having specific dynamics and specific functions. These are known as domains. An upper level exists for proteins that associate with other proteins, equal or not, to form organized protein assemblies: the quaternary-level structure. The different levels for protein structure are illustrated in Fig. 3.36, using as example insulin, whose structure was discovered by Dorothy Hodgkin, who also discovered the structure of cholesterol (see Fig. 2.10).

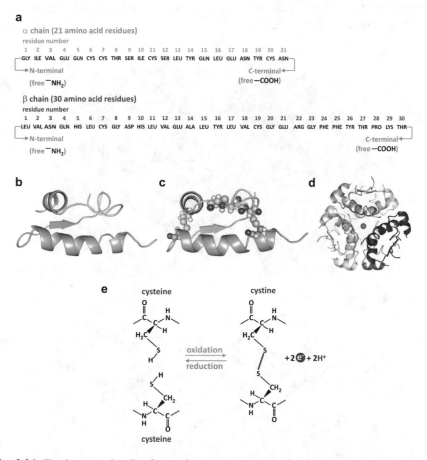

**Fig. 3.36** The hormone insulin, from primary- to quaternary-level structure. (**a**) Amino acid (three-letter code) sequence, the primary-level structure. (Note that insulin is formed by two different chains covalently connected to each other.) (**b**) Segments engaging helical secondary-level structure are represented as helical ribbons. (**c**) The protein folds into a tertiary-level structure that is stabilized by disulfide bonds (yellow in the protein structure). Disulfide bonds are the result of oxidation of two thiol (-SH) groups to form a S–S bond (**e**). It is common that Cys residues react this way in proteins. (**d**) Six insulin monomers associate forming a homohexamer, the quaternary-level structure. The quaternary-level structure is stabilized by the presence of two zinc ions (central sphere) and due to contacts between hydrophobic surfaces of monomers (entropic effect). Insulin is stored in the pancreatic beta cells and secreted into the bloodstream in the form of aggregates of these compact hexamers. Upon dilution in the blood, insulin dissociates and the active form is believed to be the monomer

Hydrogen bonding is frequently the strongest non-covalent factor in keeping the tertiary and quaternary levels of the structure of proteins. Enolase is a good example. Although there are no covalent bonds between both proteins in the dimer, hydrogen bonds are frequent (Fig. 3.37) Altogether, the sum of all hydrogen bonds creates a strong network of adhesion forces in the contact surface of the proteins. Hydrogen bonds are directional; they occur in a well-defined direction between chemical groups at a definite distance; this further contributes to maintain the structure of proteins. The extreme contribution of hydrogen bonding to polymer structure may not be intuitive, but one should bear in mind that Kevlar, an extremely resistant material used in protective items such as bulletproof vests, owes its properties in part to hydrogen bonding (Fig. 3.37).

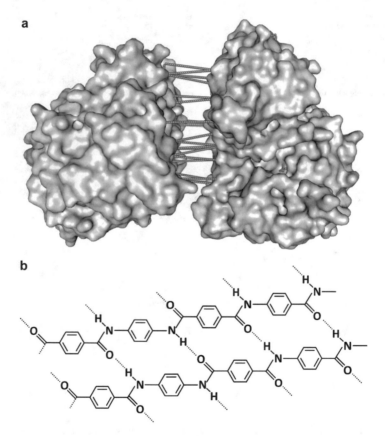

**Fig. 3.37** Examples of hydrogen bonding contribution to polymer structure. (**a**) Enzyme enolase (PDB 1IYX) is a dimer in which both subunits are attached by a dense array of hydrogen bonds in the contact surface between them. (**b**) The structure of an amide polymer (such as proteins) commercially known as Kevlar. It involves a dense network of hydrogen bonds, which confers high resistance, and Kevlar is used in protective materials such as helmets and bulletproof vests. Comparing the molecular-level details of Kevlar and β-sheets in proteins, there is a parallelism between the resistance of Kevlar and the extreme stability of aggregates formed by the juxtaposition of β-sheets in amyloid plaques (see Box 3.6)

Kevlar was named after its inventor, the chemist Stephanie Louise Kwolek, who had planned to attend a medical school but started a temporary job in chemistry and finally quit a medical career. The historic parallelism between artificial polymeric materials and biological molecules dates back to 1920, when Hermann Staudinger proposed that rubber and other polymeric molecules such as starch, cellulose, and proteins are long chains of short repeating molecular units linked by covalent bonds, a disruptive concept at that time. Staudinger used the term macromolecule ("makromoleküle") for the first time, a term now very popular among biochemists. Paul Flory, a chemist pioneer of the studies of three-dimensional organization of polymers and its relation to dynamics, also worked for the rubber industry during certain periods of his career. His work opened the field to structure–function relationships in macromolecular biochemistry. Flory was awarded the Nobel Prize in Chemistry in 1974 "for his fundamental achievements, both theoretical and experimental, in the physical chemistry of macromolecules." Thinking of natural protein fabrics, such as silk and spider webs, and artificial fabrics made of nylon and other polymers helps us realize that in the molecular world, the boundaries between nature and human artifacts are very faint.

**Hermann Staudinger**     **Paul Flory**     **Stephanie Kwolek**
**(1881 – 1965)**          **(1910 — 1985)**  **(1923 — 2014)**

Hermann Staudinger (1881–1965), Paul Flory (1910–1985), and Stephanie Kwolek (1923–2014)

As mentioned before, there are also covalent contributions to the tertiary level of structure of proteins, namely, disulfide bonds (or "bridges") and attachment of metal or other non-proteic groups to more than one amino acid residue. Disulfide bonds are formed by oxidation of two contacting Cys thiol (-SH) groups originating from an S–S bond between the Cys residues (cystine). Cell cytosol is a relatively strong reducing environment, and the contribution of disulfide bonds in cytosolic proteins is limited. However, in other circumstances, disulfide bridges form and are strong stabilizers of protein structure at the tertiary level. Insulin, a proteic hormone, is an example (see Fig. 3.36).

Metals can bind multiple ligands and covalently link different amino acid residues in a protein, therefore also contributing to stabilize a tertiary-level structure. Frequently, metals bind to the thiol group of Cys. In the electron transfer chain proteins, several metallic complexes are present, which in addition to chemical functions also contribute to the stability of the proteins (Fig. 3.38; see also Sect. 6.2.2).

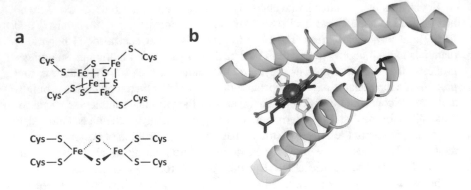

**Fig. 3.38** Contribution of metals to the structure of proteins. (**a**) Metal complexes, such as iron–sulfur centers, are common among the proteins of the electron transfer chain. Iron complexes with sulfur atoms but also with the thiol group of the side chain of Cys residues, resulting in stabilization of the structure of the proteins where they insert. The nitrogen atoms in the side chains of His are also prone to metallic complexation. Panel (**b**) shows a detail of metallic complexation in Complex IV (PDB 1OCC) of the electron transfer system (see also Sect. 6.2.3). An iron ion (*red sphere*) complexes simultaneously the N atoms of two His side chains (*blue*), stabilizing the tertiary-level structure of the protein. It also binds to a non-protein molecule, the heme a (*red organic structure*), which is also associated with the Complex IV

When one refers to quaternary-level structure, one usually refers to proteins that associate with high specificity and well-defined function, such as hemoglobin, for instance. This does not include pathological cases in which aggregation of proteins leads to loss of function and increase in toxicity. Extensive tertiary-level alterations are observed when amyloid fibers form upon aggregation of proteins, or when prions trigger conformational changes of native proteins (see Box 3.6), for instance. These are referred to as protein folding diseases as folding is the expression used to comprise secondary- and tertiary-level structure altogether.

> **Box 3.6  Amyloids and Prions: When Misfolding Turns into Disease**
> The relationship between the structure and function of proteins has been one of the main focuses of modern biochemistry for decades. Mutated proteins may have important changes in their structure and may thus be defective in their function, which is not surprising. However, the knowledge that proteins without mutations can fold in diverse forms, some of them pathogenic, is recent. Protein folding is the key to important diseases such as Alzheimer's, in which massive stacks of β-sheet-folded proteins accumulate in the brain. These stacks form plaques of insoluble protein in the extracellular tissue, which cannot be broken down by enzymes. When these plaques were found for the first time, they were described as related to saccharides and named amyloids. Although the chemical nature of the plaques is now known not to be related to saccharides, the name "amyloid" is still used, and the group of diseases is known as amyloidoses.
>
> Amyloid plaques grow with an ordered structure forming long filaments (fibrils). There are about 20 different proteins that can act as the building blocks of these fibrils, each of which is associated with a different disease. In

(continued)

**Box 3.6** (continued)

so-called systemic amyloidoses, the precursors of these plaques are transported through the bloodstream from their point of origin to their point of deposition. Localized amyloidoses are of greater clinical significance, as they mainly affect the central nervous system, the extracellular matrix of which is particularly susceptible to damage.

Transmissible spongiform encephalopathies (TSEs), which include mad cow disease (bovine spongiform encephalopathy, BSE) and Creutzfeldt–Jakob disease (CJD) in humans, are forms of amyloidoses in which the diseased brain degenerates to a porous sponge-like structure. These diseases appear when human protein called prions misfold (see below figure). The human prion (named PrPc) is a component of the membrane of healthy nerve cells that may misfold in a particular way. Amazingly, the misfolded prion may induce misfolding in a neighboring prion if contact among both molecules occurs. This has the appearance of an infection-like process in which the misfolded molecule "infects" the "healthy" molecule. "Infectious" prions can be transmitted in the diet, triggering a domino effect in healthy prions.

In Alzheimer's disease, β-amyloid plaques are formed by cleavage of the amyloid–precursor protein (APP) by two different enzymatic activities, which release peptide fragments that are 40 or 42 amino acids long. When these peptides fold into β-sheets and aggregate, fibrils are formed, surrounding neurons and causing damage. This does not happen when the same peptides fold differently. It is only in β-sheets that hydrophobic amino acids are exposed, and they rapidly bind to hydrophobic groups on other peptides due to the entropic effect. The β-sheet structure, being highly ordered, is prone to regular stacking, ultimately leading to fibrils (see below figure).

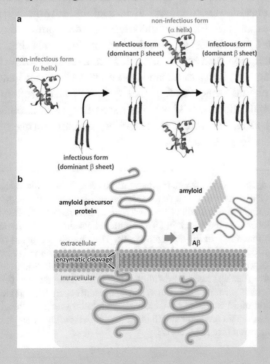

Schematic representation of structural conversion occurring in amyloidosis. (**a**) The non-infectious form of a prion protein domain PrP(121–231) (mouse, PDB 1AG2) may be converted to the infectious form, whose structure is not known in detail but dominated by β-strands (tentatively depicted; *red*). The infectious form may induce conformational changes in the non-infectious protein to produce a replica of itself. The process may amplify, and aggregates of the infectious form thus accumulate in pathological conditions. (**b**) In Alzheimer's disease, β-amyloid plaques are not formed by an infection-like propagation of the proteins dominated by β-strand domains, but by the cleavage of the APP with concomitant release of peptides (*green*) that fold into β-sheets and aggregate

## 3.3.2   Structure and Function in Proteins

Proteins can adopt many different structures at the tertiary level, from extended rods to compact globules. Extended proteins may associate in fibers, and globular proteins may have flexible domains able to bind other molecules. It is clear that proteins with extended conformation, like keratin, collagen, or silk fibroin, are good to maintain the structure of tissues or biomaterials, whereas globular proteins are good to intervene in dynamical processes, which justifies why enzymes are globular proteins. While this is generally true, it should also be acknowledged that the frontier between both is not always clear. For instance, actin is a globular protein that binds other actin molecules to form quaternary-level fibers in the cytoskeleton and muscle contractile system. Actin is an example of a globular protein having structural functionality (Fig. 3.39; see also Sect. 10.1.1). Another interesting molecule that challenges the classical dichotomy of structural/extended vs. functional/globular proteins is lung elastin. When relaxed, it is a globular protein but it stretches to encompass lung expansion. Elastin molecules are interconnected covalently by the side chain of four Lys residues: cross-linked desmosin bonds (Fig. 3.39). This way, the continuous alteration between the globular and extended conformation of elastin confers to the lung the ability to expand and contract without histological lesions. To finalize, one should stress the fact that there are also proteins that have both extended and globular domains. This is the case of myosin, another central protein in muscle contraction, in which an extended domain forms oligomeric fibers (Fig. 3.39; see also Sect. 10.1.1).

---

**Fig. 3.39** (continued) form very resistant biomaterials such as spider silk. It is important to realize that collagen has a Gly residue at every third position of its amino acid sequence. Gly is a very small amino acid because the side chain is replaced by H. This creates a line along the helix surface where two other alike helices may dock in close contact. Lys side chains are exposed in collagen and form cross-links that strength the collagen fibers. The chemical process of this cross-linking reaction between the endings of the Lys side chains depends on vitamin C (ascorbic acid). (**d**) Lys residues are also involved in cross-links of elastin, a connective tissue protein of unusual elasticity. The cross-links involve four Lys residues forming a desmosin arrangement. (**e**) Actin is a globular protein that self-associates forming fibers important in muscle contraction. (**f**) Myosin, on the other hand, is a combination of an extended domain with a globular domain ("head") having catalytic activity. The extended domain associates with other proteins forming fiber-like oligomers responsible for muscle contraction (see Sect. 10.1.2)

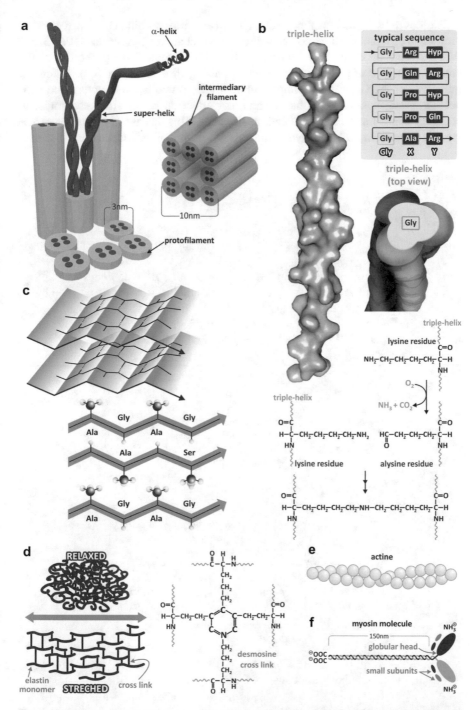

**Fig. 3.39** Typical protein structures and protein–protein interactions. Proteins such as (**a**) keratin and (**b**) collagen (PDB 1BKV) adopt string-like conformations (very extended helices) that associate to form fibers that stabilize structures like hair or nails (keratin) or the connective tissue (collagen). (**c**) Silk fibroin forms very extended β-sheets that associate antiparallel to each other to

Besides tight parallel packing in collagen fibers, proteins bind covalently to other proteins in different fibrils of collagen through side chains of Lys residues (Fig. 3.39). When this strong covalent meshing of collagen is disrupted, the properties of the connective tissue are very much affected causing diseases (Box 3.7). Likewise, mutations of the Gly residues impact dramatically on collagen structure and connective tissue function. This has a particular effect on bones since early age. In addition, bones loose resistance and break easily. This disease is known as *osteogenesis imperfecta*, Lobstein syndrome, or, more commonly, "brittle bone disease."

---

**Box 3.7  Scurvy: An Example of a Pathology Directly Associated with Protein Structure**

Scurvy is a pathology characterized by fatigue, anemia, gingivitis (gum disease), and skin hemorrhages caused by diets with a prolonged deficiency of ascorbic acid (vitamin C). It was a frequent disease in sailors on long voyages in the pioneering intercontinental discoveries of the fifteenth century. Many men died until it was discovered that scurvy could be cured and prevented by consuming citrus, such as oranges, lemons, and limes. The Portuguese sailor Vasco da Gama was the first European to lead a fleet that reached India by sea, linking Europe and Asia, connecting the Atlantic and the Indic. The drama of scurvy in his first trip to India (1497–1499) is eloquently described in an epic poem by Luís de Camões, in *The Lusiads* (1572):

> And 'twas that sickness of a sore disgust,
> the worst I ever witness'd, came and stole
> the lives of many; and far alien dust
> buried for aye their bones in saddest dole.
> Who but eye-witness e'er my words could trust?
> of such disform and dreadful manner swole
> the mouth and gums, that grew proud flesh in foyson
> till gangrene seemed all the blood to poyson:
>
> "Gangrene that carried foul and fulsome taint,
> spreading infection through the neighbouring air:
> No cunning Leach aboard our navy went,
> much less a subtle Chirurgeon was there;
> but some whose knowledge of the craft was faint
> strove as they could the poisoned part to pare,
> as though 'twere dead; and here they did aright; —
> all were Death's victims who had caught the blight.

*"The Lusiads" (Canto V, 81 and 82) version reproduced here was translated to English in 1880 by Richard Burton*

Nearly two thirds of the sailors of the entire fleet of Vasco da Gama died during the trip, although documents from that time clearly show that it was known empirically by Portuguese sailors that a diet based on fruits and other unprocessed foods was a treatment for scurvy. A manuscript from a pilot in

(continued)

**Box 3.7** (continued)

the fleet of Pedro Álvares Cabral, discoverer of Brazil in 1500, says that a diet of fresh foods, including sheep, chicken, ducks, lemons, and oranges, was used to heal scurvy.

It was only in the eighteenth century that the Scottish doctor James Lind related scurvy to diets poor in citrus on a reasonably scientific way. Ascorbic acid was discovered by the biochemist Albert Szent-Györgyi (born Hungarian, later US citizen), who was awarded the Nobel Prize in Medicine or Physiology in 1937 (Szent-Györgyi also performed important studies on muscle contraction; see Box 10.1 in Chap. 10). Ascorbic acid is part of several biochemical pathways, the synthesis of collagen being one of them. Specifically, it is mandatory in protein hydroxylation, which is a posttranslational modification in which a hydroxyl group (–OH) is added to a protein residue. Collagen is naturally hydroxylated in healthy individuals. Ascorbic acid is also mandatory in the biosynthesis of carnitine. Impaired synthesis of carnitine and collagen account for the common symptoms of scurvy.

The primary defects behind rotten or loose teeth, rigid tendons, or cartilage fragility observed in scurvy reside in the connective tissue. Without ascorbic acid, collagen is not hydroxylated and a nonfibrous, defective incomplete collagen is formed instead of fibrous collagen. The enzyme prolyl 4-hydroxylase, for instance, hydroxylates a Pro residue using an iron atom that is oxidized in the process. Ascorbic acid is needed to reduce the iron and make the enzyme active again (see below figure).

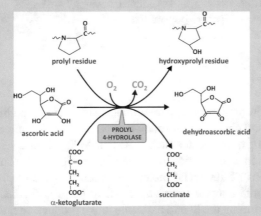

Ascorbic acid oxidation in the reaction catalyzed by prolyl 4-hydroxylase

Lysyl hydroxylases are also operative. Hydroxylation of Pro and Lys residues, both exposed in the triple helix of collagen, favors intermolecular adhesion interactions by hydrogen bonding and further chemical modifications, which are strengthened with age (this is one of the reasons why meat from young animals is more tender than from older animals).

(continued)

**Box 3.7** (continued)

Carnitine is involved in the transport of fatty acids into the mitochondria, where they are oxidized (see Sect. 7.4.3). Ascorbic acid is used by two different enzymes in the carnitine biosynthesis. Without ascorbic acid, production of carnitine declines and fatty acids cannot be used as energy source. This leads to fatigue, which is one of the symptoms of scurvy. Curiously, fatigue appears first than other symptoms. This may be explained by the fact that the enzymes in carnitine biosynthesis require higher concentrations of ascorbic acid to function (they have "lower affinity" for ascorbic acid) when compared to hydroxylases.

Other pathologies, such as the Ehlers–Danlos syndrome (EDS) and *osteogenesis imperfecta* (OI), are genetic pathologies associated with deficiencies on protein–protein interactions in collagen. A fraction of the Lys residues in collagen react with each other forming covalent cross-links in collagen fibers. In some forms of EDS, this crosslinking is impaired, rendering the skin less firm and less resistant, hyperelastic. Collagen contributes to the mechanical strength of skin, joints, muscles, ligaments, blood vessels, and visceral organs. In OI, replacement of Gly residues in the collagen amino acid residue sequence destroys the capacity of the protein to assemble in perfect triple helices because all other residues are bulkier than Gly. This leads to an extremely severe condition that is characterized by alterations in the physical properties of collagen and perturbations in the biochemical processes involving collagen homeostasis. The relationship between the collagen fibrils and hydroxyapatite crystals when bones are formed is altered, causing brittleness. For this reason, OI is also known as "brittle bone disease."

One remarkable property of many globular proteins is the ability to both bind other molecules and change conformation upon binding. Taking adenylate cyclase as example, binding of molecules such as ATP or ADP causes a very mobile domain of the protein to change position (red domain in Fig. 3.40). This often leads to dynamic distortions of the ligand molecules because of the contact of amino acid residues. The results may be such that covalent bond is formed or broken, thus chemically transforming the ligand molecule into a product. In practice, the protein action is to increase the rate of transformation of the ligand in the product. If only the ligand, not the protein, is chemically transformed in this process, this can be seen as an enzymatic catalysis, i.e., increase of chemical reaction velocity caused by proteins, the enzymes. The ligand (reactant) is named substrate in these cases.

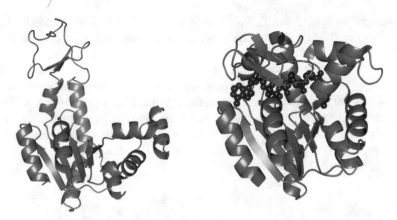

**Fig. 3.40** Example of the conformational dynamics of an enzyme. Adenylate cyclase (or adenylyl cyclase), unbound (PDB 4AKE; *left*) and bound (PDB 1AKE; *right*) to a dinucleotide analog (*red*). Upon binding to the dinucleotide analog, the mobile domains adapt by changing position

### 3.3.3  Cooperative Interplay Between Tertiary-Level and Quaternary-Level Structure

As discussed in the previous section, the tertiary-level structure of a protein is determined and maintained by arrays of spots in which attractive or repulsive forces between groups of atoms exist. This is a relatively delicate balance. When a significant number of such "force spots" is altered, the configuration of the protein adapts a different tertiary-level structure, which corresponds to the new balance of forces. Likewise, in cases in which a quaternary-level structure exists, the changes in the conformation of one protein monomer at the surface of contact with other monomer may impose alterations in the "force spots" (hydrogen bonds, electrostatic repulsion and attraction, entropic factors, etc.) so that the second monomer changes conformation to adapt. In practice, this means that conformational changes in one protein may be transmitted to a neighboring protein that is in contact with it. In other words, tertiary-level structural changes may be transmitted and amplified to other proteins throughout the quaternary-level structure. Hemoglobin is a good example. It is formed by two subunits, α and β, forming a dimer that associates with other dimer— a tetramer that is in fact a dimer of dimers. Dimers are numbered 1 and 2; therefore, hemoglobin is a tetramer of four subunits β1, α1, β2, and α2 (Fig. 3.41). α–β attractive forces are stronger than 1–2 attractive forces, but both are sufficient to transmit to neighboring monomers conformational changes. This affects the affinity of the hemoglobin monomers to oxygen. Each hemoglobin monomer is covalently associated with a non-proteic group, i.e., a prosthetic group, of the porphyrin family (Box 3.8). In this case, the porphyrin binds in its center an iron ion, forming a heme. The iron ion complexes with the heme through four bonds in the plane of the porphyrin and to a His residue side chain orthogonally to the heme plane. Other orthogonal bond, opposite to His, is established with small molecules having electron donor atoms, such as $O_2$ or CO. When an $O_2$ molecule binds to the iron in the heme, the position of the iron slightly shifts, which in turn affects the position of the His residue. When the His residue is pulled, the whole structure of the protein changes slightly. This change in conformation induces a change in conformation of the neighboring monomers. As a consequence, the neighboring monomers acquire increased affinity to bind an oxygen molecule. So, binding of $O_2$ to a monomer increases the chances that a second $O_2$ molecule binds to another monomer in the hemoglobin tetramer relative to a monomer in a tetramer without bound $O_2$ molecules. This is called positive cooperativity: several entities influencing each other making more likely a certain event to occur.

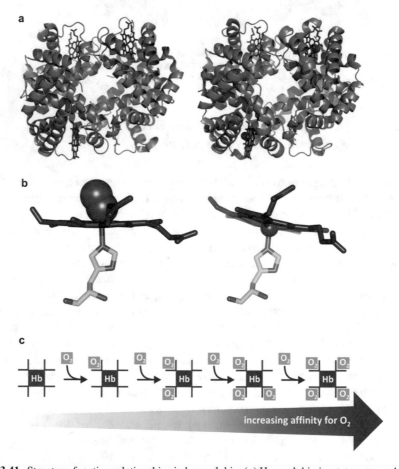

**Fig. 3.41** Structure-function relationships in hemoglobin. (**a**) Hemoglobin is a tetramer, each subunit containing a heme group (*red*). Human deoxyhemoglobin (*left*; PDB 2HHB) and oxyhemoglobin (*right*; PDB 1GZX) show subtil but important changes in conformation. (**b**, *left*) Binding of molecular oxygen to the heme group (PDB 1HHO) causes a (**b**, *right*) shift in the orientation of the heme relative to the His residue relative to the heme in deoxyhemoglobin (PDB 4HHB). This slight distortion in the position of the heme leads to a variation in the conformation of the protein, which propagates to neighboring monomers in the tetramer. (**c**) The neighboring monomers acquire higher affinity for $O_2$

**Box 3.8  The Importance of Heme Groups in Proteins**

Many natural proteins are associated with prosthetic groups of similar chemical nature called porphyrins. Porphyrins are macrocyclic compounds related to porphin (see below figure). Hemoglobins, for instance, bind porphyrin groups, such as heme B (see below figure). Hemes are porphyrins that bind iron ions in the center. The remarkable capacity to bind metallic ions of charge +2 or +3 in the center of the ring may explain the success of these molecules during natural selection and subsequent ubiquity of porphyrins in nature. The porphyrin macrocycle has 26 delocalized ($\pi$) electrons in total, being classified as aromatic from a chemical point of view. This system of delocalized electrons extends to the nitrogen atoms and is available to bind the cationic metals. It is also responsible for the intense absorption bands in the visible region of electromagnetic radiation. This is the reason why compounds with porphyrins, such as hemoglobin and chlorophyll, are deeply colored. Heme-containing proteins, hemoproteins, and other metal-containing proteins, metallo-proteins, due to their unique electronic properties, are adequate for the transient binding of diatomic gases that occurs during their transportation in blood and electron transfer (i.e., electron donation and reception to and from other compounds). It is possible that hemoproteins evolved from ancient proteic forms, whose function was electron transfer in sulfur-based photosynthesis in the ancestors of cyanobacteria before molecular oxygen existed in atmosphere.

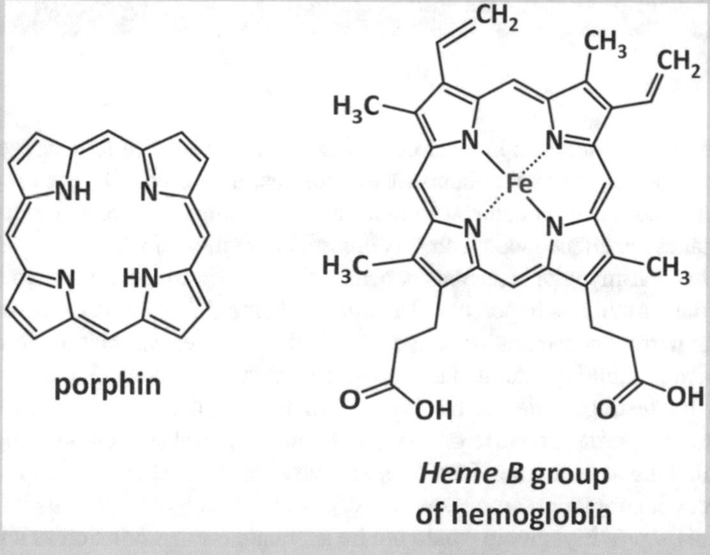

Chemical structure of porphyrins. Porphyrins, such as heme B (*bottom*), are macrocyclic compounds related to porphin (*top*)

(continued)

**Box 3.8** (continued)

In addition to the peculiar intrinsic properties of hemes and other porphyrin–metal associations, the interaction between the porphyrin and the amino acid residues of the site where it inserts in proteins is of extreme importance. Variations in the shape, volume, and chemical composition of the binding site, in the mode of heme binding and in the number and nature of heme–protein interactions, result in significantly different heme environments in proteins having different biological roles. The outcome is a fine-tuning of the heme properties. Take the 3D structure of the hemoglobin amino acids' residues sequences as example. The position of a His residue is such that its protonation interferes with oxygen release from the iron ion. Acidification of the medium causes the protonation of His, which in turn facilitates the release of oxygen. This is known as Bohr effect and is not a simple curiosity: in the pulmonary vasculature, the pH is higher than in the peripheral tissues because the pH is affected by the local abundance of $CO_2$; therefore, the Bohr effect helps in increasing the efficacy of binding oxygen in lungs and releasing in peripheral tissues.

The porphyrin groups are equally important to stabilize protein structure and resistance to proteolysis, although these properties are frequently overlooked. Even in cases in which the porphyrins are not covalently bound to the protein, the interplay between amino acid residues and the non-proteic groups is very specific.

Hemoglobin monomers interact cooperatively to bind up to four oxygen molecules. Myoglobin, a protein abundant in muscles, also binds $O_2$, but this protein occurs as a monomer, in contrast to hemoglobin. Comparing hemoglobin to myoglobin makes the effect of cooperativity clear. The fraction of myoglobin binding $O_2$ relative to total myoglobin increases nearly linearly up to saturation. To be more precise, the variation is hyperbolic. In contrast, hemoglobin binds $O_2$ critically at a triggering narrow concentration range, in which it reaches saturation. Hemoglobin changes from highly unsaturated to almost saturated in a narrow $O_2$ partial pressure interval. Interestingly, the narrow interval of transition to near saturation corresponds to the partial pressure of oxygen found in peripheral tissues (Fig. 3.42), away from lung alveoli. Therefore, cooperativity among the hemoglobin monomers enables that hemoglobin saturates with oxygen in the lungs and delivers its cargo in peripheral tissues. Myoglobin would not be adequate for this function as it is almost saturated in both situations. Myoglobin is fit for oxygen storage in muscle cells. Release of oxygen occurs when the consumption in mitochondria is such that the cell is almost depleted in oxygen (Fig. 3.42).

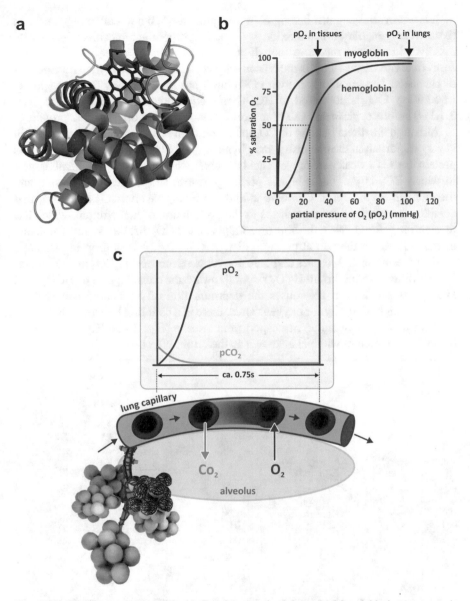

**Fig. 3.42** Bridging oxygen-binding biochemistry to physiology. (**a**) Myoglobin is a monomeric oxygen-binding protein. Like a hemoglobin monomer, it also binds oxygen through a heme group (red organic structure). (**b**) The absence of cooperativity in myoglobin when compared to hemoglobin implies distinct binding capacities at different oxygen partial pressures, $pO_2$. Hemoglobin has an abrupt transition from low to high binding when $pO_2$ changes from values typical from peripheral tissues to values typical of lungs. (**c**) As the erythrocytes pass adjacent to alveoli, $O_2$ and $CO_2$ diffuse freely across arterial and lung cells driven by partial pressure gradients

It is worth stressing that hemoglobin transports most, but not all, oxygen used in tissues. Oxygen, like carbon dioxide, is a small, hydrophilic molecule, which easily dissolves in aqueous media. Although being hydrophilic, it is very small and diffuses freely in tissues. This is the reason why cells do not need oxygen transporters or channels. The same happens with $CO_2$ but it has lower affinity for hemoglobin. In addition, $CO_2$ is converted to $HCO_3^-$ that equilibrates with $H_2CO_3$ (see Sect. 2.1.1). Therefore, plasmatic $CO_2$ transport is not dependent on a specific protein.

Although the direct binding of $CO_2$ to hemoglobin is not significant, hemoglobin is very important in the chemistry and physiology of $CO_2$ in the human body as the protein itself is a weak base or weak acid depending on pH. In the peripheral tissues, in which $CO_2$ is present at higher partial pressure, $CO_2$ diffuses to plasma and therefore in erythrocytes. Carbonic anhydrase then converts $CO_2$ to $H_2CO_3$ that acidifies the medium. Acid pH leads to the protonation of hemoglobin, which has lower affinity for $O_2$. Near the lung alveoli, plasmatic $CO_2$ diffuses to the alveoli due to the gradient in the partial pressure of $CO_2$ (Fig. 3.42). This drop in plasmatic partial pressure of $CO_2$ causes carbonic anhydrase to convert $H_2CO_3$ in $CO_2$, therefore shifting the equilibrium $HCO_3^-/H_2CO_3$ toward the consumption of $HCO_3^-$ and $H^+$. The slight drop in pH causes the deprotonation of protonated hemoglobin, which has higher affinity for oxygen. Thus, there is a coupling between pH and the efficiency of oxygen capture, transport, and release (Fig. 3.43). The coupling of hemoglobin structure with pH is known as the "Bohr effect."

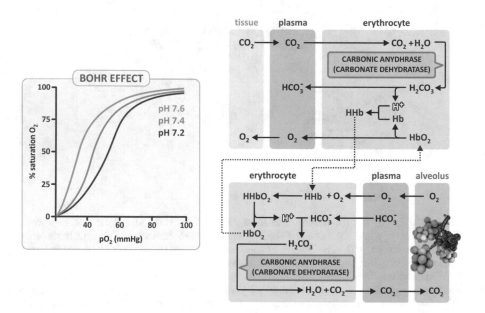

**Fig. 3.43** The Bohr effect associated with hemoglobin. $CO_2$ levels in blood influence the transport of $O_2$ by hemoglobin through plasma pH because protonation/deprotonation of hemoglobin affects cooperativity in $O_2$ binding

The same way oxygen binding, transport, and release by hemoglobin is affected by pH, it is also affected by binding of 2,3-bisphosphoglycerate (2,3-BPG) in the interface between monomers. 2,3-BPG binding to deoxyhemoglobin affects oxygen fixation in a way such that the binding curve of oxygen by hemoglobin is shifted toward higher partial pressures of oxygen (Fig. 3.44). This is far from being a simple curiosity: 2,3-BPG forms from 1,3-BPG, a metabolite of glycolysis (see Sect. 6.1.3). When glycolysis is highly active, 2,3-BPG is formed, and release of $O_2$ from hemoglobin is more effective, which is convenient for the cell as higher glycolytic activity implies, in principle, a higher demand of oxygen for the human cells. The fixation of oxygen in lungs remains unaffected.

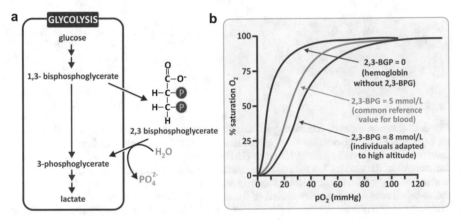

**Fig. 3.44** Coupling glycolysis to oxygen transport. (**a**) 2,3-BPG (2,3-bisphosphoglycerate) forms from 1,3-bisphophoglycerate, an intermediate metabolite of glycolysis; it is therefore a chemical signal of glycolytic activity. (**b**) 2,3-BPG binding to hemoglobin affects cooperativity in such a way that the $O_2$ binding curves are shifted in a way that $O_2$ release in peripheral tissues is facilitated but the $O_2$ capture in the lungs is not affected

### 3.3.4   Enzymes

The previous section showed how dynamic the binding of proteins to other molecules may be. In the case of hemoglobin, oxygen binds to the iron of the heme group, but it is common that more complex molecules bind directly to side chains of certain amino acid residues. Specific sets of amino acid residues in a protein may precisely locate and orient in space and have the correct physical properties (charge, polarity, hydrogen donor/acceptor groups, etc.) to specifically bind molecules that establish attractive forces with them (Fig. 3.45). The electronic clouds of these molecules are distorted by the contact with the amino acid residues, which in turn adapt their tertiary-level structure to the presence of the molecules. This mutual adaptation between protein and bound molecules frequently weakens some chemical bounds of the molecules, which may be destroyed. Likewise, formation of other bonds is possible. The result is that the molecule that bound to the protein is converted in a different molecule. If the resulting product unbinds from the protein and the protein returns to the same state as before binding the initial molecule, then the protein is an enzyme, i.e., a proteic natural catalyst, and the initial molecule that binds the enzyme to undergo a chemical reaction is said to be a substrate, as previously mentioned in Sect. 3.3.2.

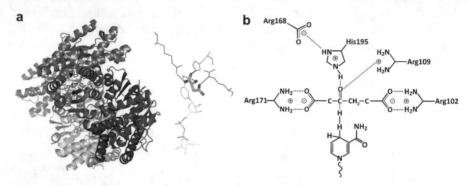

**Fig. 3.45** Determinants of substrate specificity in enzymes—an example. (**a**) Malate dehydrogenase (PDB 2DFD) is a homodimeric enzyme that catalyzes the oxidation of malate to oxaloacetate concomitantly to the reduction of $NAD^+$ to NADH (highlighted in *orange*). (**b**) A specific set of amino acids of the enzyme has the right properties (charge, H-binding ability, etc.), the right location and the right orientation to simultaneously fix the malate molecule. This set of residues form the so-called active site (or active center) of the enzyme. $NAD^+$ binds to other site, specific for it, in close vicinity to the active site and participates in the oxidation of malate facilitated by the action of the amino acids (*bottom right* structure)

Because the enzymes interact with the substrate and facilitate its conversion to products, they increase the velocity of chemical reactions enormously, typically, above $10^7$-fold. In some cases, the increase may be $10^{17}$-fold, which is a figure difficult to realize intuitively. Considering that a $2 \times 10^8$-fold increase in the velocity of a relaxed walk (~1.5 ms$^{-1}$) would leave us traveling at the speed of light (~$3 \times 10^8$ ms$^{-1}$), this intuitive perception becomes more clear. $10^{17}$-fold is more than the difference between a relaxed walk and 100 million faster than the speed of light in vacuum. One will see in Chap. 4 that enzymes accelerate reactions but cannot turn impossible reactions in possible reactions. However, enzymes turn very slow reactions (so slow that in practice reactions are as if they would not be able to occur) into fast reactions. So, in practice it is almost like if an impossible reaction was transformed into a possible reaction by the intervention of an enzyme.

### 3.3.4.1 The Importance of Studying Enzymes

Enzymes are interesting molecules because the dynamics of their tertiary-level structure implies catalytic activity, which is amazing and shaped life as it exists today. Even viruses need enzymes to be effective. Because enzymes are so proficient in speeding reactions, controlling the activity of enzymes is, in practice, controlling the course of chemical reactions in a cell. This is an essential piece to impose order in the chemistry of the cells as controlling the activity of enzymes assures that certain reactions only occur to significant extent *when* and *where* the enzyme is inside the cell. This prevents conflicting reactions in a regulated cell and enables that certain reactions can be coupled. Imagine substrate A and substrate B; now imagine that enzyme $E_A$ converts A in B and a second enzyme, $E_B$, converts B in C. The simultaneous presence of both enzymes in the same cell compartment has the practical consequence that A is converted in C. This coupling of reactions may reach considerable complexity, with many substrates, reactions, and enzymes being involved, sometimes with branched and cyclic reaction sequences (recall Figs. 1.3 and 1.4). Such sets of reactions are referred to generally as "metabolisms." Regulation of metabolism is largely dependent on enzymes. The mechanisms of metabolic regulation are extremely important and will be addressed in Chap. 5. A regulated metabolism is a sine qua non condition for the state of organisms we call "health."

Yet, the structure–activity relationship in enzymes and their significance in metabolic regulation are only part of the importance of studying enzymes. Enzymes can operate outside cells and be used in industrial processes in pharma, food, or detergent processing and manufacturing, for instance. More importantly, in biomedical sciences and clinical practice, they can be used as valuable diagnosis. When enzymes that were supposed to be confined in cellular compartments in specific tissues are found with increased levels in plasma, this is a sign of tissue lesions with rupture of cell membranes (and consequent leakage of enzymes to the plasma). The death of cells in tissues implies a constant flow of intracellular contents to the plasma, but in non-injured tissues, this occurs with a very limited extent. A severe lesion in the liver, heart, or other organ leads to unusually high increased levels in the plasma of enzymes that are specific of that organ. Prostatic-specific acid phosphatase (PSAP, Fig. 3.46), for instance, is an enzyme produced by the prostate and can be found in high amounts in the blood of men who have prostate cancer. A short number of other diseases cause moderate increased levels of PSAP, but only direct lesion of the prostate such as in tumors in this organ causes very increased levels of the protein in plasma. PSAP is then classified as a marker of prostate cancer.

Being highly irrigated (see Box 8.1), and particularly exposed to the action of drugs and other exogenous chemicals, and viruses, the liver is an organ that suffers frequent insults that lead to the presence of hepatic enzymes in plasma. Two transaminases, alanine transaminase (ALT) and aspartate transaminase (AST), are markers of lesions frequently assayed in blood samples when hepatitis, poisoning, or alcoholic liver disease is suspected. However, one should bear in mind that these enzymes are also present in other organs, albeit in small concentrations. A full diagnosis is composed of data that takes into consideration not only biochemical analysis but also the symptoms, the history, lifestyle, and other diagnostic results.

a                                             b

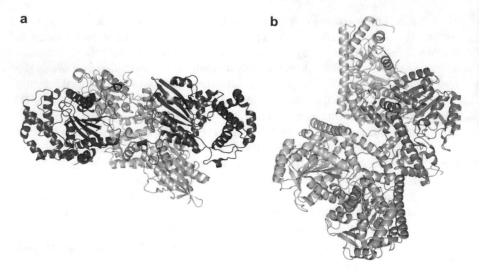

**Fig. 3.46** A marker of prostate cancer. (**a**) Human prostatic-specific acid phosphatase, PSAP (PDB 1cvi), and (**b**) aspartate aminotransferase, ASP (PDB 3II0), are markers of prostate cancer and hepatitis, respectively. Pyridoxal 5-phosphate is a cofactor that is shown bound to ASP (yellow)

The enzymes lactate dehydrogenase (LDH) and creatine kinase (CK) are of particular interest because they are markers of heart muscle lesion. Although these enzymes also exist in muscles other than the heart, there are differences in the amino acid composition that enable detection of the heart variants. Enzymes with similar activity and extensive structure homology are known as isoenzymes. For instance, three CK isoenzymes have been discovered: CK-MM or CK3, found mostly in skeletal muscle; CK-BM or CK2, found mostly in myocardium; and CK-BB or CK1, which is concentrated in lungs and the brain. Because of this distribution of CK isoenzymes, a pulmonary embolism is associated with elevated levels of CK-BB. On the other hand, an acute myocardial infarction is associated with elevated levels of CK-MB, and injuries of the skeletal muscle cause elevated levels of CK-MM. LDH isoenzymes are tetramers, both in the heart and other muscles. These tetramers may disassemble and reassemble in the form of mixed heterogeneous tetramers because the structure of the monomers is very similar. In any case, the presence of dominant heart isoenzymes can be detected in plasma in case of myocardial infarction. The plasma enzyme changes in acute myocardial infarction are shown in Fig. 3.47. CK-MB isoenzyme peaks first, AST next, and LDH last.

There are also nonenzymatic markers that are used in the diagnosis of acute myocardial infarction: myoglobin and two cardiac troponins, troponin I (cTnI) and troponin T (cTnT). CK-MB and the heart LDH isoenzyme are the most important for their heart specificity. Cardiac troponins are also important as their serum levels are frequently elevated during the first hours of acute myocardial infarction, even at a time when CK and CK-MB activities are still within the reference range but are not as consecrated as the enzymatic markers.

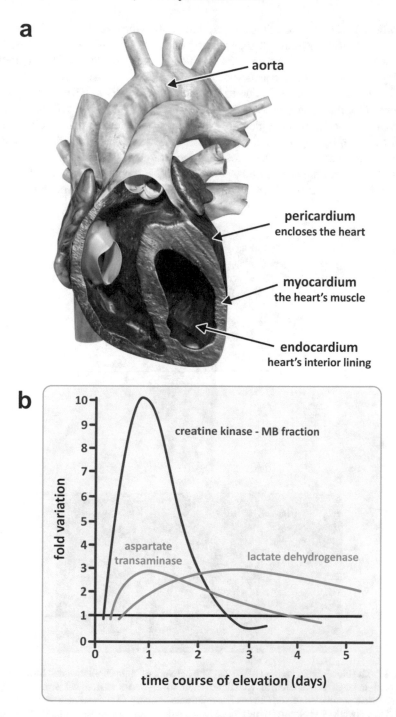

**Fig. 3.47** Biomarkers for lesions of the myocardium, the heart's muscle. (**a**) Myocardium infarction involves partial tissue death (necrosis) caused by a local deficit of oxygen supply, consequence of an obstruction of the tissue's blood flow. (**b**) Cardiac muscle enzymes, such as CK-MB, AST, and LDH, appear in the blood after the infarct. The combined information of CK-MB and LDH allow to estimate the time of the infarct, which in turn helps devising a therapeutic strategy

While the scientific and clinical discipline of studying enzymes for direct clinical diagnosis, clinical enzymology, is expanding and gaining importance, it is curious to mention that dead brain tissue does not release into the blood any significant amounts of enzymes. Despite the frequency of cerebral infarcts (strokes), no test for brain enzymes is currently available due to the blood–brain barrier (BBB, Fig. 3.48), the network of capillaries that irrigates the central nervous system. The cells of these capillaries are connected by tight junctions and adhesion molecules that severely restrict the diffusion of hydrophilic macromolecules into the cerebrospinal fluid. Small gas molecules, such as $O_2$ and $CO_2$, diffuse passively through the barrier, and some nutrients and hormones are actively transported with specific proteins (this will be revisited, for instance, in Box 9.3, which discuss the glucose transport through the BBB).

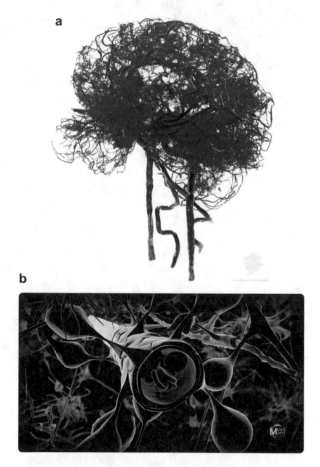

**Fig. 3.48** The blood–brain barrier (BBB). (**a**) The network of brain capillaries that forms the BBB conserved through plastination, a technique used to conserve anatomical structures (image reprinted with permission of von Hagens Plastination, Germany; © www.vonHagens-Plastination. com). The network of very thin arteries that penetrate the brain forms a very reticulated mesh. (**b**) The capillaries are associated with a thick basement membrane and astrocytic endfeet (brown cells covering the endothelium). Passage of molecules across the endothelial cells of the BBB is highly selective. Enzymes released from nerve cells upon a stroke cannot reach the blood, which is highly detrimental for diagnosis. (Image by Ben Brahim Mohammed, reproduced from Wikimedia Commons under a CC-BY license)

### 3.3.4.2   The Nomenclature of Enzymes

Because an enzyme is very specific for the substrate or for a small family of closely related molecules of very similar structure, it is very unique. In the early days of metabolic studies, enzymes were named individually, one at each time, with no concern for general rules of nomenclature. With time, the diversity of names and multitude of criteria to identify newly discovered enzymes was such that the lack of a nomenclature that could be used worldwide was detrimental to the progress of enzymology (the scientific discipline devoted to study enzymes). The International Union of Pure and Applied Biochemistry (IUPAB), an international organization emerging from the joint efforts of many national societies of biochemists around the globe, appointed a working group to propose general rules that could be used to classify and identify enzymes. The result was a nomenclature based on the kind of reaction catalyzed by the enzyme consisting of:

1. A name based on the contraction of "substrate + suffix ase" (e.g., urea + ase = urease, an enzyme that catalyzes a reaction with urea). This type of nomenclature has some flexibility.
2. A rigid four-number code preceded by EC (for "Enzyme Commission"), which is unique for each enzyme (or sets of isoenzymes). The numbers refer to a family of enzymes and three successive sub-families (Box 3.9). EC 5.2.1.3, for instance, identifies an enzyme from family 5 ("isomerases"—catalyzes an isomerization reaction), first sub-family 2 ("cis–trans isomerization"), and the total code identifies specifically retinal isomerase, an enzyme involved in vision (see Sect. 2.2). Part of the whole tree of enzyme nomenclatures is presented in Box 3.9.

**Box 3.9   Enzyme Classification and Nomenclature**

IUBMB (International Union of Biochemistry and Molecular Biology) is the organization responsible for recommendations on the nomenclature and classification of enzymes. Enzyme classification and strict nomenclature rules allow the unambiguous identification of enzymes. A working group, named "Enzyme Commission," was established in 1956 to propose a universal classification and nomenclature system. Nearly 659 enzymes were known by then and the chaos in enzyme naming was clear. Nowadays, more than 5500 enzymes are known, and it would be virtually impossible to communicate in enzymology if an official universal classification and nomenclature systems had not been established.

In 1961, the Enzyme Commission presented its report, in which enzymes are divided in six classes according to the reaction they catalyze. Classes and three levels of subclasses are numbered. The Enzyme Commission thus identified enzymes through a four-number code preceded by the letters EC to clearly identify that the numeric code corresponds to the classification set by the Enzyme Commission. The Enzyme Commission itself has been renamed but the classification system is still the same. The initials EC have remained although the commission they refer to has not.

(continued)

**Box 3.9** (continued)

Besides the numeric EC code, a name is also used because names are more intuitive and immediate than numeric codes. The most commonly used name for the enzyme is preferred, provided that it is unambiguous, but there are alternative systematic names that attempt to describe unambiguously the catalysis. Systematic names consist of two parts. The first contains the name of the substrate or, in the case of a bimolecular reaction, of the two substrates separated by a colon. The second part, ending in -ase, indicates the nature of the reaction, e.g., oxidoreductase, oxygenase, transferase (with a prefix indicating the nature of the group transferred), hydrolase, lyase, racemase, epimerase, isomerase, mutase, and ligase.

In practice, the enzyme classification and nomenclature stems from the classification of enzyme-catalyzed reactions, not from protein structures. A single protein may have two or more EC numbers if it catalyzes two or more reactions. This is the case, for example, for two proteins in *Escherichia coli*, each of which catalyzes the reactions both of aspartate kinase and of homoserine dehydrogenase. It may also happen that two or more proteins with no detectable evidence of homology catalyze the same reaction. For example, various different proteins catalyze the superoxide dismutase reaction and share a single EC number, EC1.15.1.1. This latter case is relatively rare, but it is almost universal that proteins catalyzing the same reaction in different organisms, or sets of isoenzymes in one organism, are homologous, with easily recognizable similarities in sequence.

Take class EC 1 of enzymes, oxidoreductases, as example. This class contains the enzymes catalyzing oxidation reactions. Since the oxidation of one group must be accompanied by the reduction of another, they are grouped together as oxidoreductases. The systematic enzyme name is in the form "*donor/acceptor* oxidoreductase." The substrate that is being oxidized is regarded as being the hydrogen donor. The name is commonly "*donor* dehydrogenase." Although the term reductase is sometimes used as an alternative, it is important to remember that the recommended name does not define the equilibrium position of the reaction or the net direction of flux through the enzyme in vivo. The term "*donor* oxidase" is used only when $O_2$ is the acceptor.

**Enzyme classes:**
There are six classes of enzymes:

**EC 1—Oxidoreductases** catalyze reactions in which a substrate donates one or more electrons to an electron acceptor, becoming oxidized in the process.

**EC 2—Transferases** catalyze reactions in which a chemical group is transferred from a donor substrate to an acceptor substrate.

(continued)

**EC 3—Hydrolases** catalyze reactions in which a bond in a substrate is hydrolyzed to produce two fragments.

**EC 4—Lyases** catalyze non-hydrolytic reactions in which a chemical group is removed from a substrate leaving a double bond.

**EC 5—Isomerases** catalyze one-substrate one-product reactions that can be regarded as isomerization reactions.

**EC 6—Ligases** catalyze the joining together of two or more molecules coupled to hydrolysis of ATP or an analogous molecule. These enzymes are also sometimes called synthetases, a name that was already in use before creation of the original Enzyme Commission.

In reality, all of the enzymes in classes 1–3 satisfy the definition of transferases. However, as these three classes are all large compared with the other three groups, it is convenient to break them into three classes and to reserve the name transferase for enzymes that are not oxidoreductases or hydrolases.

**Enzyme subclasses:**
Each of the six classes is divided into subclasses on the basis of the salient differences between the enzymes in the class. In **EC 1**, for example, the subclasses define the type of substrate acted on:

**EC 1.1:** Acting on the CH–OH group of donors
**EC 1.2:** Acting on the aldehyde or oxo group of donors
**EC 1.19:** Acting on reduced flavodoxin as donor
**EC 1.97:** Other oxidoreductases

This last subclass is numbered **EC 1.97** because it is provisional. In due course, the enzymes it contains may be reclassified more appropriately. The original report had two subclasses **EC 1.99** and **EC 1.98** that were removed when sufficient information was available to place the enzymes they contained elsewhere.

Classes **EC 3–5** are divided into subclasses on the basis of types of substrate, in much the same way as in **EC 1**. In **EC2**, however, it was more useful to emphasize the nature of the transferred group. So, for example, we have:

**EC 2.1:** Transferring one-carbon groups
**EC 2.2:** Transferring aldehyde or ketone residues
**EC 2.3:** Acyltransferases
**EC 2.8:** CoA-transferases

In **EC 6**, the division into subclasses is made on the basis of the type of product:

**EC 6.1:** Forming carbon–oxygen bonds
**EC 6.2:** Forming carbon–sulfur bonds
**EC 6.3:** Forming carbon–nitrogen bonds

(continued)

**EC 6.4:** Forming carbon–carbon bonds
**EC 6.5:** Forming phosphoric ester bonds

**Enzyme sub-subclasses:**
The subclasses are divided into sub-subclasses in much the same way as the way the subclasses themselves are defined. For example, **EC 1.16** (oxidoreductases oxidizing metal ions) contains two sub-subclasses:

**EC 1.16.1:** With $NAD^+$ or $NADP^+$ as acceptor
**EC 1.16.2:** With oxygen as acceptor

As with the numbering of subclasses, 99 (or a smaller number if necessary) is used for sub-subclasses containing a miscellaneous group of enzymes. For example, subsection **EC 1.6** contains oxidoreductases acting on NADH or NADPH, and within this, there is **EC1.6.99** for miscellaneous acceptors.

There are also sub-sub-subclasses so that each enzyme is identified by four different numbers. The division of sub-subclasses into sub-sub-subclasses follows the same rationale as before. An exhaustive visit of the fourth level of classes is not justified here.

*Final note:* Text based on *"Enzyme Classification and Nomenclature" by S Boyce and K Tipton (Encyclopedia of Life Sciences, 2001) and "Current IUBMB recommendations on enzyme nomenclature and kinetics" by A, Cornish-Bowden (Perspectives in Science, 2014, 1, 74–87)*

# Selected Bibliography

Boyce S, Tipton KF (2001) Enzyme classification and nomenclature. In: Encyclopedia of life sciences. Nature Pub. Group, London, pp 1–11

Cornish-Bowden A (2014) Current IUBMB recommendations on enzyme nomenclature and kinetics. Perspect Sci 1:74–87

Dobson CM, Gerard JA, Pratt AJ (2001) Foundations of chemical biology, Oxford chemistry primers. Oxford University Press, Oxford

IUPAC-IUB Joint Commission on Biochemical Nomenclature (JCBN) (1985) Nomenclature and symbolism for amino acids and peptides. J Biol Chem 260:14–42

Stevenson J, Brown AJ (2009) How essential is cholesterol? Biochem J 420:e1–e4

Westheimer FH (1987) Why nature chose phosphates? Science 235:1173–1178

# Challenging Case 3.1: Biochemistry in Dali's (he)art

## *Source*

This case is based on a biographic paper by Juan Esteves de Sagrera that describes the influence of science in general, biochemistry in particular, in the work of the surrealist artist Salvador Dali: "Dali científico. Las ciencias biomédicas en la obra del pintor ampurdanés" ("Dali scientist. The biomedical sciences in the work of the painter from Ampurdania").[1]

The emblematic Dali painting "Galatea de las esferas," from 1952, highlighted in the paper in which this case is based. (Fundación Gala-Salvador Dalí, Figueres)

## *Case Description*

Salvador Dali was not an ordinary man. Known for his eccentricity and geniality, the artist is famous for his prolific surrealistic work in painting. His inspiration for the oneiric-like landscapes and rich details in his paintings is less known but worth exploring. Surprising as it might be at first glance, science was a major inspiration for Dali, mainly chemistry, biochemistry, and psychoanalysis. While psychoanalysis would obviously inspire him for the dimensions of human nature, chemistry and biochemistry provided a repository of hidden geometries inherent to the principles of matter. With the right doses of mysticism, the organization of matter was translated into the decoding of life. The atomic implications of quantum physics and, above all, the structure of DNA had a strong fascination over Dali. DNA was regarded by Dali as the key to life. The importance of DNA in Dali's work is such that some scholars in arts name one of his creative periods as "the DNA period."

Dali used to read a lot and forge ideas from his readings. His personal home library had more than 100 books annotated by him on physics, quantum mechanics, the origin and evolution of life, and mathematics. He also subscribed to scientific magazines. This passionate avidity for scientific information was kept throughout his life. In 1985, for instance, at the age of 81, Dali was an active spectator of a

---

[1] de Sagrera JE (2004) Dali científico Las ciencias biomédicas en la obra del pintor ampurdanés. Offarm 23:122–128.

scientific meeting on culture and science that took phase in his hometown. He also read *A Brief History of Time* by Stephen Hawking.

In the 1940s, Dali was influenced by the atomic nature of matter and atomic nuclear energy. *"La desmateralizacion de la Nariz de Neron"* (1947) is an example of this influence (see next figure). In the following decade, the 1950s, Dali became fascinated by the quantum theory and praised Werner Heisenberg, the German scientist known for the "uncertainty principle" in quantum mechanics. The paintings frequently became corpuscular as if made of atomic units (e.g., *"Galatea de las esferas,"* 1952; see previous figure).

Influence of science in Dali's paintings: *"Desmaterialización cerca de la nariz de Nerón. La separación del átomo* (1947)" (Fundación Gala-Salvador Dalí. Figueres (Spain)). *"Galatea de las esferas,"* which illustrates the case source, is another example of the influence of science in Dali's paintings

In the 1960s, the structure of DNA dominates. Interestingly, DNA first appears in Dali's work in 1957, only 4 years after Watson and Crick published their note on the structure of DNA, many years before the importance of DNA structure was dully recognized and became an icon. In this painting, *"La paisaje de la mariposa,"* DNA coexists with butterflies, which also happens in other representations. Butterflies in Dali's paintings are symbols of tranquility and peacefulness. In 1971, the advertising poster of the National Congress of Biochemistry (of Spain) was painted by Dali having the structure of DNA forming Jacob's Stairway to Heaven.

In the last period of his highly creative life, Dali develops intense interest for virtual imaging and the sense of three dimensionality. In 1970, Dali used the Fresnel lens to create holograms. Dennis Gabor was awarded with the Nobel Prize in Physics for his work with lasers and development of the holographic method in

1971. Gabor's work was his source of inspiration, which demonstrates how Dali was up to date with scientific novelties. The painter's main interest however was stereoscopic painting, which is simpler and closer to his art. Stereoscopy is an elegant trick to create an optical illusion having two similar paintings and a pair of mirrors forming the arrangement depicted in the next figure. This technique consists of presenting two similar images to the left and right eye of a viewer in a way that the perception of integrated vision in the brain is three-dimensional, i.e., the sum of the two flat images in the brain creates the illusion of depth. Dali painted pairs of very similar paintings, eventually with different objects and colors in the background, which when presented to the vision of each eye individually through the reflection in mirrors resulted in a virtual 3D painting. This technique is still used in 3D viewing devices to be coupled to the screen of mobile phones. Amazingly, Dali learned about stereoscopy from the work of Gerrit Dou, in the seventeenth century. This Dutch painter probably used lenses and mirrors to create his stereoscopic paintings and was helped in this task by Van Leeuwenhoek, the seminal founder of modern optical microscopy.

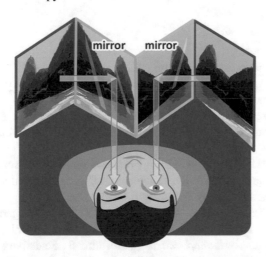

The basic principle of stereoscopy. Two similar images are placed adjacent to two planar mirrors in a symmetrical W-like setup. The viewer observes the images having the common edge of the mirrors in between the eyes at close distance so that each eye receives an individual image. The overlay composition of both images takes place in the brain

Two stereoscopic paintings of Dali have important representations of DNA: "*El pie de Gala*" (1975–1976) and "*La estructura del ADN*" (1975–1976) (shown in the next figure). The intention to bring the representation of DNA closer to the scientific standard representation of macromolecules with balls and sticks models is obvious. This path that deviates Dali from artistic standards toward the language of science is regarded by many as detrimental to his artistic creation. For them, the beginning of the end for Dali was his progressive metamorphosis from a genial painter to an academic artist and a dilettante "dandy man" of science.

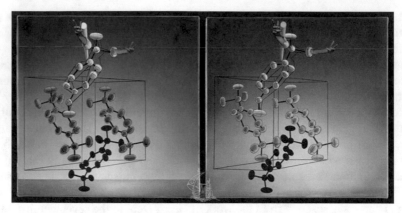

"*La estructura del ADN*" (1974–1976; diptych for stereoscopic view). (Fundación Gala-Salvador Dalí, Figueres (Spain))

## Questions

1. Take into consideration the painting "*La estructura del ADN*" (The structure of DNA). Is there a correspondence between the structure represented and the atomic-resolution structure of DNA we know today?
2. Now take into consideration the painting "*La estructura del ADN*" as perceived in the reflections of the stereoscopic setup. What are the differences relative to the painting?
3. Search for the 1957 painting "*Paisage de mariposas*" (Landscape of butterflies) in the Internet and correlate the structure of DNA as depicted with the structure proposed 4 years before by Watson and Crick.
4. In a famous painting besides "*El pie de Gala*," Dali also merged two of his main inspirations: the lifetime muse and wife, Gala, and DNA. "*Galacidalacidesoxyribonucleicacid*" (Museum S. Petersburg, Florida, USA) was the strange title given to this painting. Imagine Gala would refer to galactose and try to represent the structure of a hypothetical nucleic acid having a galactose residue replacing the ribose residue.

## Biochemical Insights

1. No. The graphics are obviously similar but there is no relation to the atomic resolution structure of DNA.

2. Reflections in the mirror result in a symmetrical but not identical structure, as is the case with enantiomers.
3. There is impressive realism in the way the blue spheres correspond to the charged phosphate backbone, and the yellow spheres' core correspond to the bases' residue stacking. The polar vs. nearly non-polar nature of the blue vs. yellow regions is the primordial physical reason for the double-helix folding of DNA.
4. Relative to ribose, galactose would have an extra hydrophilic and reactive OH group (see below figure) to accommodate in the polymeric structure of a nucleic acid imposing steric constraints (see figure). This is probably one of the reasons why evolution favored pentoses relative to hexoses.

RNA nucleotide                    DNA nucleotide

## Final Discussion

Dali used to say about himself that he was a better writer than painter. Nevertheless, his legacy in literature is largely overlooked. Dali also felt dazzled by geometric pattern in scientific disciplines other than biochemistry. The eyes of the flies and some viruses under the electronic microscope, for instance, were fascinating for him. Yet, strangely, his most obsessive natural structure besides DNA was the rhinoceros' horn, which he associated with logarithmic function in mathematics. Strangely, he found a similarity in rhino horns and DNA as both have spiral forms. DNA and rhino horns were the key to the composition of all beings. A strange association for biochemists, to whom rhino horns are little more than keratin. Coincidence or not, rhino horns are also believed to have miraculous pharmacologic properties in some cultures.

## Further Reading

"Dali y la estrutura de doble hélice". S. Grisolia (25/04/2003) ABC, hemerotace (www.abc.es/hemcrotcca/historico-25-04-2003/abc/Sociedad).

## Challenging Case 3.2: From Best with Proteins to Worst with DNA

### Source

This case is based on the biography of Linus Pauling, *"Linus Pauling. A Natureza da Ligação Química"* (*"Linus Pauling. The nature of the chemical bond."*), written by the Portuguese author Raquel Gonçalves-Maia, from the University of Lisbon.[2]

(left) Cover of the book *Linus Pauling. A natureza da ligação química* (reproduced with permission of Dr. Raquel Gonçalves-Maia, and Edições Colibri); (right) Oil portrait by Leon Tadrick with Linus Pauling and a model of the alpha-helix, 1951. (Image provided by the Ava Helen and Linus Pauling Papers, Oregon State University Libraries)

### Case Description

Biochemistry owes a lot to Linus Pauling. Not only did Linus Pauling unravel the basics of the detailed structure of helical domains of proteins, but he also was a fierce defender of the importance of biochemistry to understand life. *"All the activities of living beings are chemical in nature"* he said once. In fact, through a chemical vision of the world, he worked across physics, biology, pharmacology, and medicine. He also engaged activism for social causes. His merits in science and his efforts for a better society were properly recognized with the Nobel Prize in Chemistry (1954) and the Nobel Prize for Peace (1963). However, no one is perfect and even the very best can and do fail. Over-confident in his chemical knowledge and intuition, which led him to the structure of the protein helix, Linus Pauling rushed to a grossly erroneous structure of DNA. He could have been the discoverer of the DNA double helix, yet he proposed an unreasonable triple helix of inconceivable organization.

---

[2] Gonçalves-Maia R (2017) Linus Pauling. A natureza da ligação química. Edições Colibri, Lisbon. ISBN 978-989-689-692-8.

### 3.5.2.1 The Protein Vision

In the 1930s, several research groups were committed to unlock the mysteries of the origins of life and how cells could transmit information from generation to generation, i.e., how could cells divide and still be functional. Proteins were everywhere, so ubiquitous that they were thought by most as the molecules responsible for information storage and transmission. Pauling expertise was chemistry, yet biology was an irresistible challenge.

Pauling's *The Nature of The Chemical Bond* was the bestselling book in science in the twentieth century and undeniably the most influential book in the history of chemistry. The book was dedicated to Gilbert Lewis, who had proposed before that chemical bonds consisted of electron sharing. In a letter of August 25, 1938, Lewis reacted positively to the book: "*I have returned from a short vacation for which the only books I took were a half a dozen detective stories and yours 'Chemical bond'. I found yours the most exciting of the lot.*" Pauling knew the previous work of Hermann Emil Fischer (1852–1919), who discovered that proteins are formed by amino acid residues connected through amide bonds (also known as "peptide bonds" among biochemists). Pauling believed that proteins were long polypeptide chains. Others believed proteins would be formed by amorphous conglomerates of small peptides. The seminal work of William Astbury in X-ray crystallography seemed to support the long polypeptide hypothesis, but the evidence was not totally conclusive. Serendipitously, Astbury worked with wool, hair, and other materials rich in keratin and collagen, i.e., materials of almost purely aligned helices. The genius of Pauling's chemical input consisted in deducing the structure of polypeptides from the three-dimensional arrangement of the consecutive chemical bonds in the amino acid residue sequence. He realized that the planar arrangement of the amide bond and H bonds could be the master factors in stabilizing the overall structure. In 1936, Pauling published a general theory on the structure of proteins[3] together with Alfred Ezra Mirsky, a visiting professor at his lab. This structure was born out of chemical reasoning and intuition. Hard evidence was yet to come. It took about a dozen years … with World War II in between.

Working with paper models, Pauling considered it reasonable to admit proteins would form helices having 3.7 ($\alpha$-helix) or 5.1 ($\gamma$-helix) amino acid residues per turn. In 1950, Pauling and Robert Bohr Corey, an X-ray crystallographer, published a landmark paper titled "Two-hydrogen spiral configurations of the polypeptide chain",[4] followed by several other papers in the scientific periodical *The Proceedings of the National Academy of Sciences* of the USA,[5] commonly known as PNAS. The helical structure of proteins was finally unveiled (see next figure).

---

[3] Mirsky AE, Pauling L (1936) On the structure of native, denatured, and coagulated proteins. Proc Natl Acad Sci U S A 22:439–447.

[4] Corey RB, Pauling L (1950) Two-hydrogen spiral configurations of the polypeptide chain. *J Am Chem Soc* 72:5349–5349.

[5] Pauling L et al (1951) The structure of proteins: two hydrogen-bonded helical configurations of the polypeptide chain. Proc Natl Acad Sci U S A 37:205–211.

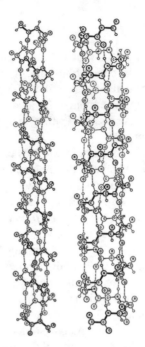

The helix with 3.7 residues per turn (*left*) and the helix with 5.1 residues per turn (*right*), as shown in the article cited in the footnote 5. Hydrogen bonds are highlighted by dashed lines

Pauling and Corey also demonstrated that H-bonding was the key structural main determinant of another stable protein structure: β-sheets[6] (see next figure).

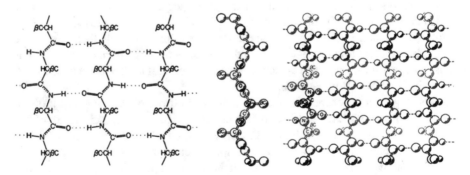

Representations of the antiparallel-chain pleated sheet structure, as shown in the article cited in the footnote 6

Globally speaking, Linus Pauling united the physical chemistry of chemical bonding to structural biochemistry of complex biological molecules first and then united molecular structure to function in biology. In this sense, he was one of the founders of modern biochemistry. He was awarded with the Nobel Prize in Chemistry for this multidisciplinary contribution. For the first time in history, a Nobel Prize was awarded for multiple contributions during a lifetime, not for a single event discovery.

[6] Pauling L, Corey RB (1951) Configurations of polypeptide chains with favored orientations around single bonds: two new plated sheets. Proc Natl Acad Sci U S A 37:729–740.

## The DNA Vertigo

Linus Pauling was very close to proposing the correct structure of DNA, but he failed. Not only he did not propose the correct structure, but he actually proposed a wrong structure. A cascade of capital errors dragged him to a proposal that was not compatible with his chemical knowledge and reasoning at all:

Error #1—Pauling underestimated the importance and complexity of nucleic acid polymers. Probably the fascination with proteins, with about 20 different natural amino acid monomers, and conformational plasticity, convinced him that proteins rather than nucleic acids, with only four different monomers, were responsible for "heritage" in cell division. This was a matter of intense debate in the 1940s, and Pauling was on the wrong side of the barricade.

Error #2—Pauling also underestimated the time and effort needed to deduce DNA structure from the fragile experimental evidence available. Astbury had very low-resolution X-ray diffraction data from DNA. Other researchers were ahead in this front. Maurice Wilkins and Rosalind Franklin, under the guidance of John Randall, were studying DNA thoroughly and methodically. None had the prestige of Linus Pauling, but they were focused and determined.

Error #3—Pauling wanted to be the first to publish DNA structure and rushed to a beginner's mistake. In the last day of 1952, Pauling and Corey submitted to PNAS a manuscript titled "A proposed structure of the nucleic acids"[7] without solid evidence for their pitch. The structure proposed consisted of a triple helix (see next figure) with the phosphate groups forming a core densely packed by a network of hydrogen bonds. The bases pointed outward.

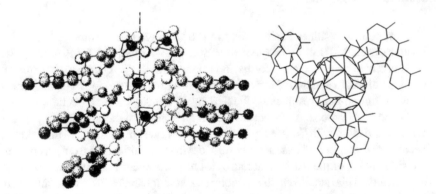

Perspective drawing of a portion of the nucleic acid structure as proposed by Pauling in the article cited in the footnote 7 (left), showing the phosphate tetrahedral near the axis of the molecule, the β-D-ribose rings connecting the tetrahedral into chains, and the attached purine and pyrimidine rings (represented as purine rings in the drawing). On the right, a plan of the proposed nucleic acid structure having several nucleotide residues is shown

---

[7] Pauling L, Corey RB (1953) A proposed structure of the nucleic acids. Proc Natl Acad Sci U S A 39:84–97.

The double-helix structure was proposed few years later, based on solid experimental evidence and generally accepted. Likewise, the hypothesis of DNA being the molecule responsible for heritable legacy became increasingly demonstrated and accepted.

## Questions

1. Other researchers, contemporary to Linus Pauling, tried to resolve the structure of protein helices assuming that the number of amino acid residues per turn would be an integer. Knowing the helix structure as proposed by Pauling, what makes this assumption unrealistic?
2. The helix with 3.6 residues per turn is named $\alpha$-helix and the helix with 5.1 residues per turn is named $\gamma$-helix. $\pi$-helices have 4.4 residues per turn. There is another helix structure named $3_{10}$-helix, in which H-bonding is established between an amino acid residue and the amino acid 3 residues earlier. What variations are expected among these types of helices in packaging density, angles between residues, and side chain projection?
3. Why were protein materials such as collagen and keratin adequate to be studied by X-ray crystallography?
4. What makes the Pauling's DNA structure unconvincing?

## Biochemical Insights

1. The amino acid residues in a $\alpha$-helix tilt $100°$ relative to each other, making 3.6 residues (i.e., $360°$) per turn (near the value of 3.7 predicted by Pauling). The N–H group of a peptide bond forms an H bond with the C=O group of the fourth residue before; in symbolic terms, this is represented by $i + 4 \rightarrow i$. The alignment between amino acid residues is dictated by H-bonding, which occurs between C=O and N–H groups. A perfect integer in all types of helices would imply that C=O would align with C=O every N (integer) residues. In addition, in the $\alpha$-helix, because the number of residues per turn deviates significantly from an integer, the side chains of the residues are also not aligned; therefore steric hindrance is minimized. This stabilizes the structure and contributes to explaining why $\alpha$-helices are dominant in nature relative to other helices. It is worth highlighting that $\alpha$-helices are dominant in nature over $3_{10}$- and $\pi$-helices. The $\gamma$-helix proposed by Pauling remains conceptual: it was never found in nature.
2. The $\pi$-helix is $i + 5 \rightarrow i$ with 4.4 residues per turn and the $3_{10}$-helix is $i + 3 \rightarrow i$ with 3.0 residues per turn. Less residues per turn imply increasing the angle between consecutive residues. Three residues correspond to $360°/3 = 120°$ tilt between residues. The packing density of the helix increases, and the side chains

of every three residues are aligned linearly with steric hindrance among them, as illustrated in the below figure, in comparison with an α-helix.

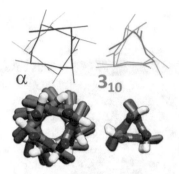

Schematic representation of an α- and $3_{10}$-helix

3. Collagen and keratin are ordered arrays of proteins almost exclusively composed of α-helices. They are thus homogenous structures repeated in space. This property is prone to the formation of crystals, a sine qua non condition to obtain an X-ray diffraction pattern from which structure can be deduced.
4. Phosphate groups are deprotonated even at neutral pH, which makes H bond between phosphate groups unrealistic. Also, the proposed structure has highly hydrophilic groups in the core shielded by not so hydrophilic groups, the bases, exposed to the solvent, which is also unrealistic. In addition, there is steric hindrance between the bases.

## Final Discussion

Linus Pauling made consistent work from the theoretical chemistry he rooted in quantum mechanics to the structure of biological macromolecules. His work highlights the importance of H-bonding to protein structure but extends much further than the unveiling of the α-helix and β-sheet. He also worked on hemoglobin pathologies and antigen–antibody structural complementarity. In this sense, he can be considered one of the founding fathers of immunochemistry. His mistake with DNA is nothing compared to the greatness of new ideas he donated to biochemistry. His beginner's mistake with DNA is an opportunity to reflect on how strongly we should rely on the power of evidence against mere intuition, the border between self-confidence and overconfidence, and how much science is modulated by competition of scientists for reputation.

## Challenging Case 3.3: London Fatberg Analysis—Impact of Urban Obesity on the Metropolitan Vasculature

### *Source*

This case is based on the documentary "Fatberg autopsy: secrets of the sewers" (Channel 4, UK, 2018), produced by BBC studios in conjunction with the UK water and wastewater services Thames Water[8, 9], which exposed the contents of one huge blockage of congealed fat, a "fatberg," trapping daily use chemicals and other human waste in their composition, discovered underneath the streets of London. The Whitechapel fatberg, for instance, would have risked raw sewage flooding onto the streets had it not been discovered during a routine inspection. In the event of a blockage, contents of the sewers could come back up through domestic pipes causing flooding; the threat to public health would be immense. Yet, the documentary reveals another face of fatbergs. The molecular contents in the chemical analyses (cleverly named "autopsy" in the documentary), being made up of materials people flush down the toilet or pour down the sink, such as cooking oil and medicines, reveal a lot about our society: rapidly changing cooking habits and obsession with body image, for instance. The documentary was produced by BBC Studios—The Science Unit; Executive Producer: Paul Overton; Series Producer: Rob Liddell; Director: Nick Clarke Powell. Another analysis followed, this time to a fatberg in Sidmouth (UK), by researchers of the team of Professor John Love, University of Exeter (UK).

Protagonists of the documentary "Fatberg autopsy: secrets of the sewers," Rick Edwards and Carla Valentine. (Courtesy of Thames Water)

### *Case Description*

The study in the documentary by Channel 4 analyzed the contents of one supersized fatberg discovered underneath the streets of South Bank in Central London. Fatbergs are part of a growing urban problem across the UK as the sewage infrastructure

---

[8] (2018) Fatberg Autopsy promises 'unforgettable sight'. Thames Waters media. https://corporate.thameswater.co.uk/media/News-releases/Fatberg-Autopsy.

[9] (2018) Fatberg autopsy: secrets of the sewers. Thames Waters media. https://corporate.thameswater.co.uk/Media/News-releases/Fatberg-Autopsy-Secret-of-the-Sewersy.

struggles to cope with the population's changing habits. The South Bank mass is thought to be larger than the first large-scale fatberg discovered under Whitechapel, East London, which weighed the same as 11 double decker buses and stretched the length of two football/soccer pitches. *"We and other water companies are facing a constant battle to keep the nation's sewers free from fatbergs and other blockages,"* Thames Water's waste networks manager, Alex Saunders, said[10] Fatbergs are so big that water companies are forced to employ teams of "flushers" to remove them. The filming crew of Channel 4 registered the activity of a "flushers" team while removing the South Bank fatberg, chunk by chunk, and the setup of an improvised lab big enough and adequate to manipulate fatberg chunks but having all necessary equipment for biochemical analysis and film shooting. A disused pumping station at Abbey Mills, dubbed "the cathedral of sewage," was the elected laboratory/studio venue (see next figure). The aim was to produce a standout documentary, uncover what was making fatbergs an ever-growing problem, and reveal what that specific fatberg could tell us about the dirty secrets of the people who contributed to it.

Thames Water fatberg fighters prepared for action (*left*) and the laboratory/studio setup at Abbey Mills, where fatberg material was crushed into small pieces and analyzed (*center* and *right*). In "Fatberg autopsy: secrets of the sewers," Rick Edwards, the presenter, discusses with Carla Valentine, anatomical pathology technologist, the interpretation of the "autopsy". (Images reproduced from the Evening Standard website (*left*)[11] and Thames Waters media (*center* and *right*)[8,9]. (Courtesy of Thames Water)

Almost every aspect of the production involved contact with material that should never really leave a sewer—from filming underground with Thames Water's team of "flushers" to setting up a makeshift studio with a crew of 78, who were all kitted out head to toe in hazmat suits. Filming professionals were told by Thames Water that due to gases building up behind fatbergs, every bit of kit needed to be "intrinsically safe," which narrowed their options to one specific model of GoPro camera, which was attached to the "flushers," without Wi-Fi connection as it equaled another explosive hazard. The "autopsy" itself (i.e., dividing samples of fatberg mass into components and identifying the components) was a different matter; no one had ever done one before, so it was harder to plan. A decontamination tunnel was set up (it looked like something from ET). In addition, there was the risk of biological

---

[10] Khomami N (2018) Fatberg 'autopsy' reveals growing health threat to Londoners. The Guardian.

[11] (2017) London's monster fatberg FINALLY defeated after two months of 'gut-wrenching' work. Evening Standard website. https://www.standard.co.uk/news/london/londons-monster-fatberg-finally-defeated-after-two-months-of-gutwrenching-work-a3675866.html.

contamination with bacteria living in the sewers. Everyone entering had to put on a hazmat suit and gloves, and some had to wear face masks. Every time anyone left, for whatever reason, they had to strip off the suit, be sprayed with a bleach-like substance, and wash their hands with alcohol gel. The hurdles of documentary production are fun and enticing but we will focus on the outcome of the so-called autopsy.

The fatberg itself was a surprise. It wasn't some gelatinous lump of fat, but solid yellow-colored matter that was so hard the "flushers" had to hack it out of the sewer with pick axes and shovels. It looked more like wax or soap than feces. Typical items found in the fatberg included condoms, sanitary towels, nappies, wet wipes, and cotton buds. But it was the in-depth analysis by the specialist team of scientists which revealed new insights into a growing urban crisis, as described in the "The Guardian" report cited in the footnote 10, and briefly transcribed with adaptations in the following text.

### Cooking Oil Forms the Matrix of the Fatberg

The fatberg samples were analyzed. A small quantity of fats come from personal hygiene and beauty products. Topically applied creams and gels, which may contain oils and fats, can make their way into the sewer from bathing and washing. However, it is fats and grease from cooking that make up the largest proportion of the fatberg. Just under 90% of the sample is comprised of palmitic acid, commonly found in cooking oil, and oleic acid, found in olive oil (see next table).

Most important parent fatty acids of lipids found in the fatberg and their typical source in human diet and hygiene products

| Trivial name and typical source | Structure | % content |
|---|---|---|
| **Palmitic acid** Palm oil, meats, and dairy products | | 80.865 |
| **Oleic acid** Olive and other vegetable oils | *trans*-Oleic acid *cis*-Oleic acid | 5.870 |
| **Stearic acid** Meats and lard, detergents, soaps, cosmetics | | 4.129 |

(continued)

| Trivial name and typical source | Structure | % content |
|---|---|---|
| **Myristic acid** Dairy products, vegetable oils | 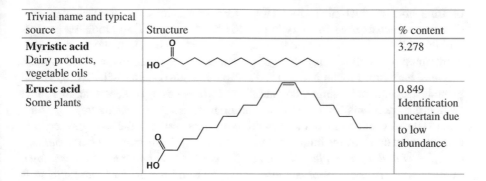 | 3.278 |
| **Erucic acid** Some plants | | 0.849 Identification uncertain due to low abundance |

Water companies continue to advise customers and commercial food outlets to not dispose of oils and fats down the drain.

## Cultivating Superbugs

Being rich in organic matter that may serve as growing media for microbials, the fatberg was tested for dangerous bacteria. The tests found potentially infectious bacteria including *Listeria*, *Campylobacter*, and *E. coli*. More importantly, the results discovered bacteria which were able to thrive in antibiotic environment in vitro. Antibiotic-resistant bacteria, sometimes known as "superbugs," are a serious concern for public health. Bacteria like these pose an immediate severe risk to the operatives who work inside the sewers and the public at large could be at risk in the event of a sewer blockage, as contents of the sewers, including these harmful superbugs, could come back up through domestic or commercial pipes causing flooding to homes and businesses.

Thames Water waste networks manager Alex Saunders, quoted above, also said: *"For the sake of our sewer workers like Vince and the other guys who feature in the show please only flush the three Ps (pee, poo and toilet paper) and don't feed the fatberg."*

## Unraveling the Hidden Dimension of Drug (Ab)Use

The fatberg analyses uncovered evidence indicating people's contact with street drugs and other pharmaceuticals, e.g., intact drugs-related items, including small plastic "baggies," needles, and syringes. Presenter Rick Edwards described the finds as *"a sobering window into the lives of people living above the sewer"*[10]. Dr. John Wilkinson, from the University of York, collaborated with a team from Cambridge using mass spectrometry to identify different chemicals inside the fatberg. The teams analyzed the sample for traces of pharmaceutical chemicals and discovered a high proportion of salicylic acid, commonly found in analgesics and topical creams for acne, and paracetamol. The tests also discovered evidence of hordenine and ostarine, both of which can be found in performance-enhancing sports supplements. Ostarine, which is used for muscle gain, is on the World Anti-Doping Agency's prohibited list and is not licensed for medical use in the UK. Hordenine and ostarine represented over half the proportion of pharmaceuticals found in the tested sample.

Other hard drugs found in the tested sample include cocaine, MDMA (3,4-meth ylenedioxymethamphetamine, commonly known as ecstasy), ketamine, morphine (which also accounts for heroin), and amphetamines (see next figure). Dr. John remarked that chemicals like these could make their way into the sewers when they aren't broken down by the human body: *"You wake up in the morning have your cup of coffee. Not all of the caffeine in that coffee you drink is going to be broken down by your body. In the case of caffeine, somewhere around 60% of that caffeine is broken down into metabolites, the remaining 40% is excreted as caffeine."* There's no way of knowing if the chemicals have been consumed or directly flushed down the toilet, but these results revealed insights into the drugs that people interact with in modern society.

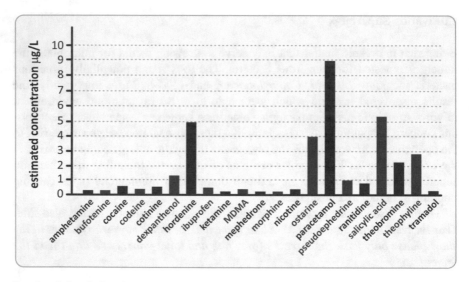

Results of chemical analyses to pharmaceuticals found in the fatberg

**Fatberg by the Sea**

A 64 m fatberg—greater in length than the Tower of Pisa—was discovered in Sidmouth, a tourist resort town situated on the English Channel coast. A team of scientists from the University of Exeter (23 km from Sidmouth) were asked to carry out an extensive "autopsy" of the fatberg to try and help solve the mystery of how it was constructed and whether it posed any environmental risks. The challenge is well elucidated in the local university news.[12] The scientists were given four samples from the fatberg, each weighing around 10 kg, as workers were removing it from the sewer. As in the London fatberg, the team found that the samples they received were mostly made of animal fats—consistent with domestic food preparation—combined with household hygiene products such as wet wipes and sanitary products, as well as natural and artificial fibers from toilet tissues and laundry (see next figure). Crucially, the team lead by Professor John Love found the fatberg contained no detectable levels of toxic chemicals—meaning its presence in the sewer, while increasing the risk of a blockage, did not pose a chemical or biological risk to the environment or human health. The Sidmouth fatberg was simply a lump of fat aggregated with wet wipes, sanitary towels, and other household products that should had never been put down the toilet. The microfibers found were probably from toilet tissue and laundry; the viruses and bacteria found were those normally associated with a sewer. At variance with London fatberg, the chemicals were those found in personal care products, rather than pharmaceuticals. The autopsy carried out at the University of Exeter was covered in an episode of BBC's Blue Planet UK, broadcasted March 2019.

Researchers at the University of Exeter performing an "autopsy" on Sidmouth fatberg to reveal its dirty secrets. (*Left*) A section of the huge fatberg in the sewer (*left*; courtesy of Southwest Water Company, UK). (*Center*) Professor John Love and Nicky Cunningham in action (courtesy of the researchers and University of Exeter). (*Right*) Wet wipes, sanitary towels, and other household products—which included incontinence pads, false teeth, and sanitary products—appeared after the fatberg samples were melted (courtesy of the researchers and University of Exeter). The researchers also used state-of-the-art equipment to study the fats, particles, fibers, and microbiological DNA included in the material

## Questions

1. Fatty acids shown in the previous table are all long chain. Propose an explanation.
2. Morphine and heroin are both accounted for as morphine in the chemical analyses shown in the previous figure. Propose an explanation.

---

[12] (2019) Autopsy reveals Sidmouth fatberg's dirty secrets. University of Exeter news. https://www.exeter.ac.uk/news/research/title_756108_en.html.

3. Lipid-driven clogging of arteries is the main cause of stroke. Bridge the findings in this case with the general biochemical determinants of atherosclerosis and stroke.
4. Many toxins accumulate in the adipose tissue of humans and other animals. Bridge the findings in this case with the general biochemical determinants of toxicology associated with fast body mass losing diets ("yo-yo diets").
5. Why is not adipose tissue commonly used in clinical biochemistry to search for toxins? Many toxins also accumulate in hair, but hair is also not commonly used to probe for toxin poisoning or drug abuse. Why?
6. Cholesterol is not part of the molecules found in fatberg. Propose an explanation.
7. Check which molecules detected in the chemical analyses shown in the previous figure can be assigned to drugs or cosmetic/hygiene products. Formulate your own opinion about illegal, prescription, and nonprescription ("over-the-counter") drug abuse.

## Biochemical Insights

1. Long-chain fatty acids have stronger adhesion forces among them and pack with each other easily. Their solubility is lower and they tend to have higher melting temperatures. In addition, fatty acid chains having 14–20 carbons are the most common in nature, therefore most abundant in food. Altogether, these factors (low solubility, high melting temperature, and abundance) concur to cause massive deposition of long-chain lipids in fatbergs. An additional important factor contributes to the compact nature of fatbergs: saponification. Ionized forms of fatty acids bind counterions forming salts that have the consistency of soaps. This soapy matter forms trapping other molecules inside.
2. Morphine and heroin have very similar chemical structures (see next figure). Depending on the analytical technique used to discriminate them, the separation between them may be total or not. If separation is not wide enough, it may not be possible to quantify them individually. This probably occurred in the reported study. Curiously, codeine, which is also very similar to morphine, could be discriminated. Small details in chemical structure sometimes dictate different outcomes for analytical procedures.

morphine                          codeine                          heroin

Chemical structures of morphine, codeine, and heroin. Morphine is an alcohol, codeine is an ether, and heroin is an ester, but the global structure is very similar

3. Atherosclerosis is addressed in Sect. 3.1. It is a pathological condition in which deposition of lipid-rich structures (atheromatous plaques) deposit in the walls of arteria. Surface fixation of lipid-rich layers facilitates deposition of additional layers; a potential clogging site slowly forms. In the myocardium, this condition may cause a heart attack (myocardial infarction); in the brain, it may cause stroke. Both are very serious life-threatening conditions. Clogging halts blood irrigation to tissues. Without oxygen, the myocardium (see next figure) and brain cells die within minutes. Stroke and myocardial infarction are both medical emergencies. Early clinical action can minimize heart and brain damage and potential complications.

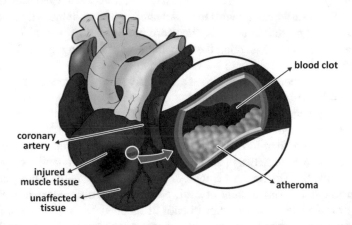

Clogging of a heart artery with lipid-rich materials may cause total interruption of blood flow and consequent supply of oxygen and nutrients to tissues. Cells die forming a lesion in the heart's muscle. Cellular contents are released in extracellular space diffusing to local blood vessels; these molecules serve as identifiers and time course markers of heart failure (Sect. 3.3.4.1)

One important difference between lipid deposition in fatbergs and atheromatous plaques is the presence of cholesterol. Atheromatous plaques are rich in cholesterol, which is present in lipoproteins together with triacylglycerols so also depositing as lipid mixtures.

4. Hydrophobic (non-polar) molecules coexisting with lipids incorporate lipid depots when formed. This is a physically driven process dominated by entropy (see Sect. 3.1). This driving force occurs in lipid assemblies of any kind, such as the adipose tissue, which consists of lipid-loaded cells. Lipid disassembly releases the hydrophobic "contaminants." In the case of adipose tissue, these "contaminants" are released in the bloodstream and distributed. Some of them are toxins and are released in significant amounts in case of severe dieting.

5. Lipophilic molecules accumulate in the adipose tissue for a long time. It is difficult to establish a robust correlation between toxin levels in the adipose and intake doses or time of intake. In contrast, in blood and urine, clearance makes this correlation trustworthy. "Contamination" of hair with toxic molecules or elements has the advantage of the straightforward correlation between height of contamination in hair and time of intake. Constant hair growth makes contaminants trapped in the protein structure of hair (see Sect. 2.2) deviate from the epidermis with time. Yet, a quantitative correlation between abundance in hair and intake doses is faint because the affinity of different "contaminants" to hair components is widely diverse. Moreover, hair protein assemblies (mainly keratin) are hard to untangle to efficiently extract the "contaminants." For these reasons, despite the advantage of simplicity in handling relative to blood or urine, hair is not, in practice, a real alternative matrix in analytical biochemistry.

6. Eventually cholesterol may have not been tested. It may also have been tested but not present in sufficient amounts to be detected by the techniques that were used. Cholesterol is present in animals, therefore not present in vegetable oil such as palm oil or olive oil. Assuming the vast majority of cooking oils is from vegetables, low levels of cholesterol are not surprising,

7. Only dexpanthenol and mephedrone are clearly part of cosmetics. Salicylic acid is formed by metabolic transformation of acetylsalicylic acid ("aspirin") in the body, but it is also an ingredient in topical anti-acne products. Theobromine is one of the products of caffeine processing in the human body and is naturally present in cocoa, therefore also present in chocolate. The relative abundance of pharmaceuticals/cosmetics in the fatberg depends on total quantity disposed and lipophilicity. So, a direct correlation between abundance in the fatberg and levels of consumption is not straightforward. Nevertheless, differences in the lipophilicity of molecules such as dexpanthenol or mephedrone and drugs in Fig. 3.6c cannot account for the huge differences in abundance; therefore, a broad conclusion on levels of consumption from abundance in the fatberg seems reasonable.

## Final Discussion

Following the discussion of this case study, one additional topic is pertinent: the impact of domestic drug waste in the environment. A letter of Michael Depledge to *Nature* journal[13] is quite eloquent:

*Low-cost pharmaceuticals are increasingly accessible to the global population, which is predicted to exceed 8 billion by 2050. Rising drug use is also driven by ageing populations. Widely used preventative medication — such as statins and anti-hypertensives — and cheap generic drugs add to the problem. The UK Office of National Statistics predicts that the country's medicine usage*

---

[13] Depledge M (2011) Reduce drug waste in the environment. Nature 478:36.

*will more than double by 2050. Agricultural soils and rivers are contaminated with a range of pharmaceuticals, including antibiotics, antidepressants, analgesics and cancer-chemotherapy agents.*[14] *The effects are already evident: they include the feminization of fish by residues of the contraceptive pill, and the deaths of millions of vultures on the Indian subcontinent following ingestion of the anti-inflammatory drug diclofenac. Antibiotic overuse has led to the emergence of resistant pathogenic bacteria in the wider environment, and not just in medical settings.*

Expired or unneeded drugs that are flushed unused down the toilet are not the biggest issue. Massive quantities of metabolites resulting from drug metabolization (i.e., chemical processing in the body) are excreted in the urine. Wastewater treatment plants are not always well equipped to deal with all kinds of pharmaceuticals and pharmaceutical by-products present in household sewage.[15]

---

[14] European Environment Agency (2005) EEA technical report No 1. https://www.eea.europa.eu/publications/technical_report_2005_1/file.

[15] Scudellare M (2015) Drugging the environment. The Scientist, p 23–28.

# Part II
# The Interplay and Regulation of Metabolism

# Chapter 4
# Introduction to Metabolism

Cells are made up of molecules, but cells are not simple mixtures or assemblies of molecules. If thrown to a test tube separately or in a random mixture, the whole set of molecules of a cell would interact physically and react chemically, but would not spontaneously form a cell. Cells form and exist because the molecular events that create and maintain them are highly ordered. The sequence and specific place of events and the flows of matter and energy are such that the cell is able to preserve stability and evolve by adaptation to a certain extent, as addressed in Sect. 1.1. In Part II, we will focus on some of the most important chemical reactions occurring in different tissues of the human body and their coordination so that one understands how the changes of matter and transfer of energy enable the human body to exist, move, adapt to external challenges, and reproduce. This ensemble of chemical reactions is called metabolism.

Because the whole metabolism is such a complex array of chemical reactions, biochemists tend to study and concentrate on subsets of reactions separately. The easiest form to understand and explain metabolism is to divide it in parts depending on the chemical nature of the molecules involved, i.e., according to the families of molecules addressed in Part I. So, for practical reasons and for the sake of simplicity, the whole metabolism, which is a single highly complex network of chemical reactions and physical events, is regarded as a sum of "metabolisms": the metabolism of saccharides, the metabolism of amino acids, the metabolism of lipids, etc.

The part of metabolism more directly related to nutrient absorption and ATP production is the "energy metabolism." The human body continuously transforms molecules, sometimes forming higher molar mass products, others breaking down molecules into smaller molar mass entities. Typically, these situations correspond to incorporate mass and energy from nutrients or consuming such mass and energy in the absence of nutrient intake, respectively. The "metabolisms" of the first kind are known as anabolic (construction), the latter being known as catabolic (degrading). So, "amino acid catabolism," for instance, refers to the breakdown of amino acids into smaller molecules, as opposed to amino acid synthesis (amino acid anabolism). Amino acids, in turn, may polymerize and form proteins: protein anabolism, in

© Springer Nature Switzerland AG 2021
A. T. Da Poian, M. A. R. B., *Integrative Human Biochemistry*,
https://doi.org/10.1007/978-3-030-48740-9_4

contrast to the breakdown of proteins, amino acids resulting therefrom (protein catabolism).

Subsets of reactions that are active in both catabolic and anabolic conditions are "amphibolic" (the prefix "amphi" referring to its dual nature). This is the case of the tricarboxylic acid (TCA) cycle, also known as Krebs cycle (see Sect. 7.2). An example of generic metabolism is schematized in Fig. 4.1.

No matter how complex the metabolism may seem at first glance, the interpretation of a scheme of consecutive reactions is simple and depends on the identification of five key factors:

1. The reactants that originate the process (so-called precursors)
2. The end products, regardless of being formed in intermediate reactions or in the final reaction
3. The branching points, i.e., steps where the sequence of reactions may follow different courses
4. The irreversible reactions
5. The specific reactions that are catalyzed by enzymes that are finely regulated and so have the ability to highly accelerate segments of the metabolism, or not

These five key factors in the interpretation of a metabolism will help the reader not to look to a metabolism as a chaotic ensemble of chemicals, enzymes, and reaction arrows. Metabolism is an appealing and richly informative text on the organization of life, should we be able to interpret it.

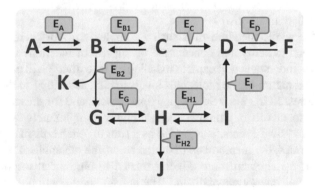

**Fig. 4.1** Hypothetical schematic metabolism involving metabolites A to K and enzymes $E_X$ (subscript X in $E_X$ represents the substrate, X = A to K). The metabolism is "fed" by A and has K, F, and J as "end products." An external source of G may also lead to the formation of J and F, but not K. An external source of D can only lead to the formation of F. However, if $E_D$ is not present or not operative, F will not be formed in any circumstance. Likewise, if $E_{B2}$ is not present or not operative, K, G, H, J, and I will not be formed even in the presence of high concentration of B. If $E_{B1}$ and $E_{B2}$ are never active or inactive at the same time, F is always formed but not J; the reactions' scheme assures the permanent formation of F but the selective formation of J when the control of the reaction course is performed by alternate states of activity of $E_{B1}$ and $E_{B2}$

Enzymes are key protagonists of metabolisms together with the metabolites (i.e., the intermediate reactants/products of the reactions of metabolisms). Not only do enzymes greatly accelerate reactions, but they are also the entities that make metabolisms happen in specific directions. Their importance and action are better illustrated if one considers first a hypothetical metabolism without enzymes (described in Sect. 4.1) and the same set of reactions with enzymes (described in Sect. 4.2).

## 4.1    Consecutive Reactions Without Enzymes

Consider the simple chemical reaction:

$$A + B \rightleftarrows C$$

which in terms of a more amenable representation to biochemists may be written as:

to highlight the conversion of A into C with the intervention of B. Nevertheless, one should not underestimate the importance of B in the balance of energy and mass of the reaction, even if B is $H_2O$ or a small ion, for instance, and A and C are big complex molecules. A, B, and C are all equally important to study the course and velocity of the reaction, i.e., the thermodynamics of the reaction. To better understand the basic principles that govern the chemical reaction, we will start with very simplistic formulations. One will assume that A and B are mixed at a specific instant in time ($t = 0$) and the reaction initiates forming C. As soon as C is formed, it will also start converting to A + B, but in the beginning, only few molecules of C exist, so the velocity of the conversion of A + B into C is larger than the opposite conversion, C to A + B. Naturally, in this condition, the concentration of C will increase until the point in which the velocities of both reactions will match: the velocity with which C is formed equals the velocity with which it is consumed. The reaction takes place because A and B are continuously transformed into C and vice versa, but the concentrations of A, B, and C present in solution do not change. This is the point of equilibrium. At this stage, it is important to dissociate the extent of the reaction (i.e., the fraction of A and B that was transformed to reach the equilibrium) and the velocity of the reaction, which is related to the time needed to reach the equilibrium.

A very extensive reaction is represented by:

$$A + B \rightarrow C$$

and named "irreversible" (therefore written with a one-way arrow), but this tells us absolutely nothing about the velocity of the reaction. In fact, the conversion of A and B into C may be so slow that in practical terms C is not formed. So, to study chemical reactions, one needs to address both the extent and the velocity of reactions. A favorable reaction has a high degree of conversion of reactants in products; a fast reaction reaches the equilibrium in a short time. Both aspects are independent of each other, and so it is not surprising that they can be modulated by separate means in metabolism. The extent of a reaction is described by the equilibrium constant, which in practical terms is calculated from the concentrations at equilibrium:

$$K_{eq} = \frac{[C]_{eq}}{[A]_{eq}[B]_{eq}}$$

It is obvious that unfavorable reactions have low $K_{eq}$ and favorable reactions have high $K_{eq}$. $K_{eq}$ varies from nil to infinity. The determinants that contribute to the value of $K_{eq}$ for a given reaction (i.e., how favorable it is) relate to the balance of both released energy and order of the whole environment in which the reaction takes place. Neither is intuitive, but both can be experimentally demonstrated. Chemical reactions tend to consume more reactants as the products are more stable ("less energetic") than reactants and the whole environment becomes more disordered with the presence of the products (i.e., closer to a random organization). The degree of disorder is generically referred to as entropy. The balance of energy and entropy is so deeply rooted in thermodynamics that they are the subject of the first and second laws of thermodynamics, respectively, as the founding principles of this discipline. The exact quantitative weighting between the balance of released energy as heat (the enthalpy, $\Delta H$) and balance of entropy ($\Delta S$) is provided by the parameter $\Delta G$, the difference of the Gibbs energy during the chemical reaction:

$$\Delta G = \Delta H - T\Delta S$$

Assuming that the temperature, $T$, does not change in the process, $\Delta G$ is related to the concentration of reactants and products:

$$\Delta G = \Delta G^0 + RT \ln \frac{[C]}{[A][B]}$$

In equilibrium $\Delta G = 0$; therefore:

$$\Delta G^0 = -RT \ln K_{eq}$$

While thermodynamics may reach complex formulations to explain the interaction of energy with matter, we shall stick to those most important in medical biochemistry. Previous equations on $\Delta G$ are useful because they provide information about the spontaneity of reactions. If $\Delta G < 0$ (exergonic reaction), the reaction will proceed spontaneously until equilibrium; if $\Delta G > 0$ (endergonic reaction), the reaction will not be spontaneous; if $\Delta G = 0$, the reaction is at equilibrium. Whether $\Delta G$ is positive, negative of nil depends largely on the concentrations of reactants and products, and the reaction will continue until these concentrations reach equilibrium ($\Delta G = 0$). $\Delta G^0$ is related to the concentrations at equilibrium specifically, and so provides information on the final extent of reaction (i.e., the extent at the equilibrium point):

$\Delta G^0 \ll 0$—highly extensive reaction (high $K_{eq}$)
$\Delta G^0 \gg 0$—low extensive reactions (low $K_{eq}$)

Box 4.1 illustrates these concepts for a simple reaction.

---

**Box 4.1  The Basic Thermodynamics of the Simplest Reaction**
In the very simple case of one single reactant (R) being transformed in one single product (P):

$$R \rightleftharpoons P$$

the equilibrium constant is:

$$K_{eq} = \frac{[P]_{eq}}{[R]_{eq}}$$

and

$$\Delta G^0 = -RT \ln K_{eq} = -RT \ln \frac{[P]_{eq}}{[R]_{eq}}$$

It is obvious that:

$$\Delta G^0 < 0 \text{ if } [P]_{eq} > [R]_{eq} \ (K_{eq} > 1)$$

$$\Delta G^0 > 0 \text{ if } [P]_{eq} < [R]_{eq} \ (K_{eq} < 1)$$

The reactions with highest $K_{eq}$ have the most negative $\Delta G^0$. If the equilibrium is not yet attained, two situations are possible:

1.
$$[R] > [R]_{eq} \ (\text{i.e., } [P] < [P]_{eq})$$

$$\Delta G = \Delta G^0 + RT \ln \frac{[P]}{[R]} = +RT \ln \left( \frac{[P][R]_{eq}}{[R][P]_{eq}} \right)$$

(continued)

**Box 4.1** (continued)

which is obviously negative. The reactions will proceed with transformation of R in P until equilibrium, i.e., $\Delta G < 0$ implies spontaneous transformation of reactants in products.

2.                                          $[R] < [R]_{eq}$ (i.e., $[P] > [P]_{eq}$)

This is the opposite case, in which $\Delta G > 0$ and the spontaneous process is the reverse reaction, i.e., the transformation of the product, P, in the reagent R.

The relationship between $\Delta G$, $\Delta G^0$, and the course of reaction is represented in the figure below.

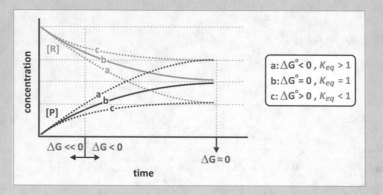

Course of the reaction R $\rightleftarrows$ P from the initial instant, in which $[P] = 0$. When $\Delta G^0 < 0$, the equilibrium is reached with P having higher concentration than R (situation a). When $\Delta G^0 > 0$, the opposite occurs (situation c). For $\Delta G^0 = 0$, R and P reach the same concentration in equilibrium (situation b)

Figure 4.2 shows two examples of how $\Delta H$ may dominate over $\Delta S$ and vice versa and be determinant for the extent of the reaction, i.e., for $\Delta G$. However, in most biological processes related to the use of nutrients, $\Delta H$ is large compared to $T\Delta S$, which makes $\Delta H$, the caloric "value" of food, an approximate measure of the total energy that can be used by the human body.

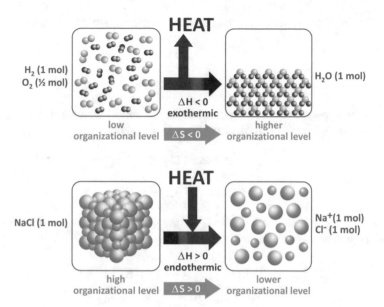

**Fig. 4.2** Role of $\Delta H$ and $\Delta S$ and molecular processes. Chemical and physical processes occur because the released energy, $\Delta H$, compensates for the decrease in $\Delta S$, or the increase in disorder compensates for the heat consumption

Isolated reactions such as the ones considered before are rarely relevant in metabolism as metabolism consists of consecutive reactions. So, let us now consider the reaction:

$$C \rightleftarrows D + E$$

coupled to the first reaction, $A + B \rightleftarrows C$, to form:

$$A + B \rightleftarrows C \rightleftarrows D + E$$

At equilibrium, it is obvious that the concentrations of A and B are related to the concentrations of D and E because C takes part in both reactions. In terms of equilibrium constants:

$$K_{eq,1} = \frac{[C]_{eq}}{[A]_{eq}[B]_{eq}}$$

$$K_{eq,2} = \frac{[D]_{eq}[E]_{eq}}{[C]_{eq}}$$

(1 and 2 refer to the first and second reactions in the sequence). Therefore,

$$K_{eq,1} K_{eq,2} = \frac{[D]_{eq} [E]_{eq}}{[A]_{eq} [B]_{eq}}$$

which is the apparent equilibrium constant of the abbreviated form ("sum of reactions") of the reactions $A + B \rightleftarrows C$ and $C \rightleftarrows D + E$ above:

$$A + B \rightleftarrows D + E$$

$$K_{eq,app,1,2} = K_{eq,1} \cdot K_{eq,2}$$

The amazing implication is that $\Delta G^0$ of the global process is simply the sum of $\Delta G$ of the two consecutive reactions:

$$\Delta G^0_{eq,1,2} = -RT \ln K_{eq,app,1,2} = -RT \ln \left( K_{eq,1} K_{eq,2} \right)$$
$$= -RT \ln K_{eq,1} - RT \ln K_{eq,2} = \Delta G^0_{eq,1} + \Delta G^0_{eq,2}$$

The consequences of $\Delta G^0_{eq,1,2} = \Delta G^0_{eq,1} + \Delta G^0_{eq,2}$ to metabolism are immense because it is implied that the conversion of C in D + E may be extremely unfavorable, but the extent to which A + B is converted to C "pushes" the reaction toward the formation of D + E through the increase in the concentration of C. This can be generalized to any set of consecutive reactions, and metabolisms are generally composed of favorable and unfavorable reactions influencing each other. In the extreme case in which an irreversible reaction follows one or more reversible reactions, such as:

$$A + B \rightleftarrows C \rightleftarrows D + E \rightarrow F$$

it is obvious that regardless of how unfavorable the first and second reactions may be, the end result is the total depletion of A and/or B and the formation of F. The same would happen in

$$A + B \rightleftarrows C \rightleftarrows D + E \quad \begin{array}{l} \nearrow F \\ \rightleftarrows G \end{array}$$

but not in

$$A + B \rightleftarrows C \rightleftarrows D + E \quad \begin{array}{l} \nearrow F \\ \searrow G \end{array}$$

in which the end products would be both F and G. Although both the formation of F and G are irreversible, they are competing reactions. Although they both occur to the total depletion of at least one of the reagents, the product of the fastest reaction dominates. One of the reactions may be much faster than the other, in which case only one product forms in practice.

In a branched chain of reactions with an irreversible conversion, for instance,

$$A + B \rightleftarrows C \rightleftarrows D + E \quad \begin{array}{l} \nearrow \quad F \rightleftarrows J + K \rightleftarrows L \\ \searrow \quad G \rightleftarrows H \rightleftarrows I \end{array}$$

the final outcome critically depends on the irreversible steps and the branching points. Thus, the result may be complex.

Given the importance of the velocity of reactions to the course of chain reactions, we shall now address kinetics in more detail as $\Delta G^0$ only accounts for how far the reaction goes in the degree of conversion of reactants into products, not kinetics.

## 4.2 Consecutive Reactions with Enzymes

The chain of reactions above is dominated by the irreversibility of reactions $D + E \rightarrow F$ and $D + E \rightarrow G$ ($\Delta G^0 \ll 0$ for both). If the latter, for instance, is much faster than the first, the metabolites G, H, and L will dominate. In the opposite case, F, J, K, and I will dominate.

Imagine now that (1) E is a key reactant of high reactivity that makes most reactions it enters irreversible ($\Delta G^0 \ll 0$—in nature ATP fulfills this role, which makes metabolism critically dependent on this molecule, Box 4.2); and (2) both reactions are catalyzed by enzymes. The combination of these two factors has high impact in the variation of the concentrations of the metabolites over time (see an example in Fig. 4.3). Importantly, the concentration of E and the control of the activity of the enzymes are ways to control the course of reactions.

The following typical situations are possible:

1. If E is depleted by a side reaction, low "level" (concentration) of E results in "accumulation" (increased concentration) of D.
2. A highly active catalysis of $D + E \rightarrow F$ with simultaneous absence of an enzyme to catalyze $D + E \rightarrow G$, or the presence of an enzyme that is inhibited, results in the production of F at high rate. If $F \rightleftarrows J + K$ is not catalyzed, a transient accumulation of F occurs.

**Box 4.2 ATP: The Quasi-universal Driver of Metabolism**
ATP is a fascinating molecule for its ubiquity. ATP hydrolysis into $ADP + PO_4^{3-}$ or $AMP + PO_3^{2-}$-$O$-$PO_3^{2-}$ has $\Delta G^0 \ll 0$, and these reactions are present in several metabolisms, coupled to unfavorable reactions. Any chemical bond splitting involves expenditure of energy and ATP is no exception. The idea that ATP has "high-energy bonds" between phosphoryl groups whose energy is liberated and used to force the occurrence of unfavorable reactions is

(continued)

**Box 4.2** (continued)

misleading albeit widely disseminated. The complete reaction of ATP hydrolysis is:

$$ATP + 3H_2O \rightarrow ADP + PO_4^{3-} + 2H_3O^+$$

but the intervention of water is frequently overlooked. The net balance of the bonds disrupted and created in this reaction is such that energy is liberated yet the chemical bonds themselves are not energy depots that release energy when broken. It is the balance of bond dynamics involved in the reaction that matters. Part of this misunderstanding comes from the fact that one tends to oversimplify the writing of chemical reactions. When written like:

$$ATP \rightarrow ADP + P$$

the involvement of water is implicit but frequently overlooked. Only one bond fission is shown, leaving the idea that the released energy is contained in the chemical bond itself, which is illusory.

See the structure of ATP in the next figure (also previously shown in Fig. 3.22). At first glance, it may seem intriguing why the hydrolysis of ATP into ADP or AMP involves different $\Delta G^0$. This happens because the phosphoryl groups are not equivalent. The electronic distribution in the molecule, for instance, is different.

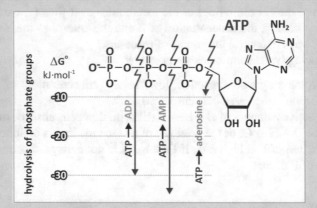

Different $\Delta G^0$ values for the hydrolysis of the phosphate groups in ATP

Having a common highly endergonic hydrolysis reaction to couple to other reactions is a big advantage for cells as only one molecule needs to be synthesized to drive metabolism, and the complexity and diversity of enzymes

(continued)

**Box 4.2** (continued)

involved are very much decreased compared to a situation where many and diverse endergonic reactions would be used. To power metabolism, the cells use almost exclusively ATP and enzymes that catalyze its hydrolysis, coupling it to a simultaneous unfavorable reaction. Assuring that ATP is always present and never completely depleted is vital for the cell.

Specific processes that need to be driven thermodynamically by the use of endergonic reactions but should not be dependent on ATP make use of similar molecules such as GTP or UTP.

It is interesting that hydrolysis of ATP $\rightarrow$ ADP, ATP $\rightarrow$ AMP, and $P_2O_7^{4-} \rightarrow 2PO_4^{3-}$ involve $\Delta G^0 \ll 0$ in any case. Nevertheless, the frequency of ATP $\rightarrow$ ADP over the others in metabolic processes is much higher. The eventual advantage may have been that, in parallel to the dedicated mechanisms cells have to convert ADP back to ATP using energy from nutrients, ATP can be readily obtained from ADP + ADP $\rightarrow$ ATP + AMP (see Sect. 8.4). The appearance of AMP may signal a state of ATP depletion that triggers regulatory events in the cell that favor ATP synthesis. This matter is still controversial and will be readdressed in Chap. 8 in connection to the use of ATP for muscle contraction.

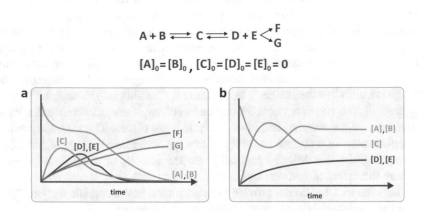

**Fig. 4.3** Evolution of the concentration of metabolites in a hypothetical reaction. Depending on the reversibility of reactions and their velocity, the concentration of the metabolites following the addition of A and B at equal concentrations changes dramatically over time, in case reactions D + E $\rightarrow$ F and D + E $\rightarrow$ G are occurring (**a**) or not (**b**). In Panel (**a**), the velocity of the formation of F is assumed to surpass the velocity of the formation of G

In most metabolic sequence of reactions ("pathways"), all reactions are catalyzed by enzymes, which generates a situation in which the concentration of metabolites may change very rapidly in high amplitude intervals, depending on the thermodynamics of the reactions involved, the enzymes present, and the eventual degree to which their activity is affected by other molecules or physical conditions such as temperature.

## 4.2.1   The Bases of Enzymatic Catalysis and Its Impact in Metabolism

As addressed in previous sections and Box 4.2, ATP or any related nucleotide triphosphate is a quasi-universal driver of metabolism for its hydrolytic dephosphorylation with $\Delta G^0 \ll 0$. Coupling this reaction to others in a metabolic pathway grants the ability to perform reactions that would otherwise occur to a very low level. As long as ATP is available in the cell, the process is assured. This simplicity contrasts with the apparent complexity of the control of enzyme activity in the cellular environment. We say apparent because this complexity is founded on simple basic principles and is not difficult to understand. The factors affecting enzyme activity are basically the ones affecting protein structure, as the essence of enzyme catalysis is the conformational dynamics of the protein with the bound substrate. Environmental factors affecting catalysis, such as pH or temperature, are not very important in regulating the activity of enzymes because most organelles in cells have buffered pH and constant temperature. The activity of enzymes changes depending on these conditions, but it is not feasible for the cells to change pH or temperature to influence the course of metabolic reactions. So, enzymes have a structure that fits optimal activity at specific temperature and pH ranges (recall the pH of different human tissues in Fig. 2.8), and their activity depends most on ligands, called effectors, that can improve activity ("activators") or decrease it ("inhibitors"). The exact mechanism these ligands use to act as activators or inhibitors of enzymes depends a lot on the structure of the enzyme and how catalysis takes places. The thermodynamics and performance of enzymes as catalysts can be studied under simple but rather abstract principles that describe generally how the velocity of catalysis is independent from the extent of reaction and how it changes with the concentration of substrate and effectors, but do not provide information on how a specific enzyme binds the substrate and distorts its electronic structure to facilitate catalysis (see Sect. 3.3.4).

The simplest thermodynamic description of catalysis is explained in Box 4.3. It illustrates that the velocity of reactions is independent of $\Delta G^0$. The simplest formulation for an enzymatic reaction was devised mainly by Leonor Michaelis and Maud Menten (Box 4.4). In their abstract model of catalysis, the generic enzyme, $E_z$, would bind the generic substrate, S, to form a complex of undetermined nature from which the generic product, P, was formed irreversibly:

$$E_z + S \rightleftarrows E_z S \rightarrow E_z + P$$

The mathematical deduction in Box 4.4 shows that in this case the velocity of the reaction (i.e., the variation in [P] per time unit) is:

$$v = \frac{V_{max}[S]}{K_M + [S]}$$

in which $V_{max}$ is the maximal possible velocity for the reaction at a given concentration of enzyme and $K_M$ is the dissociation constant of $E_zS$ (if the first reaction reaches equilibrium). $K_M$ is named the Michaelis constant and $K_M^{-1}$ reflects the extent to which $E_z$ binds S, which is usually referred to as the "affinity" of $E_z$ toward S, an intuitive but rather ambiguous concept.

The irreversible step makes the model simpler, but in many cases, the second step is reversible and the same enzyme can catalyze a reaction both ways, from S to P and vice versa. In the excess of S, the dominant reaction is S → P, and when P accumulates, P → S dominates.

---

**Box 4.3 The Thermodynamics of a Simple Catalyzed Reaction**

Isomerization is a simple process that we will use to illustrate the thermodynamics associated with the kinetics of reactions and the effect of enzymes on both. *Cis-trans* isomerization, for instance, can be represented by a simplified scheme:

in which $R_1$ and $R_2$ are any group of the molecule different from H. Depending on the temperature, solvent, and exact chemical nature of $R_1$ and $R_2$, the conformational dynamics of these molecules varies, but there is always a certain degree of flexibility that makes some molecules more distorted than others at any given instant. Some of these molecules acquire a structure that is intermediate between the *cis* and the *trans* isomer, $R_1$, $R_2$, and the H being located in unstable positions that make the molecule more energetic. Higher-energy states associated with intermediary conformations are very unstable, and few molecules acquire these conformations, although it may happen occasionally. Plotting a point for each molecule in an energy (of the total molecule) vs. degree of structural conversion plot, one would obtain at a given instant, in equilibrium, the distribution depicted in the following

(continued)

**Box 4.3** (continued)

figure. Many molecules are in the *cis* and *trans* states because they correspond to energy minima, but few molecules reach conformations in between these states because these conformations correspond to electronic distributions and nuclei localizations that are not optimal in the balance of charge distributions. The difference between the energy of the reactant (*cis* isomer in this case) and the energy associated with the molecule in the most unlikely state (energy maximum) is the so-called activation energy because evolution from this point to the products corresponds to loss of energy (decrease in $\Delta G$), i.e., the molecule spontaneously progresses to the product (*trans* isomer) after this point.

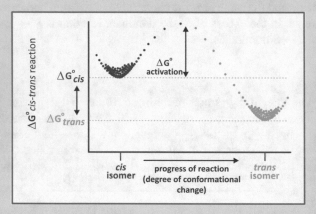

Energy of each molecule according to its degree of conversion between *cis* and *trans* conformations. Each point represents a molecule. This is a hypothetical distribution of a population of molecules in equilibrium at a certain given instant in time. Most molecules are in the *cis* or *trans* isomers conformation, which correspond to local minima of energy. The intermediate conformations correspond to molecules in high-energy states, therefore less populated. In practice, this implies that very few molecules spontaneously convert between conformations

When the activation energy is high, it is very rare that a molecule acquires the energy necessary to adopt the intermediate state that enables spontaneous conversion to products. In other words, the reaction progresses very slowly. Enzymes bind to substrates and distort the structure of the molecules in a way that the electronic distribution and nuclei interactions are more favorable. The energy of activation assigned to the reactant is decreased, and the consequence is that more molecules reach the maxima of energy and progress to products. This means that the velocity of the reaction increases (see below figure).

(continued)

**Box 4.3** (continued)

$\Delta G^{\circ}_{cis}$ and $\Delta G^{\circ}_{trans}$ remain unchanged, which means that the catalyzed reactions reach a state of equilibrium faster, but the extent of reaction (fraction of reactants converted to products) is not altered by the action of the enzyme.

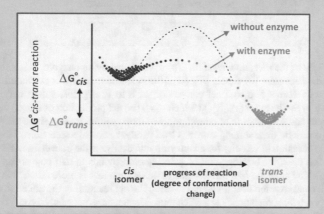

Effect of an enzyme in the molecules having an intermediate conformation between *cis* and *trans*. Enzymes distort the molecular structure of substrates turning the intermediary conformations not so energetic, therefore more likely. Because more molecules reach the intermediary conformations, a higher velocity of product formation is achieved

Box 4.4 shows that in a reversible reaction, the velocity of reaction is dependent on the relative concentrations of S and P. When the concentration of P reaches certain critical points, the velocity of the conversion of P to S surpasses the velocity of the opposite process, and the net result is the consumption of P and production of S (Fig. 4.4). This is very important in metabolism as most enzymes catalyze reactions in both directions and the dominant direction depends on the relative concentrations of the metabolites.

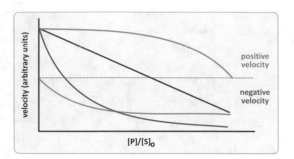

**Fig. 4.4** Velocity of the reversible reaction S ⇌ P catalyzed by an enzyme. $[S]_0$ is the initial concentration of the substrate; $[P]$ equals $[S]_0$ ($[P]/[S]_0 = 1$) when all S was converted to P. Positive values of velocity correspond to a net conversion of S to P; negative values correspond to net conversion of P to S. The exact $[P]/[S]_0$ at which $v = 0$ (no net production of P or S) depends on the specific enzymes and substrates to be considered, but the transition from positive to negative values of velocity is a generic trend among enzymes that catalyze reversible reactions. The *red line* corresponds to the variation in velocity for an enzyme that catalyzes the conversion of S to P and P to S equally. In this case, the turning point of velocity corresponds to half conversion of substrates ($[P]/[S]_0 = 0.5$). The *magenta line* corresponds to an enzyme that is more efficient in converting P into S: a small concentration of P is enough to drive the net reaction in the direction of converting P into S. The *green line* corresponds to a Michaelis–Menten enzyme: catalysis only converts S into P (always positive velocities) with maximal velocity for maximal concentration of $[S]$ ($[P] = 0$). The *blue line* corresponds to a Michaelis–Menten enzyme that only converts P into S: the velocity is always negative and maximal for maximal concentrations of P ($[S] = 0$)

---

**Box 4.4  Fundamentals of Enzyme Kinetics of Simple Reactions**

Enzymes ($E_z$) associate with substrates (S) through specific amino acid residues that distort the substrate and make the conversion to products (P) occurs faster. The enzyme interacts actively with the S molecule in the course of catalysis, but the end of reaction results in the release of P with $E_z$ being in the same state as before binding S. In abstract, this situation can be represented by:

$$E_z + S \underset{k_{-1}}{\overset{k_1}{\rightleftharpoons}} E_z S \underset{k_{-2}}{\overset{k_2}{\rightleftharpoons}} E_z + P \qquad (4.1)$$

$E_z S$ is the transitory complex in which $E_z$ and S are in contact, S being distorted. $k_x$ are the kinetic rate constants of each conversion ($x = 1, 2, -1, -2$), which are proportionality factors, between velocity and concentration of reactants. In broad terms, it may be said that $k_x$ is the intrinsic propensity of step $x$ to occur (e.g., $k_{-1}$ is the intrinsic propensity for the fast unbinding of $E_z$ and S after $E_z S$ is formed, while $k_2$ is the intrinsic propensity for the fast release of P after $E_z S$ is formed; once $E_z S$ is formed, the processes to which $k_{-1}$ and $k_2$ are associated compete with each other). The velocities of each of the four events of the reaction are:

(continued)

**Box 4.4** (continued)

$$v_1 = k_1 [E_z][S], \quad v_{-1} = k_{-1}[E_zS], \quad v_2 = k_2[E_zS], \quad v_{-2} = k_{-2}[E_z][P] \quad (4.2)$$

Assuming that the reaction proceeds in a steady-state condition, the concentration of $E_zS$ does not change with time, i.e., the velocity of creation of $E_zS$ equals the velocity of consumption of $E_zS$:

$$v_1 + v_{-2} = v_2 + v_{-1} \quad (4.3)$$

$$\left(k_1[S] + k_{-2}[P]\right)[E_z] = \left(k_{-1} + k_2\right)[E_zS] \quad (4.4)$$

$[E_z]$ and $[E_zS]$ are not practical variables to work with as it is the concentration of total enzyme ($[E_z]_0 = [E_zS] + [E_z]$) that is known or more readily measurable in an experiment. At the same time, the velocity of the reaction is the velocity at which P is produced, $v$:

$$v = k_2[E_zS] - k_{-2}[E_z][P] \quad (4.5)$$

Combining all equations, it is deduced that the velocity of product creation is:

$$v = [E_z]_0 \frac{k_2 k_1 [S] - k_{-2} k_{-1}[P]}{k_{-1} + k_2 + k_1[S] + k_{-2}[P]} \quad (4.6)$$

This is an interesting equation that describes how the velocity of a catalyzed reaction varies with the total concentration of enzyme, the concentration of the substrate, and the concentration of the product. It also describes how the velocity changes depending on the intrinsic kinetic properties of each step ($k_1, k_2, k_{-1}, k_{-2}$).

Regardless of the particular [S], [P], or kinetic constants, the equation shows that $v$ is always proportional to $[E_z]_0$. The total concentration of enzyme has a direct impact on the velocity of the reaction. In fact, in cells, increasing or diminishing the expression of enzymes is a way to directly interfere with the rate of critical steps of metabolism.

It is important to further explore the equation in particular situations that may be of significance:

(continued)

**Box 4.4** (continued)

1. [P] ≈ 0 and/or $k_{-2}$ ≈ 0 or any other conditions in which $k_{-2}[Ez][P] \ll k_2[EzS]$ (Michaelis–Menten condition)
   In this case, the reaction simplifies to:

$$E_z + S \underset{k_{-1}}{\overset{k_1}{\rightleftarrows}} E_z S \overset{k_2}{\rightarrow} E_z + P \tag{4.7}$$

and $v$ simplifies to ([P] ≈ 0 and/or $k_{-2}$ ≈ 0 in Eq. (4.6) of this box):

$$v = \frac{k_2[E_z]_0[S]}{\dfrac{k_{-1} + k_2}{k_1} + [S]} \tag{4.8}$$

$$v = \frac{V_{max}[S]}{K_M + [S]} \tag{4.9}$$

$k_2[E_z]_0$ corresponds to the velocity of reaction when all the enzymes are binding S, which is the maximal possible velocity, $V_{max}$; $(k_{-1} + k_2)/k_1$ corresponds to the equilibrium constant of the dissociation of $E_zS$ and is named Michaelis constant, $K_M$. As $K_M^{-1}$ relates to the extent of the association between $E_z$ and S in equilibrium, it is frequently taken as a measure of the enzyme–substrate affinity and is very important in biochemistry (Box 4.5).

Michaelis–Menten conditions lead to a dependence of $v$ on [S] that is hyperbolic (see below figure). Although corresponding to a very particular condition of a very simple reaction scheme, the Michaelis–Menten equation (Eq. 4.9) has incredible historical importance, and most enzymes having hyperbolic-like kinetics are described by apparent $K_M$ and $V_{max}$ even though the reaction scheme they follow is not always exactly the one in Eq. (4.7).

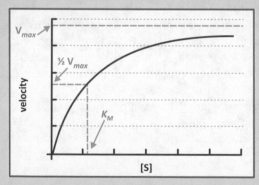

Dependence of $v$ on [S], as predicted by Eq. (4.9) (Michaelis–Menten equation). This dependence is a particular case of a rectangular hyperbole (i.e., having asymptotes perpendicular to $x$ and $y$ axes; rectangular hyperbola may be expressed as $(x - h)(y - t) = m$, in which $h$, $t$, and $m$ are constants that can be rearranged to the format of a Michaelis–Menten equation for particular cases of $h$, $t$, and $m$

(continued)

**Box 4.4** (continued)

2. $[S] \approx 0$ and/or $k_1 \approx 0$ or any other condition in which $k_1[Ez][S] \ll k_{-1}[EzS]$

This is a situation similar to the one of Michaelis–Menten but in the reversed sense:

$$E_z + S \xleftarrow{k_2}[k_{-1}] E_z S \underset{k_{-2}}{\overset{k_2}{\rightleftharpoons}} E_z + P \tag{4.10}$$

and the velocity dependence on [P] is similar, as shown in the next figure ([S] = 0 or $k_1 = 0$ in Eq. 4.6):

$$v = \frac{V_{-max}[P]}{K_{-M} + [P]} \tag{4.11}$$

(The subscripts "−" are used to refer the inverse sense of the reaction, i.e., negative velocities)

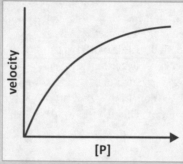

Dependence of $v$ on [P], as predicted by Eq. (4.11)

3. Intermediate values of [S] and [P]

Because we are working with an example in which the conversion between S and P follows a 1:1 stoichiometry, one may consider a situation where the reaction starts with S being at concentration $[S]_0$ in the absence of P:

$$[S] = [S]_0 - [P] \tag{4.12}$$

Combining Eqs. (4.12) and (4.6):

$$v = [E_z]_0 \frac{k_2 k_1 - (k_2 k_1 + k_{-2} k_{-1})\dfrac{[P]}{[S]_0}}{\dfrac{k_2 + k_{-1}}{[S]_0} + k_1 + (k_{-2} - k_1)\dfrac{[P]}{[S]_0}} \tag{4.13}$$

(continued)

**Box 4.4** (continued)

For a given initial concentration of S, the velocity of reaction changes as P is being produced, as depicted in the figure below. The exact curves depend on the specific values of $k_1$, $k_2$, $k_{-1}$, and $k_{-2}$ for the reactions, but the pictorial examples in the figure show that when [S] dominates over [P] ([P]/[S]$_0 \approx 0$), the reaction proceeds with net transformation of S into P with a velocity proportional to [S]. When [P] dominates over [S] ([P]/[S]$_0 \approx 1$), the reaction proceeds with net conversion of P to S with a velocity that is proportional to [P]. The negative values of $v$ in the graph reflect the net conversion of P into S (i.e., an overall course of reaction in the reverse direction). The [P]/[S]$_0$ point at which $v = 0$ (i.e., the velocity of formation of P and consumption of P is equal) is obtained setting $v = 0$ in Eq. (4.13):

$$\left( \frac{[P]}{[S]_0} \right)_{v=0} = \frac{k_2 k_1}{k_2 k_1 + k_{-2} k_{-1}}$$

(4.14)

This shows that the shifting point at which the enzyme action changes the sense of the reaction is in fact a balance between the conjugated parameters of the forward and reverse sense steps.

The practical implication of Eq. (4.13) is immense. It shows that enzymes catalyzing simple reaction described by Eq. (4.1) are able to increase the velocity of the reaction in both senses depending on if S or P is accumulating, always favoring the consumption of the metabolite (S or P) that is accumulating (see next figure). This is vital to understand metabolic regulation.

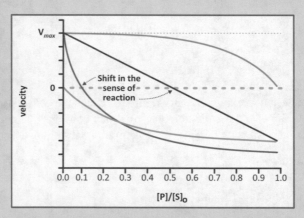

Reversible catalysis. The velocity of a reversible reaction catalyzed by a single enzyme in both senses shifts from positive (net consumption of substrate and production of P) to negative (net consumption of P and production of the substrate) depending on the balance of the kinetic constants involved and the concentration of the substrate and product. The shifting point ($v = 0$) is attained when [P] relative to [S]$_0$ is $\dfrac{[P]}{[S]_0} = \dfrac{k_2 k_1}{k_2 k_1 + k_{-2} k_{-1}}$ . The *green line* corresponds to $k_{-2} = 0$ (Michaelis–Menten condition), $k_1 = k_{-1} = k_2$; the *blue line* corresponds to $k_1 = 0$, $k_{-1} = k_{-2} = k_2$; the *red line* corresponds to $k_1 = k_{-1} = k_{-2} = k_2$; the *magenta line* corresponds to $k_1 = k_{-1} = k_2 = k_{-2}/10$

It is important to stress that a certain enzyme may catalyze the conversion of S to P irreversibly, and a second enzyme may catalyze the inverse reaction also irreversibly. In this way, S and P are interconvertible, but two enzymes are involved. Moreover, third molecules may be involved in one of the directions only. This is the case of:

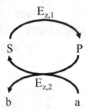

where two different enzymes are involved, and the conversion of P to S may also imply conversion of a to b, which may have $\Delta G^0 \ll 0$ to make this particular reaction more favorable. It should be noted that with this reaction scheme, the conversion of S to P may be blocked (absent or inactive $E_{z1}$) without interference in conversion of P to S, or even while $E_{z2}$ is activated. When the same enzyme catalyzes S $\rightleftarrows$ P, this possibility does not exist; this enzyme may be activated or inhibited, although both directions will be affected. In the reaction above catalyzed by $E_{z,1}$ and $E_{z,2}$, a fine-tuning of the direction of reaction is possible but in S $\rightleftarrows$ P is not. Returning to the last metabolic scheme in Sect. 4.1, now modified to include enzyme-catalyzed reactions:

$$A + B \underset{E_{z1}}{\overset{E_{z2}}{\rightleftarrows}} C \overset{E_{z2}}{\rightleftarrows} D + E \quad \begin{array}{l} E_{z3} \curvearrowright F \overset{E_{z9}}{\rightleftarrows} J + K \overset{E_{z10}}{\rightleftarrows} L \\ \curvearrowleft E_{z4} \\ \curvearrowright E_{z5} \\ E_{z6} \curvearrowleft G \overset{E_{z7}}{\rightleftarrows} H \overset{E_{z8}}{\rightleftarrows} I \end{array}$$

It is clear that $E_{z3}$, $E_{z4}$, $E_{z5}$, and $E_{z6}$ are the ones that determine the direction of the flux of the metabolism. There are mechanisms that prevent cells from having $E_{z5}$ and $E_{z6}$, and $E_{z3}$ and $E_{z4}$ simultaneously activated, so that depending exclusively on which of these key enzymes are active or inactive, this metabolic pathway may be producing L or I, or both, at the expense of consuming A and B, or it can be producing A and B at the expense of consuming I or L, or both. This is the key concept of metabolic regulation. The mechanisms used in cells to activate or inhibit enzymes are then an issue of critical importance and will be addressed in Chap. 5.

An additional factor improves the metabolic processes further. The velocities of reactions catalyzed by key enzymes such as $E_{z3}$, $E_{z4}$, $E_{z5}$, or $E_{z6}$ usually do not follow equation $v = V_{max}[S]/(K_M + [S])$. Whereas enzymes following Michaelis–Menten-like kinetics (see Box 4.4) increase the velocity of reactions proportional to the substrate concentration when [S] is low, these other enzymes change the velocity of reaction abruptly around critical concentrations of the substrate. Velocity varies sigmoidally in response to [S] (Fig. 4.5), in the same way oxygen fixation efficiency by hemoglobin depends on oxygen concentration. This means that some reactions are triggered at high velocity only when the substrate accumulates to a certain level.

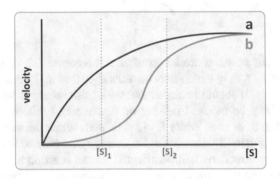

**Fig. 4.5** Two types of dependencies of the velocities of reactions catalyzed by enzymes on the concentration of substrate. Enzymes following Michaelis–Menten kinetics (a) have a hyperbolic-like dependence of velocity on [S]. The initial trend (up to $[S]_1$) is linear. In this range, accumulation of the substrate is counteracted by a proportional increase in the velocity of its conversion to product. Enzymes with sigmoidal kinetics (b) have a lag initial regime in which the velocity is low until [S] reaches a critical value, $[S]_1$, after which the velocity increases abruptly. In both cases (a and b), high substrate concentrations ($[S] > [S]_2$) correspond to almost maximal velocity of the catalyzed reaction. In this situation, the enzymes do not modulate velocity in response to substrate concentration because they are already working at maximal possible capacity

The importance of $K_M$ as the critical point of [S] under which the velocity of catalysis changes abruptly is addressed in detail in Box 4.5.

**Box 4.5 The Importance of $K_M$, the Michaelis Constant**
In strict terms, $K_M$ is only valid for an enzyme that catalyzes a reaction described by:

$$E_z + S \underset{k_{-1}}{\overset{k_1}{\rightleftharpoons}} E_z S \overset{k_2}{\rightarrow} E_z + P \qquad (4.15)$$

as explained in Box 4.4.

(continued)

**Box 4.5** (continued)

In practice, most reactions cannot be described this way. For instance, reactions involving more than one substrate do not fit in this scheme. Yet $K_M$ is a combination of all kinetic constants ($(k_2 + k_{-1})/k_1$) that reflects the extent of dissociation of $E_zS$ when near-equilibrium conditions are reached ($k_2 \ll k_{-1}$). Thus, $K_M^{-1}$ reflects the "affinity" of S for $E_z$ and is an extremely valuable tool to compare the "preference" of the same enzyme for different substrates. For this reason, $K_M$ acquired an extreme importance in the study of enzymes from the early days of enzymology, the discipline that is devoted to the study of the enzymes. Even enzymes that did not follow reaction scheme 1 but had hyperbolic-like dependences of catalytic velocity vs. [S] had apparent $K_M^{-1}$ assigned. Other enzymes were studied in conditions in which the velocity would vary in a hyperbolic-like manner with [S] and apparent $K_M^{-1}$ calculated. When more than one substrate is used, for instance, if all substrate concentrations are kept high and only one substrate concentration varies, the velocity of the reaction depends on the concentration of that substrate. The dependence is usually hyperbolic-like, and an apparent $K_M$ is estimated, which is valid for that specific substrate. Somewhat abusively, $K_M$ became the key parameter to describe the kinetic properties of enzymes, and the quest for methods on how to calculate $K_M$ for specific $E_z$–S pairs became an import part of biochemistry for many years.

It is obvious from the Michaelis–Menten equation:

$$v = \frac{V_{max}\,[S]}{K_M + [S]} \tag{4.16}$$

that when [S] equals $K_M$, $v = V_{max}/2$. The graphical interpretation is immediate, as depicted in the figure below. $V_{max}$ is the asymptotic limit of $v$ when [S] tends to infinity, and $K_M$ is the [S] in which $v$ is half $V_{max}$. This is the reason why $K_M$ is so popular among biochemists studying metabolism. When the substrate concentration drops below $K_M$, the velocity of the reaction is considerably below $V_{max}$, and the process being catalyzed loses efficacy. So $K_M$ is the reference value to estimate the metabolic impact of drops in the concentration of metabolites. A good example to illustrate the utility of $K_M$ value to explain the differences in metabolic profile of distinct tissues is the case of the isoforms of the enzyme lactate dehydrogenase (see Box 10.5).

Calculation of $K_M$ is thus of the uppermost importance. However, in an experimental plot of $v$ vs. [S] with discrete data, $V_{max}$ is not easy to identify as experimental data at very high [S] are difficult to attain (see next figure). Nowadays, a computational non-linear regression fit of Eq. (4.16) to the experimental data selects the best statistical $K_M$ and $V_{max}$ easily, quickly, and accurately, but in the early twentieth century, when Leonor Michaelis and

(continued)

**Box 4.5** (continued)

Maud Menten derived Eq. (4.15), this was not an option. In the 1930s, Dean Burk and Hans Lineweaver overcame the limitation of $V_{max}$ determination by linearizing the Michaelis–Menten equation (see next figure).

When rewritten in the form $1/v$ vs. $1/[S]$, Eq. (4.15) becomes:

$$\frac{1}{v} = \frac{K_M}{V_{max}} \frac{1}{[S]} + \frac{1}{V_{max}} \tag{4.17}$$

which describes a straight line from which both $K_M$ and $V_{max}$ are readily calculated.

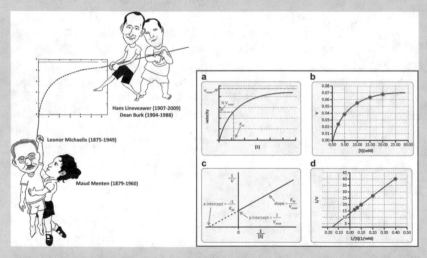

(**a**) Graphical interpretation of the Michaelis–Menten rectangular hyperbole. $K_M$ value corresponds to the substrate concentration at which the reaction velocity is half of the maximal velocity, $V_{max}$, which is the asymptotic limit of the velocity for infinite [S]. (**b**) $V_{max}$ is very difficult to be identified using experimental data as [S] cannot be extended indefinitely due to solubility or cost limitations. (**c**) Linearization of the Michaelis–Menten plot to estimate $K_M$ and $V_{max}$ graphically: the $y$ intercept is $1/V_{max}$ and the slope is $K_M/V_{max}$. $K_M$ can also be estimated from the $x$ intercept, which is $-1/K_M$. (**d**) Linearization of the data shown in Panel (**b**) allows the following calculations: $K_M = 6.39$ mM and $V_{max} = 0.085$ ms$^{-1}$. The data were extracted from a 1951 study on the hydrolysis of carbobenzoxyglycyl-L-tryptophan using pancreatic carboxypeptidase (Lumry R, Smith EL and Glantz RR. J. Am. Chem. Soc. 73:4330, 1951). The linearization of the Michaelis–Menten equation is owed to Hans Lineweaver and Dean Burk, whom used the method for the first time in 1934. With modern computational techniques and statistical methods, linearization is not mandatory as $V_{max}$ and $K_M$ can be estimated by non-linear regression methods that fit the Michaelis–Menten equation directly to experimental data (*red line* in Panel **b**)

# Selected Bibliography

Bar-even A, Flamholz A, Noor E, Milo R (2012) Rethinking glycolysis: on the biochemical logic of metabolic pathways. Nat Chem Biol 8:509–517

Kornberg A (1993) Recollections: ATP and inorganic pyro- and polyphosphate. Protein Sci 2:131–132

Newsholme P (2009) A brief history of metabolic pathways and their regulation. Mapping life's reactions. Biochemist (June):4–7

## Challenging Case 4.1: The Dark Side of Caffeine

### *Source*

On July 8, 2019, several media channels in Australia informed that the cause of death of Lachlan Foote, a healthy and athletic 21-year-old man from Blackheath, Blue Mountain, was caffeine.[1, 2, 3] The news also appeared in several media in different countries, among which we highlight the podcast "Caffeine supplement ban after overdose death," by reporter Ruby Jones, Featuring Nigel Foote, broadcasted September 20, 2019, 5:00 PM.[4] The case was surprising, not to say shocking. Caffeine consumption is so common and socially well accepted it hardly raises suspicions as a murderer. Lachlan was a talented musician saving to travel overseas and study science at a university. He had died nearly 18 months before. During all this long period, his parents believed Lachlan died of a "dodgy batch" of protein powder.

Lachlan Foote at Clarendon Concert. (Photo by Rachael Brady; reproduced with permission of Nigel Foote)

### *Case Description*

Lachlan Foote celebrated New Year's Eve 2017 having a few drinks with friends in Katoomba, Australia, and returned home by 1:45 AM. He wished his mother, Dawn, happy New Year before making himself a whey protein shake. Afterward he texted in social media to his friends:

---

[1] Burke K (2019) Young man's death prompts father's bid to ban caffeine supplement powder. 7 News. https://7news.com.au/news/health/young-mans-death-prompts-fathers-bid-to-ban-caffeine-supplement-powder-c-204573.

[2] Bruce-Smith A (2019) Tragic warning after young man's accidental caffeine powder overdose. 10 Daily News. https://10daily.com.au/news/australia/a190707awtey/tragic-warning-after-young-mans-accidental-caffeine-powder-overdose-20190708.

[3] Molloy S (2019) The father of a young man who died of caffeine toxicity speaks about his son's final hours. News Corp Australia. https://www.news.com.au/lifestyle/health/health-problems/the-father-of-a-young-man-who-died-of-caffeine-toxicity-speaks-about-his-sons-final-hours/news-story/f6698b821c9228722c4c49efce3e2c67.

[4] https://www.abc.net.au/radio/programs/pm/caffeine-supplement-ban-after-overdose-death/11534544.

Lachlan (2:07 AM, Jan 1st 2018):*"I think my protein powder has gone off. Just made an anti hangover/workout shake and it tasted awful"*
Friend: *"how"*
Lachlan: *"dunno, was kinda bitter though ... anyway night lads"*
Lachlan (2:14 AM): *"cya in the morning"*

But, as Nigel Foote, Lachlan father, said *"Morning never came for Lachlan. We found him, dead and cold on the bathroom floor on New Year's Day 2018, the day before his 22nd birthday."*[1] *"We know that Lach went off and had a shower, got into his pyjamas, turned his bed down and then made himself a shake. (...) He died not long after that, I'd say. He hadn't gotten into bed. He went to the bathroom because he felt sick. There was a small amount of vomit on the floor. (...) I put my hand on his back and he was cold as ice. I felt for a pulse but I knew there wouldn't be one. That was it. I went into a total state of shock."*[3]

The presumed direct cause of death was heart attack, but this was not totally confirmed to reporters because Nigel could not stand reading the autopsy when he spoke to the site news.com.au.[3] A toxicology report later showed there was a small amount of alcohol in his system—*"a reading of less than 0.05,"* according to the reporter—and caffeine, which puzzled investigators. *"The original toxicology (report) didn't show the level of caffeine, just that it was there, but the more thorough tests later discovered how much was in his system. It was a huge amount,"* Nigel clarified.

It was later discovered that Lachlan added caffeine powder to his late night protein shake. Many fitness enthusiasts mix whey protein with caffeine, which is widely available as a dietary supplement in many countries. *"I don't know if he made many protein shakes. I don't think he was that into it. He was athletic and a good tennis player, he did a bit of fitness stuff—weights and push ups,"* Nigel told the news.com.au reporter.

The recommended dose of pure caffeine powder supplements is about 1/6 of a teaspoon, but Lachlan was probably not aware of this limit. Chances are that he had no idea of the potency of caffeine or its potential to kill. *"We think Lachlan obtained the caffeine powder from a friend or work associate as a thorough search of his computer and bank statements, by both myself and the police, revealed no mention of caffeine powder, only related protein-powder products,"* his father told 7NEWS.com.au. *"So it appears the pure caffeine powder was bought by someone else and shared. (...) The fact that he kept the caffeine powder in our kitchen pantry—where one of us might have mistaken it for flour or sugar—proves the point. Lachlan would never have kept it there, had he known it was a threat to the family."*

Lachlan death and Nigel's tenacity launched a public discussion on safety issues related to caffeine supplement powders.[5] Before that, sodas with high caffeine content, generically known as "energy drinks," had been in the radar of food safety

---

[5] Cunningham I (2019) Warning on pure caffeine powder after death of Blackheath man Lachlan Foote. Guardian News. https://www.nambuccaguardian.com.au/story/6262877/beware-pure-caffeine-powder-warns-grieving-dad/.

agencies in many countries.[6,7] The risks associated with caffeine ingestion are a controversial matter, but one thing is for certain: with the advent and popularization of energy drinks and workout supplements, there is much more to caffeine intake than coffee drinking for social purposes or fighting somnolence. Ironically, the dark side of caffeine does not reside in black coffee but in its pure white bright powder.

## Background Information on Caffeine

Facts and Figures About Caffeine: Abundant in Nature, Used and Abused by Humans

Caffeine is 1,3,7-trimethylxanthine (also known as guaranine or theine; see next figure). It is a bitter white crystalline substance found in the seeds, nuts, or leaves of several plants native to Africa, Asia, and South America. Around 60 plants are known to contain caffeine, but the most popular sources are the seeds (beans) of coffee (both *Coffea arabica* and *Coffea canephora*), leaves of tea, nuts of kola, leaves of holly yerba mate, and seeds of guarana berries. Curiously, countries that most consume caffeine (Scandinavian countries and the Netherlands, with 400 or more mg/person/day) are not coffee-producing countries, which are among the least consumers of caffeine (e.g., Angola, Brazil, Kenya, and Tanzania, with less than 50 mg/person/day).[8]

Chemical structure of caffeine (1,3,7-trimethylxanthine)

The natural function of caffeine in plants is believed to be as a pesticide as it can kill predator insects. It may also act by preventing fungal infections. More sophisticated actions have been proposed, such as enhancing the pollinator's memory or as a reward when present in floral nectar, thus improving fertilization.[9]

---

[6] Committee on toxicity of chemicals in food, consumer products and the environment (2012) COT Statement on the interaction of caffeine and alcohol and their combined effects on health and behavior. COT statement 2012/04. https://cot.food.gov.uk/committee/committee-on-toxicity/cot-statements/cotstatementsyrs/cotstatements2012/cotstatement201204.

[7] Higgins JP (2018) Stimulant-containing energy drinks. Expert analysis. American College of Cardiology.          https://www.acc.org/latest-in-cardiology/articles/2018/02/28/10/46/stimulant-containing-energy-drinks.

[8] Fredholm BB et al (1999) Actions of caffeine in the brain with special reference to factors that contribute to its widespread use. Pharm Rev 51:83–133.

[9] Wright GA et al (2013) Caffeine in floral nectar enhances pollinator's memory of reward. Science 3339:1202–1204.

Most caffeine-containing drinks and beverages also contain plant extracts. In some cases, synthetic caffeine may even coexist with natural caffeine from such extracts, in addition to related methylxanthine alkaloids such as theophylline and theobromine. The most popular caffeine-containing drink worldwide is coffee, but caffeine is also present in other common products such as chocolate, tea, yerba mate (South America), colas, guarana sodas (soft drink made with guarana seeds), and the so-called energy drinks. Recently, coffee tablets and coffee powder are being used by athletes as supplements for "performance enhancement." These products are also used by students intensively studying for tests and exams and by people working or driving for long hours.

The typical caffeine content in foods and drinks varies greatly (see next table), which is an important factor when considering the physiological impact of the intake of this substance.

Caffeine content in foods, drinks, and drugs (adapted from the article cited in the footnote 8)

| Product | Approx. caffeine/serving (mg) | Caffeine content |
|---|---|---|
| Roasted and ground coffee | | |
| Percolated | 80–130 | 0.3–1.1 g/L |
| Drip | 110–170 | 0.4–1.2 g/L |
| Decaffeinated | 5–15 | 0.01–0.03 g/L |
| Espresso | 60–90 | 1.7–2.2 g/L |
| Instant coffee | | |
| Regular | 80–140 | 0.3–1.2 g/L |
| Decaffeinated | 5–25 | 0.01–0.05 g/L |
| Tea | | |
| Bagged | 25–40 | 0.2–0.3 g/L |
| Leaf | 25–40 | 0.2–0.3 g/L |
| Instant | 25–40 | 0.2–0.3 g/L |
| Iced | 25 | 0.2 g/L |
| Cocoa | 2–12 | 0.01–0.05 g/L |
| Chocolate bar | | |
| Milk | 10 | 0.04–0.3 mg/g |
| Dark | 50 | 0.2–1.3 mg/g |
| Baking chocolate | – | 0.6–4.2 mg/g |
| Soft drinks | | |
| Regular cola | 26–43 | 0.08–0.13 g/L |
| Caffeine-free cola | 0 | 0 |
| Diet cola | 23–53 | 0.07–0.16 g/L |
| Regular energetic drink | 75–80 | 0.3 g/L |
| Regular caffeine tablet | 100 | 100 mg/tablet |

Note: The content in gram per liter can be converted to molarity by dividing by 194.19 g/mol, the molar mass of caffeine

Pharmacokinetics of Caffeine: Absorption, Distribution, and Metabolization

Ingested caffeine is absorbed by the small intestine and distributed in the human body by blood circulation. It is soluble both in aqueous medium and lipids, which enables crossing of the blood–brain Barrier[10], the tight walls of microvessels that irrigate the brain (see Fig. 3.48). This means that caffeine reaches virtually all tissues in the human body, including the brain and the rest of the central nervous system (CNS). Moreover, the concentration of caffeine in the brain is the concentration of caffeine in the blood, which in turn depends mainly on the swollen mass of this substance. White et al. measured how ingested doses would translate into plasma levels of caffeine.[11] For this, groups of six individuals ingested 160 mg of caffeine. The dosage forms included hot coffee consumed over 20 min, cold coffee consumed over 2 min, cold coffee consumed over 20 min, sugar-free energy drink consumed over 2 min, and sugar-free energy drink consumed over 20 min. Curiously, the appearance of caffeine in plasma followed the same kinetics, with minor variations among dosage forms (next figure). The peak concentration is attained nearly 1 h after ingestion, reaching 3.5 mg/L, which is, considering the amount ingested and the body mass of the tested individuals, around 1.6 mg/L in plasma per milligram caffeine per kilogram body mass. Overall, the results from this study suggest that contrary to concerns about potential rapid absorption of caffeine from rapidly consumed cold energy drinks, caffeine absorption and exposure from instant coffee and sugar-free energy drink are similar irrespective of drink temperature or rapid versus slow administration times. Another study, using caffeinated chewing gum and caffeine capsules, showed a markedly faster rate of absorption with the gum when a high dose of 200 mg is used, presumably due to the uptake in the buccal cavity along with absorption from swallowing while chewing the gum.[12] This study also showed there is proportionality between administered dose and peak levels in the plasma.

---

[10] Fredholm BB (1980) Are methylxanthine effects due to antagonism of endogenous adenosine? Trends Pharmacol Sci 1:129–132.

[11] White JR Jr et al (2016) Pharmacokinetic analysis and comparison of caffeine administered rapidly or slowly in coffee chilled or hot versus chilled energy drink in healthy young adults. Clin Toxicol 54:308–312.

[12] Wickham KA, Spriet LL (2018) Administration of caffeine in alternate forms. Sports Med 48:S79–S91.

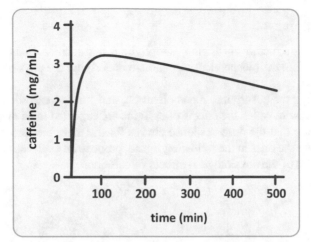

Absorption of caffeine: evolution of plasma concentration over time after ingestion of 160 mg

The maximal concentration peak in plasma is followed by a decay in concentration due to metabolization and excretion of the caffeine degradation products through urine. Although many xenobiotics are polyhydroxylated by the P450 system (see Fig. 2.6), caffeine is degraded by demethylation (i.e., removal of methyl groups, $-CH_3$), which improves solubility in plasma, therefore favoring elimination with urine. Metabolism of caffeine proceeds through three main pathways, which are illustrated in the next figure. Cytochrome P450 1A2 (abbreviated CYP1A2) and cytochrome P450 2E1 (CYP2E1), two members of the cytochrome P450 oxidase system, are involved in the metabolism of caffeine in the body[5] (see next figure). It is noteworthy that some of the metabolites of caffeine themselves, like theophylline and theobromine, have pharmacological activity.

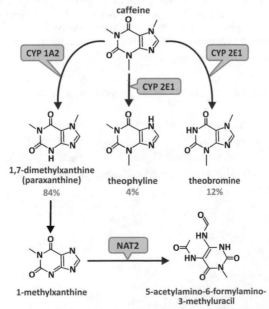

Metabolism of caffeine occurs via demethylation by CYP1A2 and CYP2E1 (percentages refer to the mean proportion of caffeine converted to each metabolite)

## Questions

1. Compare the structure of caffeine, theophylline, and adenine. Elaborate a hypothesis to explain the biochemical mechanism underlying the physiological effects of caffeine.
2. Considering the absorption, metabolization, and time to excretion, what is the time window in which the effects of caffeine are expected to be maximal?
3. Considering that the dose–response curves for the main biochemical targets of caffeine are the ones in the following figure, propose an explanation for the dose dependence of the physiological effects of caffeine.

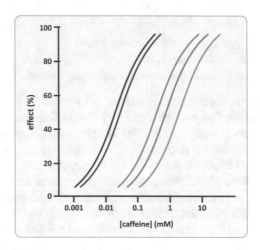

Response of biochemical targets of caffeine in relation to its plasma levels in humans. Coffee drinking causes plasma concentrations typically in the range 0.01–0.05 mM, and acute overdose victims have plasma concentrations in the range 0.2–2 mM. $A_{2A}$ receptor (*dark blue line*); $A_1$ receptor (*magenta line*); phosphodiesterase (*green line*); $GABA_A$ receptor (*orange line*); $Ca^{2+}$ release through ryanodine receptors (*cyan line*). (Figure adapted from the article cited in the footnote 8)

4. Correlate events of this case with properties and biochemical effects of caffeine.
5. Lachlan drank with friends before ingestion of caffeine. In addition, although the toxicology report refers moderate consumption of ethanol, Lachlan mentioned he had prepared an anti-hangover shake. Is it plausible that ethanol and caffeine interfere with each other? Use biochemical reasoning in your answer.

## *Biochemical Insight*

1. Adenine is one of the purines that is present in RNA (and DNA) as part of the adenosine (or deoxyadenosine) base, which in turn is also part of cyclic AMP. Caffeine, as well as its metabolization products theophylline and theobromine, is quite similar to adenine (see previous and next figures).

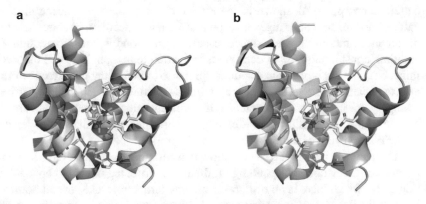

Chemical structures of adenine, adenosine, and cyclic AMP

Such similarity makes it plausible that these molecules may bind competitively and interfere with receptors in which the adenine residue plays a central role in ligand docking. This is the case of adenosine 2A ($A_{2A}$) receptors[13] (see next figure). Caffeine is an antagonist (inhibitor) of the action of adenosine with $K_M$ = 2.4 µM for binding to $A_{2A}$ receptor. The same effect is observed with $A_1$, $A_{2B}$, and $A_3$ receptors but with binding $K_M$ of 12, 13, and 80 µM. ($K_M$ is frequently named $K_D$ in pharmacology when addressing the action of drugs on receptors, but we will keep using $K_M$ for the sake of simplicity.)

Caffeine binds to $A_{2A}$ receptors (PDB 5MZP) in two different modes, shown in (**a**) and (**b**)

It is also the case for nucleotide phosphodiesterases, which comprise a group of enzymes that degrade the phosphodiester bond in the second messengers cyclic

---

[13] Cheng RKY et al (2017) Structures of human A1 and A2A adenosine receptors with xanthines reveal determinants of selectivity. Structure 25:1275–1285.

adenosine monophosphate (cAMP) and cyclic guanosine monophosphate (cGMP). They regulate the cyclic nucleotide signaling within subcellular domains (see also Sect. 5.4). Caffeine, theobromine, and theophylline are nonselective inhibitors of these enzymes, therefore interfering with cAMP-dependent signal transduction by raising intracellular concentrations of cAMP.

In addition to adenosine receptors and nucleotide phosphodiesterases, caffeine binds the $GABA_A$ receptor, which is a ligand-gated ion channel. Its endogenous ligand is $\gamma$-aminobutyric acid (GABA), the major inhibitory neurotransmitter in the CNS. The aqueous extract of coffee dose-dependently inhibits the GABA-elicited responses. Theophylline inhibits the response in a noncompetitive mechanism ($K_M = 0.55$ mM), whereas theobromine inhibit it in a competitive manner, $K_M = 3.8$ mM and 13 mM, respectively.[14] (Further details on competitive and noncompetitive mechanisms will be provided in Sec. 5.2)

Caffeine is also a voltage-independent activator of ryanodine receptors, which are a class of intracellular calcium channels in excitable animal tissues like muscles and neurons (see Sect. 10.1.3). Other targets include the enzyme acetylcholinesterase and competitive antagonism of the ionotropic glycine receptors.

The ubiquitous presence of adenosine receptors, phosphodiesterases, $GABA_A$ receptors, and ryanodine receptors implies that caffeine and their related metabolization products have a wide range of actions in different organs and systems such as the CNS, cardiovascular system, kidneys, and lungs. This poses toxicology challenges but also drug development opportunities.

2. Peak concentrations of caffeine in plasma occur around 60 min after ingestion, but the above figure may change according to the presence of other components in foods and drinks. Proteins and polyphenols, for instance, may interfere with caffeine absorption. As shown in the kinetic curve of plasma concentration of caffeine above, 10 h after ingestion, the plasma levels decrease to about half the peak concentration. The persistence of caffeine in blood is thus remarkable. In addition, considering 16% of caffeine is converted to theophylline and theobromine, both having biochemical actions similar to caffeine, the persistence of the effects of caffeine is reinforced. Theophylline is excreted unchanged in the urine, taking 8 h, in adults, to decrease its concentration by half ("half-life"). Theobromine has similar half-life, but it is further metabolized by hepatic demethylation and oxidation.

Taken together, the figures above show that caffeine may endure circulating in plasma for many hours. Depending on the initial dose, the threshold concentration under which there is no response from any target may take several hours to reach. Take the 160 mg ingestion example above; the peak concentration would be 3.5 mg/L. It takes 10 h for this concentration to drop to half this value, which

---

[14] Hossain SJ et al (2003) Effects of coffee components on the response of $GABA_A$ receptors expressed in *Xenopus* oocytes. J Agric Food Chem 26:7568–7575.

is approximately the value expected for the peak concentration of an espresso coffee.

3. The dose–response curves presented in the question are presented in log scale (i.e., the scale marks in the $x$ axis are spaced as logarithm of concentration instead of a linear scale of the concentration itself). This is the reason why dose–response curves do not appear hyperbolic as in Box 4.4, even being the same kind of relationship. The logarithmic scale facilitates representation of curves occurring at concentrations separated by orders of magnitude, as in this case. The $K_M$ of the interaction of caffeine with the different targets are easily identified (concentration of 50% effect) although interaction with adenosine receptors occurs in the tens of micromolar range and interaction with other receptors occur one or two orders of magnitude above this range of concentrations.

As explained in Sect. 4.2.1, $K_M$ is the reference value above which interaction with targets can be considered near maximal and below which rapidly approaches zero. This means that the biochemical effects of different caffeine intake doses are determined by the targets that are successively triggered by increasing concentrations of caffeine (note that intake doses translate to proportional plasma concentrations, i.e., levels of caffeine being distributed throughout the body tissues). Adenosine receptors (lowest range of binding $K_M$) respond at lower doses. At higher doses, both adenosine receptors and phosphodiesterases are inhibited, so a mix of effects caused by these two targets is observed. At even higher concentrations, in addition to inhibition of adenosine receptors and phosphodiesterases, caffeine also blocks $GABA_A$ receptors, further increasing the plethora of biochemical effects. Finally, at extreme concentrations, $Ca^{2+}$ release caused by action on ryanodine receptors leads to effects that add to the already rather complex combination of effects associated with adenosine receptors, phosphodiesterases, and $GABA_A$ receptors altogether.

It is very important to bear in mind that this escalade is determined by the binding $K_M$ values associated with each target. In the case of caffeine, there is a clear gap between doses affecting adenosine receptors and the higher doses needed for interactions with other targets to occur. Interestingly, this "biphasic" group of interactors reflects in a biphasic effect of rodents' locomotion behavior in response to caffeine doses (see next figure). The threshold effect for preference/rewarding of caffeine is 1–3 mg/kg, and the locomotion peak effect is detected at 10–40 mg/kg. Converting administration dose into plasma levels using the converting factor of 1.6 mg/L per milligram per kilogram (see background information on caffeine), and using the molar mass of caffeine (194.19 g/mol) to convert to plasma levels to millimolar, one obtains about 0.01–0.03 mM for the preference/rewarding threshold and 0.1–0.4 mM for the locomotion peak effects. (Note that it is not totally accurate because the human conversion factor may not apply exactly to rodents.) These ranges correspond to the onset of the effects on adenosine receptors and phosphodiesterases/$GABA_A$ receptor/$Ca^{2+}$ release, respectively. The effect of caffeine is shared by theophylline and paraxanthine, two metabolites of caffeine metabolization.

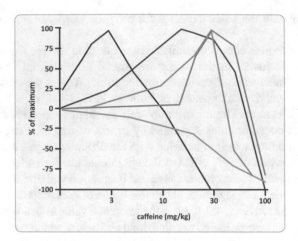

Biphasic effects of caffeine on rodent behavior in different activities: locomotor behavior in mice (*green line*); locomotor activity in rats (*magenta line*); rotation behavior in rats (*orange line*); locomotion in $A_2A$ knockout mice (*cyan line*); rewarding or aversive motivational effects of drugs in mice (*blue line*) (Figure adapted from the article cited in the footnote 8)

Caffeine is a competitive unselective antagonist of $A_1$ and $A_{2A}$ receptors, therefore counteracting the activity of the natural ligand, adenosine. While adenosine causes a decrease in heart rate upon binding to $A_1$ receptor in the heart, caffeine causes the opposite effect. Likewise, while binding of adenosine to $A_{2A}$ receptors in the heart causes coronary heart vasodilatation and in the CNS decreases dopaminergic activity and inhibits neuron excitation, caffeine causes vasoconstriction, increases dopaminergic activity, and helps neuron excitation. Adenosine causes bronchospasm due to binding to $A_{2B}$ receptors and protection in cardiac ischemia via binding to $A_3$ receptors; these effects are also counteracted by caffeine. Despite the fact that adenosine receptors are present all over the body, the effects of caffeine are particularly noticeable in the brain and heart and, to a less extent, in lungs and kidneys.

The major target of caffeine, $A_{2A}$ receptor, is a member of the G protein-coupled receptor family that activates adenylyl cyclase (to be addressed in Sec. 5.4). It is abundant in a region of the brain named basal ganglia, which is associated with control of voluntary motor movements and cognition, among other functions. In the brain, caffeine counteracts the onset of adenosine-induced drowsiness through $A_{2A}$ receptors. Wakefulness, concentration, and motor coordination are thus improved. Adenosine receptor antagonism leads to the release of neurotransmitters such as acetylcholine, enhancing the stimulant effects of caffeine. In the autonomic nervous system, caffeine increases respiratory rate and reduces heart rate. $A_{2A}$ receptors are also abundant in vasculature; they are responsible for regulating myocardial blood flow by vasodilating the coronary arteries and therefore enhancing blood flow to the myocardium at the cost of reducing arterial pressure. Not surprisingly, caffeine has a plethora of cardiovascular effects (see next table). Increase in vasoconstriction (and blood pressure) explains, at least in part, the analgesic effect of caffeine in the treatment of some forms of migraines. Other $A_{2A}$ receptor-mediated effects of caffeine may include diuresis (increase in urine output) due to the effect in proximal tubules. Yet, this effect demands high doses in acute ingestion. One death of caffeine overdose has been attributed to renal failure, but heart failure is the most common cause of deaths associated with caffeine toxicity.[15] In the gastric system, high doses of caffeine cause abdominal pain, nausea, and vomiting.

Acute and potential chronic effects of caffeine (Table adapted from the article cited in the footnote 7)

| Acute | Potential chronic |
| --- | --- |
| Increase in blood pressure | Hypertensive heart disease |
| Increase in heart rate | Coronary heart disease |
| Supraventricular arrhythmia | Atherosclerosis |
| Ventricular arrhythmia | Cerebrovascular disease |
| Coronary artery spasms | Peripheral arterial disease |
| Coronary artery thrombosis | |
| Segment elevation myocardial infarction | |
| Aortic dissection | |
| Sudden cardiac death | |
| Endothelial dysfunction | |

Adenosine receptors are also present in skeletal muscle, adipose tissue, and liver cells, where they can also inhibit adenylyl cyclase, suppressing the conversion of ATP to cAMP, thus impairing the stimuli induced by the activation of protein kinase A (PKA). In addition to the abovementioned actions of caffeine mediated by $A_{2A}$ receptors, it also triggers lipolysis, muscle glycogenolysis, and hepatic gluconeogenesis. Moreover, through the action of adenosine receptors, $Ca^{2+}$ release by the sarcoplasmic reticulum, a prerequisite for muscle contraction (see Sect. 10.1.3), also occurs. Indirectly, the series of events in muscle ends in improved uptake of glucose and fatty acids by myocytes. The description of caffeine effect over glucose and fatty acid uptake and oxidation in skeletal muscle tissue, and concomitant mobilization from the adipose tissue, is presented in the following figure.[16]

---

[15] Willson C (2018) The clinical toxicology of caffeine: a review and case study. Toxicol Rep 5:1140–1152.

[16] Silva LA et al (2017) Mechanisms and biological effects of caffeine on substrate metabolism homeostasis: A systematic review. J Appl Pharm Sci 7:215–221.

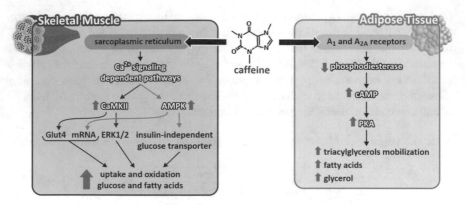

Impact of caffeine on glucose and fatty acid metabolism (figure adapted from the article cited in the footnote 16). *CaMKII* Ca$^{2+}$/calmodulin-dependent protein kinase II, *AMPK* AMP-activated protein kinase, *ERK1*/2 extracellular sign 1 and 2, *PKA* protein kinase A

Phosphodiesterases are also ubiquitous in the human body. Caffeine effects mediated by competitive nonselective inhibition of several types of phosphodiesterases are caused by increase in intracellular concentrations of cAMP and activation of PKA. In this sense, the action on phosphodiesterases exacerbates some of the effects caused by antagonism of adenosine receptors. Curiously, inhibition of phosphodiesterases also leads to TNFα and leukotriene synthesis inhibition, making caffeine an anti-inflammatory compound.

The effect of caffeine on GABA$_A$ receptors and ryanodine receptors is not well studied as their $K_M$ is well within the lethal range of caffeine doses. The effects associated with these targets are masked by the toxic effects associated with highly stimulated adenosine and phosphodiesterase receptors.

4. Some of the events described in the case can be directly correlated to caffeine:

   (a) Bitter shake; white powder, like sugar or flour. Caffeine is a bitter, white crystalline substance. The very bitter taste of the shake may be the result of a high content in caffeine. This is consistent with a "huge amount" of caffeine found in Lachlan's blood.

   (b) Lachlan died (presumably) not long after ingestion of the shake: This is also in agreement with the ingestion of a high amount of caffeine. Peak absorption of caffeine occurs 60 min after ingestion, but for Lachlan, the lethal concentration of caffeine in plasma was attained much sooner than that.

   (c) Vomit: High doses of caffeine cause abdominal pain, nausea, and vomiting.

   (d) Dead in bathroom: In addition to nausea and vomiting, caffeine can also cause diuresis. Any of these symptoms (or a combination of part of them) may have made Lachlan go to the bathroom.

   (e) Heart attack as the presumed cause of death: Caffeine has high cardiovascular impact. In high doses, it can kill by sudden cardiac death because of a supply–demand imbalance that leads to ischemia.[6] This is the most common cause of death associated with caffeine overdose and probably the direct cause of the death of Lachlan.

5. Like caffeine, ethanol is absorbed and distributed to the whole body, including the CNS as the blood–brain barrier is permeable to this small molecule that is both hydrophilic and lipophilic. In the brain, it is thought to bind to $GABA_A$ receptor, which mediates rapid inhibitory neurotransmission, where it acts as a positive allosteric modulator. The outward signs of ethanol intoxication, such as impaired sensory and motor function and slowed cognition and stupefaction, are a result of this activity.[6] Alcohol is metabolized by the enzymes aldehyde dehydrogenase and alcohol dehydrogenase, CYP2E1, and catalase. The consequences of alcohol metabolism include formation of highly reactive oxygen species (ROS), changes in the ratio of NADH to $NAD^+$ with impairment of gluconeogenesis, and medication interactions. Yet, caffeine is metabolized mainly by CYP1A2, with little interaction with CYP2E1. Moreover, Lachlan had only 0.05 (g/L, presumably) plasma content in ethanol when his blood was tested, which is very low when compared to the drinking–driving limit allowed in most countries, for instance, 0.0–0.8 g/L. His behavior upon returning home did not show evidence of drunkenness. It is very unlikely that caffeine–ethanol interactions could account for Lachlan's fatality, either at the $GABA_A$ receptor or P450 level. Ironically, Lachlan motivation for caffeine ingestion was prevention of hangover. Indeed, at least in rats, caffeine blocks some effects of acetate, which, together with acetaldehyde, is responsible for the biochemical toxicity associated with hangover.[17] In humans, according to the Committee on Toxicity of Chemicals in Food, Consumer Products and the Environment, *"There is some evidence that caffeine can ameliorate some of the neurocognitive effects of alcohol, but the findings have not been consistent in all studies, and the underlying mechanisms are unclear. (…) because of variation in the doses of caffeine and alcohol administered, the behavioural effects assessed, and other aspects of study design, it was not possible to determine whether there was a counteracting effect of caffeine. (…) Conflicting results have also been obtained in studies designed to test perceived degree of alcohol intoxication with and without caffeine. The most direct subjective ratings of intoxication were no different when alcohol was consumed with and without caffeine."*[6] Interestingly, expectancy plays a role in counteraction of ethanol-induced impairment: compensation for ethanol impairment occurs when drinkers hold expectations that caffeine will disrupt performance; when no expectation exists, no compensatory response occurs and the impairing effects of ethanol are observed.[18] Had Lachlan consumed a much smaller dose of caffeine (for instance, an espresso), the power of expectancy associated with the low quantities of ethanol ingested would have most probably sufficed to prevent hangover.

---

[17] Maxwell CR et al (2010) Acetate causes alcohol hangover headache in rats. PLoS One 5:e15963.

[18] Fillmore MT et al (2002) Does caffeine counteract alcohol-induced impairment? The ironic effects of expectancy. J Stud Alcohol 63:745–754.

## *Final Discussion: Facts and Fantasies About Energy Drinks*

From the information provided above, it is clear that caffeine has effects both on wakefulness and physical activity. Its stimulant effect makes caffeine a potential doping substance. In competitive sports, urinary concentrations above 15 mg/L are usually considered abuse. Stimulant effects have been used to drive the advertising and marketing of energy drinks. Red Bull's international campaigns, for instance, target young men mostly through extreme sports, like cliff diving, Formula 1 racing, and windsurfing. By associating the drink's image with extreme sports and energizing effects in the body, a branding identity is created.

Energy drinks' stimulant effects are almost, if not exclusively, owed to caffeine. When added to energy drinks, caffeine is typically a synthetic alkaloid rather than a naturally occurring constituent of plant-based beverages (e.g., tea or coffee). However, guarana and yerba mate, which can be contained as part of the energy blend of energy drinks, are also natural and an additional source of caffeine in these products. As such, the total amount of caffeine in energy drinks may not be accurately reflected on the label.[6] In addition, energy drinks frequently include significant amounts of glucose, taurine (an amino acid), niacin, pyridoxine, cyanocobalamin (B12), riboflavin (B2), ginseng extract, glucuronolactone (a glucose metabolite), inositol (B8), guarana (contains caffeine, theobromine, and theophylline), ephedra, yohimbine, gingko, kola nut, theophylline, vitamins, herbs, and/or L-carnitine. The health effects of these additives are not well documented.[7]

The expert analysis of John Higgins, of the American College of Cardiology[7], showed that *"the evidence for energy drinks as 'performance-enhancing' is mixed. With respect to improving aerobic/anaerobic performance and/or reaction time after energy drink consumption, studies are inconsistent. Several studies have reported improved endurance or muscle performance; however, others demonstrate no benefits, with some documenting a negative effect as evidenced by increased muscle fatigue and reduced cerebral blood flow. In 2011, there were 4854 logged calls to US Poison Control Centers regarding adverse energy drink exposures (…). The cardiac adverse events associated with acute energy drink consumption are likely related to increased heart rate and blood pressure, together with changes in conduction system associated with acute consumption of energy drinks, and include tachycardia, hypertension, supra- and ventricular arrhythmias (…). Other adverse effects associated with energy drink consumption include epileptic seizures, stroke, subarachnoid hemorrhage, pontine myelinolysis, hallucinations, anxiety, agitation, headaches, hepatitis, gastrointestinal upset, acute renal failure, rhabdomyolysis, metabolic acidosis, insulin resistance, obesity, acute psychosis, insomnia, high risk/ aggressive behavior and caffeine withdrawal. (…) Used for hydration prior to, during, or after physical activity: our research suggests an impairment in endothelial function in the hours following consumption of energy drinks, which could result in impairment in blood flow to the heart and other vital organs and complications."* A study presented in the article cited in the footnote 12 shows that there are no significant differences between conditions on any of the physical or cognitive tests performed, suggesting no stimulating effect of caffeine, taurine, or the combination with glucose on aerobic capacity, handgrip strength, jump performance, or cognitive performance.

Two consumer class action lawsuits against Red Bull were submitted to the US District Court for the Southern District of New York, USA. Consumers sued the company alleging that the brand's marketing and labeling had mislabeled the safety and functionality of the beverage.[19] A settlement was reached, and consumers of Red Bull products were entitled to either USD $10 in cash or USD $15 in Red Bull products up to a cost of USD $13 million to the company. According to Red Bull, the company settled to avoid the cost and distraction of litigation, denying any and all wrongdoing or liability.

---

[19] Picchi A (2014) Drink Red Bull? You may have $10 coming to you. CBSNews. https://www.cbsnews.com/news/drink-red-bull-you-may-have-10-coming-to-you/.

## Challenging Case 4.2: Missing the Obvious … Twice!
## The Devil Is in the Details, so Is Recognition by Peers

### *Sources*

Leonor Michaelis and Maud Menten are now famous for their pioneering work on enzyme kinetics (see Sect. 4.2.1). Notwithstanding, their work was deeply rooted on the previous studies of Victor Henri, whose contribution to enzymology is far from getting the recognition devoted to Michaelis and Menten. Henri failed in doing the obvious: developing reliable methodologies of data analysis based on equations so simple they could be easily used by non-experts. The absence of figures in his work surely did not help either. Curiously, Michaelis and Menten also failed in a similar way: they missed the obvious linearization of their working equation, skipping the opportunity of making enzyme kinetics so simple it could be used as tool by all. Hans Lineweaver and Dean Burk did it (see Box 4.5) and, in doing so, became as famous as Michaelis and Menten. Evil was in details all along the way, across three generations of the founding fathers of enzymology.

This case is based on historical facts that have been reported in several sources, among which we highlight the thesis of Victor Henri[20], the fundamental work that gave rise to the discipline of enzymology in biochemistry, and the paper by Kenneth Johnson and Roger Goody[21], having translated the original 1913 paper of Leonor Michaelis and Maud Menten[22] from German to English.

Cover page of the (second) PhD thesis of Victor Henri. (From https://archive.org/details/b28114024); public domain)

### *Description*

It all started when Victor Henri, a French researcher, studied invertase in the early twentieth century for his second PhD thesis[20] (the first was in psychology). The nature, properties, and function of enzymes were still poorly understood in this period. Scientific research on fermentation was just beginning and enzymes were

---

[20] Henri V (1903) Lois générales de l'action des diastases. Librairie Scientifique A. Hermann, Paris. https://archive.org/details/b28114024.

[21] Jonson K, Goody R (2011) The original Michaelis constant: translation of the 1913 Michaelis-Menten paper. Biochemistry 50:8264–8269.

[22] Michaelis L, Menten ML (1913) Die kinetic der invertinwirkung. Biochem Z 49:333–369.

called "ferments." The role of sugars in cellular energetics fascinated biochemists. Invertase, the enzyme that converts sucrose into its monomeric constituents (EC 3.2.1.26 in modern classification), glucose and fructose, was key. This enzyme was also popular among biochemists because it is extracellular, therefore easily accessible to be assayed. In addition, the "inversion" (i.e., hydrolysis—"inversion" of the dimerization) could be followed along time using a polarimeter because sucrose and "inverted" sugars interact differently with polarized light. Victor Henri's objective was to show that enzyme-catalyzed reactions follow physical chemistry principles, which was a matter of debate then. Vitalism was still in vogue despite that Buchner's experiments had demonstrated the opposite (a matter to be addressed in Sect. 6.1.1).

Pages 90 and 91 of Henri's (second) thesis are a landmark in biochemistry and an overlooked treasure as most of the foundations of modern enzyme kinetics are there. Using chemical kinetics reasoning from standard nonbiological physical chemistry, Victor Henri derived an equation equivalent to the one now named the Michaelis–Menten equation:

$$\text{initial rate} = \frac{K_3 a}{1 + ma} \tag{4.18}$$

($a$ is the initial amount of the substrate, sucrose; $K_3$ is a constant proportional to enzyme concentration; and $m$ is a constant). This is a particular case of a more general equation:

$$\frac{dx}{dt} = \frac{K_3 (a - x)}{1 + m(a - x) + nx} \tag{4.19}$$

($x$ is the amount of product at time $t$ and $n$ is a constant). When $x = 0$, Eq. (4.19) converts into Eq. (4.18).

He noted that the first equation predicts a hyperbolic dependence of the initial rate (velocity) on the concentration of substrate and mentioned, without showing (his thesis had no figures, only equations and data tables, etc.), that this had been experimentally confirmed. Curiously, he did not choose the obvious simplest way of analyzing the data (first equation). Instead, he preferred to be canonical and used the rather complex integrated form of the second equation.

In his approach, $K_3$, the parameter that characterizes the kinetics of the enzyme, is determined by Eq. (4.20), which contrasts with the simplicity of the approach later followed by Leonor Michaelis and Maud Menten and, even more, with the approach of Hans Lineweaver and Dean Burk:

$$K_3 = \frac{a}{t} \left[ (m - n)\frac{x}{a} + n \ln \frac{a}{a - x} \right] + \frac{1}{t} \ln \frac{a}{a - x} \tag{4.20}$$

Victor Henri was a purist. He sought a full description of the time course of the reaction for any concentration of the substrate. Leonor Michaelis and Maud Menten studied deeply the work of Victor Henri and gave him full credit for it. Their main virtue was the recognition of the advantages of using initial rates, which Henri overlooked, when analyzing kinetic data[23]:

1. Complications due to progress of the reactions vanish, for instance, the inhibition by products, the loss of catalytic activity, and, in the special case of polarimetric methods used for studying invertase, the spontaneous interconversions of α- and β-anomers of glucose (see Sect. 3.2 and Fig. 3.19d), which alter the interaction with polarized light, therefore jeopardizing the reliability of polarimeter readouts.
2. Initial rate equations are easier to rationalize in terms of molecular events and use in practice than integrated equations for the progress of reactions.
3. There is no drift in the pH, temperature, or other conditions at time zero.

Michaelis and Menten were totally aware of advantages (1) and (3) when they built their work on top of the legacy left by Henri Victor. The practical use of initial rate methods was their most important contribution and the main reason why they are still praised and remembered today, more than a century after publication of the landmark paper entitled *Die kinetic der invertinwirkung*[21, 22] ("The kinetics of invertase action").

Notwithstanding, the use of the initial rate equation also had its hurdles and difficulties. Without adequate methods for direct fitting of the hyperbola (Eq. 4.18) to the experimental data, estimation of the asymptote ($K_3$ in Eq. 4.18) was a serious limitation that had to be circumvented. To explain their approach more realistically, let's use Michaelis–Menten terminology to initial rate measurements:

$$V = \frac{[S]}{[S]+k} \tag{4.21}$$

Contrary to the meaning of the symbols that became popular much later, in this case $V$ (not to be confused with $V_{max}$) was the dimensionless fraction of maximum velocity ("V is a function that is proportional to the true starting velocity"; they wrote without explicitly mentioning the concept of maximal velocity), and $k$ is the dissociation constant of the enzyme–substrate complex. Establishing a parallel with the association curve of an acid, Michaelis and Menten proposed plotting $V$ as a function of log[s], not [S] itself. The advantage was that the maximal value $V = 1$ is the asymptote and the value of the $x$ axis for $V = 1/2$ is $k$. $k$ could then be determined. Setting s = log[S] Eq. (4.21) becomes:

$$V = \frac{10^{[S]}}{10^{[S]}+k} \tag{4.22}$$

[23] Cornish-Bowden A (2015) One hundred years of Michaelis-Menten kinetics. Perspect Sci 4:3–9.

Differentiation of this equation defines the slope of the tangent of the specified part of the curve. The "association curve" has a region whose slope is especially simple to determine, since it is practically linear over an extended stretch (see next figure). This is the middle of the curve, in particular around the region where the ordinate is ½, which corresponds to log($k$) on the abscissa. Michaelis and Menten demonstrated that the middle, almost linear, part of the curve has a slope of 0.576, i.e., forms an angle of quasi-30° relative to the abscissa as tan(30°) ≈ 0.576. So, if the experimental data reach high enough values of $V$ to allow determination of the asymptote, velocities are normalized to 1, and $k$ is directly determined from the experimental value log($k$) at $V = 1/2$. If this is not the case and the asymptote is not attainable at the highest possible values of [S] reachable in experimentation, then Michaelis and Menten proposed a recalculation of the experimental $V$ values, in arbitrary units related to the technique of detection, by adjusting $V$ values until a tangent of 0.576 (30° angle) is retrieved; $V$ values recalculated this way enable identification of log($k$) at which $V = 1/2$. This is a clever approach to solve the problem of kinetic parameters determination but is not all that straightforward.

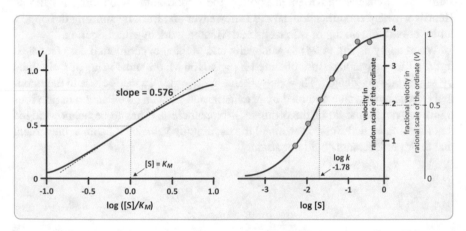

Enzyme kinetics data analyses proposed by Michaelis and Menten. Without the practical possibility of fitting the hyperbolic function of Eq. (4.21) to the experimental data, they plotted $V$ against log[S], as generally depicted in the left panel. The hyperbolic function becomes sigmoid with asymptote $V = 1$ as [S] → ∞. The slope of the tangent straight line at the data point (log($k$), $V = 1/2$) is 0.576. The right panel shows the results obtained by Michaelis and Menten[22] for the initial rate of sucrose cleavage (expressed in output units of the polarimeter—decrease in rotation in degrees per minute, referred to as "random scale" in this figure) as function of the logarithm of initial molar concentration of sucrose. The "rational scale of the ordinate" was calculated so that the slope of the tangent to the linear middle part of the curve was 0.576. At $V = 1/2$ in the "rational scale" (i.e., the true fractional velocity), the abscissa corresponds to log($k$) = −1.78, from which $k$ is determined. Figure in the right panel was adapted from the article cited in the footnote 22

Michaelis and Menten were obviously knowledgeable in mathematics, namely, differential calculus. It is amazing how they overlooked the most obvious way to transform the hyperbolic equation (Eq. 4.21) (or their starting point, Eq. 4.18) in order to calculate the kinetic parameter involved. Even more surprising, the obvious solution remained elusive for 21 more years. It was only in 1934 that Hans

Lineweaver and Dean Burk performed the linearization of the hyperbolic Henri/ Michaelis–Menten equation "upon taking the reciprocal of both sides"[24]:

$$v = \frac{V_{max}[S]}{K_S + [S]} \Leftrightarrow \tag{4.23}$$

$$\frac{1}{v} = \frac{K_S}{V_{max}} \frac{1}{[S]} + \frac{1}{V_{max}} \tag{4.24}$$

(closer to a modern notation, $v$ is the initial rate, $V_{max}$ is the maximum velocity achieved in the limit in which all enzyme is bound to substrate, $K_S$ is the equilibrium constant of enzyme–substrate binding, $K_S = [E][S]/[ES]$).

So, $1/v$ plotted against $1/[S]$ is expected to follow a straight line, easy to extrapolate, easy to plot, and easy to estimate parameters using linear regression statistics. So simple it dazzles anyone that it did not occur Michaelis or Menten to do the same. As Michaelis and Menten spotted the opportunity of building on top of Henri's work by simplification, so did Lineweaver and Burk by spotting the opportunity of building on top of Michaelis and Menten work by simplification.

When applying Eq. (4.24) to Michaelis and Menten experimental data (see next figure), Lineweaver and Burk obtained $K_S = 0.0166$ M, the same value of $k$ obtained by Michaelis and Menten. They also obtained $V_{max} = 3.94$, a value close to the maximal experimental value obtained by Michaelis and Menten (see Panel b in previous figure in "random scale of the ordinate"). Nonetheless, it has to be acknowledged that Lineweaver and Burk determined the asymptotic value of $V_{max}$, an achievement that their predecessors could not attain.

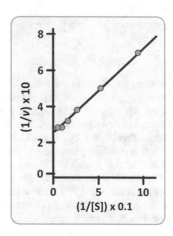

Evaluation of $K_S$ and $V_{max}$ of invertase performed by Lineweaver and Burk[24] by applying Eq. (4.24) to the data previously published by Michaelis and Menten. [S] is the molar concentration of the substrate. Figure adapted from the article cited in the footnote 24

---

[24] Lineweaver H, Burk D (1934) The determination of enzyme dissociation constants. J Am Chem Soc 56:658–666.

## Questions

1. What is the impact of the double reciprocal transformation of Eq. (4.23) (Michaelis–Menten) into Eq. (4.24) (Lineweaver–Burk) in statistical terms concerning kinetic data analysis?
2. Establish the quantitative relationship between (1) parameters in equation (4.18) ($K_3$ and $m$), (2) parameters in Eq. (4.21) ($V$ and $k$), and (3) parameters in equation (4.23) ($V_{max}$ and $K_S$), with parameters used in the contemporary formulation of Michaelis–Menten equation ($v$, $K_M$, $V_{max}$).
3. The importance of $K_M$ is praised in Box 4.5, but some researchers prefer to highlight the advantages of using the ratio $K_M/V_{max}$. Elaborate on the meaning of this ratio and its potential to unravel the role of enzymes in metabolism.
4. Once invertase acts on sucrose, fructose and glucose are available to be metabolized. Taking into account the simple sequence of reactions below, discuss the effects of fructose intolerance caused by defects of aldolase in the liver.

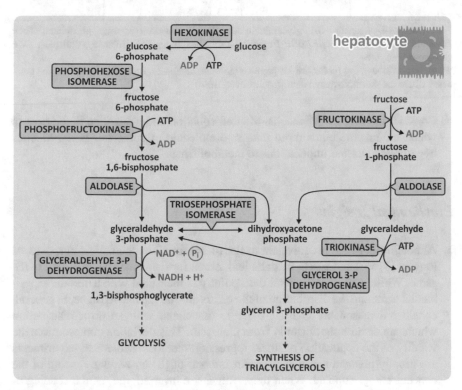

Interrelationship between hepatic glycolysis and fructolysis. Fructolysis converges to glycolysis at the level of the trioses, which follow the critical regulatory step of phosphofructokinase. This allows that fructose promotes lipid synthesis without control by insulin

5. Using the thermodynamic data in the following table and the metabolic scheme shown in Question 4, discuss the conversions of (1) sucrose into glucose and fructose and (2) the possibility of using fructose to increase glycemia.

Standard Gibbs free energy change in the chemical reactions of sucrose hydrolysis and conversion of fructose into glucose

| Reaction reactant | Products | Enzyme | $\Delta G^0$ (kcal/mol)[a] |
|---|---|---|---|
| Sucrose | Fructose, glucose | E.C.3.2.1.26/E.C.3.2.1.48 | −6.8 |
| Fructose, ATP | F1P, ADP | E.C.2.7.1.3 | −14.8 |
| F1P | GA, DHAP | E.C.4.1.2.13 | 9.5 |
| GA, ATP | GAP, ADP | E.C.2.7.1.28 | −13.5 |
| DHAP | GAP | E.C.5.3.1.1 | 1.8 |
| DHAP, GAP | F1,6BP | E.C.4.1.2.13 | −8.2 |
| F1,6BP | F6P | E.C.3.1.3.11 | 6.9 |
| F6P | G6P | E.C.5.3.1.9 | −1.4 |
| G6P | glucose | E.C.3.1.3.9/ E.C.3.1.3.58 | 7.2 |

*F1P* fructose-1-phospate, *GA* glyceraldehyde, *DHAP* dihydroxyacetone phosphate, *GAP* glyceraldehyde-3-phosphate, *F1,6BP* fructose-1,6-bisphosphate, *F6P* fructose-6-phospate, *G6P* glucose-6-phospate
[a]From https://biocyc.org (collection of pathway/genome databases, including MetaCyc, the databases where the thermodynamic data was retrieved from)

6. Consider the original Michaelis–Menten equation and generalize to the case in which an enzyme binds more than one molecule of substrate. Is the kinetics hyperbolic? Discuss implications to metabolism.

## *Biochemical Insights*

1. Although Eq. (4.23) is equivalent to Eq. (4.24) for all mathematical purposes, as long as $[S] \neq 0$ and $v \neq 0$, in statistical terms, for data analysis, is not quite the same. While each experimental data point $v$ is measured with a certain experimental error and the distribution of the errors around the mean can be, in general, considered reasonably Gaussian, $1/v$ is associated with an error distribution which can deviate significantly from Gaussian. This deviation compromises the validity of the application of linear regression methods. Moreover, experiments are usually planned to cover a certain interval of [S] by regular spacing of the concentrations sampled. When these values are inverted, the regular spacing of the interval is altered, and data points tend to cluster in the lower 1/[S] region. In the figure above exemplifying the Lineweaver–Burk plot, for instance, this effect is clear: when they plotted Michaelis–Menten data using $1/v$ vs. $1/[S]$ data points clustered at the lower 1/[S] region of the plot. This clustering impacts on the extrapolation procedure and statistical analysis. Compressing data points at high

substrate concentration into a small region emphasizes the points at lower substrate concentration (higher 1/[S] ending of the plot), which may bias the slope of the regression line.[25]

The limitations of Lineweaver–Burk method are well-known since the work carried out by … Lineweaver and Burk! As they put it in the original paper[24]: *"The relative weighting of the experimental observations alters in a definite manner when the form of an equation is altered, and if not taken into account may alter slightly the parameter constants obtained, whether graphical or analytical methods are employed. This possible disadvantage will rarely outweigh the convenience of the graphical method, where proper weighting is less easily applied."* Presently, with computer-aided non-linear regression analyses, it is possible to use Michaelis–Menten equation directly to fit the experimental data. Systematic testing of different methods to analyze enzyme kinetic data has shown that non-linear regression is the most reliable method for determining enzyme kinetic parameters. In contrast, when used without proper weighting factors, the Lineweaver–Burk plot is the less satisfactory. *"Though this is (or at least should be) known since long, this plot continues to be widely used."*[26] In fact, despite being under the fire of passionate criticism since early times, the simplicity of the Lineweaver–Burk approach has granted its endurance and resilience across several generations of biochemists. For instance, in 1965, J. Dowd and D. Riggs were peremptory: *"The marked inferiority of the Lineweaver-Burk plot strongly suggests that it should be abandoned as a method for estimating $K_m$ and $V_{max}$ from unweighted points, whether the points are fitted by eye or by the method of least squares."*[27] It never happened. The method remains popular.

2. Comparing Michaelis–Menten equation using modern nomenclature:

$$v = \frac{V_{max}[S]}{K_M + [S]}$$

with Eqs. (4.18), (4.21), and (4.23), it is possible to perform the following formal assignments:

$$m = \frac{1}{K_M}; \quad \frac{K_3}{m} = V_{max} \Leftrightarrow K_3 = \frac{V_{max}}{K_M}$$

[25] Cho Y-S, Lim H-S (2018) Comparison of various estimation methods for the parameters of Michaelis-Menten equation based on in vitro elimination kinetic simulation data. Transl Clin Pharmacol 26:39–47.

[26] Ranaldi F et al (1999) What students must know about the determination of enzyme kinetic parameters. Biochem Educ 27:87–91.

[27] Dowd JE, Riggs DS (1965) A comparison of estimates of Michaelis Menten kinetic constants from various linear transformations. J Biol Chem 240:863–869.

$$V = \frac{v}{V_{max}}; \; k = K_M$$

$$K_S = K_M$$

Using these transformations, it is clear that the hyperbolic functions used by Henri, Michaelis, and Menten and Lineweaver and Burk are all, indeed, equivalent.

3. Henri pioneered the potential use of $V_{max}/K_M$ ($K_3$ in his equation) and $1/K_M$ ($m$ in his equation) without discussing (or realizing?) their implication.

   $m$ ($1/K_M$) itself has a simple interpretation, usually named "affinity." The lower the $K_M$, the less substrate is needed for the enzyme to reach half-maximal velocity. This means that the binding of substrate to enzyme is more extensive in this case (i.e., complexation equilibrium favors the formation of the enzyme–substrate complex). The "affinity" concept quantified by $1/K_M$ arises therefrom.

   $K_3$ ($V_{max}/K_M$) is a more complete parameter. It reaches higher values for enzymes combining high $V_{max}$ and low $K_M$, i.e., highly active enzymes even at low concentrations of substrate. In this sense, $K_3$ provides a better quantification for the catalytic performance of an enzyme.

   $K_3$ is named "specificity" of an enzyme because it measures the capacity of the enzyme to discriminate between two competing substrates that are available simultaneously being thus of high relevance in studies of metabolism and toxicology. In mathematical terms, it has been demonstrated[23] that:

$$\frac{v_A}{v_B} = \frac{K_{3,A}}{K_{3,B}} \frac{[A]}{[B]} \tag{4.25}$$

where A and B refer to competing substrates of the same enzyme. For equal concentrations of A and B, the velocity at which each one is processed depends strictly on the "specificity" for each. If $K_{3,A}$ is much higher than $K_{3,B}$ (i.e., the specificity for A is much higher than for B), then $v_A$ is much higher them $v_B$, i.e., A will be processed at much higher rate than B.

Taking the analogous example of the enzyme AMPK (AMP-activated protein kinase), which binds both AMP and ADP, with $K_M$ of 80 μM and 50 μM, respectively, and an estimated ratio of corresponding maximal velocities[28] of reaction of $V_{max,AMP}/V_{max,ADP} = 1.5$, the relative $K_3$ values are $K_{3,AMP}/K_{3,ADP} = 1.5 \times 50/80 = 0.94$. Specificity for both ligands is thus similar, meaning that AMPK does not clearly

---

[28] Gowans GJ et al (2013) AMP is a true physiological regulator of AMP-activated protein kinase by both allosteric activation and enhancing net phosphorylation. Cell Metab 18:556–566.

distinguishes ADP from AMP in terms of binding. Discussion on the metabolic implication of this property will be carried out in Chap. 10 (Box 10.4).

4. Invertase present in small intestine cells converts sucrose into glucose and fructose. Also, at the small intestine cells, fructose is converted to glucose (see next figure) in a process that in the end is coincident with the last steps of gluconeogenesis in the liver (see Sect. 9.3).

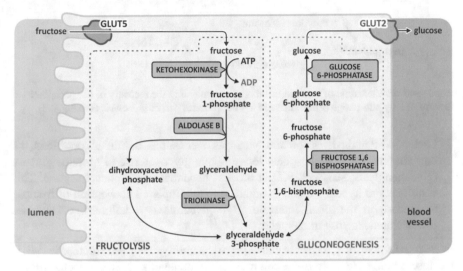

Metabolism of fructose in the small intestine (enterocytes) highlighting the branches in common with fructolysis and gluconeogenesis (see Sect. 9.3)

However, depending on the diet, the influx of fructose may be such that not all fructose is converted to glucose in the intestine and fructose reaches the blood. In this case, fructose is processed in the liver through the metabolism shown in Question 4. Fructose is converted to metabolites of glycolysis and triacylglycerol synthesis. In both cases (liver and intestine cells), impairment of aldolase B causes depletion of free phosphate ions as they are sequestered in F1P, which accumulates in the cells. The immediate consequence is loss of capacity to convert ADP into ATP and consequently a negative impact on all ATP-dependent metabolic processes. While in the full fructose metabolism the phosphate is recovered, aldolase B impairment abrogates such recovery. This is better illustrated in the following figure:

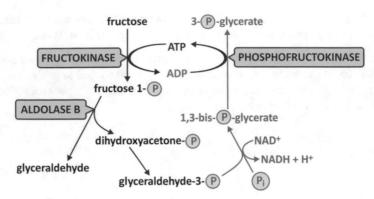

Condensed representation of fructose metabolism highlighting the recovery of the phosphate ion to form ATP by phosphoglycerate kinase in a process that is part of glycolysis (see Fig. 6.5)

Lack of ATP in the liver potentially causes liver failure. In milder conditions, the unavailability of free phosphate ions impairs glycogen use to produce glucose (glycogenolysis; see Sect. 9.2), a process dependent on a phosphorylase. This contributes to hypoglycemia, in addition to the impaired conversion of fructose to glucose in the small intestine. Hypoglycemia is a hallmark of aldolase B-associated fructose intolerance.

In essential fructosuria, a disease caused by the inability to convert fructose into F1P, fructose is neither retained in the cell interior nor metabolized. The concentration of fructose in blood increases, which is accompanied by an increase of the concentration in the urine. Not surprisingly, in contrast to fructose intolerance associated with aldolase B, essential fructosuria is an asymptomatic and not dangerous condition.

5. The hydrolysis of sucrose into fructose and glucose has negative standard Gibbs free energy change. It is therefore considered a spontaneous process.

   The table provided above has the standard Gibbs free energy change for each reaction of the fructose to glucose conversion. Using the additivity of $\Delta G^0$ for consecutive reactions, it is possible to determine the $\Delta G^0$ for the overall process:

(a) For the fructose $\rightarrow$ glucose pathway using the GA (glyceraldehyde) metabolite:
   $\Delta G^0 = -14.8 + 9.5 - 13.5 - 8.2 + 6.9 - 1.4 + 7.2 = -14.3$ kcal/mol

(b) For the fructose $\rightarrow$ glucose pathway using the DHAP metabolite:
   $\Delta G^0 = -14.8 + 9.5 - 8.2 + 6.9 - 1.4 + 7.2 = -0.8$ kcal/mol

$\Delta G^0$ is clearly negative in the first case, meaning that conversion of fructose in glucose is favorable and powered by the GA + ATP $\rightarrow$ GAP + ADP step, which makes it possible that fructose can be used to increase glycemia.

(Note: The DHAP–GAP conversion was not considered, as its influence is only significant in case there is an extra source of any of these metabolites besides

F1P. F1P originates equimolar quantities of GA and DHAP in case of full conversion of GA into GAP—supported by a very negative $\Delta G$ in this case, $-13.5$ kcal/mol—there will be equimolar quantities of DHAP and GAP to react, making DHAP–GAP conversion not so important in practical terms).

6. Michaelis and Menten considered the case in which the fractional velocity ($V$ in their nomenclature) was dependent on [S] by:

$$V = \frac{[S]}{[S] + k}$$

$k$ being the dissociation constant of the enzyme–substrate complex, $k = \frac{[E][S]}{[ES]}$. $V$ can be represented by:

$$V = \frac{v}{V_{max}} = \frac{[ES]}{[E] + [ES]}$$

as the velocity of reaction is proportional to the concentration of the complexed enzyme, [ES], and the maximum velocity is attained when all enzyme is complexed, i.e., maximal velocity is proportional to the total concentration of enzyme, [E] + [ES].

Considering an enzyme having multiple binding sites ($n$) for substrates:

$$E + nS \leftrightarrows ES_n$$

The ratio of bound to total enzyme is proportional to $V$, which in this case is:

$$V = \frac{v}{V_{max}} = \frac{[ES_n]}{[E] + [ES_n]}$$

The association constant is:

$$K_a = \frac{[ES_n]}{[E][S]^n}$$

Combining both equations:

$$\frac{v}{V_{max}} = \frac{K_a [E][S]^n}{[E] + K_a [E][S]^n}$$

Which simplifies into:

$$v = \frac{V_{max}[S]^n}{K_d + [S]^n}$$

$K_d$ is the dissociation constant ($1/K_a$).

This is equivalent to the Michaelis–Menten equation in the special case in which $n = 1$, i.e., an enzyme that binds only one substrate molecule. $K_d$ corresponds to $[S]^n$ at which $V = \frac{1}{2}$ (i.e., $v = V_{max}/2$).

   As in the Lineweaver–Burk approach, this equation can also be linearized by inversion of both sides and rearrangement:

$$\frac{1}{v} - \frac{1}{V_{max}} = \frac{K_d}{V_{max}} \times \frac{1}{[S]^n}$$

If $v \ll V_{max}$, this equation is simplified to:

$$\frac{1}{v} = \frac{K_d}{V_{max}} \frac{1}{[S]^n} \Leftrightarrow$$

$$\log v = n \log[S] + \log\left(\frac{V_{max}}{K_d}\right)$$

Plotting $\log(v)$ vs. $\log[S]$ in the low $v$ regime yields a straight line having slope $= n$ and intercept $= \log(V_{max}/K_d)$. $V_{max}$ and $K_d$ cannot be individually known but the specificity ($V_{max}/K_d$) of the enzyme can. $v$ vs. $[S]$ no longer follow and hyperbolic variation (except for $n = 1$), being a sigmoid instead (see next figure). The higher $n$ is, the steeper is the sigmoid. Higher $n$ means that the function becomes closer to a step function: at $[S]^n$ close to $K_d$, $v$ changes suddenly from very low to very high, i.e., the enzyme becomes close to a switch: for $[S]$ below $\sqrt[n]{K_d}$, there is no catalysis, in practice, while for $[S]$ above the critical value of $\sqrt[n]{K_d}$, catalysis occurs at maximal velocity. It is an all-or-nothing mechanism that may switch on or switch off metabolic pathways using only one key enzyme for regulation.

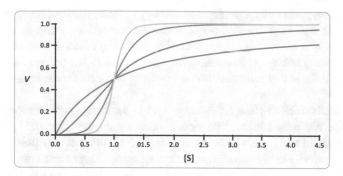

Example of kinetic curves expected for an enzyme (or enzyme cluster) having $K_d = 1$ that binds 1 (blue curve), 2 (orange curve), 5 (gray curve), or 10 (yellow curve) substrate molecules

## *Final Discussion*

Victor Henri's biography is very rich and interesting, described in a short and eluci-dative biosketch by Athel Cornish-Bowden[23]: *"Henri himself pursued an extremely varied career. Before his studies of enzymes he worked in experimental psychology, and was the first collaborator of Alfred Binet, the pioneer of intelligence testing. He received his second doctorate in 1903 on the basis of his thesis on diastases, but he appears to have done no further work on invertase. He was Professor of Physiology in Paris, and afterwards was responsible for the organisation of the chemical indus-try of Russia for defence. After the First World War he spent 10 years at the University of Zürich, after which he was to be in charge of a planned great institute of petro-chemistry at Berre L'Étang (near Marseilles). However, he moved to Science Faculty in Liège before this was finished. His later work was mainly in physical chemistry, with a particular interest in the use of absorption spectra as a source of information about molecular structures."*

Leonor Michaelis was born in Berlin (1875), where he graduated as a physician and became (unpayed) professor at the local university. His major motivation, like Henri's, was to put studies of enzymes on the framework of physical chemistry.[23] His work on enzyme inhibition distinguishing between competitive inhibition, char-acterized by the increase in $K_M$, and noncompetitive inhibition, characterized by a drop in $V_{max}$, remains the hallmarks of enzyme inhibition despite modern refine-ments on this topic.[23] Leonor Michaelis spent 4 years in Japan, Nagoya, as Professor of Biochemistry, where he helped the development of biochemistry as a scientific discipline. After that, he moved to the USA (John Hopkins University, Baltimore, and Rockefeller Institute, New York). In the USA, he contributed to establish that radial semiquinones are intermediates in some biological redox reactions, which is now well accepted but was highly controversial when Michaelis proposed it.

Maud Menten had a more discrete path. She moved away into other areas and left enzymology behind, as explained by Athel Cornish-Bowden[23]: *"As Menten's primary interest was in experimental pathology (…) she rather faded from the view of biochemists. Among her various important contributions one can mention her development of a method of histochemical detection of alkaline phosphatase in the kidney that was considered by a major textbook of the 1950s to have revolutionised the field, and the use of sedimentation and electrophoresis for detecting haemoglobin variants."*

The contributions of Hans Lineweaver and Dean Burk to enzymology were circumstantial. When the landmark paper of both was published,[24] in 1934, Lineweaver was 26 years old and Burk was 30. Burk got his PhD in plant nutrition and chemistry. They were not exactly experienced enzymologists. Their contribution was methodological. Lineweaver showed Burk, his boss, the new plot and Burk immediately realized its value. They submitted a paper on the plot to the *Journal of the American Chemical Society, JACS*, the reference journal of the prestigious American Chemical Society, but the referees were opposed to its publication for reasons that "nowadays, if not then, make entertaining reading"[29], in the words of Dean Burk himself. Some reviewers considered the paper "just a mathematical exercise and not really chemistry at all."[30] Actually, a total of six referees in two rounds of review gave the manuscript thumbs-down but the editor of *JACS* Arthur Lamb, using editorial privilege, published it anyway. The editor's judgment was correct as the paper went on to become by far the most highly cited paper ever to appear in *JACS*.[30]

Fifty-one years after the publication of *JACS* paper, Lineweaver was characteristically humble about his contribution[31]: *"Why the many citations? The paper revealed no new fundamental concepts or profound results. It did describe, with examples, a simple treatment of enzyme kinetic data that yielded straight-line plots if the data are consistent with a postulated mechanism, and these can be extrapolated easily to yield characterizing constants of the enzyme. (…) It was perhaps serendipity that the paper was timely, having appeared just after interest in enzyme research was increased by the Nobel Prize-winning proof of John H. Northrop, Wendell M. Stanley, and James B. Sumner that enzymes are proteins rather than some phantom substances."*

Lineweaver went on to obtain his PhD in 1936 from John Hopkins University. *"Then I got a better job."*[31] Burk later became a leading authority on photosynthesis, receiving the American Chemical Society's Hillebrand Prize in 1952 for his work in this area. During more than 30 years at the National Cancer Institute, he has made important contributions to other areas of biochemistry, including cancer research.

---

[29] Burk D (1984) Enzyme kinetic constants: the double reciprocal plot. Trends Biochem Sci 9:202–204.

[30] Dagani R (2003) Straightening out enzyme kinetics. Chem Eng News 81:27.

[31] Lineweaver H (1985) This week's citation classic. Curr Cont 11:19.

## Challenging Case 4.3: Building Models in Metabolism, Descriptive, Predictive … Useful!

### *Source*

Exceptionally, this case does not have a specific source from the literature, art or cinema. A simple model having similarities with saccharose metabolism is devised. The model is able to describe some features of experimental observations despite the blunt contrast between the simplicity of the model and the complexity of metabolism.

### *Case Description*

Consider the schematic metabolism with metabolites from A to L shown in the following figure:

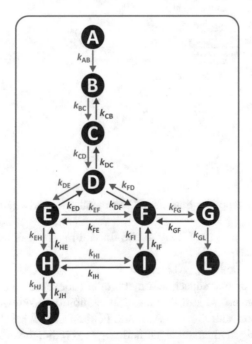

Scheme of reactions for a prototypical metabolism. $k_{xy}$ are the rate constants for the conversion of $X$ into $Y$, which represent any metabolite A–L

You will learn in Chaps. 6 and 7 that the schematic set of reactions shown in the previous figure can approximately describe the metabolism of glucose in glycolysis and/or the metabolism of fructose. The scheme can also be used to gain insight into the chemical processing of saccharose in cells. Saccharose is a dimer formed by one monomer of glucose and one monomer of fructose. If one's attempt is to study glycolysis and fructolysis, the metabolites A–L represent the compounds listed in the next table. Yet, we are not so ambitious at this point. Our goal is to illustrate, for the sake of didactics, how elaboration on simple representations of reality may provide clues to explain important experimental observations.

Correspondence of metabolites to establish a parallelism with the scheme of reactions shown in the previous figure and the metabolism of glucose and/or fructose

| Metabolite | Corresponding metabolite in glycolysis or fructolysis |
|---|---|
| A | Glucose |
| B | Glucose-6-phosphate |
| C | Fructose-6-phosphate |
| D | Fructose-1,6-bisphosphate |
| E | Dihydroxyacetone phosphate |
| F | Glyceraldehyde-3-phosphate |
| G | Pyruvate |
| H | Frutose-1-phosphate |
| I | Glyceraldehyde |
| J | Fructose |
| L | Acetyl-CoA |

## Questions

1. What are the metabolites that "feed" the metabolic pathway represented in the previous figure?
2. The end product is L, but L may not be formed in all possible conditions. What happens if $k_{GL} = 0$?
3. In practice, what does $k_{GL} = 0$ means?
4. Imagine that L or one subsequent metabolite (not represented) impacts on the enzyme that converts C into D, causing its inhibition. How would this situation impact on the concentrations of A, J, and L, assuming $[A] = [J]$ at $t = 0$?
5. Develop a simple mathematical methodology to predict the evolution of the concentrations of the metabolites for a specific set of rate constants ($k$'s). Suggestion: start by computing velocities of metabolite consumption or formation followed by approximate numerical integration. Assume all metabolite concentrations are small enough for enzymatic catalysis in the linear velocity regimen.

## *Biochemical Insight*

1. A and J are the starting points. In theory, any metabolite can have a sudden increase in concentration, and the metabolism is responsive. However, metabolisms are usually represented systematically to have starting points with metabolites expected to have concentration fluctuation caused by exogenous factors, such as the uptake of nutrients following digestion and absorption in the gastro-intestinal tract, such as in glycolysis.

2. If $k_{GL} = 0$, this means G is not converted in L (in mathematical terms, the velocity of G → L conversion is zero). If all other rate constants ($k$'s) are not nil, all other metabolites, except A, accumulate. The final concentration of metabolites will depend on the relative value of $k$'s. A does not accumulate because A → B is irreversible.

3. $k_{GL} = 0$ means, in practice, that the conversion of G to L is not operative. The enzyme that catalyzes this conversion is inhibited or has not been expressed.

4. At a certain point, L may reach a concentration that slows considerably the consumption of A. In contrast, J consumption is not affected, which leads to continuous production of L. If L is not cleared at the same or higher rate, it will accumulate and further inhibit consumption of A. Overall, J is consumed at a higher rate than A due to retroinhibition of the reactions starting in A, but not in J.

5. The velocities at which metabolites are converted correspond to changes in concentration per unit of time. The net velocity corresponds to the balance between velocity of formation and velocity of consumption. For metabolite B, for instance, the net velocity is:

$$v_B(t) = \frac{\partial [B](t)}{\partial t} = k_{AB}[A](t) - k_{BC}[B](t) + k_{CB}[C](t) \qquad (4.26)$$

$k_{AB}[A](t)$ and $k_{CB}[C](t)$ are the velocities of formation of B from A and C, respectively, and $k_{BC}[B](t)$ is the velocity of consumption of B to form C. $\dfrac{\partial [B](t)}{\partial t}$ is the velocity of conversion of B at instant $t$, $v_B(t)$. For short, albeit not infinitesimal, time intervals ($\Delta t$), $\dfrac{\partial [B](t)}{\partial t} \approx \dfrac{\Delta [B](t)}{\Delta t}$, and so:

$$\frac{\Delta [B]}{\Delta t} \approx v_B(t) \Leftrightarrow \frac{[B](t + \Delta t) - [B](t)}{\Delta t}$$
$$= v_B(t) \Leftrightarrow [B](t + \Delta t) = v_B(t)\Delta t + [B](t) \qquad (4.27)$$

This means that the concentration of B at instant $t + \Delta t$, $[B](t + \Delta t)$, can be calculated from the concentration of B at instant $t$, $[B](t)$, inasmuch as $\Delta t$ is small. Specifically, for metabolite B, combining Eqs. (4.26) and (4.27):

$$[B](t + \Delta t) = \left(k_{AB}[A](t) - k_{BC}[B](t) + k_{CB}[C](t)\right)\Delta t + [B](t) \qquad (4.28)$$

Starting from $t = 0$ with an initial concentration set of the metabolites, step by step ($\Delta t$, $2\Delta t$, $3\Delta t$, etc.), it is possible to predict how concentrations vary. Computing the equivalent of Eq. (4.28) to all metabolites in a spreadsheet calculator helps. Velocities are:

$$\frac{\partial[A](t)}{\partial t} = -k_{AB}[A](t)$$

$$\frac{\partial[B](t)}{\partial t} = k_{AB}[A](t) - k_{BC}[B](t) + k_{CB}[C](t)$$

$$\frac{\partial[C](t)}{\partial t} = k_{BC}[B](t) + k_{DC}[D](t) - k_{CB}[C](t) - k_{CD}[C](t)$$

$$\frac{\partial[D](t)}{\partial t} = k_{CD}[C](t) + k_{ED}[E](t) + k_{FD}[F](t) - k_{DC}[D](t) - k_{DE}[D](t) - k_{DF}[D](t)$$

$$\frac{\partial[E](t)}{\partial t} = k_{DE}[D](t) + k_{FE}[F](t) + k_{HE}[H](t) - k_{ED}[E](t) - k_{EF}[E](t) - k_{EH}[E](t)$$

$$\frac{\partial[F](t)}{\partial t} = k_{DF}[D](t) + k_{EF}[E](t) + k_{GF}[G](t) - k_{FD}[F](t) - k_{FE}[F](t) - k_{FG}[F](t) + k_{IF}[I](t) - k_{FI}[F](t)$$

$$\frac{\partial[G](t)}{\partial t} = k_{FG}[F](t) - k_{GF}[G](t) - k_{GL}[G](t)$$

$$\frac{\partial[H](t)}{\partial t} = k_{EH}[E](t) + k_{JH}[J](t) + k_{IH}[I](t) - k_{HE}[H](t) - k_{HJ}[H](t) - k_{HI}[H](t)$$

$$\frac{\partial[I](t)}{\partial t} = k_{HI}[H](t) + k_{FI}[F](t) - k_{IH}[I](t) - k_{IF}[I](t)$$

$$\frac{\partial[J](t)}{\partial t} = k_{HJ}[H](t) - k_{JH}[J](t)$$

$$\frac{\partial[L](t)}{\partial t} = k_{GL}[G](t)$$

The following graph represents three sets of results, obtained with three different sets of initial metabolite concentrations and $k$'s. Other parameters would lead to different outcomes. The realism of a model is very much dependent on the accuracy of the estimates of concentrations and $k$'s.

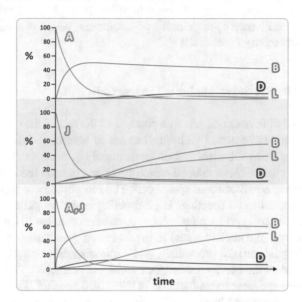

Variation in four selected metabolites' concentrations (% relative concentration to total sum of concentrations) for three initial distinct conditions. *Top panel*: all concentrations set to zero, at $t = 0$, except A, and all $k$'s equal to 1, except $k_{CD} = 0.1$, and $k_{HE} = k_{IF} = k_{EH} = k_{IH} = k_{JH} = k_{FI} = k_{HI} = k_{HJ} = 0$ (vaguely, a glycolysis-like situation). *Middle panel*: all concentrations set to zero at $t = 0$, except J, and all $k$'s equal to 1, except $k_{AB} = k_{BC} = k_{CD} = k_{HJ} = 0$ (vaguely, a fructolysis-like situation). *Lower panel*: all concentrations set to zero at $t = 0$, except A and J (equal concentrations), and all $k$'s equal to 1, except $k_{CD} = 0.1$ and $k_{HJ} = 0$ (vaguely, a saccharose metabolism-like situation)

## *Final Discussion*

### Fructose, Obesity, and Insulin Resistance

The conditions set in the graph are such that they are illustrative of the impact of the metabolism of fructose in obesity. Glycolysis is regulated at the level of metabolite C to D conversion. L, which stands for acetyl-CoA or any generic product of glycolysis, never reaches high levels because conversion of C is slowed down. Instead, B accumulates and becomes available to other processes. In this case, B stands for glucose-6-phosphate, which can be used in glycogen synthesis. If glycogen stores are full, glycemia remains high for longer times, and insulin resistance arises as a long-term consequence. At variance, in the middle panel of the graph, which illustrates a fructolysis-like metabolism, fructose (J) will cause a high increase in the end products (L) because no retroinhibition mechanism is operative. This situation will favor lipid synthesis and the occurrence of obesity in the long term. Saccharose metabolism (modeled in the lower panel of the graph) is a combination of glucose and fructose metabolisms. L and B reach high levels. Lipid synthesis and insulin resistance are then favored by high intake of saccharose, which is a matter of debate in medical biochemistry literature.[32, 33]

## A Suitable Energetic Substrate for a Tireless Cell

One special kind of human cell acts independent of the rest of the cells of the body; is highly responsive to external stimulus, but not to hormones from its own body; and has an energetic metabolism that depends mainly on fructose, not glucose. We are talking about the only human cell which functions outside the body in which it was created: the spermatozoon or sperm cell. The morphology of this kind of cells is extremely adapted to its function. High mobility and minimal cargo dictate the architecture of the cell and its metabolism. Architecture is based on the storage a nearly single cargo (the nucleic acids) in the so-called "head" region (see figure), a powerful uniflagellar tail for propelling, and a mitochondria-rich "power plant" in the mid-piece where the energy needed for mobility is processed. The advantages of such architecture are clear when function is taken into account. The same does not hold for metabolism. The advantage of having fructose as main nutrient, instead of glucose, used by all cells in the same organism, remains puzzling.

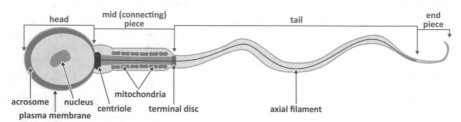

Sperm cell structure. Sperm cells have evolved under strong selective pressure. They became small and highly motile, lacking endoplasmic reticulum, Golgi apparatus, and ribosomes and having scarce amounts of cytoplasm. Sperm cells are specialized transporters of the nucleic acid carried in their head

---

[32] Elliott JJ et al (2002) Fructose, weight gain, and the insulin resistance syndrome. Am J Clin Nutr 76:911–922.

[33] Tappy L et al (2013) Effects of fructose-containing caloric sweeteners on resting energy expenditure and energy efficiency: a review of human trials. Nutr Metab 10:54.

In 1946, T. Mann[34] has shown that the most abundant saccharide present in the seminal fluid of the bull, ram, rabbit, boar, and man is fructose, not glucose. Glucose is also present but to a lesser extent. Fructolysis is the dominant process rather than glycolysis, a specialization for which the advantage is not obvious. One possible explanation is that fructose (rather than glucose) present in the seminal fluid is less susceptible to competitive use by microorganisms present in the path to the ovum. The other, more plausible may be that fructose usage is not limited by the feedback control mechanism inherent to glycolysis at the phosphofructokinase level (conversion of metabolic C to D in the first figure), so fructose can be used freely at the high rate demanded by the strenuous effort of racing to reach the ovum. Citrate and ATP resulting from fructolysis may inhibit glycolysis but not fructolysis, taking to the maximum the power to convert the precursor nutrient (fructose in this case) into energy (ATP) to be used in flagellar movement. The role of glucose could then be the production of NADPH through the pentose phosphate pathways (see Sect. 8.3.1.3).[35] This pathway assures the protection against harsh oxidant species produced at a high rate by the high density of active mitochondria during the sperm race.

*This case had the collaboration of Prof. Cláudio M. Soares (Instituto de Tecnologia Química e Biológica António Xavier, Universidade Nova de Lisboa, Portugal).*

[34] Mann T (1946) Studies on the metabolism of semen. Biochem J 40:481–491.
[35] Visconti PE (2012) Sperm bioenergetics in a nutshell. Biol Reprod 87:72.

# Chapter 5
# Regulation of Metabolisms

In a cell, the fluxes of matter and energy are highly controlled so that cells can maintain their organization and multiply when needed. We have discussed in the previous chapter that consecutive reactions can be driven through coupling favorable reactions to unfavorable reactions, many of those benefiting from the $\Delta G^0$ of ATP hydrolysis. Synthesizing ATP involves using energy associated with the chemical processing of nutrients or molecules stored for this purpose (catabolic metabolism). When in excess, the nutrients tend to engage a series of reactions whose end products are the storage molecules (anabolic metabolism) for later use. This shift implies a complex network of metabolisms that must be inhibited or activated. Inhibition and activation occur selectively at specific reactions which in turn occur in specific locations and precise timings inside the cells. In addition, the shift requires that different cells in the same tissue or in different tissues operate coordinately. Liver, adipose tissue, muscle, and brain, for instance, need to be coordinated so that when the brain and muscle require specific nutrients to operate, this process does not conflict with processes in other organs and no failure of the body function as a whole occurs.

Only specific selected sets of reactions take place at a given time in each organelle of a cell. Chemical entities (hormones) circulate in the body and are captured by cell receptors that trigger short chemical reaction sequences, generically named "signal transduction pathways" (see Box 5.1 and Fig. 5.1) upon binding. In the end, these short sequences of chemical reactions modify enzymes stimulating or inhibiting the catalysis of metabolic reactions (Fig. 5.1). The same hormone may be sensed by different cells in different organs. The metabolic events triggered in different organs are not necessarily the same but are coordinated. For instance, during prolonged aerobic exercise, muscles are consumers of fatty acids, which must be mobilized from the adipose tissues; in this situation both the muscle and the adipose tissue are not activating the same metabolic pathways, but they are certainly coordinated. A drop in glycemia (glucose concentration in blood) leads to the release of glucagon, a peptide hormone synthesized in the pancreas that binds to receptors in hepatocytes (liver cells) and triggers events that activate reactions leading to the

© Springer Nature Switzerland AG 2021
A. T. Da Poian, M. A. R. B., *Integrative Human Biochemistry*,
https://doi.org/10.1007/978-3-030-48740-9_5

production of glucose (see Chap. 9). Glucose is then released in the blood through the optimized mesh of capillaries throughout the whole liver. This matter will be revisited later in Box 8.1.

---

**Box 5.1  Biosignaling, the Communication Among Cells and Inside Cells**

In a complex organism, such as the human body, having specialized systems, with specialized organs, specialized tissues, and specialized cells, coordination requires fine-tuning and reliability. The homeostasis of the human body requires that the action of different organs is not conflicting. Imagine a situation such as prolonged starvation. The liver synthesizes glucose and releases it in the blood to keep the glycemia within safe levels for the brain to operate. What would happen if other organs such as striated muscle were subtracting glucose from the blood to synthesize glycogen, for instance? This conflict between liver action and striated muscle action would be fatal or, at least, result in a paramount waste of energy and matter. There are mechanisms that prevent conflicts of this kind. These mechanisms coordinate the action of cells that may be in contact with each other in the same tissue or in very remote locations relative to each other, such as different organs. There are molecules that serve as "signals" that are released and trigger synchronized and compatible events in different cells. This is known as cellular communication or biosignaling.

Naturally, biosignaling, such as any communication process, requires that there is a source for the signal (e.g., hormone), the means for the signal to disseminate, and a receptor, or receptors. The signal is a molecule that is synthesized, so the steps of biosignaling are (1) synthesis of the molecule that will serve as signal; (2) release of the signal molecule; (3) transport/dissemination to the target cells, i.e., cells with the receptors that bind and are responsive to the signal molecule; (4) interaction with the receptor, usually a protein, in the target cell; (5) triggering intracellular chemical or physical events that result directly from signal–receptor interaction ("signal transduction"); and (6) generation of other events, frequently a series of events ("cascade" or "signaling pathway"), which constitutes the final response of the cell to the signal.

Hormones, neurotransmitters, prostaglandins, growth factors, and cytokines, such as interleukins and interferons, serve as biochemical communication signals. These signals may act at short range or long range (see below figure). Short-range communication involves diffusion of signals in the extracellular medium in the immediate vicinity of the cell that secretes the signal molecule. This is named paracrine signaling. Synaptic signaling is an example of paracrine signaling: neuronal termini release neurotransmitters that bind to receptors in the postsynaptic cells. Long-range signaling is known as endocrine signaling. In this case, specialized cells in specialized organs synthesize hormones, which are usually released into the bloodstream and dis-

(continued)

**Box 5.1**  (continued)

tributed throughout the body. Hydrophilic hormones are soluble in blood and are easily distributed; hydrophobic hormones might need transporters and/or have severe limitations in the concentrations they can achieve in blood. Steroid hormones, for instance, are hydrophobic hormones derived from cholesterol, and they need to be associated with carrier proteins to be distributed in the body through the blood. A third form of signaling exists in addition to endocrine and paracrine: autocrine. However, this signaling is mainly found in pathological conditions such as cancers. Tumor cells release growth factors that bind to receptors in the surface of the same cell that releases them, stimulating cell growth and cell division. These events become unbalanced, and the tumors grow uncontrolled.

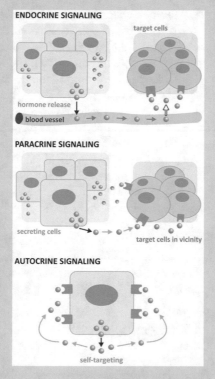

Outline of cellular communication through hormones (represented by green spheres) in endocrine, paracrine, and autocrine signaling

Most molecular signals have very high selectivity and affinity for their receptors. High selectivity means that the receptor is responsive only for a very precise structure of the ligand, i.e., the molecule that serves as signal. High affinity means that the binding equilibrium of the ligand signal to the receptor is very extensively shifted toward the bound ligand–receptor complex, which in practice means that very low concentrations of the ligand

(continued)

**Box 5.1** (continued)

generate an integrated high response from the receptor. Nanomolar ($10^{-9}$ M) concentration of hormones usually suffices to trigger physiological responses because the high affinity for the receptor compensates the low concentration of the ligand. Depending on whether the hormones may translocate through the plasma membrane or not, the receptors may be located on the surface of cells, or inside the cells, in the cytoplasm or nucleus. Adrenalin, insulin, and glucagon, for instance, have surface receptors; in contrast, testosterone and progesterone, for instance, bind to receptors intracellularly.

Upon binding of the ligand to receptor, the signal transduction process may be of three main different types depending on the functionality of the receptor:

1. When the receptor is an ionic channel, the ligand may activate or inhibit the flux of ions through the channel. This is the case of the receptor of acetylcholine in the neuromuscular junction.
2. When the receptor is coupled to G protein (guanosine nucleotide-binding protein), binding of the ligand causes conformational changes in the receptor that indirectly activate the G protein, which detaches it from the receptor and binds to adenylate cyclase (also known as adenylyl cyclase) or a phospholipase or another enzyme that catalyzes the formation of the molecules so-called second messengers. Second messengers then initiate series of reactions that will interfere with metabolic processes. Cyclic AMP, cyclic GMP, inositol triphosphate, and diacylglycerol are examples of second messengers, and the receptors of adrenaline and serotonin are examples of G protein-coupled receptors. This subject is further developed in Sect. 5.4.
3. When the receptor has catalytic tyrosine kinase activity, binding of the ligand causes conformational changes that make the enzyme active and thus able to phosphorylate proteins in Tyr residues using ATP as a source of phosphate. The insulin receptor is of this kind.

The intracellular signaling pathways that follow the binding of the signal molecule to the receptor are diverse. The main signaling pathways associated with the regulation of metabolism are addressed individually in the main text.

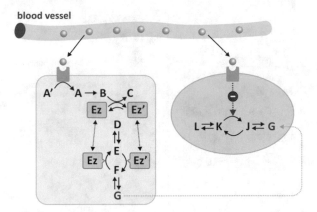

**Fig. 5.1** Schematic representation of signal transduction pathways. When the hormone (green sphere) binds to its receptor, which may be located on the membrane or may be intracellular, this translates into a short series of chemical reactions (exemplified as compounds A' to C) named "signal transduction." One of the consequences may be that one enzyme is modified ($E_z$ to $E_{z'}$ in the picture). This transformation may be phosphorylation, for instance ($E_{z'}$ being in this case $E_z$ with a covalently bound phosphate group). In this hypothetical situation in which $E_z$ and $E_{z'}$ catalyze opposing reactions irreversibly (F to E and E to F), the practical effect of the hormone is to dictate the sense of metabolic reactions. In the case depicted in this figure, the presence of the hormone in blood generates metabolite G into the cell on the left. The same hormone may be acting at the same time in different cells (colored light blue and salmon) in different organs triggering different metabolic pathways. The hormone may stimulate G-consuming pathways in cells from other organs (e.g., leading to inhibition of the enzyme that converts K in J)

Glycemia is a key factor in energetic metabolism as the brain uses almost exclusively glucose as an energy source for ATP production (the only known exception occurs in long-term starvation; see Sect. 9.3.4), and there is no mechanism to keep a higher concentration of glucose in the central nervous system than in blood because glucose transporters across the blood–brain barrier do not operate against a concentration gradient. Therefore, the coordination of metabolisms occurring in different organs simultaneously is such that glucose concentration in the blood is kept higher than a specific threshold. Values under this threshold produce loss of conscience; coma and death may occur (Fig. 5.2). Given the importance of glycemia control and the investment represented by regulatory pathways, it is not surprising that biochemists tend to overemphasize the importance of glucose as a nutrient compared to amino acids and lipids. Yet, lipids and proteins are the most important energetic reserves in the human body (see Table 9.1). Energetic metabolisms are often but erroneously associated with carbohydrate metabolism only. In reality, fatty acid, amino acids, and glucose metabolisms are interconnected. But despite the connections, the metabolites pertaining to each pathway are not necessarily interconvertible. Lipids are typically storage molecules, and they cannot be converted to glucose or proteins; it can only be converted in metabolites that in principle will follow reaction routes leading to ATP synthesis (Fig. 5.2).

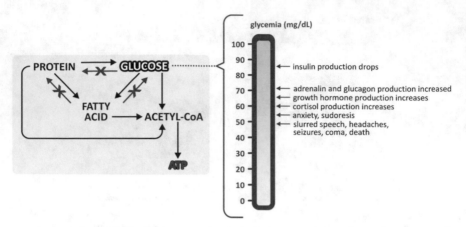

**Fig. 5.2** Allowed and forbidden conversion routes in human metabolism. Glucose and fatty acids cannot be converted to proteins. Fatty acids cannot be converted to proteins or glucose. This implies that all excess nutrients from food intake end in lipids, stored in adipose tissue. Glycemia (the concentration of glucose in blood) needs to be kept above a certain threshold, even in the absence of food intake. In this case, synthesis of glucose is possible using amino acids from muscle protein degradation

In summary, metabolisms need regulation both inside the cells and in different tissues, frequently in different organs. Therefore, different levels of regulation exist. They are associated with mechanisms with different efficacies, time scales, and areas of impact.

## 5.1  Levels of Regulation: Impact and Time Scale

Think about your simplest daily routines, like eating, moving, and sleeping. They seem extremely banal and simple, almost unnoticeable in our lives, but they are demanding challenges from the metabolic point of view. They involve relatively fast alternation between states of nutrient abundance (meals) and nutrient absence (fasting), rest (e.g., sleeping), moderate exercise (e.g., walking), exercise bursts (e.g., short run to catch a bus) or intense enduring exercise (e.g., athletic running or swimming), and all possible combinations of feeding state with exercise. At the same time, there are processes that constantly consume energy such as brain activity (any basal activity, not only mental work), heartbeat, or keeping body temperature. Metabolic adaptation to fluctuating conditions on top of basal permanent activities requires mechanisms of metabolic regulation that are (1) fast, (2) efficient, and (3) reliable. In fact, these factors are not independent: reliability comes from redundancy of mechanisms as having more than one mechanism to assure the same effect decreases the chances of failure. Moreover, redundancy is a cooperative combination of similar mechanisms, a fast one, albeit not so efficient, and a very efficient one, albeit not very fast. Usually fast regulation mechanisms consist in controlling the availability of substrates and/or the activity of enzymes locally by intracellular processes, while very efficient regulation mechanisms consist in controlling the presence of enzymes, i.e., their genetic expression, which is naturally a slower process. Genetic expression and covalent activation/inactivation of enzymes are usually dependent on hormones, which guarantee coordination among the metabolism in different organs, all under the influence of the same circulating hormones. This interplay between mechanisms is illustrated in Fig. 5.3.

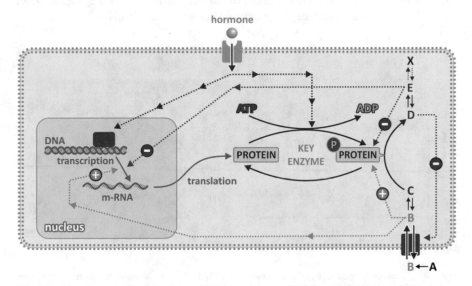

**Fig. 5.3** Generic representation of the regulation of the metabolisms. The different levels of regulation are (1) substrate availability through control of transport across membranes (metabolite B); (2) enzyme activation by upstream metabolites in the metabolic pathway (the role performed by B, for instance) and/or enzyme inhibition by the product of downstream metabolites in the metabolic pathway (the role performed by E, for instance); (3) activation or inactivation of enzymes by covalent attachment of a phosphate group (represented by the pink circle labeled with P), which is usually the end effect of a cascade of events caused by the binding of a hormone to its receptor (signal transduction); and (4) regulatory enzymes' expression/translation controlled by hormones through signal transduction inside cells. Hormone-dependent mechanisms are slower than the direct effect of metabolites on enzymes (activation or inhibition) because enzymes coexist with regulatory metabolites (named effectors) in the same cell compartment. Among the hormone-dependent mechanisms, those involving covalent modification of enzymes are faster than those involving the regulation of gene expression

## 5.2  Inhibition and Activation of Enzymes by Ligands

As mentioned in the previous section, controlling the activity of enzymes (i.e., changing their kinetic characteristics such as $K_M$ or $V_{max}$) is a fast way to influence the rate and course of metabolic pathways. Physical factors, such as temperature and pH, have a direct effect on the structure of proteins because they alter the intramolecular forces that stabilize protein folding. Therefore, they affect the activity of enzymes. Enzymes have optimal temperature and pH ranges to operate (Fig. 5.4). Below and above the optimal ranges, the reaction velocity decreases. Human enzymes are adapted to body temperature and have optimal activity around 37 °C. They are also adapted to the pH of their microenvironments. Pepsin and other stomach enzymes, for instance, have optimal activity at acidic pH.

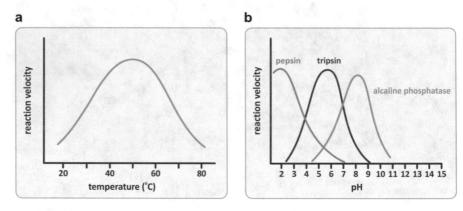

**Fig. 5.4**  Effects of temperature and pH on enzyme activity. Enzymes undergo structural alteration in their folding when physical factors such as temperature (**a**) or pH (**b**) change, which in turn cause alterations in the kinetics of catalysis. The reaction velocity is maximal in a limited interval of temperature or pH, decreasing for higher or lower values. Enzymes are adapted to the local pH of the different microenvironments of the human body from very acidic (such as pepsin) to alkaline (such as alkaline phosphatase), as illustrated in Panel (**b**)

There are microorganisms living in extreme environments such as hot springs, at low pH and high temperatures. They are called extremophiles for this reason. Nevertheless, even in these cases, there is adaptation of the enzymes to the local pH and temperature. These molecules are very appealing to the biotechnological industry because they can be used in industrial processes that combine extreme pHs with high temperature; however, they are not in the realm of human biochemistry and will not be further discussed here.

In principle enzymatic activity could be modulated by changing pH or temperature, but this is not an option as these factors do not affect specifically a single enzyme in a specific metabolic pathway and would impact on many biochemical processes inside cells. The inhibition or activation of an enzyme has to be very selective. Shutting down a metabolic pathway demands inhibition and/or activation of very specific enzymes (recall Sect. 4.2.1), which is not compatible with the manipulation of pH or temperature. Instead, having small molecules that bind specifically to unique binding sites of enzymes, affect their conformation, and influence their kinetics is a much better way to specifically modulate the activities of selected enzymes. These small molecules are called ligands, and they can slow (inhibitors) or accelerate (activators) the catalytic process.

In the case of an enzyme obeying Michaelis–Menten kinetics, ligands may in principle affect $V_{max}$, $K_M$, or both. The molecular mechanisms behind the influence of ligands on $V_{max}$ or $K_M$ may be very diverse, even if the end result is the same. Two different inhibitors may impact on $V_{max}$, for instance, binding to different sites of the same enzyme and producing different effects on enzyme structure. In other words, the alterations ligands cause on kinetic parameters tell nothing about how ligands and enzymes interact. To relate kinetics with mechanism of interaction, one has to resort to models, i.e., hypothetical arrangements and postulated events, and deduce on the final effect in kinetics. This means that the examples in Box 4.4 can be extended to the case in which an additional molecule, the inhibitor or the activator, interacts with the enzyme $E_z$, in addition to S (the substrate) and P (the product). Let's work on a simple example. Postulating that there is a ligand that binds to $E_z$ and prevents S from binding to $E_z$ (in this case the ligand acts as an inhibitor and will be represented by I), the reaction scheme is composed of:

$$E_z + S \leftrightarrows E_z S \rightarrow E_z + P \ \left(\text{The unperturbed catalysis itself}\right) \tag{5.1}$$

$$E_z + I \leftrightarrows E_z I \ \left(\text{The action of the inhibitor}\right) \tag{5.2}$$

The action of I is to prevent a fraction of $E_z$ to take part in the reaction. The effective concentration of $E_z$ available to interact with S is decreased, and, naturally, $K_M$ appears to be increased. Having less extensive binding of $E_z$ to S appears to be a decreased affinity of $E_z$ to S. In reality, the intrinsic binding of $E_z$ to S is not affected in its nature; it is the "sequestration" of $E_z$ by I that causes the alteration in $K_M$. For very high S concentration, S is so abundant relative to I that, in practice, the influence of I on binding of $E_z$ to S is not relevant and so $V_{max}$ remains constant. $V_{max}$ is $k_2 \cdot [E_z]_0$ (Box 4.4) and I does not change $k_2$ or $[E_z]_0$. The reaction velocity vs [S] plot shows a decreased slope at small [S] but the same asymptotic value ($V_{max}$) in the presence of I when compared to the plot in the absence of I (Fig. 5.5).

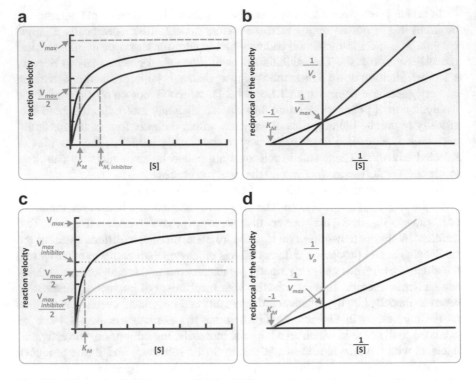

**Fig. 5.5** Effects of an inhibitor on the Michaelis–Menten kinetics of an enzyme. Changes in the enzyme velocity caused by the presence of the inhibitor may be detected in regular velocity vs [S] (**a**, **c**) or Lineweaver–Burk (**b**, **d**) plots. (**a**, **b**) $K_M$ is increased with no alteration in $V_{max}$. (**c**, **d**) $V_{max}$ is decreased with no alteration of $K_M$. Simultaneous alterations on $V_{max}$ and $K_M$ are also possible (not shown), which are the most frequent situations in practice

If I binds to $E_z$ irreversibly ($E_z + I \rightarrow E_zI$), part of enzyme population is permanently blocked by I, so the effective concentration of $E_z$ free to interact with S is decreased even at high concentrations of S. In this case $V_{max}$ is decreased (recall again $V_{max} = k_2 \cdot [E_z]_0$—Box 4.4).

In the reaction Schemes (5.1) and (5.2) shown above, $E_z$ was assumed to bind to S or I but not to both simultaneously. This is expected to occur in cases in which S and I compete for the active catalytic site of the enzyme. This mechanism is thus called "competitive." There are other reaction schemes that consider simultaneous binding of I and S to $E_z$ (which implies that they are not competing to the same binding site—"noncompetitive" inhibition):

$$E_z + S \leftrightarrows E_zS \rightarrow E_z + P \ \ \left(\text{The unperturbed catalysis itself}\right) \tag{5.3}$$

$$E_z + I \leftrightarrows E_zI \ \ \left(\text{The action of the inhibitor}\right) \tag{5.4}$$

$$E_zI + S \leftrightarrows E_zSI \ \left(\text{Enzyme binding S after binding I}\right) \tag{5.5}$$

$$E_zS + I \leftrightarrows E_zSI \ \left(\text{Enzyme binding I after binding S}\right) \tag{5.6}$$

In this case, binding of I to $E_z$ hinders the catalytic process but not the binding of S to the catalytic site itself, so $V_{max}$ is decreased but $K_M$ is not altered (Fig. 5.5). Even at high [S], a fraction of $E_z$ is bound to I. The total amount of $E_z$ available for catalysis is decreased, and $V_{max}$ is also decreased (again recall that $V_{max} = k_2 \cdot [E_z]_0$—Box 4.4). In practice, it is not frequent to find inhibitors that perturb the structures of proteins with a selective effect on $V_{max}$ leaving $K_M$ unchanged, but noncompetitive inhibition retains didactic interest.

It is interesting to note that $V_{max}$ is also decreased in cases in which $k_2$ is decreased. This may happen when $E_zSI$ in Reactions (5.5) and (5.6) above retains catalytic activity ($E_zSI \leftrightarrows E_zS + P$) but at slower rate when compared to free $E_z$. The catalytic rate constant ($k_2$) is decreased in the fraction of enzyme associated with I ($E_zI$), and so $V_{max}$ decreases.

It may also happen that $E_z$ bind ligands that have the opposite effect of inhibitors on $E_z$: to increase $k_2$, thus increasing the velocity of the catalytic process. In this case the ligand is called an "activator," as opposed to "inhibitor." There are also activators that decrease $K_M$. The binding of S to $E_z$ may be affected when the structure of the catalytic site of $E_z$ is altered by the conformational changes caused by the binding of the activator to its specific binding site in the protein.

Depending on the characteristics of the reaction schemes, namely, stoichiometries of interactions, diversity of interacting molecules, reversibility of the reactions, etc., the expected kinetics may vary greatly. Figure 5.6 has simple illustrative examples based on Michaelis–Menten kinetics.

When performing work on enzymology, one has to bear in mind that there is no direct unequivocal correspondence between the experimental kinetics, reaction schemes, and molecular mechanisms of action. One may conceive mechanisms, infer the corresponding underlying reaction scheme, deduce mathematically the associated kinetics, and study the match of the deduced kinetics to the experimental data, but the inverse is not possible. It is not possible to deduce unambiguously mechanism from experimental kinetic curves. In this way, it is precipitated to deduce a competitive mechanism in a situation in which experimental data show higher $K_M$ and equal $V_{max}$ as other mechanisms may result in similar kinetics. However, for most practical purposes in health sciences, the exact mechanisms of action are not all that relevant, and this reasoning of enzymology research practice will not be further developed.

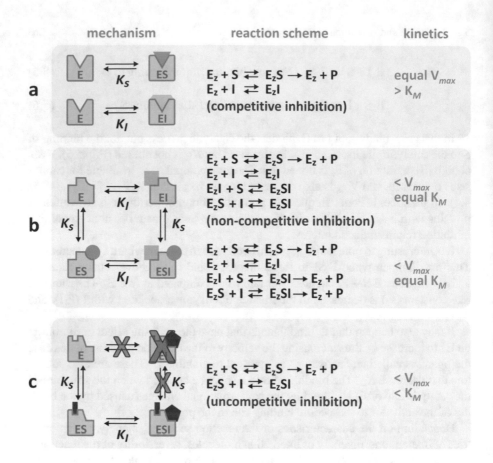

|  | mechanism | reaction scheme | kinetics |
|---|---|---|---|

**a** — $E_z + S \rightleftarrows E_zS \rightarrow E_z + P$
$E_z + I \rightleftarrows E_zI$
(competitive inhibition)

equal $V_{max}$
$> K_M$

**b** — $E_z + S \rightleftarrows E_zS \rightarrow E_z + P$
$E_z + I \rightleftarrows E_zI$
$E_zI + S \rightleftarrows E_zSI$
$E_zS + I \rightleftarrows E_zSI$
(non-competitive inhibition)

$< V_{max}$
equal $K_M$

$E_z + S \rightleftarrows E_zS \rightarrow E_z + P$
$E_z + I \rightleftarrows E_zI$
$E_zI + S \rightleftarrows E_zSI \rightarrow E_z + P$
$E_zS + I \rightleftarrows E_zSI \rightarrow E_z + P$

$< V_{max}$
equal $K_M$

**c** — $E_z + S \rightleftarrows E_zS \rightarrow E_z + P$
$E_zS + I \rightleftarrows E_zSI$
(uncompetitive inhibition)

$< V_{max}$
$< K_M$

**Fig. 5.6** (**a–c**) Four examples of reaction mechanisms and the associated kinetic alteration in Michaelis–Menten or Michaelis–Menten-like enzymes. Depending on the mechanisms and respective reaction scheme, the resulting apparent $V_{max}$ and $K_M$ may differ or not from $V_{max}$ and $K_M$ in the absence of the inhibitor, I. The first mechanism is named "competitive inhibition," but the other mechanisms bear names that are not always consensual. The second and third mechanisms are usually referred to as "noncompetitive" and the fourth as "uncompetitive"

It is worth stressing that all that has been mentioned to enzymes having a Michaelis–Menten (or a Michaelis–Menten-like) kinetics is valid with adaptations for enzymes having more complex kinetics such as sigmoidal (Fig. 5.7). Thus, the general concept of inhibiting or activating enzymes using small molecules as ligands may potentially apply to all enzymes in a metabolism. The fundamentals of sigmoidal kinetics have been addressed in discussing question 6 of challenging Case 4.2 and Fig. 4.5.

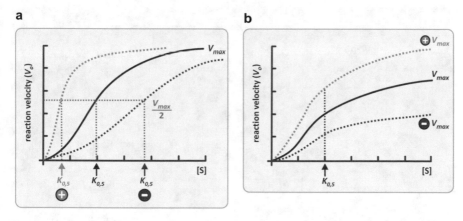

**Fig. 5.7** Effect of activation and inhibition in sigmoidal kinetics of catalysis. The impact may be on the [S] needed to achieve 50% of the maximal velocity ($K_{0.5}$) (**a**), or the maximal attained velocity, $V_{max}$ (**b**)

The importance of enzyme inhibitors in drug discovery and development is huge. Many of the drugs in current medicines target enzymes (Table 5.1) and were conceived to block processes in which those enzymes are essential. Inhibiting HMG-CoA reductase, for instance, results in the inhibition of cholesterol synthesis. One of the most common strategies to create inhibitors is to synthesize molecules that are similar enough to the substrate to bind the catalytic site but incapable of being catalyzed. Figure 5.8 shows several examples of natural substrates and similar molecules that act as inhibitors of their enzymes. In some cases, binding of the inhibitor to the enzymes is reversible, but in other cases, inhibitors react covalently with the enzyme, and the binding is thus irreversible in practice.

**Table 5.1** Examples of drugs, targeted enzymes, and field of therapy

| Drug | Target enzyme | Field of therapy |
|---|---|---|
| Aspirin | Cyclooxygenase | Anti-inflammatory |
| Captopril and enalapril | Angiotensin-converting enzyme (ACE) | Antihypertension |
| Simvastatin | HMG-CoA reductase | Lowering of cholesterol levels |
| Desipramine | Monoamine oxidase | Antidepression |
| Clorgyline | Morpramine oxidase A | Antidepression |
| Selegiline | Morpramine oxidase B | Treatment of Parkinson's disease |
| Methotrexate | Dihydrofolate reductase | Anticancer |
| 5-Fluorouracil | Thymidylate synthase | Anticancer |
| Viagra | Phosphodiesterase enzyme | Treatment of male erectile dysfunction |
| Allopurinol | Xanthine oxidase | Treatment of gout |
| U75875 | HIV protease | AIDS therapy |
| Ro41-0960 | Catechol-$O$-methyltransferase | Treatment of Parkinson's disease |
| Omeprazole | $H^+/K^+$ ATPase proton pump | Ulcer therapy |
| Organophosphates | Acetylcholinesterase | Treatment of myasthenia gravis, glaucoma, and Alzheimer's disease |
| Acetazolamide | Carbonic anhydrase | Diuretic |
| Zileuton | 5-Lipoxygenase | Anti-asthmatic |

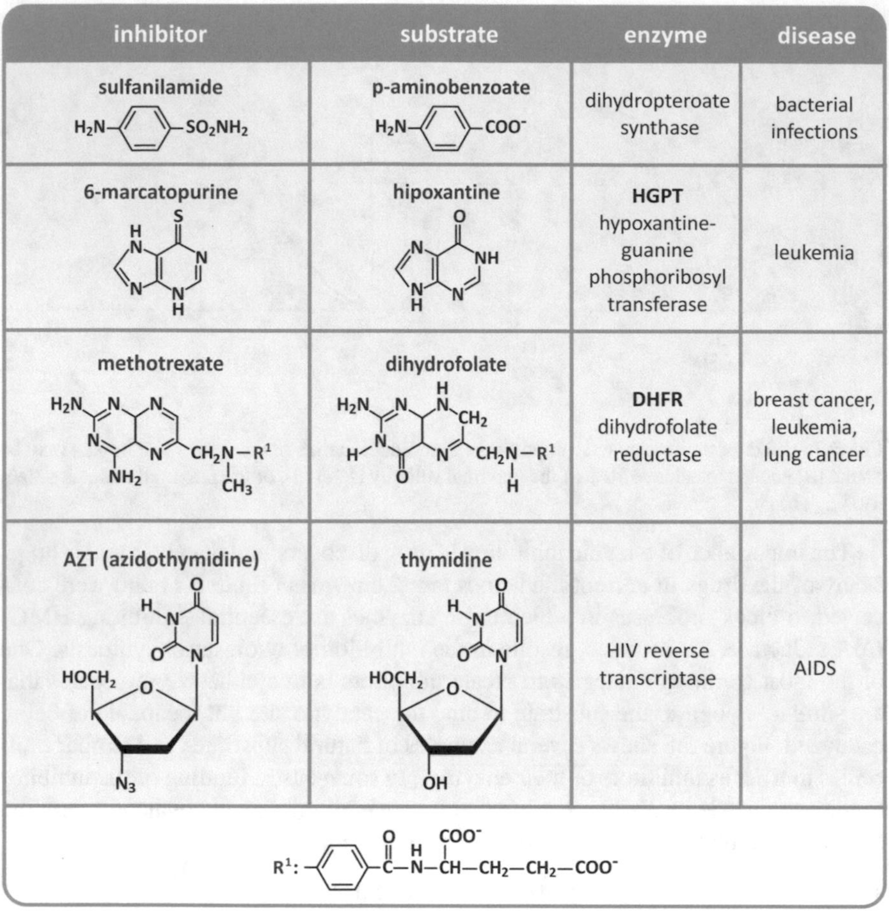

| inhibitor | substrate | enzyme | disease |
|---|---|---|---|
| sulfanilamide<br>$H_2N-\!\!\bigcirc\!\!-SO_2NH_2$ | p-aminobenzoate<br>$H_2N-\!\!\bigcirc\!\!-COO^-$ | dihydropteroate synthase | bacterial infections |
| 6-marcatopurine | hipoxantine | **HGPT** hypoxantine-guanine phosphoribosyl transferase | leukemia |
| methotrexate | dihydrofolate | **DHFR** dihydrofolate reductase | breast cancer, leukemia, lung cancer |
| AZT (azidothymidine) | thymidine | HIV reverse transcriptase | AIDS |

**Fig. 5.8** Examples of enzyme inhibitors that are very similar in chemical structure to the natural substrates

## 5.2.1 Nomenclature of Ligands

Ligands other than the substrates that have a significant effect on the velocity of catalysis are generally called effectors. They can be activators (speed catalysis up) or inhibitors (slow down catalysis). In principle, effectors bind and dissociate from enzymes without undergoing chemical modification, in contrast to substrates. For historical reasons, some ubiquitous substrates, such as NADH or NADPH, are often named separately as coenzymes. These so-called coenzymes are small molecules that transfer or accept groups from another substrate. This should not be confused with prosthetic groups. Prosthetic groups are chemical entities other than amino acid residues covalently attached to enzymes, such as heme groups of iron–sulfur centers (Fig. 5.9). They may have a direct role in catalysis, but they are not considered substrates as they are part of the enzyme itself in chemical terms.

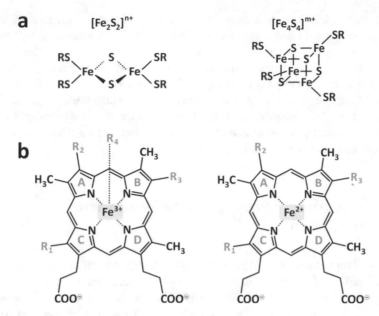

**Fig. 5.9** Examples of prosthetic groups of enzymes. (**a**) Iron–sulfur centers and (**b**) heme groups are considered prosthetic groups as they are not amino acid residues but bind covalently to enzymes. The A–D multiring structure (tetrapyrrole) in Panel (**b**) is the basic structure of a family of molecules named porphyrins. The central metal ion varies among porphyrins. Porphyrins conjugating iron ions are hemes. Both iron oxidation states (+3, *left*, or +2, *right*) are accommodated by the tetrapyrrole, meaning the central iron ion can engage redox reactions while inserted in the heme. The organic groups $R_1$–$R_4$ vary among hemes from different sources. In protoporphyrin IX, for instance, $R_1$ is $CH_3$, and $R_2$ and $R_3$ are $C=CH_2$. See also Box 3.8 and Fig. 3.38

The rules of nomenclature are not well defined. ATP, for instance, is often referred to as a coenzyme, but this is not a general rule. The same happens with CoA. NADH and NADPH are more consensually classified as coenzymes. FADH is often referred to as coenzyme, but it occurs in nature covalently bound to proteins, so it is in reality a prosthetic group. While it is important to have precise nomenclature for the purpose of efficient and unbiased communication, one should not overemphasize name over action, and one shall not further discuss semantics.

## 5.3   The Availability of Primary Precursors in a Metabolic Pathway

In a metabolic pathway such as the one depicted in Fig. 5.1, the velocity of reactions is not only controlled by the kinetic characteristics of the enzymes involved but also by the accumulation of G in the cell where it appears as the end product and availability of G in the cell where it is the primary precursor. To leave one cell and enter another, metabolite G needs to cross the membranes of both cells. Assuming that G is polar, it will not diffuse across the lipid bilayer of the membrane. It will need to be carried by specialized molecules or molecular assemblies. Take glucose as an example. It may be produced by hepatocytes in a metabolic pathway called gluco-neogenesis and be consumed in neurons. Glucose is a polar molecule, soluble in plasma. To leave the hepatocyte and enter the neurons, glucose uses transmembrane proteins that assist in the process of translocating glucose across membranes.

Given the importance of transport across membranes to metabolisms, we shall address this issue in some detail in the next section.

### 5.3.1   Transport of Metabolites and Effectors Across Membranes

Figure 5.10 represents the six typical alternative routes molecules use to cross membranes. Small relatively hydrophobic molecules such as ethanol or caffeine may not need transporters as they are able to simply diffuse across the lipid bilayer (Fig. 5.10a). Small ions have hindered diffusion across the bilayer due to their polarity, but they are able to cross membranes if encapsulated by chelating agents that are soluble in lipids (Fig. 5.10b). Ions can also cross membranes through ion channels, proteins that connect both sides of the membranes and are selectively permeable to specific ions (Fig. 5.10c). Protein channels may also transport small molecules, which have thus a facilitated diffusion across the membrane (Fig. 5.10d). All these processes, from simple diffusion to facilitated diffusion, operate with net translocation of molecules in one direction, from the higher electrochemical gradient to lower electrochemical gradient. If the concentration and charge distribution is the same on both sides of the membrane, there is no net mass movement across the membrane, as the velocity of transfer on one sense equals the velocity in the opposite sense.

Transporting molecules against electrochemical gradients is not energetically favorable, needing an external source of energy to occur. Some transporters use the hydrolysis of ATP as the energy source (primary active transport—Fig. 5.10e); others couple the transport of the solute to the transport of an ion along its electrochemical gradient (secondary active transport—Fig. 5.10f).

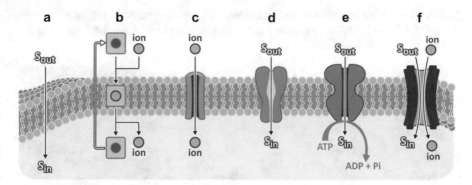

**Fig. 5.10** The six different routes that molecules and ions may use to translocate lipid membranes: (**a**) simple diffusion, (**b**) ionophore-mediated, (**c**) ion channels, (**d**) facilitated diffusion, (**e**) primary active transport, and (**f**) secondary active transport. Routes (**a–d**) are driven by electrochemical potential gradient (difference of concentration and charge distribution between both sides of the membrane)—passive transport. Routes (**e**, **f**) use ATP hydrolysis or the dissipation of an ion gradient, respectively, as sources of energy—active transport

The coupling between ion and solute transport may be such that both ion and solute are co-transported in the same direction (symport) or opposed directions (antiport) (Fig. 5.11b, c). Primary active transport and co-transport mechanisms may also be coordinated. For example, active transport may create a transmembrane gradient that is used to transport a solute independent of its gradient (Fig. 5.11d, e).

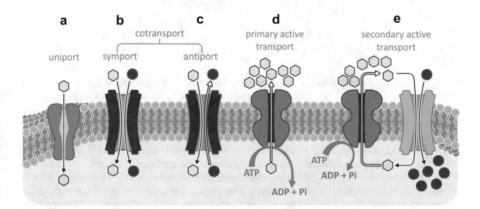

**Fig. 5.11** Types of solute transport through protein channels: (**a**) uniport, when a single solute is transported unidirectionally; (**b**, **c**) co-transport, when two solutes (or a solute and an ion) are transported simultaneously, either in the same direction (symport—**b**) or in opposite directions (antiport—**c**); (**d**) primary active transport, when the hydrolysis of ATP is used as the energy source to transport an ion or a solute against its gradient; and (**e**) secondary active transport, when an active transport is coupled to a co-transport by the creation of a gradient of ions or small molecules (*yellow hexagons*) that is used to transport the solute (*red circles*) independent of its gradient

The molecular structure of two examples of passive transport channels is shown in Fig. 5.12: the K$^+$-specific channel and aquaporin, a channel for the facilitated diffusion of water across membranes.

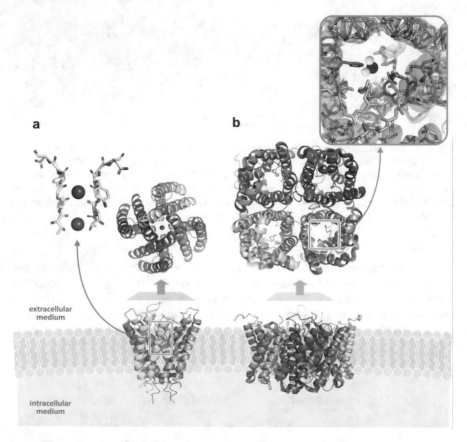

**Fig. 5.12** Examples of passive transport channels. (**a**) Molecular structure of the K$^+$ channel of *Streptomyces lividans* (PDB 1BL8) in longitudinal (*bottom*) and perpendicular (*top*) views in relation to the membrane. K$^+$ channels are the most widely distributed ion channel in living organisms. Four identical subunits, containing two transmembrane helices each, are arranged in a conic structure in which the central backbone carbonyl oxygens form K$^+$-selective pores that fit the ion precisely (see detail in the *left*, PDB 1J95), removing the hydration shell from the ion when it enters the channel. The channel accommodates four K$^+$ sites that are occupied alternately. (**b**) Structure of the spinach aquaporin (PDB 2B5F) in longitudinal (*bottom*) and perpendicular (*top*) views in relation to the membrane. Aquaporin selectively transports water molecules in and out of the cells. The protein is a tetramer of identical subunits, each of them containing a transmembrane pore (see detail in the *right*). All aquaporins contain a conserved sequence of Asn-Pro-Ala (highlighted in *yellow*) as part of the water channel. Water channel also contains a conserved His that narrows pore diameter limiting the passage of molecules larger than water and a conserved Arg (both His and Arg residues highlighted in *orange*) that repels cations, including H$_3$O$^+$. The positioning of the water molecule is tentative for illustrative purposes only

An example of simple diffusion operating in coordination with antiport active transport is shown in Fig. 5.13.

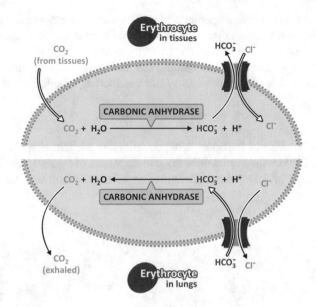

**Fig. 5.13** Transport of ions and solutes across the erythrocyte membrane. $CO_2$ is a small molecule able to diffuse directly across the membrane according to its concentration gradient (in lungs, in → out; in respiring tissues, out → in). Hydrogen carbonate (bicarbonate) is co-transported with chloride using an antiport mechanism. The direction of transport depends on $CO_2$ concentration because $CO_2$ equilibrates with $HCO_3^-$

Glucose transport in intestinal epithelial cells is depicted in Fig. 5.14 as example of a transport assembly that can be found in human epithelia.

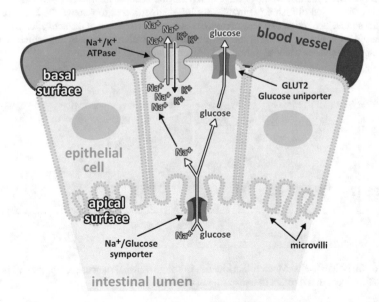

**Fig. 5.14** Glucose transport in intestinal epithelial cells. The glucose transport system in intestinal epithelial cells is complex and includes symport of $Na^+$ and glucose from the intestinal lumen to the cytoplasm and antiport of $Na^+$ and $K^+$ between the cytoplasm and blood. $Na^+$ has a central role by coupling both steps. Glucose is then transported to blood by a uniporter

Channels and transporters operate under the same thermodynamic principles as enzymes. Diffusion through membranes mediated by transporters has a decreased activation energy (Fig. 5.15). Moreover, transporters are proteins whose actions share similarities with enzymes. Assuming that the transport of a solute, S, from the outer to the inner side of a membrane by the transporter, T, is described by:

$$S_{out} + T \leftrightarrows ST \rightarrow S_{in} + T$$

the associated kinetics of transport is hyperbolic, like a Michaelis–Menten process (Fig. 5.16). Likewise, there are inhibitors and activators of transporters. Thus, transporters may also be key points of regulation of metabolism and therefore drug targets.

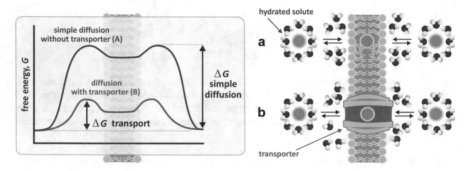

**Fig. 5.15** Thermodynamics of the passage of a polar molecule or ion (*green circle*) across a lipid membrane. Simplified diffusion (**a**) implies a high energetic cost to remove the hydration shell around the solute. The action of a transporter (**b**) leads to the reduction of the activation energy (the $\Delta G$ represented in the *left panel*) of the membrane translocation process by the molecule or ion. The transporter replaces water by forming hydrogen bonds with the solute

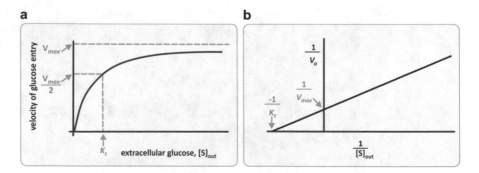

**Fig. 5.16** (**a, b**) Michaelis–Menten-like kinetics in solute transport across membranes. $K_t$ is equivalent to $K_M$, except that it refers to transport, not catalysis

Another way to control the action of transporters is to control their availability on the surface of the cells. Transporters are embedded in the bilayer matrix so that they can be removed from the surface of the cell by vesiculation. This happens with glucose transporters such as glucose transport 4 (GLUT4). GLUT4 is responsible for insulin-stimulated glucose uptake in muscle and adipose tissue. Binding of insulin to its receptor on the surface of membranes triggers the fusion of intracellular vesicles having membranes loaded with GLUT4 with the cell membrane, exposing the transporters (see Sect. 8.4.3). Glucose transport may thus occur, and this molecule becomes available inside the cells (Fig. 5.17). From that moment on, metabolic pathways having glucose as precursor, such as glycolysis, may initiate.

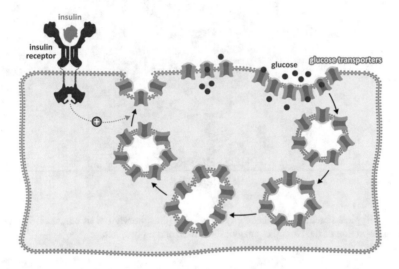

**Fig. 5.17** Insulin-stimulated exposure of GLUT4 on the surface of muscle and adipose cells. Vesicles bearing GLUT4 are recruited and fused with the plasma membrane. GLUT4 is automatically exposed. In the absence of insulin, the process is reversed

## 5.4   Slower (But Efficient!) Mechanisms of Controlling Enzyme Action

Enzyme response to binding of effectors is very fast. Concomitant to the effector's accumulation or depletion, the binding or dissociation, respectively, of the effector to the enzyme is nearly immediate. Yet, the efficacy of the process of inhibition or activation is concentration dependent. At extreme concentrations, the effect is also extreme, but at intermediate concentrations of the effector, the inhibition or activation is only partial. In contrast to non-covalent binding of effectors, there are two mechanisms of enzyme action that are highly efficient, although not immediate. These are the covalent modification of enzymes using phosphates, which may take up to minutes, and the synthesis or degradation of the enzymes themselves, which may take up to hours or days.

   Phosphorylation of enzymes occurs at the –OH groups of the exposed Ser, Thr, or Tyr residues. This process frequently involves ATP and is catalyzed by a second enzyme, a kinase. The inverse process, dephosphorylation, is also catalyzed by enzymes, called phosphatases (Fig. 5.18). Phosphate is the ubiquitous group used for covalent modification of enzymes in nature, but there is no rule on whether the active forms of the enzymes are the ones phosphorylated or not. Table 5.2 shows examples of enzymes that are active or inactive when phosphorylated.

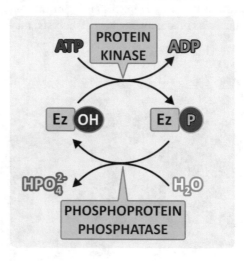

**Fig. 5.18** Typical reactions for phosphorylation or dephosphorylation of enzymes. The process involves the action of the two other enzymes: a kinase and a phosphatase

**Table 5.2** Examples of enzymes regulated by phosphorylation

| Metabolic pathway | Enzyme | Phosphorylated form | |
|---|---|---|---|
| | | Active | Inactive |
| Glycogenolysis | Glycogen phosphorylase kinase | ✓ | |
| | Glycogen phosphorylase | ✓ | |
| Glycogenogenesis | Glycogen synthase | | ✓ |
| Glycolysis and gluconeogenesis | 6-Phosphofructo-2-kinase | | ✓ |
| | Fructose 2,6-bisphosphatase | ✓ | |
| | Pyruvate kinase | | ✓ |
| Lipolysis (conversion of triacylglycerols to glycerol and fatty acids) | Lipase | ✓ | |
| Lipogenesis | Citrate lyase | | ✓ |
| | Acetyl-CoA carboxylase | | ✓ |
| | 3-Hydroxy-3-methylglutaryl-CoA reductase | | ✓ |

The presence or absence of a phosphate group bound to the protein may have a high impact on its structure (Fig. 9.7 is an elucidative example). Local changes in structure may propagate through the 3D architecture of the protein and trigger or block its enzymatic activity. The reason why phosphates are the only choice of nature to perform this task is intriguing. This is believed to be related to the abundance of phosphates and its chemical nature. Phosphates bind to organic molecules, forming esters, keeping their anionic charge, which protects them from hydrolysis. Nucleophiles, such as the hydroxide ion, are repelled by negative charges and therefore react less rapidly with anions than with neutral substrates. Furthermore, the same is generally true with respect to electrically neutral nucleophiles such as water. The rate constant for the attack of hydroxide to the dimethyl phosphate anion is less than that for the attack of hydroxide ion to the trimethyl phosphate by a factor of more than $10^5$ (Table 5.3).

**Table 5.3** Rates of reaction of esters with $OH^-$ at 35 °C

| Ester | $k$ ($M^{-1} s^{-1}$) | Fold increase in $k$[a] |
|---|---|---|
| $(CH_3O)_2PO_2^-$ | $2 \times 10^{-9}$ | 1.0 |
| $(CH_3O)_3P=O$ | $3.4 \times 10^{-4}$ | $1.7 \times 10^5$ |
| $CH_3CO_2C_2H_5$ | $1.0 \times 10^{-2}$ | $5.0 \times 10^6$ |

[a]Relative to $(CH_3O)_2PO_2^-$

The phosphorylation/dephosphorylation switch to turn on or turn off enzymes is found for key enzymes of metabolisms. The primary triggering of the events that result in phosphorylation or dephosphorylation of enzymes is the interaction of a hormone with its receptor, as illustrated before in Fig. 5.3. This figure shows that binding of a hormone to its receptor triggers a series or transformations whose end result is the phosphorylation/dephosphorylation of the key enzyme. These intermediary steps are generally referred to as signal transduction and the chemical species involved as second messengers, as mentioned before in Box 5.1. We shall not address the theme of signal transduction in detail because, in this section, we intend to remain focused on the basics of metabolic regulation (signal transduction triggered by specific hormones will be discussed in detail in the chapters that explore the regulation of metabolism in different physiological situations—Chaps. 8–11). Nonetheless, it is worth mentioning that there are two main classes of receptors that trigger signal transduction: G protein-coupled and kinase-linked receptors.

In the case of G protein-coupled receptors, binding of the ligand (in pharmacology, usually named "agonist") to the receptors causes unbinding of GDP with concomitant binding of GTP and the release of a complex of proteins associated with the receptor (named G protein) that will associate with other molecular targets, such as adenylyl cyclase, which converts ATP into cyclic AMP (cAMP) (Fig. 5.19). The increase in cAMP concentration may in turn stimulate other chemical reactions in a cascade-like manner. The amplitude of the effects triggered by G protein-coupled receptors is controlled not only by the ligand binding to the receptors but also by the activity of the cyclic nucleotide phosphodiesterases. These enzymes break the phosphoester bond of cAMP, generating AMP, which does not act as a second messenger. There are different known inhibitors of phosphodiesterases, such as caffeine or theophylline (see Challenging Case 4.2). These inhibitors prevent the degradation of cAMP, thus prolonging the physiological effects G protein-coupled receptors' ligands.

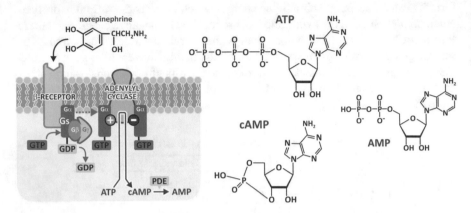

**Fig. 5.19** Example of the signal transduction triggered by G protein-coupled receptor. Upon binding of the ligand to the receptor (e.g., norepinephrine), G protein exchanges GDP to GTP and dissociates from the receptor and binds to adenylyl cyclase, activating or inhibiting (depending on the type of G protein) the production of cAMP

Besides adenylyl cyclase, there are other membrane-associated enzymes, such as phospholipase C, that mediate the action of the G protein-coupled receptors. In this case, the G proteins associate with phospholipase C and leads to hydrolysis of the phosphatidylinositol bisphosphate in the membranes into diacylglycerol and inositol triphosphate, the latter serving as second messenger.

Kinase-linked receptors are transmembrane proteins with extracellular domains that bind agonist ligands, including hormones, and intracellular domains having amino acid residues that undergo phosphorylation, triggering a cascade of events ("kinase cascade").

The most important signal transduction pathways for metabolic dynamics and regulation will be shown throughout the following chapters, together with the metabolic pathway for which they are relevant. However, it is important to highlight that the end effect of some hormones that bind to cell surface receptors may take place in the nucleus, involving, for instance, the recruitment of transcription factors (Fig. 5.20). This enables hormones to control the activity of enzymes at a higher level: regulating enzyme biosynthesis.

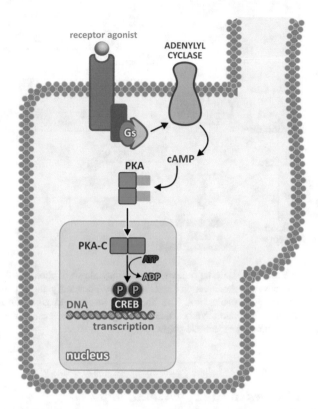

**Fig. 5.20** Impact of a receptor agonist on transcription of a given gene via adenylyl cyclase. Hormones may interfere in enzyme biosynthesis using similar mechanisms. cAMP produced after agonist binding to its G protein (Gs)-coupled receptor binds to the cAMP-dependent protein kinase (PKA), causing the dissociation of its catalytic subunits (PKA-C, *blue squares*) from the PKA regulatory subunits (*pink squares*). PKA-C phosphorylates a transcription factor (cAMP response element-binding protein—CREB) that induces the expression of different genes

Not only hormones may activate/inhibit enzymes by phosphorylation or dephosphorylation, but they can also control the availability of the enzymes through the control of their biosynthesis. This redundancy assures high efficacy in a wide time range as covalent modification of enzymes is operational in the seconds to minutes time scale and the control of synthesis is operational in hours to days. Cholesterol synthesis is shown as example in Fig. 5.21.

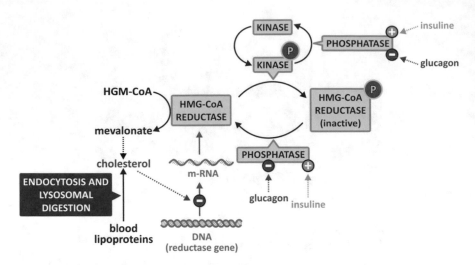

**Fig. 5.21** HMG-CoA reductase is the key enzyme in the biosynthesis of cholesterol. It is regulated by slow mechanisms: covalent modification (inactivation by phosphorylation) and transcription inhibition by the end product, cholesterol. The hormones insulin and glucagon have opposite effects: insulin stimulates cholesterol synthesis, and glucagon stimulates the inhibition of HMG-CoA reductase, shutting down the synthetic pathway

## 5.5   Key Molecules in Energy Metabolism

In the same way ATP is frequently used in metabolism because its favorable hydrolysis coupled to an unfavorable reaction makes the whole set of chemical reactions favorable, NADH (see NADH chemical structure in Fig. 3.23) is also a molecule frequently used to drive redox processes. The process can be represented by:

$$NAD^+ + 2H \leftrightarrows NADH + H^+$$

but this is not elucidative for its role in redox reactions. A better representation is:

$$NAD^+ + RH_2 \leftrightarrows NADH + RH^+ \left( or\ NADH + R + H^+ \right)$$

in which $RH_2$ is a generic molecule. In this reaction the reduction of $NAD^+$ is coupled to the oxidation of $RH_2$. This is the case of:

$$NAD^+ + \underset{\text{Lactic acid}}{\text{(structure)}} \leftrightarrows NADH + \underset{\text{Pyruvic acid}}{\text{(structure)}} + H+$$

For the sake of simplicity, in biochemical literature the reaction stoichiometries and the involvement of $H_2O$ or $H^+$ (or $H_3O^+$) are not always represented, and this process can simply be written as:

$$\text{Lactic acid} \underset{NAD^+ \quad NADH}{\overset{NAD^+ \quad NADH}{\rightleftharpoons}} \text{Pyruvic acid}$$

but one must not forget the chemical details implied albeit not explicit.

NADH is so ubiquitous in cells that it can be considered as a "universal" electron carrier. NADH is used in cells due to its strong reducing (electron-donating) power; $NAD^+$ is converted back to NADH by oxidation of other molecules. Most reactions involving $NADH/NAD^+$ in metabolism are catalyzed. The enzymes that catalyze the oxidation of a substrate with the transfer of one or more hydrides ($H^-$) to an electron acceptor such as $NAD^+$ (or equivalent molecules such as $NADP^+$, or FAD, or FMN; see Fig. 3.23) are named dehydrogenases.

There are other examples of ubiquitous molecules in metabolism, although not as striking as NADH or ATP. Biotin, for instance, is a non-proteic "prosthetic" (not an amino acid residue) group that exists in carboxylases and participates in the addition of $CO_2$ groups from $HCO_3^-$ to organic molecules, as exemplified in Fig. 5.22 (in which $E_z$ is the enzyme):

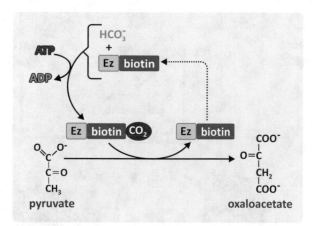

**Fig. 5.22** Example of the role of biotin in carboxylation reactions. Biotin occurs in carboxylases ($E_z$-biotin) and participates in the addition of $CO_2$ groups from $HCO_3^-$ to organic molecules such as pyruvate

Acetyl-CoA (acetyl bound to coenzyme A, CoA) is an amazing molecule and a very common metabolite. Acetyl-CoA is important in glucose metabolism, fatty acid degradation, amino acids metabolism, and cholesterol synthesis. Its ubiquity is not as simple to explain as the universality of ATP (hydrolyses with $\Delta G^0 \ll 0$), NADH (strong reducing power; $\Delta G^0 \ll 0$ in reduction), or biotin (specific functionality in carboxylation processes). Acetyl groups seem to be the universal "carbon transfer unit" and CoA their carrier for this purpose. Acetyl (or ethanoyl) is a two-carbon group with the following chemical structure:

$$
\begin{array}{c}
\mathsf{R} \\
\mid \\
\mathsf{C}\!=\!\mathsf{O} \\
\mid \\
\mathsf{CH_3}
\end{array}
$$

In this formula, R represents the generic molecule to which acetyl group is attached to (i.e., the acetylated molecule). In acetyl-CoA, R is CoA. Degradation of glucose, fatty acids, or some amino acids results in acetyl groups being attached to CoA. The acetyl groups can then be used to generate NADH from $NAD^+$, which in turn will help in transforming ADP to ATP. Yet, under different circumstances, acetyl groups in acetyl-CoA may also be used to synthesize molecules, serving as a carbon supplier for the growth of their carbon backbones. CoA concurs with biotin-containing enzymes for this purpose.

A generic scheme of the metabolisms built around the "universal molecules" ATP, NADH, and acetyl-CoA is presented in Fig. 5.23.

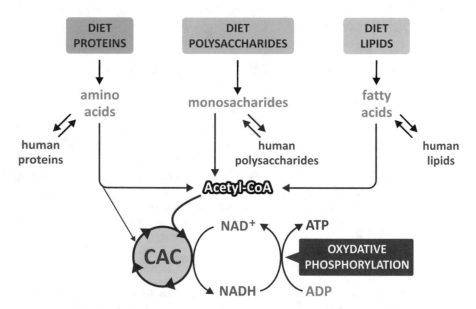

**Fig. 5.23** Very broad and abbreviated view of human energy metabolism. Intermediary metabolites are not shown for the sake of simplicity. The main interplay of NADH, acetyl-CoA, and ATP is highlighted. CAC: citric acid cycle (also known as tricarboxylic acid cycle or Krebs cycle)

It is very curious that the fraction of the structure of the ATP, NADH, or acetyl-CoA molecules involved in chemical reactivity itself is very small when compared to the whole molecule (Fig. 5.24). Moreover, at first glance it may seem puzzling that in some reactions in metabolism, ATP is replaced by GTP or UTP, and NADH is replaced by FADH$_2$ or NADPH (Fig. 5.24). GTP or UTP do not differ from ATP in the phosphodiester bond broken to form ADP (which is analogous to GDP or UDP). Phosphorylation of NAD to form NADPH occurs at a site that does not affect the electron donor group. So, what is then the advantage of having relatively large molecules to perform simple chemical reactions that involve only small groups in their structures? What is the advantage in having analogous molecules with same reactivity that differ only in the "inert" part of these molecules? The answers are simple: having molecules of a significant dimension facilitates recognition and specificity of the binding sites of enzymes, and small variations in the chemical structure of these molecules determine that other enzymes should be used instead. In reactions in which NADPH intervenes, the associated enzyme is specific for NADPH, not NADH. The occurrence of these reactions is an advantage in the sense of allowing the donation of electrons without affecting NADH pools in the cell. The reaction only occurs if NADPH is present. Therefore, the use of NADPH in addition to NADH enables that specific redox reactions in the cell may be regulated regardless of the metabolic state of cell at the moment, i.e., regardless of NADH levels. Protective mechanisms against reactive oxygen species (ROS), for instance, are very much dependent on the reductive action of NADPH. The capacity to repair the chemical damage imposed by ROS is thus maintained regardless of low concentrations of NADH (see Sect. 6.2.5). Likewise, the use of GTP or UTP in some reactions instead of ATP may have the same advantage.

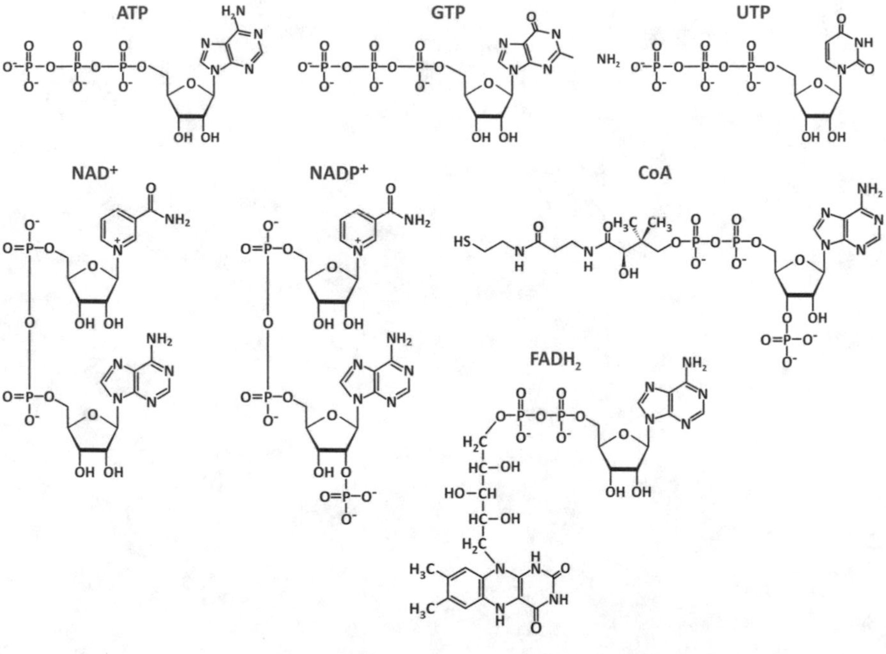

**Fig. 5.24** Molecular structure of reactive molecules commonly involved in metabolic processes. The reactive groups of ATP, NAD$^+$, or CoA are relatively small compared to the whole molecule. Analogous molecules to ATP or NAD$^+$ differ in the "nonreactive" part of the molecule. Although not involved directly in reactivity, these parts of the molecules are important to grant specificity to different enzymes

In the following chapters, we will devote our attention to the synthesis of ATP, which in turn is dependent on the NADH production, which in turn is dependent on acetyl-CoA that is obtained from nutrients: fatty acids, monosaccharides, and amino acids.

## Selected Bibliography

Dagani R (2003) Straightening out enzyme kinetics. Lineweaver and Burk's 1934 paper showed biochemists a better way to plot their data. Chem Eng News 81:27

Laskowski RA, Gerick F, Thornton JM (2009) The structural basis of allosteric regulation in proteins. FEBS Lett 583:1692–1698

Lineweaver H, Burk D (1934) The determination of enzyme dissociation constants. J Am Chem Soc 56:658–666

# Challenging Case 5.1: Mushroom Poisons Starring on Stage and Science

## Source

**AGRIPPINA**
DRAMA
*Per Mufica.*

Da Rapprefentarfi nel Famofif-
fimo Teatro Grimani di
S.Gio:Grifoftomo

L'Anno M.DCCIX.

IN VENEZIA , M. DCCIX.
Appreffo Marino Roffetti in Merceria,
all' Infegna della Pace .
*Con Licenza de' Superiori , e Privilegio.*

Poisoning frequently shows up both in actual history and in works of fiction. The present case deals with mushroom poisoning, using as background the rumors around the death of Roman Emperor Claudius, reported by Roman historians, which inspired the opera Agrippina, by George Frideric Handel with the libretto by Cardinal Vincenzo Grimani. Mushroom poisoning also inspired the fiction film Phantom Thread, which is a historical drama, written and directed by Paul Thomas Anderson, about a fictional fashion designer of the 1950s London haute couture scene.

Cover of the libretto for the first performance of Agrippina, in the Teatro San Giovanni Grisostomo, at Venice, in 1709. (Public domain)

## Case Description

Murder by poisoning follows the history of mankind and has inspired many crime fictions. Among the causes of poisoning, mushrooms appear prominently. Mushrooms grow in forests and meadows and are used commonly in the culinary. However, some varieties are very toxic and may be lethal, making them a very effective murder weapon, not only due to the difficulty to detect the poisoning substance in the victim (especially in the past) but also because it is not easy to prove that the killing was deliberate.

A classic example that combines history and fiction is the case of the Roman Emperor Claudius, who died at the age of 64, in 14 October AD 54. Although the cause and the circumstances of Claudius death are controversial, the versions of at least two major Roman historians, Dio Cassius (author of *Roman History*) and Suetonius (author of *De Vita Caesarum*), support he was murdered by his fourth wife, Agrippina, who served him up a portion of poisonous mushrooms during a banquet. The described symptoms that led Claudius to die in about 12 h include excruciating abdominal pain, vomiting, diarrhea, excessive salivation, low blood pressure, and difficulty breathing, agreeing with the historic version of poisoning. Interestingly, the murder of Claudius was the subject of the 2001 historical diagnosis conference at the University of Maryland's annual Clinicopathologic Conference,[1]

---

[1] Roberts L, Smith M (2001) Murder by mushroom - homicide is confirmed in the death of the Roman Emperor Claudius. NewsWise. University of Maryland. Article ID: 22443. https://www.newswise.com/articles/murder-by-mushroom-homicide-is-confirmed-in-the-death-of-the-roman-emperor-claudius.

where William A. Valente, clinical professor of medicine at the University of Maryland School of Medicine, defended Claudius' poisoning as strongly likely. However, this case is far from being closed as there are doctors who, based on the features of Claudius' death described in other sources, dispute this conclusion.[2]

In any case, being history or fiction, which reasons would lead Agrippina to kill her husband Claudius? Well, Claudius was the Roman Emperor from AD 42–54. In AD 14, when Augustus, the first Emperor of Roman Empire, died, Claudius was the fourth in the line of succession, after Tiberius (son of the first marriage of Livia, adopted by Augustus), Germanicus (son of Drusus, the second son of the Livia's first marriage), and Caligula (son of Germanicus) (see below figure). Tiberius became Emperor after Augustus, and his reign lasted until AD 37. Germanicus died in AD 19, so Caligula succeeded Tiberius. But the 4 years of his reign were of despotic horror, leading to his assassination in AD 41, when Claudius ascended to the throne. Claudius had four wives, Plautia, Paetina, Messalina, and Agrippina. His marriage with Agrippina was troubled, starting with the need of the Roman Senate to authorize their illegal incestuous union (Agrippina, sister of Caligula, was Claudius' niece). Even in the Handel's opera, in which the murder is not explicit, a net of intrigues clearly shows Agrippina obsession to put Nero, her son from the previous marriage, in the line for succession, rather than the Claudius's own son by Messalina, Britannicus (see below figure), who was close to reaching his adulthood.

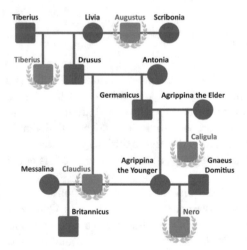

The Julio-Claudian Roman dynasty: Augustus, the first Emperor of the Roman Empire (ascension in 27 BC), married Livia after she has divorced Tiberius. He adopted Livia's son Tiberius, who ascended to the throne in AD 14. Germanicus, nephew of Tiberius, died in AD 19, 18 years before Tiberius death, so Germanicus' son Caligula became the third emperor of the Roman Empire in AD 37. Caligula was assassinated in AD 41, being succeeded by his uncle Claudius, the younger brother of Germanicus. Britannicus, the natural successor of Claudius, was younger than Claudius' great-nephew and stepson Nero. Claudius adopted Nero and died before Britannicus reached adulthood, so Nero ascended to the throne in AD 54. The suicide of Nero in AD 68 ended the reign of the Julio-Claudian dynasty, as he lacked natural or adopted sons

---

[2] Marmion VJ, Wiedemann TEJ (2002) The death of Claudius. J R Soc Med 95:260–261.

Claudius death changed the course of Roman history as it paved the way for Nero, a notoriously cruel leader, to take Roman Empire throne.

If the ingestion of mushrooms had been indeed the actual cause of Claudius' death, the main candidate to be the poisoning weapon would be a specimen of the *Amanita phalloides* species, known as the "death cap" mushroom (see below picture).

Picture of the death cap mushroom *Amanita phalloides*, characterized by yellowish- or olive-green cap over a white stipe. They resemble several edible species commonly consumed by humans. (Image by Archenzo, reproduced from Wikimedia Commons, under a CC BY license)

The death cap causes the majority of fatal mushroom poisonings, leading to fatal liver damage 1–3 days after ingestion. α-Amanitin, a modified bicyclic octapeptide (see below figure), is its main toxin.

Chemical structure of α-amanitin, consisting in a highly modified octapeptide arranged in a conserved macrobicyclic motif

α-Amanitin, as other amatoxins, is lethal in even small doses. It acts as a potent and selective inhibitor of the enzyme RNA polymerase II, which catalyzes the synthesis of RNA molecules, including the messenger RNAs. This would deeply interfere with cellular metabolism as without messenger RNAs, protein synthesis is abrogated.

Being or not the cause of Claudius' death, more than 2000 years later, the death cap and its poisoning properties are again on stage, now in a 2017 cinema film, the "Phantom Thread." Set in the 1950s London, the main character of the film is Reynolds Woodcock, a renowned elite dressmaker, whose geniality matches with an obsessive and controlling personality. Reynolds' sister Cyril manages his professional issues as well as his personal and romantic life, helping him to get rid of each lover of whom he inevitably tires.

The main focus of the plot is a strange relationship established between Reynolds and the waitress Alma, who becomes his muse and lover. When Alma revolts against Reynolds' strict routine, fed up with his sickly likes and dislikes, they establish a strange codependent relationship. Alma repeatedly feeds Reynolds with poisonous mushrooms in precise doses to make him severely ill but not enough to kill him. Then, she devotedly takes care of him until he gets recovered. This morbid relationship seems to be the way Alma found to show her power over Reynolds, but, intriguingly, being poisoned also became almost a necessity to Reynolds, maybe because it makes him feel as fragile as a baby in the hands of the mother. Indeed, Reynolds is haunted by the death of his mother. No wonder he has hallucinations about his mother during the most critical phase of the illness.

## Questions

1. You learned in Sect. 5.2 that ligands, which are molecules that bind specifically to unique sites of enzymes, affect their conformation and, thus, influence their kinetics properties. This is the case of the toxic component of the death cap, α-amanitin, which acts as a ligand of RNA polymerase II, inhibiting the reaction this enzyme catalyzes, the DNA transcription (see below figure).

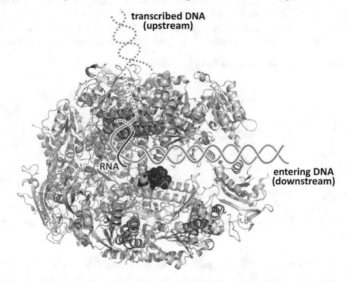

Crystallographic structure of polymerase II (PDB 1K83) complexed with α-amanitin (shown in *red spheres*), showing a schematic representation of the putative positions of the entering and the transcribed DNAs as well as the nascent RNA chain

   (a) The reaction catalyzed by RNA polymerase II is not directly involved in the production of a metabolic intermediate. Nonetheless, this enzyme controls the synthesis of the enzymes that, in turn, will regulate the availability of cellular metabolites. With this in mind, discuss which would be the effects of RNA polymerase II inhibition by α-amanitin on the different levels of metabolic regulation depicted in Fig. 5.3.
   (b) The symptoms observed after intoxication with α-amanitin are characteristics of hepatic failure. Propose a reason to explain why the liver is the main organ affected by α-amanitin ingestion.

2. α-Amanitin is the most potent member of the naturally occurring amatoxins, a group of toxic compounds found in several genera of poisonous mushrooms. A series of articles published in the 1970s explored the mechanisms of RNA polymerase II inhibition by the amatoxins. The following questions are based on results extracted from one of these articles.[3]

---

[3] Cochet-Meilhac M, Chambon P (1974) Animal DNA-dependent RNA polymerase. II. Mechanism of the inhibition of RNA polymerases B by amatoxins. Biochim Biophys Acta 353:160–184.

(a)  Based on the analyses of the effect of α-amanitin on (1) DNA binding to
     RNA polymerase (first table) and (2) RNA chain initiation (second table),
     discuss which step of polymerase reaction is inhibited.

Effect of α-amanitin on polymerase binding to DNA

| Expt. number | Incubation conditions | Control | + amanitin |
|---|---|---|---|
| 1 | Enzyme + α-amanitin (10 min) + DNA (10 min) | 517 cpm | 503 cpm |
| 2 | Enzyme + DNA (10 min) + α-amanitin (10 min) | 820 cpm | 786 cpm |

In this experiment a radioactively labeled DNA was used. After the incubation period, the reaction
mixture was filtered, and the radioactivity retained in the filter was measured. The composition of
the reaction medium is detailed in reference [3]. cpm: radioactivity counts per million

Effect of α-amanitin on chain initiation and elongation

| | Incorporation of labeled nucleotides (μmol) | | | |
| | Expt. 1 | | Expt. 2 | |
| Incubation conditions | [γ-$^{32}$P]GTP | [$^3$H]UMP | [γ-$^{32}$P]ATP | [$^3$H]UMP |
|---|---|---|---|---|
| Control (no α-amanitin) | 0.547 | 121 | 0.263 | 130.5 |
| + 0.002 μg α-amanitin | 0.342 | 81.6 | 0.166 | 78.3 |
| + 1 μg α-amanitin | 0 | 0 | 0 | 0 |

In this experiment radioactively labeled nucleoside triphosphates were used. The enzyme was
preincubated with α-amanitin in ice and then incubated with the nucleoside triphosphates for
20 min at 37 °C; the incorporation of radioactively labeled nucleotides in the synthesized RNA
chain was measured. The composition of the reaction medium is detailed.[3] cpm: radioactivity
counts per million

(b)  As you learned in Sect. 5.2, kinetics studies may tell us about the mechanism
     by which the ligand (for instance, the inhibitor) interacts with the enzyme.
     With this in mind, analyze the Lineweaver–Burk plot obtained for RNA
     polymerase activity in the presence and in the absence of α-amanitin (next
     figure), and discuss the possible mechanism of enzyme inhibition.

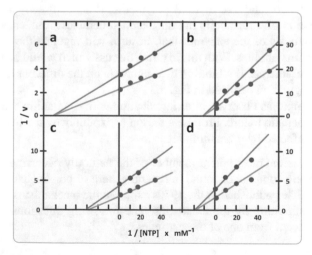

Effect of α-amanitin on the kinetics of the reaction catalyzed by RNA polymerase II: Lineweaver–
Burk plot with varying concentrations of nucleoside triphosphate (NTP) (a, CTP; b, UTP; c, ATP;
d, GTP). Empty symbols (no α-amanitin added); filled symbols (α-amanitin added at a concentration of 4.35 nM). Figure reproduced with permission from the article cited in the footnote 3

3. In order to further study the binding of amatoxins to RNA polymerase, the authors of the article described above developed a membrane filter assay. This assay uses a radioactively labeled analog of amanitin and is based on the fact that RNA polymerase-amatoxin complexes are retained on nitrocellulose filters while free labeled compound can be washed through nitrocellulose membrane. So, measuring radioactivity in the filter allows one to probe the amount of amatoxin complexed with the enzyme. Using this method, they determined the dissociation and the association constants for the interaction between amatoxin and the enzyme, at different temperatures, as shown in the next table.

| Temperature (°C) | Amanitin analog | |
|---|---|---|
| | $K_D$ (M) | $K_A$ (M$^{-1}$) |
| 0 | $1.66 \times 10^{10}$ | $6.01 \times 10^9$ |
| 5 | $1.72 \times 10^{10}$ | $5.81 \times 10^9$ |
| 10 | $2.16 \times 10^{10}$ | $4.63 \times 10^9$ |
| 15 | $3.89 \times 10^{10}$ | $2.57 \times 10^9$ |
| 20 | $6.66 \times 10^{10}$ | $1.50 \times 10^9$ |
| 30 | $2.82 \times 10^9$ | $3.54 \times 10^8$ |
| 37 | $6.42 \times 10^9$ | $1.56 \times 10^8$ |

(a) Using the value of the association constant at 37 °C, calculate the $\Delta G^0$ of amatoxin binding to RNA polymerase.
(b) Search the literature and learn about the so-called van't Hoff plots. Then, using the equilibrium constants measured in the different temperatures, construct a van't Hoff plot to determine the $\Delta H$ and $\Delta S$ of the binding reaction, and comment the possible molecular interactions involved in the stabilization of the complex.

4. In an article published in 1971, naturally occurring and chemical modified amatoxins were studied in respect with their in vitro inhibition of RNA polymerase as well as in vivo toxicity in mice.[4]
Observe general chemical structure of the amatoxins (next figure) and the respective R groups of four of the compounds tested (see below table), and correlate their toxicity with the dissociation rate constants ($k_2$) for their binding to RNA polymerase and ability to inhibit the enzyme.

---

[4] Buku A et al (1971) Inhibitory effect of naturally occurring and chemically modified amatoxins on RNA polymerase of rat liver nuclei. FEBS Lett 14:42–44.

| Peptide | $R_1$ | $R_2$ | $R_3$ | $k_2$ (s$^{-1}$) | Conc. (nM) | Inhibition of RNA pol (%) | i.p. LD$_{50}$ in mouse (mg/kg) |
|---|---|---|---|---|---|---|---|
| α-Amanitin | CHOH-CH$_2$OH | NH$_2$ | OH | $1.2 \times 10^{-4}$ | 5 | 50 | 0.3 |
| O-Methyl-α-amanitin | CHOH-CH$_2$OH | NH$_2$ | OCH$_3$ | $1.8 \times 10^{-4}$ | 5 | 44 | 0.2 |
| O-Methyl-demethyl-γ-amanitin | CHO | NH$_2$ | OCH$_3$ | $9.8 \times 10^{-4}$ | 5 | 33 | 3 |
| O-Methyl-aldoamanitin | CH$_2$OH | NH$_2$ | OCH$_3$ | ND | 100 | 0 | >50 |

Chemical structures of distinct amatoxins and their in vitro inhibition of RNA polymerase and in vivo toxicity in mice, as well as their inhibition constant ($K_i$) and dissociation rate constants ($k_2$) for their binding to the enzyme. Data from the articles cited in the footnotes 4 and 5. ND = not determined

## Biochemical Insights

1. RNA polymerases are enzymes that open the double-stranded DNA, exposing the nucleotides of one strand and allowing their use as a template for the synthesis of an RNA molecule, a process called transcription. Different types of RNA polymerases are responsible for synthesis of distinct subsets of RNA molecules. RNA polymerase I synthesizes the larger ribosomal RNA; RNA polymerase II synthesizes messenger RNAs and most the regulatory small RNA molecules (snRNAs and microRNAs); and RNA polymerase III synthesizes transporter RNAs, 5S ribosomal RNA, and other small RNAs. α-Amanitin specifically inhibits RNA polymerase II.[5]

---

[5] Lindell TJ et al (1970) Specific inhibition of nuclear RNA polymerase II by α-amanitin. Science 170:447–449.

(a) At first, α-amanitin would directly affect the level (4) of metabolic regulation depicted in Fig. 5.3, i.e., the expression/translation of enzymes: blocking RNA polymerase II, α-amanitin impairs the transcription of the genes expressed in the cell, lowering the amount of the respective proteins and enzymes codified by them. This would then affect the other levels of metabolic regulation as proteins and enzymes with different functions in the cell metabolism would have their amount lowered. For instance, if the expression of cellular transporters is inhibited, availability of substrates needed to be uptaken would be affected, lowering the production of metabolites downstream the respective pathways. If cell depends on the activation of a given enzyme by phosphorylation, but the enzyme itself or the respective kinase has its expression blocked, this activation will not occur, impairing the production of the required metabolite. Also, if the expression of a given enzyme depends on the production of a specific transcriptional factor or effector whose expression is committed by the inhibition of RNA polymerase, cellular metabolism would also be deregulated.

(b) The liver is the main organ affected after the ingestion of amatoxins because it is the first organ to be in contact with the substances absorbed from the gastrointestinal tract through the hepatic portal vein system (see Box 8.1).

2. In their article, Cochet-Meilhac and Chambon deeply analyzed the effects of α-amanitin on calf thymus RNA polymerase II.[3]

(a) To evaluate the effect of α-amanitin on DNA binding to RNA polymerase, the enzyme was either incubated with α-amanitin and then a labeled DNA molecule was added or incubated with the labeled DNA molecule and then α-amanitin was added. The formation of the enzyme-DNA complex was measured by passing the reaction mixture through a filter in which the enzyme is retained, while free DNA is not.

To evaluate the effect of α-amanitin on RNA chain initiation and elongation, the incorporation of radioactively labeled GTP and ATP (initiation nucleotides) and subsequent incorporation of UMP, respectively, was measured.

The results showed no significant differences in the amount of DNA bound to the enzyme, demonstrating that the α-amanitin neither does not inhibit DNA binding to the enzyme nor releases DNA release from a pre-formed complex. On the other hand, incorporation of radioactively labeled nucleotides to the nascent RNA chain was inhibited by α-amanitin in a dose-dependent manner, showing that α-amanitin affects both initiation of RNA synthesis and chain elongation.

(b) The Lineweaver–Burk plot obtained for RNA polymerase activity in the presence and in the absence of α-amanitin shows that the reaction $V_{max}$ is decreased while $K_M$ value is not affected, independently of whether the nucleoside triphosphates participate in initiation of RNA synthesis (ATP and GTP) or not (UTP and CTP). This pattern is typical of noncompetitive inhibition and reveals that α-amanitin does not affect the affinity of the enzyme for any of the four nucleotides (it does not compete with the nucleotides for binding to the enzyme consistent with discussion of question 2a), but decreases the amount of enzyme free to interact with the substrate.

3. As discussed in Chap. 4, the $\Delta G$ of a given reaction corresponds to the balance of released energy as heat (the enthalpy, $\Delta H$) and changes in entropy ($\Delta S$):

$$\Delta G = \Delta H - T\Delta S$$

The $\Delta G$ of a reaction is calculated using (see section 4.1):

$$\Delta G = \Delta G^0 + RT \ln K_{eq}$$

In the equilibrium, $\Delta G$ value is 0. So:

$$\Delta G^0 = -RT \ln K_{eq}$$

(a) Thus, the $\Delta G^0$ of α-amanitin binding to RNA polymerase, at 37 °C, is:

$$\Delta G^0 = -1.98 \times 10^{-3} \text{ kcal K}^{-1} \text{ mol}^{-1} \times 310 \text{ K} \times 18.87 = -11.6 \text{ kcal mol}^{-1}$$

Being a negative value, it is demonstrated that α-amanitin binding to RNA polymerase, at 37 °C, is extensive, which is the reason behind the potency of the toxin.

(b) The graph correlating $\ln K_{eq}$ (on the $y$-axis) and $1/T$ (on the $x$ axis) is called the van't Hoff plot. It is used to calculate the enthalpy and entropy changes involved in a chemical reaction. From this plot, $-\Delta H/R$ is the slope, and $\Delta S/R$ is the intercept of the linear fit as in equilibrium:

$$\Delta G^0 = -RT \ln K_{eq} \Leftrightarrow \Delta H - T\Delta S^0 = -RT \ln K_{eq} \Leftrightarrow \ln K_{eq} = -\frac{\Delta H^0}{R} \times \frac{1}{T} + \frac{\Delta S^0}{R}$$

Using the experimental data for α-amanitin binding to RNA polymerase described in the article cited in the footnote 4, it is possible to construct the van't Hoff plot.

| Temperature (°C) | Temperature (K) | $1/T \times 10^3$ (K$^{-1}$) | $K_A$ (M$^{-1}$) | $\ln[K_A$ (M$^{-1}$)] |
|---|---|---|---|---|
| 10 | 283 | 3.53 | $4.63 \times 10^9$ | 22.26 |
| 15 | 288 | 3.47 | $2.57 \times 10^9$ | 21.67 |
| 20 | 293 | 3.41 | $1.50 \times 10^9$ | 21.13 |
| 30 | 303 | 3.30 | $3.54 \times 10^8$ | 19.68 |
| 37 | 310 | 3.23 | $1.56 \times 10^8$ | 18.87 |

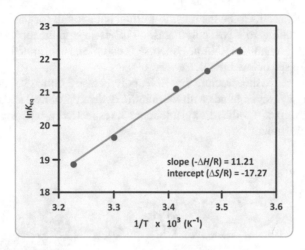

van't Hoff plot for the binding of α-amanitin to RNA polymerase

$\Delta H^0 = -22.19$ kcal mol$^{-1}$
$\Delta S^0 = -34.19$ cal mol$^{-1}$ K$^{-1}$

Enthalpy and entropy changes are both negative, thus having opposing effects on $\Delta G$. In this case, the negative $\Delta G$ (i.e., the spontaneity of α-amanitin binding to RNA polymerase) is driven by a negative enthalpy change ($\Delta H < 0$; heat release), overcompensating the drop in entropy.

The contributions of $\Delta H$ and $\Delta S$ to $\Delta G$ are closely related, as explained. For instance, the tight binding resulting from multiple favorable non-covalent interactions between association partners will lead to a large negative enthalpy change, but this is usually accompanied by a negative entropy change due to the restriction of the mobility of the interacting partners. Similarly, a large entropy gain is usually accompanied by an enthalpic penalty (positive enthalpy change) due to the energy required for disrupting non-covalent interactions. This phenomenon is called the enthalpy–entropy compensation.

The enthalpy–entropy compensation may be rooted in the formations and disruptions of the weak non-covalent interactions. Multiple factors seem to influence the compensation behavior, including (1) the structural and thermodynamic properties of the solvent (hydrophobic effect, solvation, desolvation, and local water structure—see next point on structural data obtained for α-amanitin–RNA polymerase complex); (2) the flexibility of the ligand-binding site/pocket or of the regions in the surrounding of the localized site; (3) the molecular structure of the ligand; and (4) the changes in intermolecular forces during the binding process.

4. Amatoxins' in vivo toxicity seems to directly correlate with their in vitro inhibition of RNA polymerase, and this is also related to their affinity to the enzyme. Indeed, the nontoxic amatoxin $O$-methyl-aldoamanitin neither inhibits RNA synthesis in vitro nor binds to RNA polymerase.

Additionally, $R_1$ seems to be the most important substituent for the stabilization of
   amatoxin binding to RNA polymerase. This is in agreement with the crystal
   structure of α-amanitin–RNA polymerase II complex, determined by X-ray crystallography (see below figure).
It is possible to see in the structure that RNA polymerase 2 Glu-A822 and Gln-A768
   form strong hydrogen bonds with α-amanitin hydroxyproline 2 and the backbone
   carbonyl group of 4,5-dihydroxyisoleucine 3, respectively, both located adjacent
   to the $R_1$ substituent.

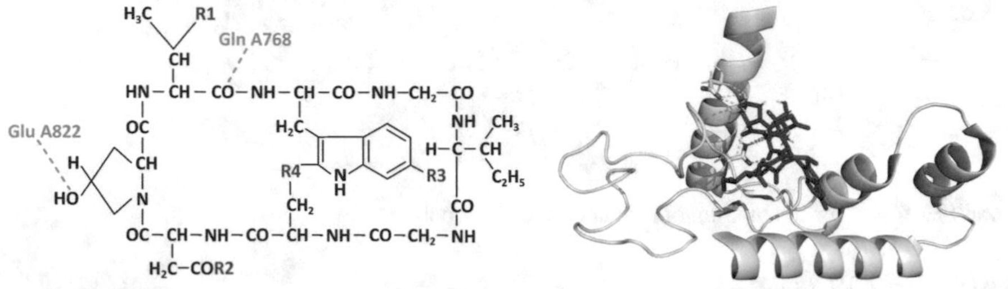

(*Left*) Chemical structure of α-amanitin, pointing out the chemical groups involved in hydrogen
bonding with RNA polymerase II (residues indicated in blue). (*Right*) α-Amanitin binding pocket
in RNA polymerase II showing the possible hydrogen bonds with Glu A822 and Gln A768 (*dotted
yellow lines*)

## Final Discussion

As stated by a Portuguese chemistry teacher João Paulo André, in an article published in *Journal of Chemistry Education*,[6] "The storyline of operas, with historical
or fictional characters, often include potions and poisons." In his text, he classifies
the operas that have potions and poisons in their plot in four categories: (1) apoth-
ecary operas, inspired in chemists and pharmacists of the eighteenth and nineteenth
centuries; (2) operas in which poisoning by natural products is a crucial part of the
story; (3) operas of the great poisoners of antiquity, which include historical characters known as experts in manipulating poisons; and (4) arsenic operas, in which
arsenic compounds are used as poison in the plot.
   The content of these operas may be useful to inspire chemistry classes and moti-
vate the students. The following table, reproduced from the article, shows the examples analyzed in the article.

---

[6]André JP (2013) Opera and poison: a secret and enjoyable approach to teaching and learning
chemistry. J Chem Educ 90:352–357.

| Materials and compounds | Opera (composer) | Aria or scene |
|---|---|---|
| Rhubarb (oxalic acid); manna (mannose; mannitol) | Der Apotheker (Haydn) | "Per quel che ha mal di stomaco" |
| Antimony(III) chloride; mercury(II) sulfide; sulfur; manna; castor oil (ricinoleic acid) | Il campanello (Donizetti) | Prescription duet |
| Oleander (oleandrin); Cherry laurel (amygdalin); hemlock (coniine); *belladonna* (scopolamine and hyoscyamine) | Suor Angelica (Puccini) | "Amici fiori" |
| *Solanaceae* (scopolamine and hyoscyamine) | Tristan and Isolde (Wagner) | "Am Obermast die Segel ein" |
| Henbane (scopolamine and hyoscyamine) | Hamlet (Thomas) | |
| Mandrake (scopolamine and hyoscyamine) | Romeo and Juliet (Gounod) | Poison aria |
| *Datura stramonium* (scopolamine and hyoscyamine) | Lakmé (Delibes) | "Viens Mallika" (Flower duet) |
| *Hippomane mancinella* (physostigmine) | L'Africaine (Meyerbeer) | "O Paradis" |
| Curare (tubocurarine) | Il Guarany (Gomes) | |
| Venom of snakes (neurotoxins) | Antony and Cleopatra | "Give me my robe" |
| ? | Mitridate, Re di Ponto (Mozart) | "Ah ben ne fui presaga … Pallid' ombre" |
| Arsenic trioxide | Simon Boccanegra (Verdi) | Poison scene (act II) |
| Arsenic trioxide; BAL and DMSA (agents for As(III) chelation therapy) | Lucrezia Borgia (Donizetti) | End of act I |
| Scheele's green (copper(II) arsenite) trimethylarsine | Patience (Gilbert and Sullivan) | "Am I alone and unobserved?" |
| Radium(II) chloride | Madame Curie (Sikora) | |

## Challenging Case 5.2: The Biochemistry of Zombies

### Source

This case is based on the book *The Serpent and the Rainbow: A Harvard Scientist's Astonishing Journey into the Secret Societies of Haitian Voodoo, Zombies, and Magic*,[7] by Wade Davis, a Harvard anthropologist who got interested in news clips reporting the certified deaths of several Haitian individuals who later on were found by their relatives to be alive and working as slaves.

Wade Davis. (Image reproduced from Wikimedia Commons, under a CC0 1.0 license)

### Case Description

According to the *Encyclopaedia Britannica*, "*A zombie is regarded by voodooists as being either a dead person's disembodied soul that is used for magical purposes, or an actual corpse that has been raised from the grave by magical means and is then used to perform agricultural labour in the fields as a sort of a will-less automaton. In actual practice some voodoo priests do appear to create zombies by administering a particular poison to the skin of a victim who enters a state of profound physical paralysis for a number of hours.*"

Being particularly interested in ethnobotanics, a speciality that deals with the psychoactive properties of plants, the anthropologist Wade Davis was inspired to investigate whether reports about the existence of zombies in Haiti were authentic or just folklore. He published his findings in the 1985 book *The Serpent and the Rainbow*, which described his quest and actually brought a degree of credence to the legend. Davis began his work in Haiti by consulting several voodooists. After winning their trust, Davis analyzed their zombie-making recipes seeking in them the occurrence of active principles displaying pharmacological actions compatible with the zombie profile.

---

[7] Davis W (1985) The serpent and the rainbow: a Harvard scientist's astonishing journey into the secret societies of Haitian voodoo, zombies, and magic. Simon & Schuster, New York, NY. ISBN: 978-0-684-83929-5.

Voodoo ceremonial artifacts. (*Left*) Voodoo flag exhibited at the Nationaal Museum van Wereldculturen, the Netherlands, representing "Dambala," the rainbow snake god who bears the cosmos, and his female counterpart the rainbow snake "Ayida Wédo," with eggs (wealth and fertility) at their heads. (*Right*) Voodoo mirror exhibited at the Ethnological Museum of Berlin, Germany, which in Haiti's voodoo culture is considered as a doorway into the world of the dead. (Images reproduced from Wikimedia Commons)

The potions usually involved the production of crude extracts from different species of animals, insects mostly, but occasionally toads and fish were also present. Davis reasoned that the common ingredients that could contain some pharmacologically active compounds were the puffer fish (fugu) and the cane toad *Rhinella marina* formerly known as *Bufo marinus*. Soon Davis narrowed down his search to the toxin tetrodotoxin (TTX), which was originally discovered in 1909 by Dr. Yoshizumi Tahara[8] and found to occur in several species of endosymbiotic bacteria (bacteria that lived in these animals). In other words, the animals that contain TTX do not make it themselves, but through bioaccumulation of the TTX producing microorganisms end up by becoming poisonous to other animals that consume them.

[8] Bane V et al (2014) Tetrodotoxin: chemistry, toxicity, source, distribution and detection. Toxins 6:693–755.

Chemical structure of tetrodotoxin and the cane toad *Rhinella marina*. *(Image reproduced from Wikimedia Commons)*

TTX is a powerful neurotoxin (a thousand times more toxic than cyanide) that acted very specifically by binding to integral membrane proteins whose function was that of a sodium channel. The mode of action of TTX is to temporarily disable the function of the sodium ion channel, thereby blocking the passage of sodium ions across the plasma membrane. This blocks the firing of action potentials in neurons, which then prevents the nervous system from propagating pulses along the axons, inhibiting muscular activity.

When reading about the properties of TTX, Wade Davis was convinced that zombies were in fact people who ingested or absorbed nonlethal doses of TTX. Davis went on to conclude that the main zombifying mechanism was the muscle paralysis mediated by TTX that produced a near-death situation. He also suggested that in order to produce useful zombies, the toxin had either to be ingested or rubbed onto wounded flesh, but understandably the dosage had to be very carefully estimated, or else the potential zombies would simply die. The voodooist also had to be sure that the TTX-treated victim was not buried for too long, or else resuscitation would not occur. The narrative went on to describe that among the side effects of the TTX-intoxicated zombies was loss of memory, a feature that actually helped unscrupulous voodooists to convince them that they were slaves and had to carry on their servitude for the rest of their lives.

As much as Wade Davis's book caused an impact at the time of its publication, his zombie hypothesis was later dismissed by the scientific community on the basis of careful chemical analysis and remains today as a bizarre curiosity. Notwithstanding, Wade Davis's tale inspired yet another account that appeared in 1985. This was the novel *Bufo & Spallanzani* by the Brazilian author Rubem Fonseca[9] which made use of the TTX effect to narrate a very well-spun story involving a health insurance scam. Regardless of the present lack of the conflicting information about the true origins of bona fide zombies, rest assured that if they do actually exist, the final scientific interpretation will most certainly be biochemical in essence.

---

[9]Fonseca R (1985) Buffo & Spalanzanni. Editora Nova Fronteira, Rio de Janeiro. ISBN: 9788520926512.

## Questions

1. TTX is known to be a very potent neurotoxin, and, as pointed out in the text, it is thousands of times more toxic than cyanide. In practical terms, how would you devise an assay that would allow for quantification of the potency of toxins?
2. Recently TTX has been used as an adjuvant for the treatment of cancer. In your view, how is this neurotoxin medically useful for the treatment of a chronic degenerative disease such as cancer?
3. Why don't the animals that contain TTX-producing bacteria suffer any effects from the toxin?

## Biochemical Insights

1. In order to classify toxins in terms of their potency, one has to take into account the avidity with which they bind to their receptors, in this case, the proteins that make up the sodium channels. In biochemistry, this avidity is better referred to as affinity, which can be translated as the tendency of a compound to combine with, or bind to, other chemical substances. The affinity can then be quantified by measuring a certain effect (in the case of TTX, this could be muscular paralysis or death) and correlate it to the concentration of the toxin that produces it. To measure the potency of TTX, one could use a standard assay, in which several concentrations of TTX would be injected into experimental animals (caterpillars, for instance) and then record the concentration of TTX that produces the death of half the population studied. This is also called the lethal dose 50, or $LD_{50}$ for a certain compound. The higher the potency, the lower the concentration of TTX required to cause the death of the animals. Affinity can be measured in a number of ways, but always looking at the correlation between concentration of a ligand and its overall effect. Affinity has been explained in detail in Chap. 4 within the context of the enzyme–substrate interaction and illustrates well the general concept (recall the molecular interpretation given to the $K_M$).
2. In many chronic diseases, one of the major problems which significantly affect the patient's quality of life is pain. This is definitely the case with cancer, in which the tumors grow to such a size that they start compromising the functions of different organs and also cause intense pain. Physiologically, pain is the sensation one has when certain nerves are stimulated, and the signals are then transmitted to the brain. Transmission occurs via propagation of the impulses along the nerves' axons. In turn, transmission depends on the function of the sodium channels. As mentioned above, TTX blocks muscular contractions by binding to and inhibiting the action the sodium channels, thereby blocking not only muscular contraction but also pain, since this signal is also transmitted via the same sodium channel proteins. As with any drug, the analgesic effect of TTX has to be carefully monitored or else death ensues.

3. The animals that contain TTX-producing bacteria do not suffer any effects from the toxin because the proteins that constitute the sodium channels of animals like the puffer fish have a very discrete change in their amino acid composition. Thus, these mutated sodium channel proteins have very little affinity for TTX, which render them resistant to it. By the way, to date there is no known antidote to TTX.

## Final Discussion

This case study exemplifies situations that are common in the history of science. Scientists heed tales from the folklore that narrate events that at first seem fantastic but that later prove to have some basis in reality, both in medicine and basic sciences. Apart from the pharmacological origin of the zombies described above, there are other instances along the same vein. For example, the tale of vampires (the original story of Dracula, by Bram Stoker), those night creatures who feed on blood and who for centuries have instilled fear in the minds of people. Real-life vampires have been described in the realm of psychiatry. They are represented by people who suffer from the so-called Renfield's syndrome and who derive sexual pleasure by drinking the blood from their victims. Curiously Dracula's legend is loaded with sexual undertones, and the Count is usually presented as an incorrigible womanizer. Another psychiatric disorder is clinical lycanthropy, in which the affected individuals believe they can turn into animals, most commonly wolves, especially in moonlit nights. This is the basis of the legend of werewolves. Obviously, in both cases, Renfield's syndrome and lycanthropy, the extra powers described in the tales are not real. People affected by vampirism cannot fly, become invisible, and turn into bats. The sufferers from lycanthropy do not really become all powerful wolves and, contrary to legend, would die if hit by simple bullets as opposed to the silver bullets described in the novels.

*This case is a contribution of Prof. Franklin D. Rumjanek (Instituto de Bioquímica Médica Leopoldo de Meis, Universidade Federal do Rio de Janeiro, Brazil). In memoriam (1945–2020).*

# Chapter 6
# Energy Conservation in Metabolism: The Mechanisms of ATP Synthesis

In living organisms, different types of energy are always interconverting into one another within the cell enabling the distinct cellular functions to be performed. This can be illustrated by several examples, such as the conversion of the energy of light into chemical energy in photosynthetic organisms or the chemical energy into mechanical energy in muscle contraction. Also, virtually, all the cells need to use chemical energy to transport ions and other compounds across a membrane, generating a concentration gradient and thus converting chemical into osmotic energy. Finally, chemical energy is continuously converted in other forms of chemical energy during the biosynthesis of new molecules in cellular metabolism.

In the beginning of the twentieth century, a set of experiments carried out by Arthur Harden and William J. Young, in which they showed that phosphate is essential for yeast alcoholic fermentation, started a new era for the understanding on how energy is obtained from the environment and stored within the cells for later use. This discovery was the first association between phosphate and energy transformations in living cells, paving the way for the subsequent identification of ATP, more specifically its phosphoanhydride bond, as the main cellular energy carrier.

In this chapter, we will discuss the principles and the most relevant steps of the main processes of ATP synthesis in heterotrophic cells. Heterotrophic organisms conserve the energy of nutrient molecules by coupling the breaking of their chemical bonds to the synthesis of ATP, which occurs through two distinct mechanisms.

The first mechanism of ATP synthesis to be identified is known as substrate-level phosphorylation. It does not depend on oxygen and thus may occur in anaerobiosis. The general principle involved in ATP synthesis through this mechanism is the formation of a phosphorylated molecule that presents a so-called high-energy

© Springer Nature Switzerland AG 2021
A. T. Da Poian, M. A. R. B., *Integrative Human Biochemistry*,
https://doi.org/10.1007/978-3-030-48740-9_6

phosphate bond or, in a more precise term coined
by Fritz Lipmann, a high potential of transferring
its phosphoryl group, which is used to phosphorylate ADP, generating ATP.

The second mechanism of ATP synthesis is
known as oxidative phosphorylation. It depends on
organized membranes and, in eukaryotic cells,
takes place in mitochondria. The unraveling of this
mechanism was possible due to someone who
could see beyond the paradigm of the energy of the
chemical bonds. This man was Peter Mitchell, who
formulated the revolutionary chemiosmotic theory,
which states that the synthesis of ATP from ADP
and inorganic phosphate (Pi) is driven by the pH
gradient across the mitochondria inner membrane
and the consequent formation of a transmembrane
electrical potential.

Peter Mitchell     Fritz Lipmann
(1920-1992)       (1899-1986)

## 6.1  ATP Synthesis by Substrate-Level Phosphorylation

When the first forms of life arose on Earth, the atmosphere contained no oxygen. In
this scenario, the anaerobic use of glucose appeared as the main metabolic pathway
of energy conservation. This pathway was termed fermentation (see Sect. 6.1.1 to
understand the origin of the term), and involves the synthesis of ATP through the
transfer of the phosphoryl group from an intermediate of the pathway directly to
ADP, a process known as substrate-level phosphorylation.

Although in humans the oxidative phosphorylation is the main pathway for ATP
synthesis, for some cell types or in some situations, the fermentation is of crucial
importance. Mature erythrocytes, cells from the crystalline, and some cells of the
retina lack mitochondria, the organelles in which the oxidative phosphorylation
takes place. There are also situations in which some cells, even being highly oxidative, may experience limited oxygen availability. This is the case of muscle cells in
intense contraction activity working in low oxygen conditions due to the adrenaline-
induced contraction of peripheral blood vessels (see Chap. 10). In some pathological conditions, fermentation overcomes oxidative metabolism even in the presence
of oxygen. This is the case of cancer cells, which undergo the called "Warburg
effect" (see Box 6.4).

It is important to note that among the nutrient molecules, only saccharides can be
used as energy source in anaerobic conditions. The degradation of the other nutrients, the lipids and the proteins, depends on metabolic pathways that occur in mitochondria and culminate with ATP synthesis through oxidative phosphorylation,
which requires oxygen as the final substrate (see details in Sect. 6.2).

### 6.1.1 A Historical Perspective of the Fermentation Process Discovery

Since ancient times, humans have dealt with processes involving fermentation, although only in the nineteenth century it has been understood as a metabolic pathway associated with the reactions of energy transformation essential for life.

Fermentation is the basis of the production of bread, beer, and wine, activities that accompanied humans since the early civilizations. The term fermentation arose from these practices, coming from the Roman word *fermentare*, which is related to the formation of bubbles. Thus, fermentation was firstly associated with a process in which a gas was produced. Although now we know that, depending on the organism, fermentation leads to different end products, not always including a gas, in the alcoholic fermentation, which occurs during the production of bread, beer, and wine, sugars from fruits or cereals are converted into ethanol and $CO_2$, the gas that is released.

Despite several advances in the scientific studies on fermentation during the nineteenth century, Louis Pasteur was the first to connect living organisms to fermentation, creating the basis for its understanding as a biological process (Box 6.1).

---

**Box 6.1 The Impact of Pasteur's Ideas About Fermentation**

The concept that fermentation was performed by living organisms was not readily accepted when proposed by Pasteur, especially because it seemed to reinforce the theory of vitalism, which assumed that organic substances in living organisms could be formed only under the influence of a mysterious "vital force." Jacob Berzelius, Justus von Liebig, and Friedrich Wöhler, important chemists from that time, were strongly against the vitalistic ideas, and, although they were right in their concepts on the chemical processes, their convictions made them very reluctant to acknowledge the valuable contributions of Pasteur's observations. This can be illustrated by the ironic text published by Wöhler and von Liebig, in 1839, in the *Annals of Chemistry*, regarding the participation of yeast in the transformation of sugars in ethanol and $CO_2$ in fermentation. "Beer yeast, when dispersed in water, breaks down into an infinite number of small spheres. If these spheres are transferred into an aqueous solution of sugar, they develop into small animals. They are endowed with a sort of suction trunk with which they gulp the sugar from the solution. Digestion is immediate and clearly recognizable because of the discharge of excrements. These animals evacuate ethyl alcohol from their bowels and carbon dioxide from their urinary organs. Thus, one can observe how a specifically lighter fluid is extruded from the anus and rises vertically, whereas a stream of carbon dioxide is ejected at very short intervals from their enormously large genitals."

Pasteur observed microscopic globules in a wine sample (Fig. 6.1a) and sustained that these globules were living microorganisms responsible for the sugar transformation into ethanol and $CO_2$. He also demonstrated that each type of fermentation is linked to a specific microorganism or, as he named, a specific "ferment." Interestingly, similar globules have been described in a sample of beer very earlier, in 1680, by Anton van Leeuwenhoek, the pioneer of the use of the microscope (Fig. 6.1b), although there is no evidence that van Leeuwenhoek had associated these globules with living organisms.

In addition to associate fermentation to the presence of living organisms, Pasteur also carried out crucial experiments showing that it occurred in the absence of oxygen. These experiments discredit the view of fermentation as a chemical process resulting from the reaction between oxygen and sugars, reinforcing the idea that fermentation was a biological process. Pasteur's findings also allowed the formulation of other two important biological concepts derived from his experiments. The first one was that some forms of life can exist without oxygen, in what he referred in one of his most famous articles, published in 1861, as *La vie sans l'air* (the life without air). The second concept came from an intriguing observation that he made in those experiments: more $CO_2$ was produced by yeast in the absence than in the presence of oxygen, a phenomenon known as the "Pasteur effect" (Fig. 6.1c), which is now explained by the fact that the number of ATP molecules synthesized per glucose molecule in fermentation is much lower than that in oxidative phosphorylation; therefore, much more glucose are consumed.

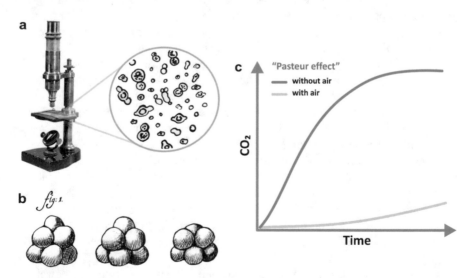

**Fig. 6.1** Louis Pasteur's seminal contributions to the understanding of the fermentation process. (**a**) The microscope used by Pasteur to observe the wine samples and his drawing representing the "globules" he had observed in the sample. (Figure adapted from Pasteur's book *Études sur la bière*, 1876). (**b**) Drawing by van Leeuwenhoek, reporting, in 1680, the observation of a beer sample in the microscope that he developed. (**c**) Representation of the "Pasteur effect," in which the production of $CO_2$ by yeast is much higher in the absence than in the presence of oxygen

The next significant advance in the understanding of fermentation came from an unexpected observation. Eduard Buchner, working with his brother Hans in the preparation of a yeast extract to treat patients with tuberculosis, discovered that even when all the yeast cells were completely disrupted, fermentation continued to proceed normally. This finding marks the end of the vitalism theory by showing that the presence of living cells was not necessary for fermentation, and can be regarded as the dawn of biochemistry, introducing the concept that biological reactions are catalyzed by molecules, which Buchner called zymases, the term firstly used to refer to what we now call enzymes. Due to the great impact of his discoveries, Eduard Buchner was the first scientist to be awarded the Nobel Prize in Chemistry, in 1907.

Finally, we must comment the experiment by Harden and Young, mentioned in the beginning of this chapter. Using the cell-free system developed by the Buchners, they observed that the fermentation rate, which decreased with time, could be restored with the addition of salts of phosphoric acid (Fig. 6.2). This was the first evidence that phosphate was involved in cell metabolism, opening the way in the subsequent years to the recognition of the ubiquity and importance of the biological phosphorylations in life.

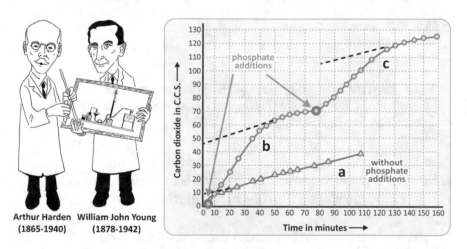

**Fig. 6.2** Classical experiment by Harden and Young. Representation of the original figure from the article published in 1906, showing the production of $CO_2$ by yeast fermenting glucose in the absence of any addition (curve **a**) and when salts of phosphoric acid were added at the indicated times (curves **b** and **c**). (Figure adapted from Harden and Young. Proc. Royal Soc. Lond. B 77:405–520, 1906)

Few years later, Otto Meyerhof, studying muscle contraction in different organisms, showed that contraction occurred in the absence of oxygen, a similar phenomenon to that observed by Pasteur for yeast alcoholic fermentation (Fig. 6.3a). However, in the case of muscle cell fermentation, the end product was lactate. The molecular similarity of the end products of alcoholic and lactic fermentations suggested that both processes could be equivalent (Fig. 6.3b).

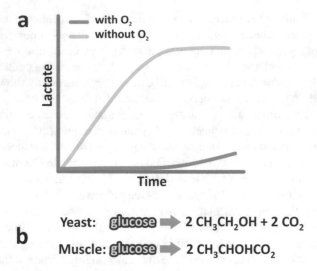

**Fig. 6.3** Muscle fermentation experiments by Otto Meyerhof. (**a**) Schematic representation of the results obtained by Meyerhof when a muscle fiber was incubated with glucose. (**b**) Similarity of the end products of alcoholic fermentation performed by yeast and lactic fermentation performed by the muscular fibers

Additionally, the same requirement of phosphate was observed for fermentation in the muscle cells. Now we know that indeed the processes are identical, except for the last reaction (see next section).

### 6.1.2   An Overview of the ATP Synthesis by Substrate-Level Phosphorylation During Fermentation

The general mechanism of ATP synthesis that occurs during fermentation consists in a series of reactions that rearranges the molecular structure of an initially phosphorylated monosaccharide, which is a hexose-phosphate, in such a way as to form phosphorylated compounds with high potential of transferring their phosphoryl group (see Box 6.2). These high-energy intermediates are triose phosphate molecules originated from a cleavage step and whose phosphoryl group is transferred to ADP, generating ATP (Fig. 6.4a).

The triose phosphates with high potential of transferring their phosphoryl groups are 1,3-bisphosphoglycerate (1,3BPG) and phosphoenolpyruvate (PEP) (Fig. 6.4b). The anhydride bond that links 1,3BPG's carboxyl group to the phosphate and the double bond between carbons 2 and 3 of PEP show $\Delta G^0$ values of hydrolysis of

−49.4 kJ mol⁻¹ and −61.9 kJ mol⁻¹, respectively. These values are higher than that of the hydrolysis of ATP phosphate group, which is −30.5 kJ mol⁻¹, allowing the transfer of the phosphoryl groups of 1,3BPG and PEP to ADP. This is the molecular basis of the substrate-level phosphorylation mechanism.

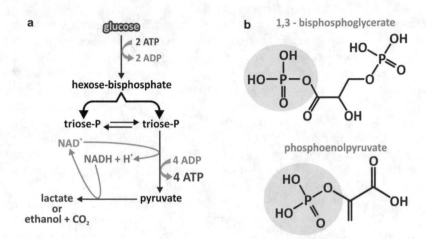

**Fig. 6.4**   General overview of the essentials of the fermentation process. (**a**) The pathway initiates with a hexose (usually glucose). After two steps of phosphorylation with ATP as the phosphate donor, the resultant hexose bisphosphate is cleaved into two triose phosphates, which may inter-convert. One of them follows the pathway, which includes one NAD⁺-dependent oxidation reaction and two transfers of the phosphate group of a high-energy triose phosphate to ADP, generating ATP and, at the second of those reactions, pyruvate. Pyruvate must be reduced to allow NADH reoxida-tion. The main end products of fermentation are lactate or ethanol and CO₂, depending on the organism. (**b**) Chemical structures of the high-energy triose phosphate molecules 1,3-bisphosphoglycerate and phosphoenolpyruvate

The initial substrate for fermentation is usually glucose, the most abundant monosaccharide derived from a regular human diet. The two initial steps of phos-phorylation use ATP as the phosphoryl group donor (Fig. 6.4a). This may seem nonsensical at a first glance, if we think that this is a metabolic pathway for the synthesis of ATP. However, as it will become clear through the analysis of the stoi-chiometry of the entire pathway, although two ATP molecules are used for each hexose that enters the pathway in its initials steps, four ATP molecules are synthe-sized at the end, yielding a positive balance of two ATP molecules for each hexose molecule that is metabolized. Moreover, consumption of ATP is mandatory to "force" the first steps to occur in terms of thermodynamics (see Box 4.2).

Additionally, the pathway includes an oxidative step in which the oxidation of a three-carbon intermediate is coupled to the reduction of the coenzyme nicotinamide adenine dinucleotide (NAD⁺), generating NADH. Since the typical amount of NAD⁺ in the cytoplasm is much lower than the amount of glucose metabolized, fermenta-tion should end with a step leading to the reoxidation of the NADH molecule. The reaction responsible for this synthesizes the end product of the pathway, which in human cells is lactate (Fig. 6.4a).

**Box 6.2  The Concepts of "Energy-Rich" Phosphate Compounds**
In his classical article from 1941, Fritz Lipmann classified the phosphate compounds as "energy rich" and "energy poor." This classification was based on the conception that the energy released by the hydrolysis of phosphorylated compounds was essentially determined by the nature of the chemical bond that links the phosphate to the molecule. Phosphoanhydride linkages generate "energy-rich" compounds, while phosphoester linkages generate "energy-poor" compounds. According to this concept, the free energy of hydrolysis of a phosphate compound would be determined essentially by the contribution of enthalpy. Due to the experimental approaches used at that time, the contribution of the environment in which the reaction takes place could not be taken into consideration, but new data obtained after 1970 introduced the idea that the energy of the hydrolysis of phosphate compounds would be also determined by the differences in their solvation energy and thus would vary greatly depending on the medium. This was found valid for compounds containing phosphoanhydride bonds, such as ATP, pyrophosphate, or acyl phosphates, but not for compounds in which the phosphate group is linked to the molecule through a phosphoester bond (see figure comparing the energy of hydrolysis of the phosphoanhydride bond of pyrophosphate with the phosphoester bond of glucose-6-phosphate as a function of water activity). Based on these observations, a new conception was formulated by the Brazilian scientist Leopoldo de Meis, according to which molecules containing phosphoanhydride bonds are susceptible to changes in the entropic energy (dependent on the solvation energy and, thus, on the environment), which contributes to the free energy involved in the reaction.

Leopoldo de Meis
(1938-2014)

Effect of water activity (Wa) on the energy of hydrolysis of pyrophosphate (*yellow circles*) and glucose-6-phosphate (*blue circles*). (Graph reproduced from de Meis, in Calcium and Cellular Metabolism: Transport and Regulation, chapter 8. Springer Science+Business Media, NY, 1997)

With this in mind, it is possible to understand the mechanism of ATP synthesis by ATP synthase presented in Sect. 6.2.4, in which the energy required in the reaction is not needed for the condensation of ADP and Pi, but for the ATP release from the catalytic site of the enzyme.

## 6.1.3  Glucose Fermentation Reactions

The first ten reactions of the fermentation process, starting with glucose and ending with the formation of two pyruvate molecules, are the same in many organisms that synthesize ATP anaerobically. The last step, which involves NADH reoxidation, is different depending on the organism, resulting in either lactate or ethanol and $CO_2$ as the end products. Furthermore, the reactions from glucose to pyruvate are also the same that occur when carbohydrates are used aerobically (see Sect. 7.4). The pathway from glucose to pyruvate is named glycolysis and may be divided in two parts.

In the first part, two phosphorylation steps using ATP generate a hexose with two phosphoryl groups linked, the fructose-1,6-bisphosphate, which was the compound discovered by Harden and Young (see Sect. 6.1.1), and the first intermediate of fermentation to be identified, known as the Harden–Young ester. In the first phosphorylation step, the enzyme hexokinase catalyzes the conversion of glucose in glucose-6-phosphate, which is then isomerized to fructose-6-phosphate by the phosphohexose isomerase. The second phosphorylation step is catalyzed by the phosphofructokinase and occurs at the hydroxyl group on carbon 1 of fructose-6-phosphate, generating fructose-1,6-bisphosphate. Fructose-1,6-bisphosphate is then cleaved by the aldolase into two triose phosphate molecules, glyceraldehyde-3-phosphate, and dihydroxyacetone phosphate. These triose phosphates may interconvert in a reaction catalyzed by the triose phosphate isomerase, with dihydroxyacetone phosphate forming glyceraldehyde-3-phosphate, the next intermediate of the pathway. Thus, in the end of this first phase, one glucose molecule is converted into two glyceraldehyde-3-phosphate molecules (Fig. 6.5). In the light of the points discussed in the previous section, it is important to note that all the phosphorylated compounds of this first part of glycolysis are phosphoesters.

In the second part of the pathway, each glyceraldehyde-3-phosphate molecule is oxidized in a $NAD^+$-dependent reaction followed by a phosphorylation step, in a reaction catalyzed by the glyceraldehyde-3-phosphate dehydrogenase, which forms 1,3-bisphosphoglycerate. Note that this phosphorylation step uses inorganic phosphate directly and not ATP as the phosphate donor. 1,3-Bisphosphoglycerate is the first high-energy phosphorylated compound generated in the pathway. The oxidation of the aldehyde group of glyceraldehyde coupled to the phosphorylation reaction forms, instead of a free carboxyl group, an acyl-phosphate, from which the phosphoryl group is transferred to ADP, generating ATP and 3-phosphoglycerate, in a reaction catalyzed by the phosphoglycerate kinase. Phosphoglycerate mutase converts 3-phosphoglycerate into 2-phosphoglycerate, which is then dehydrated by the enolase, forming the second compound in the pathway with high potential of transferring the phosphoryl group, the phosphoenolpyruvate (PEP). Then, another substrate-level phosphorylation step occurs, with the phosphoryl group of PEP transferred to ADP, generating ATP and pyruvate, a reaction catalyzed by the pyruvate kinase (Fig. 6.5).

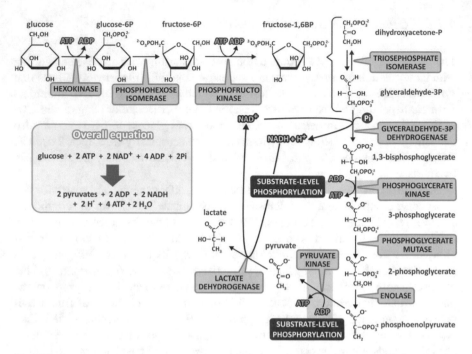

**Fig. 6.5** Reactions of the fermentation of glucose to lactate. The first ten reactions consist in the metabolic pathway named glycolysis, which is also the pathway for the aerobic metabolization of carbohydrates. The last step occurs when pyruvate is not oxidized through the aerobic metabolism and NADH must be reoxidized. It should be noticed that each fructose-1,6-bisphosphate molecule leads to two glyceraldehyde-3-phosphate molecules, as dihydroxyacetone phosphate will be converted to glyceraldehyde-3-phosphate. The names of the enzymes are highlighted in *yellow boxes*. The steps that involve substrate-level phosphorylation are also indicated

The last step in the fermentation of glucose in human cells is the reduction of pyruvate to lactate, catalyzed by the lactate dehydrogenase, allowing the reoxidation of NADH (Fig. 6.5).

When glycolysis is the means of carbohydrate utilization with concomitant use of oxygen, instead of pyruvate being reduced in the last step of fermentation, it is completely oxidized to $CO_2$ in mitochondria (see Sect. 7.4).

## 6.2   ATP Synthesis Through Oxidative Phosphorylation

Oxidative phosphorylation accounts for 95% of ATP synthesis in the human organism. This metabolic pathway is compartmentalized in mitochondria, the organelles responsible for most of the bioenergetic functions within the eukaryotic cell. The accepted hypothesis for mitochondria origin proposes that these organelles were originally prokaryotic cells that became endosymbionts living inside the ancient eukaryotic cells (Box 6.3). Indeed, the mitochondria are unique organelles, comprising an own circular genome that shows substantial similarity to bacterial genomes and codifies ribosomal RNAs and the 22 tRNAs necessary for the translation of mRNAs into proteins. Moreover, like bacteria, mitochondria are composed of two lipid membranes (Fig. 6.6). The outer mitochondrial membrane contains several porins, proteins that make membrane permeable to ions and small molecules (with molecular mass lower than 5 kDa). The inner membrane has a very high protein–lipid ratio and is rich in an unusual phospholipid, cardiolipin. It is very impermeable, with the transport of molecules and ions, including $H^+$, occurring only through specific proteins. This membrane has a very large surface area provided by several convolutions called mitochondrial *cristae*. The inner membrane encloses the mitochondrial matrix, which comprises most of the enzymes of the oxidative metabolism, the mitochondrial ribosomes, tRNA, and several copies of the mitochondrial DNA.

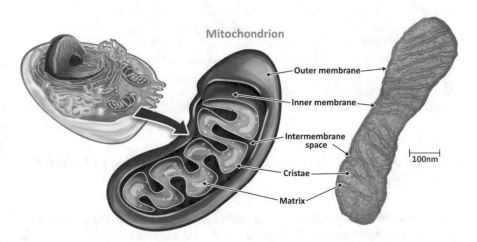

**Fig. 6.6** The structure of a mitochondrion. Mitochondria are organelles formed by a permeable outer membrane and a very impermeable inner membrane with several convolution named cristae. The intramitochondrial medium forms the mitochondrial matrix. (Mitochondrion transmission electron micrograph: courtesy from Prof. Marlene Benchimol)

**Box 6.3 The Endosymbiotic Theory for the Origin of Mitochondria**
The incorporation of the mitochondria was an important event in the evolution of the eukaryotic cells. It is believed that it occurred more than 1.5 billion years ago through the invasion of a heterotrophic anaerobic cell by an aerobic bacterium. This is known as the endosymbiotic theory, which was postulated in the beginning of the twentieth century but was revived and better argued by Lynn Margulis, in the 1960s. Several genetic evidences suggest that the ancestral symbiont was an aerobic α-proteobacterium that consumed oxygen through a respiratory chain. The role of the anaerobic host in the symbiosis would be to make pyruvate accessible to the endosymbiont metabolism. On the other hand, evidence indicate that the role of the ancestral symbiont in the initial phase of the evolution of the mitochondria was to protect the anaerobic cell components from the toxic effects of oxygen through the activity of the last respiratory chain component, the enzyme cytochrome oxidase, which converts oxygen to water (see next sections of this chapter). In fact, the sharp increase in the oxygen tension around two billion years ago introduced a great threat for the anaerobic ancient cells, which did not have the detoxifying enzymes peroxidases, catalases, or superoxide dismutases that protect modern cells against the toxic effects of reactive oxygen species. Thus, the aerobic symbiont would function as an oxygen scavenger inside the host cell. With time, evolution of host genomes probably contributed with new functions to the symbiont, including the ATP/ADP transporter, transforming it into an organelle with an ATP-exporting function.

## 6.2.1   A Historical Perspective of the Understanding of Cellular Respiration

In the eighteenth century, Lavoisier's classical experiments correlated respiration to the combustion of organic matter. He demonstrated that both animal respiration and organic matter combustion involved $O_2$ consumption, $CO_2$ release, and heat production, establishing the concept that aerobic metabolism proceeds with the direct reaction between $O_2$ and the organic compounds in the body. However, the connection between metabolic oxidative reactions and $O_2$ consumption in respiration remained elusive for a long time.

In the first decades of the twentieth century, there was a serious controversy regarding the mechanisms of biological oxidations in aerobic metabolism (Fig. 6.7). On one side, Heinrich Wieland postulated, together with Torsten Thumberg, that in the biological oxidations, the reactions catalyzed by the dehydrogenases "activated" some hydrogen atoms of the metabolic intermediates, making them labile to be transferred to a hydrogen acceptor. It is important to mention that their hypothesis, although incomplete by not taking into account the role of oxygen, correctly supported the concept that free oxygen does not directly combine with carbon to form

$CO_2$. On the opposite side was Otto Warburg (Box 6.4), who defended the oxygen-activating hypothesis, in which he argued that the dehydrogenase concept was unnecessary. In this hypothesis, Warburg postulated that the oxidation of all metabolites was catalyzed by an iron-containing enzyme, which he named *Atmungsferment* (meaning oxygen-transferring enzyme or respiratory enzyme) and in which the iron atom was oxidized to its ferric state ($Fe^{3+}$) by oxygen and reduced back to its ferrous form ($Fe^{2+}$) after reaction with organic substances. The key to solve the Wieland–Warburg controversy was provided by the entomologist and parasitologist David Keilin, who changed biochemistry with his findings (Fig. 6.7).

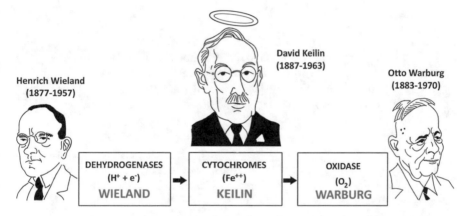

**Fig. 6.7** Solution of Wieland–Warburg controversy with the discovery of the cytochromes by Keilin. Wieland postulated that the reactions catalyzed by the dehydrogenases "activated" some hydrogen atoms of the metabolic intermediates, making them labile to be transferred to a hydrogen acceptor, while Warburg defended the oxygen-activating hypothesis, in which biological oxidations were catalyzed by an iron-containing enzyme. Keilin proposed that the cytochromes connected the dehydrogenases and the oxidase, being alternately reduced by the dehydrogenases and oxidized by oxygen through the Warburg enzyme

---

**Box 6.4 The Diversity of Otto Warburg's Contributions for Science**
Otto Warburg was a very active and interdisciplinary scientist, giving outstanding contributions to different fields in science, including respiration, photosynthesis, and cancer cell metabolism. Warburg strongly defended the use of quantitative methods and worked on the improvement of the instruments to get reliable measurements. Using manometric techniques, he developed an apparatus, the Warburg respirometer (see next figure), to quantify $O_2$ uptake by thin slices of a tissue through the changes in the chamber pressure, measured by the connection of a manometer to the end of the glass joint (H). The apparatus also allowed the measurements of $CO_2$ emission by adding, for example, potassium hydroxide in the chamber (E) to precipitate $CO_2$.

(continued)

**Box 6.4** (continued)

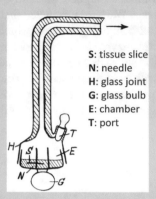

S: tissue slice
N: needle
H: glass joint
G: glass bulb
E: chamber
T: port

Due to his pioneering studies on cellular respiration in the beginning of the twentieth century, in which he proposed the existence of an iron-containing respiratory enzyme (the Warburg *Atmungsferment*), he was awarded the Nobel Prize in Physiology or Medicine in 1931.

Warburg also dedicated much time of his life investigating cancer cell metabolism. Studying different types of cancer cells in the decade of 1920, Warburg made a very intriguing observation: he found a behavior that was the opposite of the Pasteur effect (the inhibition of fermentation by $O_2$; see Sect. 6.1.1). Warburg showed that cancer cells produced lactic acid from glucose even under aerobic conditions, which is known as the Warburg effect (see next figure). This was firstly interpreted as a consequence of a damaged respiration in cancer cells, but now we know that the Warburg effect occurs due to alterations in the regulation of glycolysis in tumorigenesis.

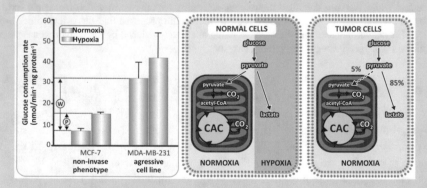

Pasteur (P) and Warburg (W) effects; CAC: citric acid cycle (see Sect. 7.2). (Graph in the left reproduced by permission from Macmillan Publishers Ltd.: Gatenby & Gillies. Nat. Rev. Cancer 4:891–899, 2004)

Studying the physiology of insects, Keilin rediscovered a substance that has been described more than a century before by Charles A. MacMunn. This substance was forgotten after it has been considered a hemoglobin contaminant present in MacMunn's preparations (see Box 6.5). Keilin proved that it was not hemoglobin, naming it "cytochrome" (meaning cellular pigment) due to its ubiquity and ability to absorb visible light. He showed that cytochrome absorption spectrum was very characteristic, containing four distinct bands, which he named *a*, *b*, *c*, and *d*, and identified this substance in cells of many different organisms, such as insects, worms, yeast, and plants. But how Keilin's findings correlate to the Wieland–Warburg controversy? Keilin proposed that this pigment acted as the link between the Wieland dehydrogenases and the Warburg oxidase, being alternately reduced by the dehydrogenases and oxidized by oxygen through the Warburg enzyme.

---

**Box 6.5 The Discovery of the Cytochromes by David Keilin**

The basic features of the respiratory chain were established between the 1920s and 1930s by the entomologist and parasitologist David Keilin. In 1925, he published his first paper in this field, entitled "On cytochrome, a respiratory pigment, common to animals, yeast, and higher plants" (Keilin. Proc. R. Soc. Lond. B 98:312–339, 1925), marking a new phase in the studies of the biological oxidations. It is interesting to revisit the reflections of E. F. Hartree, who worked with Keilin on this subject, published in the periodic *Biochemical Education*, in 1973, in which he comments the history of the discovery of the cytochromes: "In these days, when the world of science is under pressure to organize for the pursuit of practical ends, when the scale of scientific endeavour is making the lone furrow an anachronism, it is salutary to recall the simple-handed achievements of men of science stimulated solely by an urge to understand the living world. The discovery of cytochromes is a notable example of such achievements, not only because the disarming simplicity of the experimental approach by the discoverer, David Keilin, but also because the discovery came at a critical moment. It resolved a serious dilemma that was impeding the evolution of biochemistry from an untidy and rather primitive branch of chemistry into a major scientific discipline: more succinctly the transformation of Bio-Chemistry into Biochemistry."

At that time, Keilin was interested in studying the adaptations of a fly larva that parasitizes the stomach of horses. Using a microspectroscope to compare the absorption spectra of the larvae and the adult thoracic muscle, he found the presence of four sharp absorption bands in the spectrum of the adult muscle, which were very different from the bands seen for hemoglobin and oxyhemoglobin (see in the next figure the results published in the 1925's paper with a schematic representation of the experiment). Intrigued by this absorption pattern, he examined many other organisms, including different insects, yeast, and other microorganisms, plants, and animal tissues, finding always the same absorption spectra with the four bands.

(continued)

**Box 6.5** (continued)

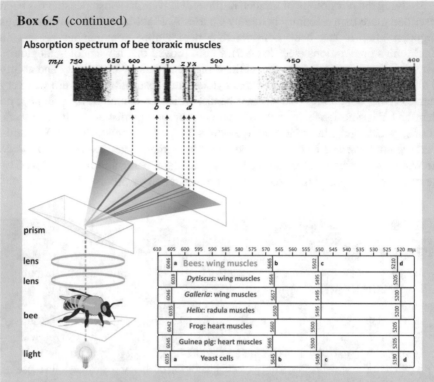

Schematic representation of the classical experiment by Keilin, showing the original results published in the Proc R Soc Lond 98:312–339, 1925, reproduced with permission of the Royal Society UK

A more intriguing observation was that when a yeast suspension was submitted to vigorous shaking, the bands disappeared, appearing again after a period of rest. Keilin's interpretation of this phenomenon was that the pigment, which he named cytochrome, became oxidized during the shaking in the air, leading to the disappearance of the bands. After interruption of the shaking, yeast respiratory activity led to reduction of the pigment, which in the reduced form displayed the four-banded absorption spectrum. Using insects, Keilin observed strikingly similar results: the absorption band pattern observed in the thoracic muscle was strong when the wings were stimulated to vibrate but disappeared after the movement stopped. Thus, Keilin concluded that the cytochrome was acting as a respiratory carrier linking oxygen and the oxidizable metabolic intermediates.

Keilin's experiments provided a great advance in the comprehension of the energy transformations in aerobic organisms, establishing the concept of what we now know as the respiratory chain, a series of membrane-associated redox carriers that transfer electrons from the metabolic substrates to molecular oxygen. However, the process by which this sequence of redox reactions is coupled to energy storing into the cells was still unknown at that time.

One of the landmarks in the discovery of this missing link was the article published by Belitser and Tsybakova in 1939, showing a direct correlation between oxygen consumption in cellular respiration and phosphate esterification, as it can be seen in Fig. 6.8.

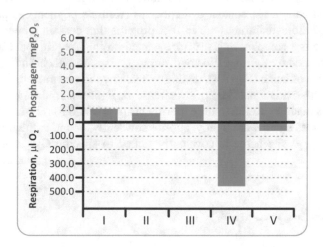

**Fig. 6.8** Classical experiment that correlated $O_2$ consumption to the formation of high-energy phosphate compounds. The authors used a pigeon muscle preparation in the following conditions: (I) before any incubation; (II) in the absence of $O_2$ (in a $N_2$ atmosphere) without addition of a respiratory substrate; (III) in the presence of pyruvate but in the absence of $O_2$; (IV) in the presence of pyruvate and $O_2$; and (V) in the presence of $O_2$ but in the absence pyruvate. (Figure adapted from Belitser and Tsybakova. Biokhimiya 4:516–535, 1939)

After this, years and years were spent searching a high-energy intermediate, which, in analogy to the substrate-level phosphorylation process already known to be the mechanism of ATP synthesis in fermentation (see Sect. 6.1.2), would couple the redox process to ATP synthesis. This intermediate had never been found, and how oxidative phosphorylation occurred remained one of the most challenging questions of biochemistry for a long time. The basic principles behind this process could only be understood after the proposal of the chemiosmotic hypothesis by Peter Mitchell (see next section).

## 6.2.2   An Overview of Oxidative Phosphorylation Process

In the oxidative reactions of catabolism, the electrons removed from the metabolic intermediates are transferred to two major electron carrier coenzymes, nicotinamide adenine dinucleotide (NAD$^+$) and flavin adenine dinucleotide (FAD), which are converted to their reduced forms, NADH and FADH$_2$ (see Chap. 7). These reactions are catalyzed by dehydrogenases, as firstly postulated by Wieland and Thumberg (see Sect. 6.2.1).

Oxidative phosphorylation depends on the electron transport from NADH or FADH$_2$ to O$_2$, which is reduced to H$_2$O. Electron transport occurs through a number of protein complexes associated with the inner mitochondrial membrane. Some of these protein complexes contain cytochromes (Keilin's pigment) as part of their structures. These cytochromes are in fact proteins that contain a heme prosthetic group. Heme is a complex ring structure named protoporphyrin that binds an iron atom. As detected by Keilin, there are different classes of cytochromes that contain different types of heme (Fig. 6.9), distinguishable by their characteristic absorption spectra. Heme oxidation leads to a decrease in light absorption, explaining the observations made by Keilin in his experiments (see Box 6.5). The heme groups in the cytochromes $a$ and $b$ are tightly, but not covalently, bound to the protein, while in cytochrome $c$, the heme is covalently linked to specific Cys residues of the protein.

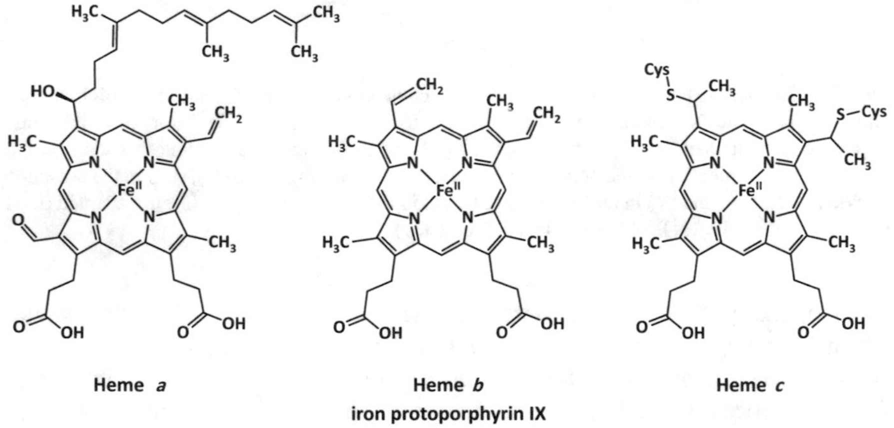

Heme $a$          Heme $b$          Heme $c$

iron protoporphyrin IX

**Fig. 6.9** The distinct types of heme prosthetic groups of the cytochromes: heme $a$, which has a long isoprenoid tail attached to the porphyrin ring; heme $b$, which is a protoporphyrin IX; and heme $c$, which is covalently bound to Cys residues in the polyprotein chain of the cytochrome $c$

The structures of respiratory protein complexes are now known with detail, as it will be explained in the next section, but one can safely say that the concept and the principles of their role in electron transport were established by Keilin after the discovery and the study of the cytochromes, which he defined as "oxidation–reduction catalysts" (see Sect. 6.2.1).

The mechanism by which the electron transport is coupled to ATP synthesis was proposed by Peter Mitchell, in 1961, in his revolutionary chemiosmotic theory. Initially, it was difficult to be accepted, since most of the scientists in the field believed that a high-energy intermediate would link oxidation to phosphorylation reactions (as occurs in the substrate-level phosphorylation process; see Sect. 6.1.2). However, the chemiosmotic hypothesis was proven correct, and Mitchell was awarded the Nobel Prize in Chemistry in 1978.

The chemiosmotic theory proposes that the electron transfer through the respiratory protein complexes is coupled to proton ($H^+$) pumping across the proton-impermeable mitochondrial inner membrane, from the mitochondrial matrix to intermembrane space. $H^+$ pumping generates what Mitchell called the proton-motive force, the simultaneous effect of the pH gradient across the membrane and transmembrane electrical potential that drives the ATP synthesis from ADP and Pi (Fig. 6.10). The $H^+$ transport occurs through specific protein segments that are part of some of the respiratory complexes. The ATP synthesis is catalyzed by another protein complex in the mitochondrial membrane, the ATP synthase, through which the $H^+$ ions return to the matrix (Fig. 6.10).

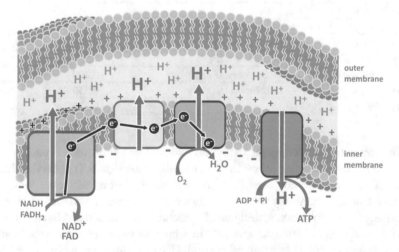

**Fig. 6.10** Schematic representation of the general mechanism of ATP synthesis in oxidative phosphorylation. Electrons are transferred from NADH or FADH$_2$ through a number of protein complexes associated with the inner mitochondrial membrane to the final acceptor $O_2$, which is reduced to $H_2O$. Electron transport is coupled to $H^+$ pumping across mitochondrial inner membrane, generating a pH gradient and a transmembrane electrical potential, which are the driving forces for ATP synthesis by ATP synthase. Note that this is a simplified scheme in which the NAD- or FAD-associated complexes are represented as one entity, although they are different protein complexes. It is important to point out that electron transport through FAD-associated complexes are not coupled to $H^+$ pumping across mitochondrial membrane, as detailed in the next section

## 6.2.3   The Electron Transport System

The electron transport complexes are integral membrane proteins that contain attached chemical groups or metal ions (flavins, iron–sulfur groups, heme, or copper ions) capable of accepting and donating electrons (see detailed structures in the next section).

### 6.2.3.1   The Sequence of Electron Transfer Between the Electron Carrier Groups

A form to deduce the sequence in which the electrons are transferred between the carriers is to compare the reduction potential of each individual electron carrier group (Table 6.1). The reduction potential is a measure of how "easy" is for a molecule to accept an electron.

**Table 6.1**  Reduction potential of respiratory electron carriers

| Reaction | Reduction potential ($V$) |
|---|---|
| $2H^+ + 2e^- \rightarrow H_2$ | −0.414 |
| $NAD^+ + H^+ + 2e^- \rightarrow NADH$ | −0.320 |
| NADH dehydrogenase (FNM) + $2H^+ + 2e^- \rightarrow$ NADH dehydrogenase (FNMH$_2$) | −0.300 |
| Ubiquinone $+2H^+ + 2e^- \rightarrow$ ubiquinol | 0.045 |
| Cytochrome $b$ ($Fe^{3+}$) + $e^- \rightarrow$ cytochrome $b$ ($Fe^{2+}$) | 0.077 |
| Cytochrome $c_1$ ($Fe^{3+}$) + $e^- \rightarrow$ cytochrome $c_1$ ($Fe^{2+}$) | 0.220 |
| Cytochrome $c$ ($Fe^{3+}$) + $e^- \rightarrow$ cytochrome $c$ ($Fe^{2+}$) | 0.254 |
| Cytochrome $a$ ($Fe^{3+}$) + $e^- \rightarrow$ cytochrome $a$ ($Fe^{2+}$) | 0.290 |
| Cytochrome $a_3$ ($Fe^{3+}$) + $e^- \rightarrow$ cytochrome $a_3$ ($Fe^{2+}$) | 0.350 |
| $\frac{1}{2} O_2 + 2H^+ + 2e^- \rightarrow H_2O$ | 0.817 |

The sequence deduced by the reduction potential was confirmed experimentally. Keilin in his first experiments on the cytochromes (see Box 6.5) observed that the bands do not appear nor disappear simultaneously. From this observation, he deduced that what he firstly named simply cytochrome was a mixture of pigments containing three components, designated cytochromes $a$, $b$, and $c$. Afterward, it was possible to distinguish two components in what was considered as cytochrome $a$, whose absorption bands were superimposed. One of them, the cytochrome $a_3$, is the last component in the chain, being directly oxidized by oxygen. Furthermore, the use of specific inhibitors of the electron transport allowed the complete sequence to be determined, confirming that the sequence follows the order of increasing reduction potential, as expected.

### 6.2.3.2 The Organization of the Respiratory Complexes in the Inner Mitochondrial Membrane

Although Keilin's experiments allowed the determination of the sequence through which the electrons flow in the respiratory chain, nothing was known at that time about how the electron-transferring groups were organized. Keilin always worked with a grinded heart muscle preparation that later was identified as sub-mitochondrial particles, which are actually vesicles of the inner mitochondrial membrane.

At the end of the 1940s, the group of David Green was able to fractionate four components of the respiratory chain, which they named Complexes I, II, III, and IV. These components corresponded to four integral membrane protein complexes: the NADH/ubiquinone oxidoreductase (or NADH dehydrogenase), the succinate/ubiquinone oxidoreductase (or succinate dehydrogenase), the ubiquinone/cytochrome $c$ oxidoreductase (or cytochrome $bc1$ complex), and the cytochrome $c$ oxidase, respectively (Fig. 6.11). Now we know that, in addition to these four protein complexes, there are two other proteins that participate in the transfer of electrons from metabolic substrates to the respiratory chain, the acyl-CoA dehydrogenase and the glycerol-phosphate dehydrogenase (Fig. 6.11).

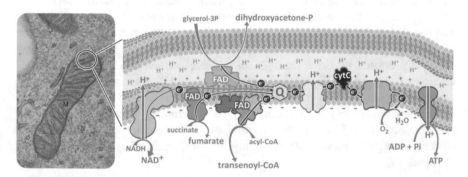

**Fig. 6.11** Schematic representation of the electron transfer complexes in the inner mitochondrial membrane. The electrons flow from NADH to $O_2$ through three protein complexes: NADH/ubiquinone oxidoreductase (*light green*), ubiquinone/cytochrome $c$ oxidoreductase (*light blue*), and cytochrome $c$ oxidase (*pink*). The electron transference from NADH/ubiquinone oxidoreductase to ubiquinone/cytochrome $c$ oxidoreductase is mediated by the membrane-soluble molecule ubiquinone (*green hexagon* labeled with a Q), and the transference from ubiquinone/cytochrome $c$ oxidoreductase to cytochrome $c$ oxidase is mediated by the small protein cytochrome $c$ (*dark blue* labeled with cytC). $O_2$ is also reduced by electrons coming from $FADH_2$, which is linked to FAD-dependent dehydrogenases, such as succinate/ubiquinone oxidoreductase (*dark green*), the acyl-CoA dehydrogenase (*orange*), and the glycerol-phosphate dehydrogenase (*blue*). The electrons from $FADH_2$ are transferred to ubiquinone and then flow to $O_2$ through ubiquinone/cytochrome $c$ oxidoreductase and cytochrome $c$ oxidase, as described for the electrons transferred from NADH. The electron transferring through NADH/ubiquinone oxidoreductase, ubiquinone/cytochrome $c$ oxidoreductase, and cytochrome $c$ oxidase is coupled to $H^+$ pumping to the intermembrane space. The $H^+$ returns to the matrix through the enzyme ATP synthase, driving ATP synthesis. The transmission electron micrograph in the left shows a cell thin section with a mitochondrion (courtesy from Prof. Marlene Benchimol)

Electron transport between the respiratory complexes occurs through two more mobile electron carriers, the ubiquinone (also called coenzyme Q), a very hydrophobic molecule with high mobility in the mitochondrial membrane (independently on its protonation state), and cytochrome $c$, a small protein associated with the outer face of the inner mitochondrial membrane (Fig. 6.11).

NADH formed in metabolic NAD$^+$-dependent oxidative reactions is a water-soluble electron carrier that reversibly associates with specific dehydrogenases. The electrons are transferred from NADH to O$_2$ through three protein complexes: NADH/ubiquinone oxidoreductase, ubiquinone/cytochrome $c$ oxidoreductase, and cytochrome $c$ oxidase (the Complexes I, III, and IV, respectively).

The electron carrier FAD is usually covalently attached to a FAD-dependent dehydrogenase. This type of enzyme includes the succinate dehydrogenase (or Complex II), the acyl-CoA dehydrogenase, and the glycerol-phosphate dehydrogenase. In the reactions catalyzed by these three FAD-dependent enzymes, FAD is reduced to FADH$_2$, whose electrons are then transferred to O$_2$ through ubiquinone/cytochrome $c$ oxidoreductase and cytochrome $c$ oxidase, as described for NADH/ubiquinone oxidoreductase electrons (Fig. 6.11).

The numbering of the four firstly identified electron transport complexes from I to IV and the concept that they form a chain, as inferred by the usually used name of "electron transport chain" or "respiratory chain," give rise to a not entirely correct idea that the electron transport complexes are arranged sequentially and that the electron transport occurs through a linear pathway. A more appropriated terminology emerges from the concept of a convergent "electron transport system," as proposed by Erich Gnaiger, in which electrons either from NADH via Complex I or from FADH$_2$ through three different FAD-associated complexes, the Complex II, the electron-transferring flavoprotein (ETF), or the glycerol-phosphate dehydrogenase (GpDH), converge to ubiquinone, in what he named the "Q-junction" (Fig. 6.12). After this point of convergence, the electrons flow to oxygen through a "linear" pathway composed of Complex III, cytochrome $c$, and Complex IV.

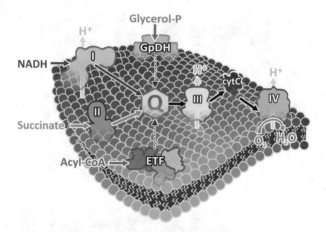

**Fig. 6.12** Convergent electron transport system. Ubiquinone receives the electrons from four different protein complexes: NADH/ubiquinone oxidoreductase (or Complex I), succinate/ubiquinone oxidoreductase (or Complex II), ETF/Q oxidoreductase (ETF), and glyceraldehyde dehydrogenase (GpDH). NADH comes from NAD-dependent dehydrogenases, mainly glyceraldehyde-3-phosphate dehydrogenase, pyruvate dehydrogenase, isocitrate dehydrogenase, α-ketoglutarate dehydrogenase, and malate dehydrogenase. $FADH_2$ is linked to succinate/ubiquinone oxidoreductase, GpDH, or acyl-CoA dehydrogenase, which in turn is associated with ETF. Then, electrons are transported to $O_2$ through a "linear" pathway composed of ubiquinone/cytochrome $c$ oxidoreductase (or Complex III) and cytochrome $c$ and cytochrome $c$ oxidase (or Complex IV)

### 6.2.3.3 The Structure of the Electron-Transferring Components

A detailed description of each electron-transferring component of the respiratory chain is presented below as these are paradigmatic cases of structure/function correlation.

NADH/ubiquinone oxidoreductase (or Complex I) is composed of 45 polypeptide chains associated with several electron-transferring groups: a flavin nucleotide (FMN) and many iron–sulfur (Fe–S) centers. The complete atomic structure of NADH/ubiquinone oxidoreductase was initially determined for the prokaryotic enzyme (Fig. 6.13a), being now also available for mammalian mitochondrial complexes. The high degree of sequence and folding conservation makes the bacterial enzyme a model for the human Complex I. The electrons removed from NADH flow through FMN and then to the Fe–S groups to finally reduce ubiquinone to ubiquinol (Fig. 6.13b). Dramatic conformational changes around the quinone binding site couple the redox reactions to proton translocation from the matrix to the intermembrane space, which occurs through protonation and deprotonation of the side chains of amino acid residues along the enzyme structure.

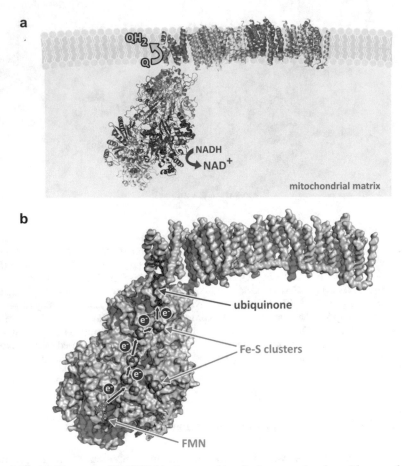

**Fig. 6.13** (**a**) Structure of NADH/ubiquinone oxidoreductase complex from *Thermus thermo-philes* (PDB 3M9S), with each subunit colored differently. This complex catalyzes the oxidation of NADH to NAD⁺, with the reduction of ubiquinone (Q) to ubiquinol (QH₂). (**b**) The transfer of electrons (e⁻) from NADH to ubiquinone (shown in *magenta sticks*), flowing through FMN (shown in *green sticks*) and iron–sulfur clusters (*orange* and *yellow spheres*), is represented by the *red arrows* over the protein surface map. Electron transfer is coupled to the translocation of four H⁺

Succinate/ubiquinone oxidoreductase (or Complex II) is an FAD-dependent mitochondrial membrane enzyme that catalyzes the oxidation of succinate to fumarate, a reaction of the citric acid cycle, the pathway that accounts for the complete oxidation of acetyl-CoA, which, in turn, is the convergent product of the degradation pathways of sugars, lipids, and some amino acids (see Sect. 7.2). This enzyme is usually referred as succinate dehydrogenase, but since the oxidation of succinate to fumarate is coupled to electron transference to ubiquinone, succinate/ubiquinone oxidoreductase is a more precise denomination. The enzyme contains four polypeptide chains, a catalytic heterodimer composed of subunit A, containing a covalently bound FAD, and subunit B, containing three iron–sulfur clusters, and two transmembrane polypeptides that anchor the enzyme in the mitochondrial membrane and where a heme *b* group is bound (Fig. 6.14).

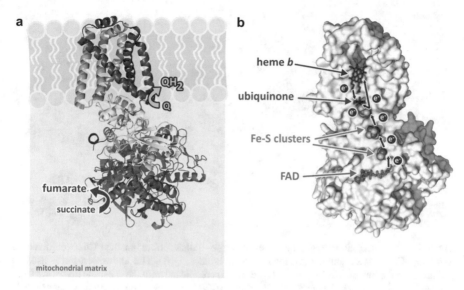

**Fig. 6.14** (a) Structure of succinate/ubiquinone oxidoreductase complex from porcine heart (PDB 1ZOY). The FAD binding protein, or subunit A, is shown in *blue*; the iron–sulfur protein, or subunit B, is shown in *light yellow*; and the transmembrane proteins are shown in *pink* and *orange*. The enzyme catalyzes an FAD-dependent oxidation of succinate to fumarate, with the concomitant reduction of ubiquinone (Q) to ubiquinol (QH₂). (b) The electron (e⁻) transfer pathway from FAD (shown in *orange sticks*) to heme *b* (shown in *red sticks*), flowing through the iron–sulfur groups (*orange* and *yellow spheres*) to finally reduce the ubiquinone (shown in *magenta sticks*), is represented by the *red arrows* over the protein surface map

Acyl-CoA dehydrogenase (ACAD) catalyzes the first step of mitochondrial fatty acid oxidation: the conversion of an acyl-CoA to trans-2,3-enoyl-CoA with the reduction of the enzyme-bound FAD coenzyme (see Sect. 7.4.4). ACADs associate with the electron-transferring flavoprotein (ETF), which reoxidizes ACAD-bound FADH₂. ETF then transfers the electrons to the ETF/Q oxidoreductase that in turn reduces ubiquinone after electron transport through the Fe–S centers (Fig. 6.15). There are five isoforms of this enzyme showing distinct specificity for the fatty acyl chain length. The very-long-chain acyl-CoA dehydrogenase (VLCAD) forms homodimers of 67 kDa subunits bound to the inner mitochondrial membrane through the 180 last residues of the C-terminal. These residues are lacking in the long-, medium-, and short-chain acyl-CoA dehydrogenases (LCAD, MCAD, and SCAD, respectively), isoforms that are soluble homotetramers with 45 kDa subunits.

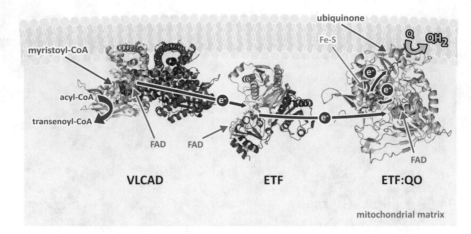

**Fig. 6.15** Representation of the multistep electron transfer from an acyl-CoA (oxidized to transenoyl-CoA) to ubiquinone (Q, reduced to ubiquinol, $QH_2$). The known structures used as examples are the human VLCAD dimer (PDB 3B96) complexed with the substrate myristoyl-CoA (shown in *pink spheres*); the human ETF (PDB 1EFV); and the ETF/QO from pig liver (PDB 2GMH) complexed with ubiquinone (shown in *magenta sticks*). The electrons ($e^-$) are transferred through each enzyme-bound FAD (shown in *orange sticks*) to the iron–sulfur center (shown in *orange* and *yellow spheres*) in ETF/QO to finally reduce the ubiquinone

Glycerol-phosphate dehydrogenase (GpDH) is a dimeric enzyme associated with the outer face of the inner mitochondrial membrane that oxidizes glycerol-3-phosphate to dihydroxyacetone phosphate (DHAP) with reduction of the enzyme-bound FAD that mediates the transfer of the electrons to ubiquinone (Fig. 6.16).

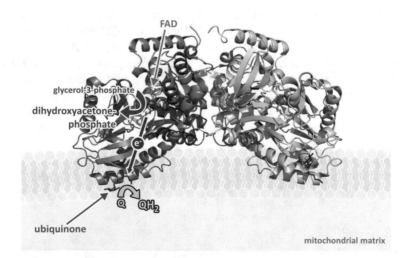

**Fig. 6.16** Structure of the GpDH dimer from *E. coli* (PDB 2QCU), showing the enzyme-bound FAD (in *orange sticks*), which transfers electrons ($e^-$) to ubiquinone (Q, reduced to ubiquinol, $QH_2$; molecular structure shown in *magenta sticks*)

Ubiquinone is a lipid-soluble benzoquinone with an isoprenoid tail, which in mammals is composed of ten isoprenyl units, making the molecule very hydrophobic and thus soluble in the membranes (Fig. 6.17). It is reduced to ubiquinol after accepting two electrons and two $H^+$ from NADH/ubiquinone oxidoreductase, succinate/ubiquinone oxidoreductase, ETF/Q oxidoreductase, or GpDH and diffuses in the membrane, reaching ubiquinone/cytochrome $c$ oxidoreductase to which the electrons are transferred.

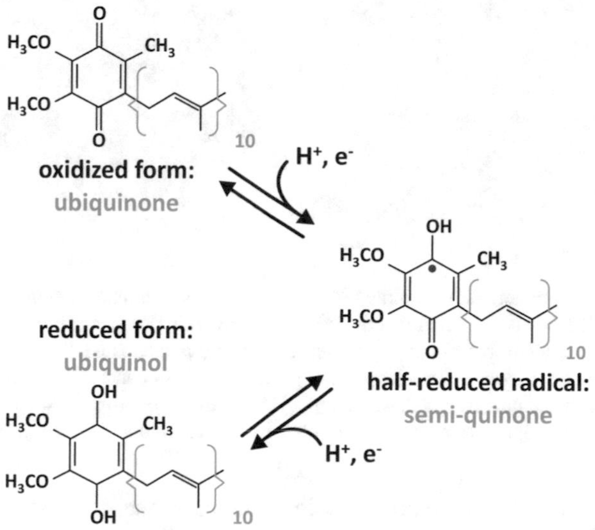

**Fig. 6.17** Structures of the redox forms of coenzyme Q: ubiquinone is the oxidized form that is completely reduced to ubiquinol by accepting two electrons and two $H^+$ in the reactions catalyzed by NADH/ubiquinone oxidoreductase, succinate/ubiquinone oxidoreductase, ETF/Q oxidoreductase, or GpDH. Ubiquinol is reoxidized by the ubiquinone/cytochrome $c$ oxidoreductase. The reduction of ubiquinone and the oxidation of ubiquinol involve an intermediate step in which a half-reduced semiquinone radical ($\bullet Q^-$) is formed. The implications of this will be discussed in Sect. 6.2.5

Ubiquinone/cytochrome $c$ oxidoreductase (or Complex III) transfers the electrons from ubiquinol to cytochrome $c$ with the coupled transport of four $H^+$ from the matrix to the intermembrane space. This protein complex is a dimeric structure with each monomer being a complex assembly of 11 polypeptide chains. The electrons are transported through three functional groups associated with each monomer: cytochrome $b$ containing two $b$-type heme groups, a Fe–S center, and cytochrome $c_1$ containing one $c$-type heme group (Fig. 6.18).

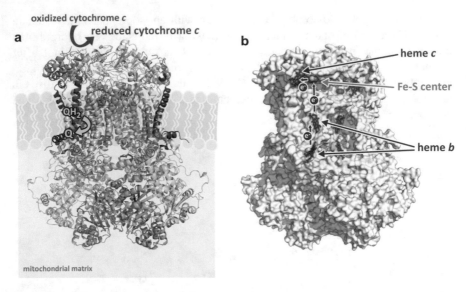

**Fig. 6.18** (**a**) The complete dimeric structure of the bovine ubiquinone/cytochrome *c* oxidoreductase (PDB 1BE3) with the 11 subunits colored differently. (**b**) The electron (e⁻) transfer pathway from ubiquinol to cytochrome *c*, flowing through the heme *b* groups (shown in *red sticks*) of the cytochrome *b* subunit (shown in *light purple* in A), the iron–sulfur center (shown in *orange* and *yellow spheres*), and the heme *c* group (shown in *red sticks*) of the cytochrome $c_1$ subunit (shown in *pink* in A), to finally reduce cytochrome *c*, is represented by the *red arrows* over the protein surface map

Cytochrome *c* is a monomeric protein containing one heme group (Fig. 6.19). This protein is located in the intermembrane space in close association with the inner mitochondrial membrane. It is reduced by ubiquinone/cytochrome *c* oxidoreductase and reoxidized by cytochrome *c* oxidase.

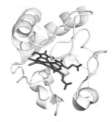

**Fig. 6.19** Structure of human cytochrome *c* (PBD 1HCR), showing its heme group in *red sticks*

Cytochrome *c* oxidase (or Complex IV) is a dimeric complex of 13 subunit monomers that transfers electrons from cytochrome *c* to $O_2$, forming $H_2O$. It contains two Cu ions associated with the SH groups of two Cys of the subunit that receives the electrons from cytochrome *c* and transfers them to an *a*-type heme group that in turn transfers the electrons to another heme *a* group, heme $a_3$, this one associated with another Cu ion, which finally transfers the electrons to $O_2$ (Fig. 6.20).

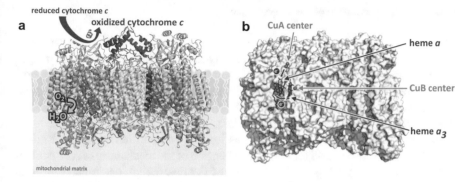

**Fig. 6.20** (**a**) Dimeric structure of the bovine cytochrome $c$ oxidase (PDB 1OCC), with the 13 subunits of each monomer colored differently. The enzyme catalyzes the oxidation of cytochrome $c$ with the reduction of $O_2$ to $H_2O$. (**b**) The electrons ($e^-$) are sequentially transferred through the CuA center (shown in *brown spheres*), to heme $a$ (shown in *red sticks*), and heme $a_3$ (shown in *red sticks*)-CuB center (shown in *brown spheres*) to finally reduce the $O_2$ to $H_2O$, as represented by the *red arrows* over the protein surface map

## 6.2.4 The ATP Synthesis Through Oxidative Phosphorylation

The synthesis of ATP is catalyzed by a large protein complex, the ATP synthase, located, as the respiratory complexes, in the inner mitochondrial membrane. The enzyme uses as substrates ADP and Pi in a reaction dependent on the flow of $H^+$ from the intermembrane space to the mitochondrial matrix. In electron microscopy imaging, ATP synthase is seen as characteristic head connected to the membrane by a long stalk. Through studies using the technique of electron cryotomography, it was possible to observe that the ATP synthase dimers are arranged in long rows along the highly curved mitochondrial cristae edges, an organization similar in mitochondria from mammals, fungi, or plants (Fig. 6.21).

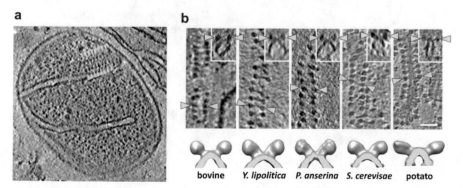

**Fig. 6.21** (**a**) Tomographic slices showing the arrays of $F_1$–$F_o$ ATP synthase dimers (*yellow arrows*) in whole mitochondria of *Podospora anserina*. (**b**) Rows of $F_1$–$F_o$ ATP synthase in mitochondrial membranes from bovine heart, *Yarrowia lipolytica*, *Podospora anserina*, *Saccharomyces cerevisiae*, and potato. A side view of each array with the dimers in relation to the membrane is shown in the inset. *Yellow arrow heads* indicate $F_1$ heads of one dimer. Scale bar, 50 nm. The surface representation of each dimer is shown in the *bottom*. (Figures reproduced with permission from Davies et al. Proc. Natl. Acad. Sci. USA 108:14121–14,126, 2011)

### 6.2.4.1   The Structure of ATP Synthase

The overall structure of ATP synthase may be separated in two components: the $F_1$ portion, a peripheral membrane protein clearly seen in electron microscopy images as projections in the inner mitochondrial membrane (see Fig. 6.21), and the $F_o$ portion (whose denomination comes from its sensitivity to oligomycin), an integral mitochondrial membrane protein, through which $H^+$ flow from the intermembrane space to the mitochondrial matrix (Fig. 6.22).

The $F_1$ portion of the ATP synthase was isolated and purified by Efraim Racker, whose studies were decisive for the comprehension of the mechanism of ATP synthesis reaction. The crystallographic structure of $F_1$ (Fig. 6.22a), determined by John E. Walker, revealed that it has nine subunits of five different types, three α-subunits, three β-subunits, and one of each γ-, δ-, and ε-subunits (Fig. 6.22). $F_1$ δ- and ε-subunits interact with the membrane-embedded transmembrane helices of $F_o$ portion of the enzyme. The catalytic sites of ATP synthesis are located in each of the β-subunits, whose conformations in the enzyme structure are different from each other due to differences in their interactions with the other subunits of the enzyme. This is essential for the mechanism of ATP synthesis, as it will be detailed in the next section.

The $F_o$ complex is composed of three types of subunits: one a-subunit; two b-subunits, which associate with $F_1$ α- and β-subunits; and 10–12 small c-subunits, which are hydrophobic polypeptides consisting of two transmembrane helices that form a membrane-embedded cylinder that interacts with the δ- and ε-subunits of $F_1$ complex (see schematic representation in Fig. 6.22b).

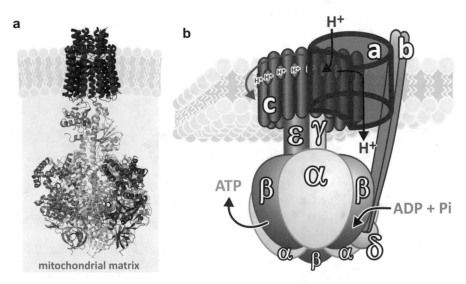

**Fig. 6.22** (**a**) Structure of the $F_1$ and part of the $F_o$ portions of ATP synthase from *Saccharomyces cerevisiae* (PDB 2XOK). The $F_1$ α-, β-, γ-, and ε-subunits are represented in *cyan, purple, light green*, and *pink*, respectively. The c-subunits of the $F_o$ portions are shown in *red*. (**b**) Schematic representation of the entire $F_1F_o$ -ATP synthase, showing the δ-subunit of $F_1$ portion as well as the a- and b-subunits of $F_o$ portion, which were not yet determined and thus are not represented in the crystallographic structure shown in (**a**)

### 6.2.4.2   The Mechanism of ATP Synthesis by the ATP Synthase

The synthesis of ATP from ADP and Pi is a very endergonic reaction in aqueous solution. However, one important point to understand the mechanism of ATP synthesis by ATP synthase is that when occurring in $F_1$ environment, it is readily reversible, with a free energy change close to zero (see Box 6.2). For the reaction catalyzed by $F_1$, the energy barrier consists in the step of ATP release from the enzyme. This energy barrier is overcome by the energy input from the H⁺ gradient, since flow through $F_o$ promotes conformational changes in the β-subunit, leading to the loss of its affinity to ATP.

This view of ATP synthesis was formulated by Paul D. Boyer. From his kinetics studies, two main new concepts emerged. The first was that the three catalytic sites of the ATP synthase participate sequentially and cooperatively in the catalytic cycle; the second was that the catalytic mechanism would be seen as "a rotational catalysis" mechanism, as Boyer named, in which the three catalytic sites alternate the reaction catalysis (see Box 6.6). One clue for this proposal was given by the crystallographic structure of $F_1$, which revealed that the three β-subunits were differentially occupied during the catalytic cycle, one having ADP bound, the other having ATP bound, and the third being empty (see Fig. 6.23a).

The rotational catalysis mechanism may be summarized in the model shown in Fig. 6.23b. ADP and Pi from the medium bind to the β-subunit catalytic site that is in the β-ADP conformation. The conformation of this β-subunit changes to the β-ATP conformation due to enzyme rotation driven by H⁺ flow through $F_o$ portion.

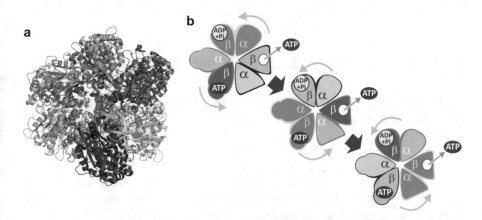

**Fig. 6.23** (**a**) Structure of mitochondrial bovine $F_1$ showing ADP bound to one of the β-subunits, in *yellow*, and a non-hydrolyzable ATP analog (phosphoaminophosphonic acid-adenylate ester) bound to another β-subunit, in *orange* (PDB 1BMF). (**b**) Schematic representation of the different conformations assumed by $F_1$ subunits: ADP and Pi bind to the catalytic β-subunit, which is in the β-ADP conformation. The enzyme rotation driven by H⁺ flow through $F_o$ portion promotes a conformation change in the β-subunit that acquires the β-ATP conformation, which stabilizes ATP in the active site. Then, another $F_1$ rotation leads the subunit to its empty conformation, which loses the affinity to ATP, releasing it to the medium

In this conformation, the β-subunit stabilizes ATP, which is in equilibrium in the active site with ADP and Pi. Then, another $F_1$ rotation occurs, leading this β-subunit to change conformation again, now to the empty conformation, which loses the affinity to ATP, releasing it to the medium. Another round starts when another $F_1$ rotation leads the β-subunit again to the β-ADP conformation. This rotational movement frequently justifies the label of "molecular machine" to ATP synthase.

Due to his great contribution to the understanding of the mechanism of ATP synthesis, Boyer shared the Nobel Prize in Chemistry, in 1997, with John Walker, who determined the crystallographic structure of the $F_1$ portion of the enzyme, an essential step for the comprehension of the catalytic mechanism.

**Box 6.6  The Confirmation of Boyer's Model by Real-Time Microscopy**
$F_1$ rotation could be directly seen in an ingenious experiment performed by the research groups of Masasuke Yoshida and Kazuhiko Kinosita (published in Nature 386:299–302, 1977), in which they attached to the γ-subunit of $F_o$ a long fluorescent actin filament and observed its movement as ATP was hydrolyzed, in real time in a microscope, in relation to the $α_3β_3$ core immobilized in the microscope slide. They also observed that the rotation occurred in three discrete steps of 120°, completely confirming Boyer's model. You can watch a movie with the experiment result in following link: http://www.k2.phys.waseda.ac.jp/F1movies/F1Prop.htm.

## 6.2.5   Regulation of Oxidative Phosphorylation

Oxidative phosphorylation is generally limited by the availability of ADP, so that the major control of ATP synthesis by oxidative phosphorylation is the cellular ATP requirement.

When respiratory substrates are freely available, ATP is synthesized so that the ratio ATP/ADP increases. If the levels of ADP become very low (when ATP synthesis overcomes its utilization in cellular metabolism), but the substrates are still available, the $H^+$ gradient reaches the maximum level. This prevents electron transport and hence respiration, since the gradient cannot be dissipated through ATP synthase, which cannot work due to the absence of its substrate, ADP. This situation is called state 4 respiration (Fig. 6.24) and illustrates the low permeability of the inner mitochondrial membrane to $H^+$ and the coupling of electron transport to ATP synthesis.

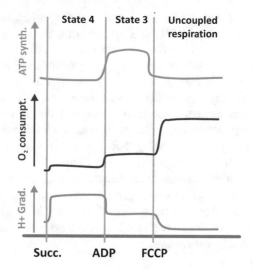

**Fig. 6.24** Representation of the extent of the proton gradient, oxygen consumption, and ATP synthesis when a respiratory substrate (e.g., succinate), ADP, and an uncoupler (e.g., $p$-trifluoromethoxycarbonyl cyanide phenylhydrazone, FCCP) are added to intact mitochondria in vitro

When ADP becomes available, the ATP synthase phosphorylates it to ATP, and $H^+$ gradient is reduced to such an extent that respiration is allowed to proceed, while ADP is available. In this situation, respiration occurs at the same rate of ATP synthesis, in what is called state 3 respiration (Fig. 6.24). However, even in state 4, or when ADP phosphorylation is inhibited (e.g., using oligomycin), some oxygen consumption is observed, demonstrating that the coupling of respiration to ATP synthesis is imperfect, and part of the energy will be normally dissipated as heat. This occurs to some extent due to $H^+$ leaking through the inner mitochondrial membrane. This phenomenon can be reproduced by the use of substances called uncouplers (e.g., $p$-trifluoromethoxycarbonylcyanide phenylhydrazone—FCCP), whose effect on $O_2$ consumption, $H^+$ gradient, and ATP synthesis is shown in Fig. 6.24. In the presence of uncouplers, $H^+$ readily move back into the matrix, bypassing the ATP synthase and collapsing the gradient, which causes respiration to be accelerated but uncoupled from ADP phosphorylation.

### 6.2.5.1 Uncoupling Proteins: The Physiological Uncouplers

Physiologically, uncoupling of electron transport to ADP phosphorylation is provided by a family of uncoupling proteins (UCP). The best studied of them, and the first to be identified, UCP1, is expressed exclusively in a specialized tissue called brown adipose tissue (BAT). In contrast to white adipocytes, brown adipocytes contain several lipid droplets and a much higher number of mitochondria (Fig. 6.25a), which confer the brown color to the tissue.

In humans, BAT is mostly found in the newborn, regulating thermogenesis through the expression of the UCP1. This protein occurs in abundance in the inner mitochondrial membrane and provides an alternative route, bypassing ATP synthase, for H⁺ to return to the mitochondrial matrix, dissipating the gradient and accelerating respiration, resulting in heat production in a regulated manner (Fig. 6.25b). Until recently, this tissue has been considered to have no physiological relevance to adult humans, but some findings suggest that reminiscent BAT cells may proliferate in response to cold exposure (Box 6.7). This effect seems to be more pronounced in lean subjects, suggesting that regulated uncoupled respiration may also be a way to control energy expenditure (see Chap. 11).

Two other UCP isoforms have been characterized, UCP2 and UCP3. They show a more ubiquitous tissue distribution (Fig. 6.25c) and probably play a role in protecting cells against reactive oxygen species, which may be produced in excess in mitochondria in some situations, such as when an excessive amount of substrates is supplied to the cells (see next section).

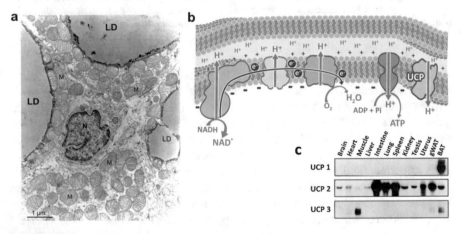

**Fig. 6.25** (**a**) Transmission electron micrograph of a BAT adipocyte thin section showing lipid droplets (LD) and a high number of mitochondria (M); N: nucleus. (Courtesy from Prof. Marlene Benchimol). (**b**) The UCPs are integral inner mitochondrial membrane proteins that allow H⁺ flow bypassing ATP synthase, dissipating the gradient and accelerating respiration. (**c**) Western blotting showing the expression pattern of the three UCP isoforms in mouse tissues. (Figure reproduced with permission of Portland Press from Ricquier and Bouillaud, Biochem J. 345:161–179, 2000)

## Box 6.7 BAT in Adult Humans

Positron emission tomography combined with computed tomographic (PET-CT) scans, with $^{18}$F-fluorodeoxyglucose ($^{18}$F-FDG) as a tracer, is generally used to diagnose neoplasms and their metastases, since tumor cells present a much higher glucose uptake when compared to other cells (see Box 6.4 about the Warburg effect). In these tests, a high glucose uptake in the supraclavicular tissue is usually observed, sometimes confusing the diagnosis. It was speculated that these highly glycolytic cells in this region would be brown adipose tissue (BAT). In a set of studies performed to investigate this issue, healthy volunteers were exposed to 16 °C for 2 h before the tests. Comparative PET-CT scans revealed a great increase in $^{18}$F-FDG uptake in the supraclavicular region upon cold exposure (see below figure). Furthermore, tissue biopsies were used for immunostaining with UCP1-specific antiserum, confirming the presence of substantial amounts of metabolically active BAT in adult humans. A systematic examination of the presence and distribution of BAT in lean and obese men during exposure to cold temperature showed that BAT activity is inversely correlated to body mass index (BMI). Additionally, analysis of 3640 consecutive PET-CT scans performed for various diagnostic reasons in 1972 patients showed substantial BAT depots in regions extending from the anterior neck to the thorax for 76 of 1013 women (7.5%) and 30 of 959 men (3.1%), with also a larger mass in women.

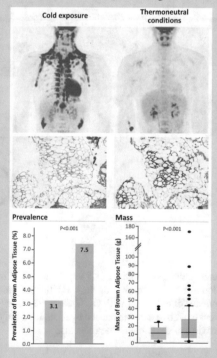

Presence of BAT in humans detected by PET-CT scans after exposure to cold (*top left*) and under thermoneutral conditions (*top right*), and histologic images of biopsy specimens showing UCP1 staining (*middle right*), and the quantification of prevalence (*bottom left*) and amount (*bottom right*) of BAT in men (*blue*) and women (*pink*). (Figures reproduced with permission from Lichtenbelt et al. New Engl. J. Med. 360:1500–1508, 2009, and Cypess et al. New Engl. J. Med. 360:1509–1517, 2009)

### 6.2.5.2  Production of Reactive Oxygen Species in Mitochondria

In physiological or pathological situations in which the input of electrons into the respiratory chain overcomes their transfer to oxygen, such as in hypoxia, the formation of reactive species of oxygen (ROS) is increased. This occurs because the passage of electrons from Complex I to ubiquinone and from ubiquinone to Complex III involves the formation of a partially reduced ubiquinone radical ($\cdot Q^-$) as an intermediate (see Fig. 6.17). When the electron flow through the respiratory chain is impaired, the probability of this radical to react with cellular components before being completely reduced or oxidized greatly increases. $\cdot Q^-$ can react with oxygen, generating the superoxide radical ($\cdot O_2^-$). This radical is very reactive, and its formation may lead to the production of an even more reactive radical, the hydroxyl radical ($\cdot OH$). These ROS can react and damage enzymes, lipids, and nucleic acids. They can also alter cellular gene expression, leading to several modifications in cellular functions.

Cells have different enzymatic systems to prevent oxidative damage caused by ROS. This includes the enzyme superoxide dismutase that converts $\cdot O_2^-$ in hydrogen peroxide ($H_2O_2$), which in turn may be used by glutathione peroxidase to reduce glutathione or by catalase to form $H_2O$ (Fig. 6.26).

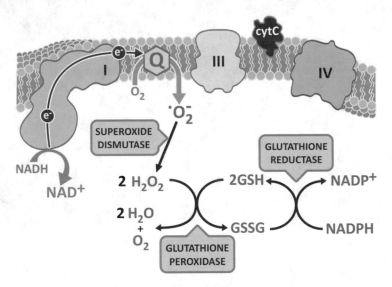

**Fig. 6.26**  A partially reduced ubiquinone radical ($\cdot Q^-$) is formed as an intermediate in the reduction of ubiquinone to ubiquinol by Complex I or the oxidation of ubiquinol to ubiquinone by Complex III. If $\cdot Q^-$ accumulates, it may react with $O_2$, forming $\cdot O_2^-$ and $\cdot OH$. Cells have different enzymatic systems to prevent oxidative damage caused by ROS, including the enzymes superoxide dismutase, glutathione peroxidase, and glutathione reductase

# Selected Bibliography

Berche P (2012) Louis Pasteur, from crystals of life to vaccination. Clin Microbiol Infect 18(Suppl 5):1–6

Boyer PD (1997) Energy, life, and ATP. Nobel lecture. http://www.nobelprize.org/nobel_prizes/chemistry/laureates/1997/boyer-lecture.html

Buchner E (1907) Cell-free fermentation. Nobel lecture. http://www.nobelprize.org/nobel_prizes/chemistry/laureates/1907/buchner-lecture.html

de Meis L (1997) The conception of phosphate compounds of high and low energy. In: Sotelo JR, Benech JC (eds) Calcium and cellular metabolism: transport and regulation, vol 8. Springer Science+Business Media, New York, pp 85–103

Efremov RG, Baradaran R, Sazanov LA (2010) The architecture of respiratory complex I. Nature 465:441–445

Gatenby RA, Gillies RJ (2004) Why do cancers have high aerobic glycolysis? Nat Rev Cancer 4:891–899

Gnaiger E (2007) Mitochondrial pathways through complexes I + II: convergent electron transport at the Q-junction and additive effect of substrate combinations. http://www.oroboros.at/fi leadmin/user_upload/O2k-Publications/O-MiPNet-Publ/2007-3_MitoPathways_p21-50.pdf

Iwata S, Lee JW, Okada K, Lee JK, Iwata M, Rasmussen B, Link TA, Ramaswamy S, Jap BK (1998) Complete structure of the 11-subunit bovine mitochondrial cytochrome bc1 complex. Science 281:64–71

Keilin D (1925) On cytochrome, a respiratory pigment, common to animals, yeast, and higher plants. Proc R Soc Lond 98:312–339

Kohler RE (1973) The background of Otto Warburg's conception of the Atmungsferment. J Hist Biol 6:171–192

Koppenol WH, Bounds PL, Dang CV (2011) Otto Warburg's contributions to current concepts of cancer metabolism. Nat Rev Cancer 5:325–337

Manchester KL (1998) Albert Szent-Gyorgyi and the unravelling of biological oxidation. Trends Biochem Sci 23:37–40

Mitchell P (1978) David Keilin's respiratory chain concept and its chemiosmotic consequences. Nobel lecture. http://www.nobelprize.org/nobel_prizes/chemistry/laureates/1978/mitchelllecture.html

Ricquier D, Bouillaud F (2000) The uncoupling protein homologues: UCP1, UCP2, UCP3, StUCP and AtUCP. Biochem J 345:161–179

Stock D, Gibbons C, Arechaga I, Leslie AG, Walker JE (2000) The rotary mechanism of ATP synthase. Curr Opin Struct Biol 10:672–679

Sun F, Huo X, Zhai Y, Wang A, Xu J, Su D, Bartlam M, Rao Z (2005) Crystal structure of mitochondrial respiratory membrane protein complex II. Cell 121:1043–1057

Tsukihara T, Aoyama H, Yamashita E, Tomizaki T, Yamaguchi H, Shinzawa-Itoh K, Nakashima R, Yaono R, Yoshikawa S (1996) The whole structure of the 13-subunit oxidized cytochrome c oxidase at 2.8 A. Science 272:1136–1144

Watmough NJ, Frerman FE (2010) The electron transfer flavoprotein: ubiquinone oxidoreductases. Biochim Biophys Acta 1797:1910–1916

## Challenging Case 6.1: The Tragic Matter in Energy Management

### Source

Gottfried Schatz wrote a series of high-quality essays to the journal *FEBS Letters* that were later collected in a book titled *Jeff's View on Science and Scientists*.[1] In the Chapter 5 of Schatz's book some of his thoughts were on bioenergetics emphasizing the role of oxygen. The original essay was titled "The tragic matter" and was published in the scientific journal *FEBS Letters*[2]

Picture of Gottfried Schatz in 2001, reproduced from Wikimedia Commons, under a CC-BY license

### Case Description

A human sitting quietly consumes a considerable quantity of energy. Intuitively, one tends to think that without exercise the human body uses only traces of energy but this is not correct. Body functions such as constant temperature regulation and basal functioning of cells demand about 4.184 kJ for each kg body mass per hour. Joule is the SI (International System of Units, abbreviated from the French *Système International*) unit of energy, but it is not a very friendly unit in health sciences. Calorie (cal), a more intuitive alternative, is more popular. 4.184 kJ $kg^{-1}$ $h^{-1}$ is 1 kcal $kg^{-1}$ $h^{-1}$ (or 1 Cal $kg^{-1}$ $h^{-1}$, as Cal—with capital C—is the way kcal is named in US food labels), which is about 1.2 mW $g^{-1}$ (see next table). This value compares to 0.2 $\mu W$ $g^{-1}$ of energy generated by the sun. A 5000-fold difference! Impressive... But not as impressive as the performance of some bacteria such as *Azotobacter*, with 10 W $g^{-1}$, i.e., a 50 million-fold difference. Energy consumption in humans is needed mainly for ATP synthesis and heat production. Mitochondria are in charge of energy management in both cases. *Azotobacter* has no mitochondria at all and

---

[1] Schatz G (2006) Jeff's view on science and scientists. Elsevier, Amsterdam. ISBN-13: 978-0-4444-52133-0.

[2] Schatz G (2003) The tragic matter. FEBS Lett 536:1–2.

does not conserve inner temperature: in total, each gram of the bacterium can make 7 kg of ATP per day!

Average energy rate consumption of the human body expressed in several units

| Energy rate (production or consumption) | Comment |
|---|---|
| 4.184 kJ kg$^{-1}$ h$^{-1}$ (1.2 mW g$^{-1}$) | The most rigorous units in scientific terms, following the SI (International System of Units) |
| 1 kcal kg$^1$ h$^{-1}$ | Kilocalorie is a popular unit in nutrition and related areas |
| 1 Cal kg$^{-1}$ h$^{-1}$ | Cal, instead of kcal, is used in US food labels |
| 1 MET | MET—metabolic equivalent—corresponds to the relative fold-increase of the energy consumption rate relative to basal metabolism (in practical terms, when sitting quite) |

The numbers are supportive of the hypothesis that animals deviated from photosynthesis throughout evolution, associating with symbiotic bacteria that were transformed into mitochondria inside their cells (see Box 6.3). On a sunny day, an average human body receives at best 500 W of solar energy, which corresponds to 7 mW g$^{-1}$. Because it is impossible to have 100% efficiency in conversion to ATP, in practice photosynthesis would not allow much more than keeping the basal metabolism operating. Locomotion and physical exercise would be very much impaired. Running at 5 mph (8 km h$^{-1}$), a modest effort, demands an eightfold increase in metabolic rate relative to the basal value (i.e., a so-called metabolic equivalent, MET, of 8). Higher velocities demand much more (see below figure). Complex forms of life cannot cope with the power shortage of photosynthesis. The same applies to fermentation.

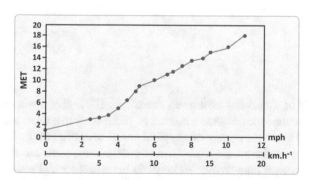

MET dependence on velocity in walking (up to 5 mph = 8 km h$^{-1}$) and running. Data taken from the article cited in the footnote[3]

---

[3] Ainsworth BE et al (2000) Compendium of physical activities: an update of activity codes and MET intensities. Med Sci Sports Exerc 32(9 Suppl):S498–S504.

Evolving from photosynthesis to respiration had probably started as a battle for survival. Photosynthesis was extremely important to life on Earth because it assured that a sustainable long-life power source could be used to feed the food chain of living beings, but it came at a very high cost: the release of poisonous molecular oxygen gas in the environment. Without specific defense mechanisms, the unsaturated lipids, nucleic acids, and proteins, among other biomolecules, were very sensitive to oxidation by $O_2$. Yet, the spread of $O_2$ was slowed down by the presence of divalent iron solubilized in the oceans, which scavenged and sequestered $O_2$ as precipitates of $Fe^{3+}$ oxides. For this reason, $Fe_3O_4$ and $Fe_2O_3$ are abundant in Earth's soil. It was only when the capacity of oceans to scavenge $O_2$ reached exhaustion that molecular oxygen invaded the atmosphere, slowly and drastically. Mass extinction must have occurred until biochemical armory was developed to deal with the effects of generalized oxidation produced by $O_2$: enzymatic reversal of disulfide bridge formation by controlled reduction, enzymatic transformation of "reactive oxygen species" into much less reactive molecules (Sect. 6.2.5.2), and the use of antioxidants such as vitamin E ($\alpha$ tocopherol) and carotenoids to prevent other molecules from reacting with oxidative agents. Throughout evolution, $O_2$ was tamed and the surviving species thrived. But this was not all: $O_2$, the former menace, was transformed into an asset. Its amazing capacity to receive electrons (i.e., oxidize other molecules) enabled the assembly of electron chains that culminated in final delivery to $O_2$ and ATP synthesis with high efficiency. The power shortage problem was finally solved and life could evolve into more sophisticated forms. Accumulation of lipids and glycogen in animals and the presence of oxygen in the atmosphere to breathe granted animals independence from direct sunlight: day or night, sunny or cloudy, animals could move at high speed and high performance, and maintain complex tissues demanding high metabolic rates, such as the ones of the central nervous system of humans.

## Questions

1. For the sake of simplicity, assume an unrealistic 100% efficacy in energy conversion and calculate the distance a human of 70 kg body mass can walk at 3.5 mph (5.6 km h$^{-1}$) using the energy of a typical cereal bar (150 kcal or Cal). How long would it take for the basal metabolism to consume the same amount of energy?
2. Consider a typical man having 70 kg body mass and 15 kg lipids. Calculate the mass of glycogen needed to synthesize the same quantity of ATP. If all storage was made of glycogen instead of lipids, what would be the body mass of the same man?
3. Given the advantages of using lipids as "fuel" for metabolism, what are the advantages of having storages of glycogen in the liver and muscles together with lipids stored in the adipose tissue?
4. Based on the reasoning above, hypothesize why big animals tend to move slower.

## Biochemical Insight

1. At rest (MET = 1), 1 kcal kg$^{-1}$ h$^{-1}$ energy consumption is expected. For 70 kg body mass, energy consumption is expected to be 70 kcal h$^{-1}$. So, at rest, it would take approximately 2 h to consume the calories in the cereal bar. A walk at 3.5 mph corresponds to nearly 4 MET (see previous figure), so the 150 kcal of the cereal bar would be exhausted in 1/4 of the time, i.e., 30 min. A 30-min walk at 5.6 km h$^{-1}$ covers 2.8 km (approximately 1.75 miles).

2. The yield in the conversion of glucose and fatty acids to ATP in catabolism is not straightforward as glucose and most fatty acids do not have the same number of carbons or molar mass. So, the first step is to calculate the relative yield of conversion of glucose and a selected fatty acid, such as palmitate, per unit mass:

$$\frac{n_{ATP}}{m} = \frac{n_{ATP}}{molec} N_A \frac{1}{M}$$

$n_{ATP}$ is the number of molecules of ATP, $m$ is mass, *molec* stands for one molecule, $N_A$ is Avogadro's number, and $M$ is molar mass ($M$ = 180 g mol$^{-1}$ for glucose; $M$ = 256 g mol$^{-1}$ for palmitate). Because one molecule of glucose converts to nearly 30 ATP and one molecule of palmitate converts to nearly 131 ATP, the mass ratio of glucose and palmitate associated with equimolar production of ATP is:

$$\frac{m_G}{m_P} = \frac{131}{30} \cdot \frac{256}{180} \approx 3$$

$m_G$ and $m_P$ are the masses of glucose and palmitate, respectively. In practice, this means a threefold higher mass of glucose relative to palmitate is needed to produce the same quantity of ATP. (You should bear in mind the contribution OH groups have to the molar mass of glucose.)

Therefore, 15 kg lipids needed to be replaced by 3 × 15 = 45 kg of glucose (i.e., nearly 45 kg of glycogen). However, glucose/glycogen is highly hygroscopic and retains threefold its own mass of water, which is 3 × 45 = 135 kg water. In total, the mass of glycogen storage would sum 45 + 135 = 180 kg.

Overall, the body mass would be the initial 70 kg with 15 kg lipids removed and 180 kg hydrated glycogen added: 70–15 + 180 = 235 kg. Think about this figure twice before cursing your adipose tissue in front of the mirror.

3. Lipids are certainly an efficient way to store molecules that can be used for body heating or to synthesize ATP, which can in turn be used as "energy source" to many different biochemical events such as metabolic processes or muscle contraction during exercise (see Chap. 10). The very low hydration level of lipids is advantageous for storing, as explored in question 2, but it has a dark side: lipids have very low solubility and are therefore difficult to mobilize. When exercise has to be started suddenly and kept very rapidly (the "fight or flight" burst), lipid uptake in the adipose tissue and transport in the body is too slow. In contrast,

muscle glycogen is on site, and its branched architecture guarantees that multiple simultaneous glucose monomers are released at a high pace. Liver glycogen is used to rapidly counteract hypoglycemia whenever needed. Hypoglycemia is potentially a very dangerous condition that demands rapid contingency measures, degradation of glycogen in the liver, and release in blood being the most important (see Sect. 9.2).

4. Energy conversion in physical activity is far from being ideal. Nutrients are not converted to energy and energy is not converted to work with 100% efficiency. Heat is released and overheating becomes a problem for large body mass animals. To avoid overheating, large animals respire less intensely and do not exercise constantly. As G. Schatz put it: *"It is respiration that keeps me going. There is no better example of its power than the flight muscles of insects. They are the Ferraris of muscles and packed with mitochondria. Thanks to them, a dragonfly can lift twice its own weight, beating our best helicopters by a factor of seven. But as with any high-performance engine, heat is a problem"*[1]. In the case of Francisco Lázaro, the carpenter who ran the marathon in the Olympics of 1912 in Stockholm, the problem ended in tragedy (see Challenging Case 11.2).

## Final Discussion

An interesting paper by Rolf Luft's group at Karolinska Institute in Stockholm, Sweden, reported the case of a woman who arrived in 1959 at Luft's Clinic with puzzling symptoms.[4] She was 30 years old but had been suffering from abnormal physiological symptoms since the age of 7. Her symptoms comprised enormous perspiration combined with increased fluid intake (without polyuria), extremely high caloric intake (>3000 kcal per day) but at a stable body mass of 38 kg (1.59 m height), and general weakness. BMR, the basal metabolic rate (measured from oxygen consumption), was 180% the reference value, which was claimed to be the highest BMR ever found in a human. No hormonal dysfunctions were reported. Luft and his collaborators focused their study in morphologically abnormal muscle mitochondria found in biopsies and abnormal ATP production concomitant with uncontrolled muscle metabolism. The proposed explanation was that the cause of the disease was a short circuit of the flow of protons in the inner membranes of mitochondria, affecting ATP production but preserving electron transport, an explanation that holds and gained further experimental evidence afterward.[5, 6] The disease

---

[4] Luft R et al (1962) A case of severe hypermetabolism of nonthyroid origin with a defect in the maintenance of mitochondrial respiratory control: a correlated clinical, biochemical and morphological study. J Clin Invest 41:1776–1804.

[5] Sjöstrand FS (1999) Molecular pathology of Luft disease and structure and function of mitochondria. J Submicrosc Cytol Pathol 31:41–50.

[6] Luft R (1994) The development of mitochondrial medicine. Proc Natl Acad Sci U S A 91:8731–8738.

was therefrom named after Luft as "Luft disease." Unfortunately, no cure was found in due time and the patient committed suicide in the 1970s. There is still no cure for the disease. Symptomatic treatment is the only option. Vitamins C, K, and E, along with coenzyme Q10 and a high caloric diet, are recommended. Protection from heat is obviously advised.

The reader is encouraged to use the knowledge learned from Chap. 6 to explain both the symptoms of the disease based on the proposed electron chain uncoupling and the reasons behind the symptomatic treatment.

## Challenging Case 6.2: The Force Comes from Within

### *Source*

"May the Force be with you" is an iconic expression that is immediately associated with the Star Wars saga, a US space film series, created by George Lucas. The first Star Wars film premiered in 1977 and was followed by other two (in 1980 and 1983), forming the original Star Wars trilogy. Then, a subsequent prequel trilogy was produced, followed by three more films that compose the final sequel trilogy. Star Wars films inspired several productions in other media, such as television shows, video games, books, and comics, resulting in a Guinness World Records award as the "Most successful film merchandising franchise."

Star Wars' logo, created by Suzy Rice in 1976, reproduced from Wikimedia Commons, under a CC-BY license

### *Case Description*

George Lucas' concept of the "Force" in the Star Wars fictional universe is related to a metaphysical power or a kind of spirituality, for him the essence of all religions, which would be used either in its "light side" or "dark side."[7] But Lucas also gave a more biological aspect to the "Force" when he associated it to symbiotic microscopic organisms that he called "midi-chlorians."

The idea of symbiotic creatures living inside the cells and having as their main function to provide power (i.e., energy) for the organism is a clear reference to the mitochondria (see Box 6.3, about Lynn Margulis' endosymbiotic theory for the origin of mitochondria). Indeed, this issue is largely explored in Star Wars-related media.[8]

The origin of the mitochondria as the inspiration for the concept of the midi-chlorians may be perceived in the dialogue between the Jedi Master Qui-Gon and the young Anakin Skywalker in the first film of the prequel trilogy ("Star Wars: Episode I – The Phantom Menace"):

> Qui-Gon: "*Midi-chlorians are microscopic life-forms that reside within the cells of all living things and communicate with the Force.*"
> Anakin: "*They live inside of me?*"
> Qui-Gon: "*In your cells. We are symbionts with the midi-chlorians.*"
> Anakin: "*Symbi-what?*"

---

[7] The Mythology of 'Star Wars'. Documentary; Available in https://vimeo.com/groups/183185/videos/38026023.

[8] https://starwars.fandom.com/wiki/Midi-chlorian.

> Qui-Gon: "*Symbionts. Life-forms living together for mutual advantage. Without the midi-chlorians, life could not exist, and we would have no knowledge of the Force. Our midi-chlorians continually speak to us, Annie, telling us the will of the Force.*"

The association of the "Force" to the midi-chlorians is probably the most criticized aspect of the series. The controversy seems to lie in the fact that the midi-chlorians in some way demystify the "Force," shifting its "metaphysical" essence to a much more "physical" nature. This is sustained by the direct relationship between the intrinsic power of the characters and the number of midi-chlorians in their cells. This idea appears when the midi-chlorians count in Anakin's blood revealed a number that had never been recorded so far.

> Qui-Gon Jinn: "*I need a midi-chlorian count.*"
> Obi-Wan Kenobi: "*The readings are off the chart. Over 20,000. Even Master Yoda doesn't have a midi-chlorian count that high.*"
> Qui-Gon Jinn: "*No Jedi has.*"

The introduction of the midi-chlorians as a kind of materialization of the "Force" made several people very upset when the "The Phantom Menace" came out. An example of the bad repercussion of this idea is the text by Evan Narcisse in the Gaming & Culture section of *Time* magazine: "*One word ruined Star Wars for me, and probably for a generation of fans, too. That word wasn't Jar Jar or Watto. It wasn't a character. It was 'midi-chlorians'. With that one word, the mechanisms of the Force became less spiritual and more scientific. Major bummer. The draw of the concept of the Force in the Original Trilogy is that it comes across as a low-maintenance religion. It's kinda like Unitarianism that also gives you psychic powers and enables you to jump, fight and stare better than other members of your respective species.*"[9]

But it seems that George Lucas intention was not to restrict the conception of the "Force," but to use the idea of symbiosis to expand it: "*… it's really a way of saying we have hundreds of little creatures who live on us, and without them, we all would die. There wouldn't be any life. They are necessary for us; we are necessary for them. Using them in the metaphor, saying society is the same way, says we all must get along with each other.*"[10]

Interestingly, this philosophical or ecological way to understand the meaning of symbiosis in nature has also been a subject of Lynn Margulis' work. In the 1970s, together with the chemist James Lovelock, she developed the Gaia hypothesis[11], which suggested that the Earth could be considered a "macroorganism," in which all level of interactions depends on mutualism and symbiosis.

---

[9] Narcisse E 20,000 Per Cell: Why Midi-chlorians Suck. Gaming & Culture, Time. http://techland.time.com/2010/08/10/20000-per-cell-why-midi-chlorians-suck/. Accessed 10 Aug 2010.

[10] Knight C Midi-Chlorians: physiology, physics, and the force. TheForce.net; http://www.the-force.net/midichlorians/midi-what.asp.

[11] Lovelock JE, Margulis L (1974) Atmospheric homeostasis by and for the biosphere: the Gaia hypothesis. Tellus 26:2–10.

## Questions

1. Putting aside the "science vs religion" controversy, discuss the pertinence of the association between the midi-chlorians and the "Force?"
2. One of the main functions of the mitochondria is to convert the electrical potential of redox reactions into the chemical energy of the ATP phosphate bond through the oxidative phosphorylation process. This idea, proposed by Peter Mitchell in his chemiosmotic hypothesis (see Sect. 6.2.2), was initially difficult to be accepted as most of the scientists in the field were searching for a high-energy intermediate as the key piece in ATP synthesis. Indeed, in biological sciences, it is usually easier to conceive energy as that involved in the breaking and formation of chemical bonds than that associated with the redox potential. To explore this subject, a simple experiment is proposed to measure the energy involved in oxidation–reduction reactions and to correlate it with the reactions that occur in the electron transport system in the mitochondrial membrane.

Materials you will need:

- 3 LED bulbs (LED Christmas lights work well)
- 2 m of thin electrical wire (0.14 mm)
- 3 1.5 V batteries
- 50 cm of rigid copper electrical wire (4 mm)
- 3 lemons
- 3 zinc screws
- A cutting plier
- A knife

Procedures:

- Cut the wire into pieces of approximately 20–25 cm.
- Peel the ends of the wires (approximately 1 cm from each end).
- Use two pieces of these wires to connect the positive and negative poles of the battery in the LED bulb. If the lamp does not light up with a single battery, try connecting two batteries in series and observe the result.

   (a) How many batteries are needed to light up the LED? Check the nominal voltage of the battery you used and calculate how many volts are necessary to light up the LED.
   (b) How should the + and - poles of the batteries be arranged for the lamp to light up?
   (c) Oxidation reactions occurring inside the battery generate voltage that allowed the LED to be lit. Using copper and zinc as the electron transfer pair and lemons as the conducting medium reproduce the previous experiment following the procedures described below and discuss the results obtained.

- Remove the rubber cover that surrounds the wire, cut a small piece of the exposed copper, and insert it into the lemon.

- Insert the zinc screw into the lemon.
- Use two pieces of wire to connect the screw to one end of the LED lamp and another wire to connect the copper wire to the other end of the LED lamp. Check whether the lamp was lighted. If it has not lighted up, add more lemons until it works. In this case, remember how you needed to arrange the batteries to turn on the LED lamp in the previous experiment.

(d) Based in the results obtained with the homemade battery (for which you know exactly the electron transfer pair), use the information in the next table and the following equation to calculate the voltage necessary to light up the LED.

| Reduction potential | Reduced state | | Oxidized state | Oxidation potential |
|---|---|---|---|---|
| −3.04 | Li | ↔ | $Li^+ + e^-$ | +3.04 |
| −2.92 | K | ↔ | $K^+ + e^-$ | +2.92 |
| −2.90 | Ba | ↔ | $Ba^{2+} + 2e^-$ | +2.90 |
| −2.89 | Ca | ↔ | $Ca^{2+} + 2e^-$ | +2.89 |
| −2.87 | Na | ↔ | $Na^+ + e^-$ | +2.87 |
| −2.71 | Mg | ↔ | $Mg^{2+} + 2e^-$ | +2.71 |
| −2.37 | Al | ↔ | $Al^{3+} + 2e^-$ | +2.37 |
| −1.66 | Mn | ↔ | $Mn^{2+} + 2e^-$ | +1.66 |
| −1.18 | $H_2 + 2(OH)^-$ | ↔ | $H_2O + 2e^-$ | +1.18 |
| −0.83 | Zn | ↔ | $Zn^{2+} + 2e^-$ | +0.83 |
| −0.76 | Cr | ↔ | $Cr^{3+} + 3e^-$ | +0.76 |
| −0.74 | $S^{2-}$ | ↔ | $S + 2e^-$ | +0.74 |
| −0.48 | Fe | ↔ | $Fe^{2+} + 2e^-$ | +0.48 |
| −0.44 | Co | ↔ | $Co^{2+} + 2e^-$ | +0.44 |
| −0.28 | Ni | ↔ | $Ni^{2+} + 2e^-$ | +0.28 |
| −0.23 | Pb | ↔ | $Pb^{2+} + 2e^-$ | +0.23 |
| −0.13 | $H^2$ | ↔ | $2H^+ + e^-$ | +0.13 |
| 0.00 | $Cu^+$ | ↔ | $Cu^{2+} + e^-$ | 0.00 |
| +0.15 | Cu | ↔ | $Cu^{2+} + 2e^-$ | −0.15 |
| +0.34 | $2(OH)^-$ | ↔ | $H_2O + ½O_2 + 2e^-$ | −0.34 |
| +0.40 | Cu | ↔ | $Cu^+ + e^-$ | −0.40 |
| +0.52 | $2I^-$ | ↔ | $I_2 + 2e^-$ | −0.52 |
| +0.54 | $Fe^{2+}$ | ↔ | $Fe^{3+} + e^-$ | −0.54 |
| +0.77 | Ag | ↔ | $Ag^+ + e^-$ | −0.77 |
| +0.85 | Hg | ↔ | $Hg^+ + e^-$ | −0.85 |
| +1.09 | $2Br^-$ | ↔ | $Br_2 + 2e^-$ | −1.09 |
| +1.23 | $H_2O$ | ↔ | $2H^+ + ½O_2 + 2e^-$ | −1.23 |
| +1.36 | $2Cl^-$ | ↔ | $Cl_2 + 2e^-$ | −1.36 |
| +2.87 | $2F^-$ | ↔ | $F_2 + 2e^-$ | −2.87 |

Crescent oxidation action ↓

Crescent reduction action ↑

$$\Delta E^0 = E^0_{red(higher)} - E^0_{red(lower)}$$

(e) Correlate your results with the electron transfer occurring in the mitochondrial respiratory complexes.

3. In the Star Wars series, George Lucas concept that the "Force" may manifest
   through its "light side" or "dark side" may illustrate the dual role of reactive
   oxygen species (ROS) production in mitochondria. ROS are formed when the
   input of electrons into the respiratory chain overcomes their transfer to oxygen
   (see Sect. 6.2.5.2). Its production is essential for the adaptation to different physiological states, but its imbalance may be the cause of pathological situations.

   (a) An uncontrolled increase in ROS production may damage different cellular
       components and structures, a condition known as oxidative stress. Describe
       the ROS-induced molecular alterations on the main biomolecules that can be
       associated with cellular damage.
   (b) Diabetes mellitus, a metabolic dysfunction associated with high plasma glu-
       cose concentrations, is one condition known to involve high production of
       ROS. The next figure shows the result of an experiment in which ROS pro-
       duction was quantified after cell incubation with two glucose concentrations
       simulating normoglycemia (5 mM) or hyperglycemia (30 mM).[12] Discuss the
       results, explaining the effects of the overexpression of the antioxidant enzyme
       superoxide dismutase (Mn-SOD) in the cells; the addition of rotenone, an
       inhibitor of Complex I; thenoyltrifluoroacetone (TTFA), an inhibitor of
       Complex II; carbonyl cyanide m-chlorophenylhydrazone (CCCP), an uncou-
       pling agent that facilitates the return of $H^+$ from intermembrane space to mito-
       chondrial matrix; or the overexpression of the uncoupling protein 1 (UCP1).

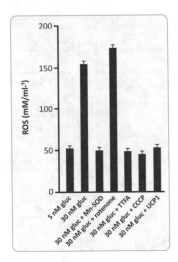

ROS production by cells incubated with 5 or 30 mM glucose alone or after the overexpression of
the enzyme superoxide dismutase (Mn-SOD) or UCP1 gene, or after the addition of either rotenone, TTFA, or CCCP. (Figure adapted with permission from the article cited in the footnote 12)

---

[12] Nishikawa T et al (2000) Normalizing mitochondrial superoxide production blocks three pathways of hyperglycaemic damage. Nature 404:787–790.

(c) Although much more attention has been given to the "dark side" of ROS, body health maintenance requires a balance between production and removal of ROS to ensure homeostasis. If ROS are produced in levels higher than those that body can handle, oxidative stress occurs, while low levels of ROS causes reductive stress that can also be deleterious to body function. Discuss the beneficial role of ROS in cellular metabolism.

## *Biochemical Insights*

1. Georges Lucas created the midi-chlorians as the source of the "Force" inspired by the mitochondria, organelles originated from symbiosis and responsible for the energy management in the organism. He conceived the "Force" as the "inner energy that powers all living things and is manifested as an intracellular symbiotic life form called midi-chlorians."[13] The word "midi-chlorian" combines mitochondria and chloroplasts (endosymbionts analogs to mitochondria that perform photosynthesis in plants). The most important function of both organelles is energy conversion in the cells. Through the mechanisms of oxidative phosphorylation, the energy release from oxidative metabolization of the nutrients is converted into ATP (known as the "molecular currency" of intracellular energy transfer) in the mitochondria. The same occurs for the photophosphorylation process that takes place in chloroplasts: the energy of light is used to synthesize ATP.

2. The experiment provides a simple way to realize that oxy-reduction reactions generate energy that can be used in other processes.

(a) To light up the LED, we need two batteries of 1.5 volts lined up in series setup, so that 3 volts are necessary (see the following diagram).

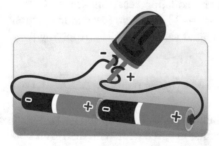

(b) The batteries are composed of metals and an electrolytic solution that allows the electrons to be transferred from the metal that is oxidized to the

---

[13] Hom M (2015) The real biology of "Star Wars" midichlorians. Elsevier SciTech Connect, Biomedicine & Biochemistry, New York. http://scitechconnect.elsevier.com/real-biology-star-wars-midichlorians/.

metal that will be reduced. This oxy-reduction reaction generates the electrical current necessary to light up the LED. The metal with greater capacity to donate electrons, i.e., a lower reduction potential, acts as the anode, or the negative pole of the battery, while the metal with the higher tendency to receive electrons, i.e., a higher reduction potential, act as the cathode, or the positive pole of the battery. The potential difference between the anode and the cathode generates the electrical energy that allows the LED to light up.

(c) As occurred with the commercial batteries in the previous experiment, the homemade battery built using copper and zinc as the electron transfer pair and lemons as the conducting medium provided energy to light up the LED. In this case, three pairs of zinc and copper were necessary (see next figure). It is important to have in mind that the energy responsible for lighting the LED comes from the electron transfer from zinc to copper and not from the metabolic reactions occurring in the lemon cells. The lemon acts just as the conducting medium, providing the electrolytes for electron transfer.

(d) The electron transfer pair that provides the electrical energy that lit up the LED are copper (oxidation reaction: $Cu \leftrightarrows Cu^{+2} + 2e^-$, with a reduction potential value of +0.34) and zinc (oxidation reaction: $Zn \leftrightarrows Zn^{+2} + 2e^-$, with a reduction potential value of −0.76). Applying these values in the equation $\Delta E^0 = E^0_{red(higher)} - E^0_{red(lower)}$, we have:

$$\Delta E^0 = +0.34 - (-0.76)\text{volts} \Leftrightarrow \Delta E^0 = +1.1\text{volts, per pair}$$

Since we needed three pairs to light up the LED, we conclude that the voltage necessary is about 3 volts, in agreement with the result obtained with the commercial batteries.

(e) The mitochondrial electron transport chain works similarly as the batteries. Several oxy-reduction pairs are distributed with a crescent reduction potential along the protein complexes, generating an electric potential difference that determines the electron flow. For instance, in Complex I, many iron–sulfur (Fe–S) centers transport the electrons from FMN to ubiquinone (See Fig. 6.13). In this case, although the oxy-reduction pairs are the same (iron and sulfur), they occupy different environments in the protein structure, which slightly alter their reduction potential, determining the electron flow direction. The electric energy is used to promote conformational changes in Complexes I, II, and IV that allow these proteins to transport protons from the mitochondrial matrix to the intermembrane space (see Sect. 6.2.2). Thus, in mitochondria, the electrical energy is used to pump the protons, generating a pH gradient across the membrane, which, together with the transmembrane electrical potential, defines the proton-motive force that will finally drive ATP synthesis by ATP synthase (see Sect. 6.2.4).

3. ROS are natural by-products of oxygen metabolism by multiple processes performed by the cells. The major sources of ROS are the activity of the enzyme NADPH oxidase (NOX) and the mitochondrial respiratory chain.

(a) ROS may react with lipids, DNA, and proteins, damaging cellular membranes, causing chemical changes in DNA bases, thus causing mutations, and oxidizing proteins, which may affect their activities and functions.

Cell membranes can be damaged due to lipid peroxidation caused by ROS, as illustrated in the following figure. The most affected lipids are the polyunsaturated fatty acids, which contain multiple double bonds that cause the methylene bridges to have especially reactive hydrogen atoms. This results in the formation of fatty acid radicals that in turn reacts with molecular oxygen generating a peroxyl-fatty acid radical. This unstable molecule starts a chain reaction that produces a different fatty acid radical and a lipid peroxide, in a continuous cycle that ultimately leads to membrane permeabilization and/or disruption.

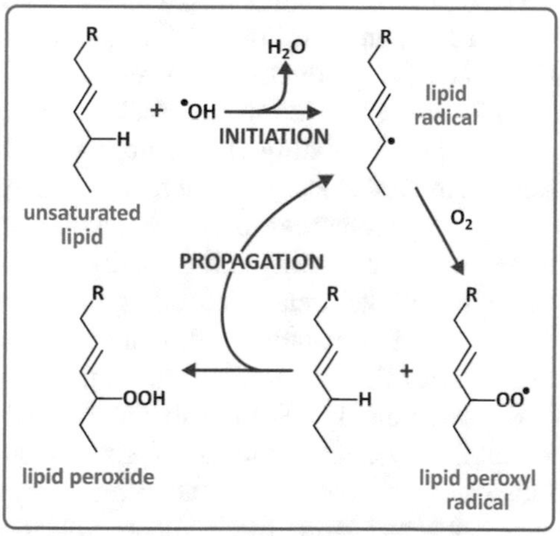

Mechanism of lipid peroxidation

Due to the lowest one-electron reduction potential of guanine when compared to the other nucleosides in DNA, the most common oxidative damage occurring in DNA molecule is the formation of 8-oxo-2'-deoxyguanosine (see below figure). Although repair enzymes remove oxidized bases from DNA, remaining modified bases result in mutations, which may cause carcinogenesis and disease.

8-oxo-2'-deoxyguanosine

ROS can promote loss of protein function by causing amino acid residues' oxidation or fragmentation as well as inducing protein–protein cross-linkages. This affects the proper protein folding, and thus its biological activity, by altering the physical interactions that maintain protein tertiary or quaternary structure.

(b) The results show that hyperglycemia leads to overproduction of mitochondrial ROS, which has been shown to be the common link for several key pathogenic pathways involved in microvascular complications of diabetes. Overexpression of the antioxidant enzyme superoxide dismutase in the cells restores the normal ROS levels, showing that superoxide is the main species produced.

There are two main sites of superoxide generation in the electron transport system (ETS), Complex I and the interface between ubiquinone and Complex III. Rotenone, an inhibitor of Complex I, has no effect on ROS production, while TTFA, an inhibitor of Complex II, restores ROS production, suggesting that ROS production is a result of ETS feeding through Complex II. CPPP- or UCP-1-induced mitochondrial uncoupling restores the normal ROS levels, indicating that a high mitochondrial membrane potential induces an excessive ROS production in hyperglycemia.

(c) ROS can act as signaling molecules that control a wide variety of physiological functions, especially regulating cell cycle by determining cell survival or inducing programmed cell death depending on the cell type and/or the physiological condition. For instance, different components of the mitogen-activated protein kinase (MAPK) cascades are activated through the interaction with ROS. Additionally, ROS also inhibit protein phosphatases, preventing the inhibitory actions of these proteins on MAPK signaling. Finally, ROS also play a pivotal role in the immune response. For instance, phagocytes synthesize and store free radicals, releasing these species to destroy invading pathogenic microbes.

## *Final Discussion*

Mitochondria are dynamic organelles that undergo continuous fusion and fission events that are necessary to maintain mitochondrial quality. These processes are mediated by a number of proteins from the dynamin family: dynamin-related protein 1 (Drp1), fission 1 protein (Fis1), and mitochondrial fission factor (Mff) regulate fission, resulting in small and rounded mitochondria, while optic atrophy 1 (Opa1) and mitofusin (Mfn) 1 and 2 regulate fusion, generating elongated, tubular interconnected networks (see next figure).

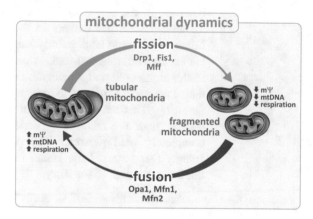

Different proteins regulate mitochondria dynamics, controlling the equilibrium between fission and fusion, which result in rounded or elongated mitochondria, showing lower or higher membrane potential (mψ), DNA content (mtDNA), and respiration taxes, respectively. (Figure adapted from the article cited in the footnote 14, with permission of SAGE Publishers)[14]

Mitochondrial fusion mixes the contents of partially damaged organelles, replacing damaged components, preserving mitochondrial membrane potential and oxidative phosphorylation, and exchanging mitochondrial DNA, while fission creates new mitochondria and is associated with apoptosis, allowing removal of damaged mitochondria. If you would like to see the fusion and fission processes in real time, watch the following video: https://youtu.be/xhruE64XzY4.

To go deep in several new aspects about mitochondria, including mitochondrial dynamics, we recommend the lecture by Jodi Nunnari, from the Department of Molecular and Cellular Biology at the University of California, Davis, which is available at https://youtu.be/PhdLh_rEnXY.

*This case had the collaboration of Dr. Wagner Seixas da Silva (Instituto de Bioquímica Médica Leopoldo de Meis, Universidade Federal do Rio de Janeiro, Brazil).*

---

[14] Balog J et al (2016) Mitochondrial fission and fusion in secondary brain damage after CNS insults. J Cereb Blood Flow Metab 36:2022–2033.

## Challenging Case 6.3: A Curious Girl Has Fallen into a Mad World

### Source

The (Mad) Hatter is an emblematic character of Lewis Carroll's 1865 novel *Alice's Adventures in Wonderland*. Although Carroll had never used the adjective "mad" for his character, the Hatter behavior in the story, together with the madness associated with hat makers in the 1800s, made Carroll's Hatter to be popularly known as the Mad Hatter. Indeed, mad hat-makers were not unusual in the eighteenth and nineteenth centuries due to their long-term exposure to mercury in hat manu-facturing. This may be the origin of the English phrase "mad as a hatter" to describe an insane or completely mad person.[15] The American actor Johnny Depp starred the Hatter in Tim Burton's 2010 film version of the story.

Cover of the 1907's edition of the *Alice's Adventures in Wonderland*, written by Lewis Carroll and illustrated by Charles Robinson; public domain

### Case Description

*Alice's Adventures in Wonderland* was written in 1865 by the mathematician Charles Dodgson under the pseudonym Lewis Carroll. The story is about a 7-year-old girl named Alice who, sitting on a riverbank with her sister, suddenly sees a talking, clothed white rabbit with a pocket watch. Attracted by the fascinating situation, Alice followed the fantastic animal and fell down into the rabbit hole, reaching a nonsense world populated by very strange creatures. *Alice's Adventures in Wonderland* is a surreal story that in fact symbolizes the difficulties we face in growing up, leaving childhood behind to become an adult, when we must deal with a variety of dramatic changes, both physical and psychological, that occur during adolescence. Growing up includes the need for adapting to a different logic based on distinct rules, including another way to deal with authority and time, necessary for the building of an own identity. In the story, this is represented, for instance, by Alice's discomfort of never being the right size: she finds the ways to grow or shrink but she can't control her growing, being too big to get into the garden, or too small to reach the key that would help her get in.

---

[15] https://www.phrases.org.uk/meanings/mad-as-a-hatter.html.

The Hatter appears in the book's Chap. 7, which describes the "Mad Tea-Party" scene. The Tea-Party takes place in a table outside the March Hare's house, where the Hatter, the Dormouse, and the March Hare are taking tea, which, as explained by the Hatter, is endless, since the "Time" stuck them at 6 pm forever. During this endless tea, they switch places at the table, ask unanswerable riddles, and recite nonsensical poetry.

In the Tea-Party, the conventions are broken, especially the very strong conventions for British people, such as time, or punctuality (the Hatter's watch tells the month and not the hour), or etiquette (the Hatter, the Hare, and the Dormouse insist there is no room for Alice in the tea table, but she sits down anyway). The rules Alice is supposed to follow are completely different from those she is used to. Also, during all the dialogues, the logic is inverted, but in this nonsense, a new sense, which is also logical in the new context, is created, challenging the value of the real-world conventions. The logic redefinition makes one to reflect on what indeed defines how mad you are, or you may be, as illustrated in the following dialogue between Alice and the Hatter.

> Mad Hatter: *"Have I gone mad?"*
> Alice: *"I'm afraid so. You're entirely bonkers. But I'll tell you a secret. All the best people are."*

Regardless of the symbolism and the interpretations of Alice's Wonderland, the choice of the characters does not seem to be by chance. During the Victorian era (1837–1901), madness was frequently observed in hat industry workers, being the likely origin of popular expressions like "mad as a hatter" and "the hatters' shakes." In the nineteenth century, the hat industry employees manipulated the toxic substance mercurous nitrate to treat animals' furs to produce the felt used in hat manufacturing. Nevertheless, it took some time for medical doctors to fully correlate hatters' madness with mercury. It was only in 1864, the year before the novel publication, that the toxicologist Alfred Swaine Taylor reported, in Britain, mercury poisoning in a hat maker.

Mercury poisoning affects mainly the central nervous system leading these workers to exhibit a variety of physical and mental disturbances, including emotional instability, changes in personality, difficulties in talking and thinking clearly, and tremors or shaking due to uncontrollable muscle twitching.

Mercury was first identified as a toxic compound in 1823 by the physician William Burnett: "It has long been known that in the vacuum barometer mercury rises in a vaporous state at the usual temperature of this climate, and that persons employed in the mines from whence this metal is procured, as well as those who are employed in gilding and plating, have suffered paralytic and other constitutional effects from inhaling the air saturated with mercurial vapors: had any doubt remained of mercury existing in the state alluded to, it would be effectually removed from the experiments conducted by Mr. Faraday, detailed in the 20th number of the Journal of Science, &c."[16]

---

[16] Burnett W (1823) An account of the effect of mercurial vapours on the crew of His Majesty's Ship Triumph, in the year 1810. Philos Trans R Soc Lond A 113:402–408.

Although being studied for a long time, the exact mechanism by which mercury causes its effects on the nervous system remains unclear. Most of the studies point to an effect on neurons' redox homeostasis leading to mitochondrial dysfunction that may culminate with cellular death. Other clues on the molecular mechanisms involved in mercury-induced neurotoxicity came from studies that evidenced that mercury binds covalently to sulfhydryl (thiol) groups from proteins and non-protein molecules[17], such as glutathione (GSH; γ-glutamyl-cysteinylglycine; see next figure).

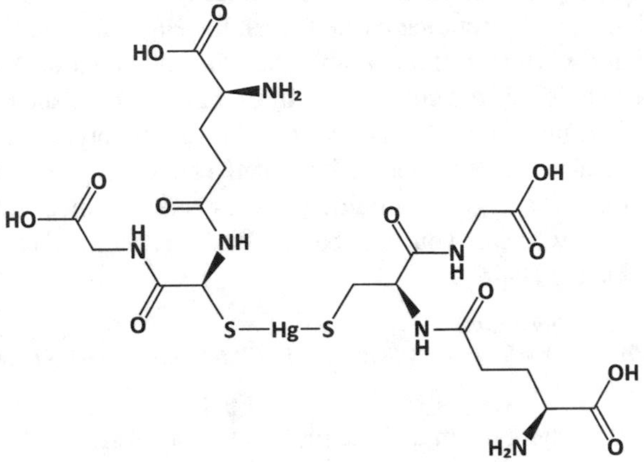

Chemical structure of bis-glutathionyl-mercury (II)

GSH is the most abundant low molecular weight thiol compound present in the body, playing a crucial role in protecting cells against oxidative damage. This occurs through the action of the enzyme glutathione peroxidase (GPx), which converts $H_2O_2$ into $H_2O$, reducing oxidized glutathione (GSSG), and glutathione reductase (GR), which reduces GSSG into GSH using NADPH as the electron donor (see Sect. 6.2.5.2), ensuring the maintenance of the proper cellular GSH/GSSG ratio, necessary to avoid oxidative damage. Mercury binding depletes GSH from the cells, impairing the GPx/GR system to work.

## Questions

1. Mercury effect on cellular GSH/GSSG ratio exemplifies how a redox imbalance may deeply affect cellular functions. In this context, antioxidant enzymes, as the system Gpx/GR, play a crucial role in maintaining cellular redox status. Together with Gpx/GR, superoxide dismutase (SOD) and catalase (CAT) form the main enzymatic systems that control the levels of reactive oxygen species (ROS) in the

---

[17] Rubino FM (2015) Toxicity of glutathione-binding metals: a review of targets and mechanisms. Toxics 3:20–62.

cells. Besides these enzymatic antioxidants, ROS levels may also be controlled by nonenzymatic antioxidant systems, which consist of low molecular mass compounds able to rapidly inactivate radicals. Choose an example of a nonenzymatic antioxidant and discuss its role in protecting cells against oxidative damage.

2. Neurons and glial cells are the main cells composing the nervous system. Neurons are electrically excitable cells highly specialized for the processing and transmission of cellular signals. Astrocytes, a subtype of glial cells, perform many functions, such as providing nutrients and protecting neurons against oxidative stress, tissue repair, and the maintenance of extracellular ion balance. Compare the energy metabolism profile and the antioxidant enzyme system in neurons and astrocytes in the context of the cross talk between these cell types.


## Biochemical Insights

1. ROS can cause cell damage by starting chemical chain reactions such as lipid peroxidation (see question 3 discussion in the Challenging Case 6.2). Ascorbic acid (or vitamin C) terminates the lipid peroxidation chain reactions by donating an electron to the lipid radical forming a relatively stable ascorbate radical, which is then converted to dihydroascorbate. Aside from this direct action on aqueous peroxyl radicals, vitamin C may act indirectly by restoring the antioxidant properties of the lipid-soluble vitamin E.

   Additionally, besides its role as an antioxidant molecule, vitamin C also acts as a cofactor in many enzymatic reactions. One example is the enzyme prolyl 4-hydroxylase, which is essential for making collagen to acquire the fibrous conformation, as discussed in Box 3.7.

2. The brain is a high-energy-demanding organ, accounting for more than 20% of the energy consumption of the organism. Considering the heterogeneity of the brain cells, one may expect that the different cell types have distinct metabolic profiles. These differences are better studied between neurons and astrocytes.

   Although neurons and astrocytes have similar mitochondria content, neurons are highly oxidative cells, able to oxidize most of the energy substrates, including glucose, pyruvate, lactate, glutamine, and glutamate. Neurons can also use ketone bodies, compounds formed in metabolic conditions where there is intense oxidation of lipids such as fasting, diabetes, and exercise (see Sect. 9.3.4). In this cell type, pyruvate coming from glycolysis will be fully oxidized in the tricarboxylic acid cycle within the mitochondria (see Chap. 7). On the other hand, astrocytes show a more glycolytic profile, in which the formation of ATP occurs mainly through glycolysis with lactate as a final product. Lactate is released from the astrocytes to be oxidized in neurons (see next figure). Thus, glycolysis in astrocytes, in addition to providing energy to these cells, is a source of lactate for neurons, in which it acts as an important substrate for aerobic ATP production.

The cross talk between neurons and astrocytes in respect to the energy metabolism can be illustrated with the astrocyte–neuron lactate shuttle model[18] (see below figure).

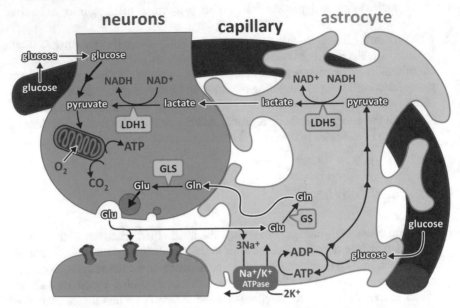

Energy metabolism cross talk between neurons and astrocytes: the lactate shuttle model. Lactate shuttle model accounts for the interrelationship between astrocyte and neuron energy metabolism. Glutamatergic neurotransmission in active neurons increases extracellular levels of glutamate, which is taken up by astrocytes through the $Na^+$-dependent glutamate transporters. The resulting increase in the intracellular $Na^+$ concentration activates the $Na^+/K^+$-ATPase, increasing ATP consumption, which in turn stimulates glycolysis in astrocytes, leading to lactate production. Lactate is released and used by neurons as an oxidative energy substrate

The very oxidative profile of the neurons results in high ROS production in these cells, which, on the other hand, are highly susceptible to oxidative stress. In this context, astrocytes' protective functions become evident.

A major antioxidant agent in the brain is glutathione (GSH). This molecule is a tripeptide (γ-L-glutamyl-L-cysteinylglycine) synthesized through the action of two enzymes: γ-glutamylcysteine synthetase and glutathione synthetase (see next figure).

[18] Bélanger M et al (2011) Brain energy metabolism: focus on astrocyte-neuron metabolic cooperation. Cell Metab 14:724–738.

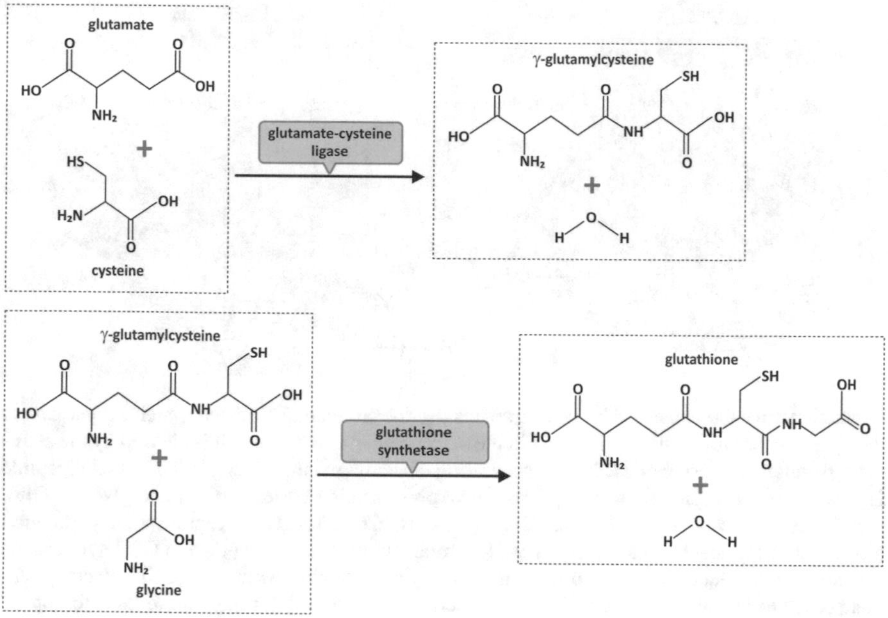

GSH synthesis. Firstly, the enzyme γ-glutamylcysteine synthetase or glutamate-cysteine ligase (GCL) covalently links glutamine and cysteine through a γ peptide linkage, forming γ-glutamylcysteine. Then, glutathione synthctasc links a glycine to the cysteine carboxyl group of the γ-glutamylcysteine to form GSH

Neurons and astrocytes have the enzymes to synthesize GSH, but neurons have a poor capacity to uptake the precursor amino acids for GSH synthesis (especially cysteine, which usually circulates in its oxidized form, cystine), being the astrocytes the main source of these precursors (see below figure). GSH is cleaved by the astrocytes to glutamate and cysteinylglycine through the action of their ectoenzyme γ-glutamyl transpeptidase. Cysteinylglycine is then cleaved by the extracellular neuronal aminopeptidase N, forming glycine and cysteine, which then may be used by the neurons. Glutamate is provided to neurons in the form of glutamine, which is synthesized in the astrocytes by the glutamine synthetase. Glutamine is then converted back to glutamate by neuronal glutaminase.

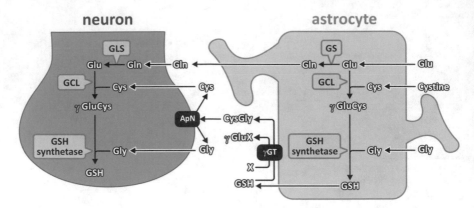

Glutathione metabolism in the brain. Astrocytes use the circulating oxidized cysteine (cystine) for the synthesis of GSH, which is released in the extracellular space, where it is cleaved by the astrocytic γ-glutamyl transpeptidase (γGT), generating cysteinylglycine (CysGly). Then, this dipeptide is cleaved by the neuronal aminopeptidase N (ApN), forming cysteine (Cys) and glycine (Gly), which serve as precursors for neuronal GSH synthesis. X represents an acceptor for the γ-glutamyl also produced in reaction catalyzed by γGT. Glutamate necessary to complete GSH synthesis is provided by the glutamate–glutamine cycle. Astrocytic glutamine synthetase (GS) converts glutamate (Glu) to glutamine (Gln), which is released in the extracellular medium to be taken up by neurons, where it is converted to glutamate by glutaminase (GLS)

## *Final Discussion*

The Hatter became one of the most popular Carroll's characters in the *Alice's Adventures in Wonderland* book. This is probable due to his eccentric behavior, characterized by madness and hyperactivity, as illustrated by the following quotes:

> "I knew who I was this morning, but I've changed a few times since then."
> "Sometimes I've believed in as many as six impossible things before breakfast."
> "Trust me. I know a thing or two about liking people, and in time, after much chocolate and cream cake, 'like' turns into 'what was his name again?'."

In terms of scientific research, the hat makers' madness called the attention to the neurotoxic effects of mercury. Indeed, there are a number of studies in the scientific literature reporting the effects of mercury exposure on cognitive functions. In these studies, the researchers usually perform behavior tests using animal models. An example is shown in the following figure, which shows the results of memory and anxiety tests performed in mice continuously exposed to mercury.[19] Memory was evaluated using the *novel object recognition test*. In this test, an animal is placed in an arena and allowed to explore two identical objects placed in the arena center (training session). Then, one of the objects is replaced by a new one, and the amount

---

[19] Malqui H et al (2018) Continuous exposure to inorganic mercury affects neurobehavioral and physiological parameters in mice. J Mol Neurosci 66:291–305.

of time spent exploring familiar and novel objects is measured. If the animal is able to recognize the familiar object as such, the exploration time of the novel object will be higher than 50% of the total time. Anxiety was evaluated using the *elevated plus maze test*. The plus maze apparatus consists of an elevated platform comprising two open and two closed arms connected by a central square. An animal is placed at the center of the platform facing one of the closed arms, and the amount of time each animal spent in the open or closed arms of the maze is determined. Anxiogenic behavior is expected to decrease in the number of open arm entries as well as the time spent on open arms and can be estimated by the proportion of open arm entries divided by the total number of arm entries and the time spent on open arms relative to the total time spent on both arms.

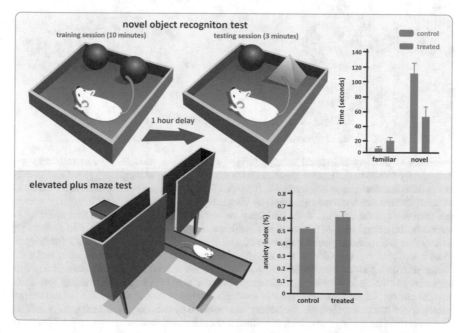

Schematic representation of the novel object recognition and elevated plus maze tests, with the respective results of experiments comparing control to mercury-exposed (treated) animals. (Results adapted from the article cited in the footnote 19, with permission)

There are several other behavioral tests used to evaluate locomotor and cognitive functions in animals. Some examples are illustrated in the following figure.

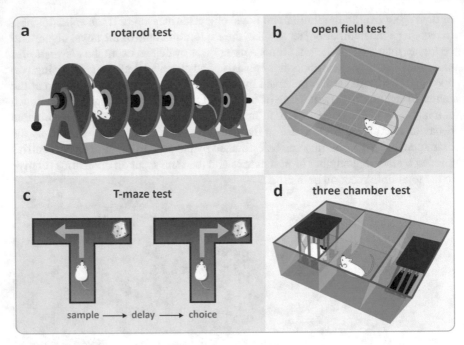

Examples of different behavioral tests performed using animal models. (**a**) *Rotarod:* mice are individually placed in the rotarod apparatus with the rod rotating at increasing speed and latency to fall from the rotating rod is determined. (**b**) *Open field test*: the animal is placed at the center of an arena divided into nine equal quadrants by imaginary lines on the floor. Total locomotor activity and time spent at the center or at the periphery are recorded, and the amount of time spent by an animal in the central area of the arena is considered to be inversely proportional to anxiety. (**c**) *T-maze test*: the animal is placed at the base of a simple maze shaped like the letter T (which provides a straightforward choice) with a reward at one or both arms. The animal must make the choice of which path to take. (**d**) *Three-chamber test*: the animal is placed in the middle of an apparatus consisting in a rectangular transparent acrylic box with two walls dividing it in three equal chambers. Cages are used to contain a stranger mouse (a mouse with which the test animal has never had any contact before). Sociability was assessed by allowing the animal to fully explore the apparatus after placing one cylindrical aluminum cage containing the stranger mouse (a mouse with which the test animal has never had any contact before) in one chamber and one empty cage in the opposite chamber

*This case had the collaboration of Dr. Lorena O. Fernandes-Siqueira (Instituto de Bioquímica Médica Leopoldo de Meis, Universidade Federal do Rio de Janeiro, Brazil).*

# Chapter 7
# Catabolism of the Major Biomolecules

Catabolism is the group of metabolic pathways in which the chemical energy contained in the nutrient molecules is transferred to other compounds, especially ATP, which in turn provides the energy for different cellular processes, such as the synthesis of new biomolecules, ion transport across cellular membranes, and muscle contraction.

In Sect. 6.2, we discussed in detail the mechanism of ATP synthesis through oxidative phosphorylation, starting with the electron transfer from the reduced coenzymes NADH and $FADH_2$ to the components of the respiratory system. In this chapter we will present the metabolic pathways along which the oxidation of the nutrient molecules results in the reduction of $NAD^+$ and FAD, generating the reduced coenzymes that feed the electron transport system.

## 7.1 An Overview of Catabolism

Carbohydrates, lipids, and proteins are the major constituents of food and serve as fuel molecules for the human body. The net energy yielded by the metabolism of each of these nutrients was determined for humans in the end of the nineteenth century, by the pioneer work of Wilbur O. Atwater, who developed a human respiration calorimeter (see Box 7.1) and obtained the values of 4.0, 4.0, and 8.9 kcal/g for carbohydrate, protein, and fat, respectively, values that are still considered valid today.

© Springer Nature Switzerland AG 2021
A. T. Da Poian, M. A. R. B., *Integrative Human Biochemistry*,
https://doi.org/10.1007/978-3-030-48740-9_7

**Box 7.1  Wilbur O. Atwater and the Studies on Human Nutrition**
Atwater started his scientific career in agricultural chemistry, studying fertil-
izers, but his major contributions for the advancement of science came when
he changed his scientific interest to human nutrition after he spent a period in
Carl von Voit's laboratory, in Germany. Atwater developed systematic studies
on the composition and the digestibility of foods as well as on the kind of
foods people from different ethnic groups in different areas were eating. At
that time, part of his interest was focused on finding out how poor people
could make more economical choices of food maintaining its nutritional
value. In 1893, Atwater's group developed a respiration calorimeter for studies of human metabolism, known as the Atwater–Rosa–Benedict's human
calorimeter (see below figure). This system allowed them to measure the
metabolizable energy contributed by each of the nutrient molecules.

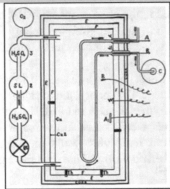

Inside view of the Atwater–Rosa–Benedict's human respiration calorimeter (image from
http://vlp.mpiwg-berlin.mpg.de/references?id=lit15620&page=p0213) and the diagrammatic representation of its components (reproduced from the book The Elements of the
Science of Human Nutrition, by Lusk G, 1917). The studied subject enters a copper chamber with a capacity of 1123 L, lay on a bed, and the chamber is hermetically closed. The air
within the chamber passes through three bottles previously weighed containing (1) sulfuric
acid, which removes the water; (2) moist soda lime, which removes the $CO_2$; (3) and sulfuric acid again, which absorbs the moisture taken from the soda lime. The gain in weight of
bottle 1 represents water absorbed, and the gain in weight of bottles 2 and 3 equals the $CO_2$
absorbed. $O_2$ was automatically fed into the system by cylinder, and $O_2$ consumption was
calculated by weighting the cylinder before and at the end of the experiment

The main products of carbohydrates, lipids, and protein digestion (monosaccharides, fatty acids, and amino acids, respectively) reach the bloodstream and enter the
cells, where they are firstly metabolized through specific catabolic routes that converge to a central pathway in the metabolism, the citric acid cycle (CAC; Fig. 7.1),
also referred to as the Krebs cycle (see Sect. 7.2.3) or tricarboxylic acid cycle.

The formation of a molecule of acetyl-coenzyme A (acetyl-CoA) is the convergence point of the degradation pathways of monosaccharides, fatty acids, and some
of the amino acids (Fig. 7.1). Then, acetyl-CoA is completely oxidized to $CO_2$ in

CAC, to which the products of the degradation of other amino acid also converge. In the oxidative reactions of the CAC, electrons are transferred from the CAC intermediates to $NAD^+$ or FAD, generating NADH and $FADH_2$ that in turn will transfer the electrons to $O_2$ through the respiratory system (see Sect. 6.2.2). Thus, in the overall process, $O_2$ is consumed and $CO_2$ is produced. This process is called cellular respiration, in analogy to the physiological term respiration, which is attributed to the gas exchanges that occur in the lungs.

The specific pathways of fatty acids and glucose degradation, the β-oxidation and glycolysis, respectively, also have oxidative steps in which FAD and $NAD^+$ are reduced (Fig. 7.1). On the other hand, the reactions of amino acid metabolism that generates the intermediates of the central metabolic pathway consist basically in transamination or deamination reactions that remove the amino group from the amino acid molecules, followed in some cases by carbon skeleton rearrangements.

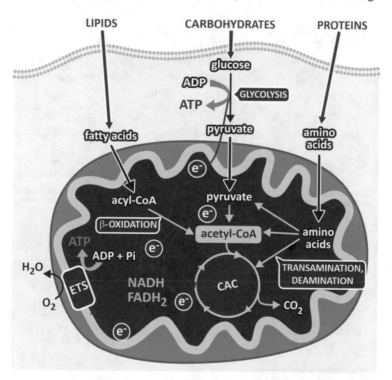

**Fig. 7.1** Schematic representation of a generic cell showing the catabolic pathways. Carbohydrate digestion generates monosaccharides, mainly glucose, which is degraded through glycolysis into two pyruvate molecules. In this pathway, an oxidative step is coupled to $NAD^+$ reduction, generating NADH that feeds the respiratory chain (electron transport system, ETS) with electrons ($e^-$) that ultimately reduce $O_2$ to $H_2O$. Pyruvate is converted to acetyl-CoA in the mitochondrial matrix in a reaction involving an oxidative decarboxylation with concomitant $NAD^+$ reduction to NADH. Degradation of fatty acids occurs through the pathway known as β-oxidation, which sequentially removes from the fatty acid chain a two-carbon fragment, in the form of acetyl-CoA, with concomitant reduction of FAD and $NAD^+$. Amino acids, the products of protein hydrolysis, undergo transamination or deamination reactions that remove their amino group and generate intermediates of the central metabolic pathway. The circles in the electron representation are used to indicate that they are not free in the cellular medium but are transferred by the compounds of each pathway

In this chapter we will first focus on the central pathway of acetyl-CoA oxidation, the CAC, to which all the nutrient degradation products ultimately converge to feed the electron transport system. Then, we will describe each specific pathway for the degradation of monosaccharides, fatty acids, and amino acids.

It is important to bear in mind that the use of each nutrient molecule as a source of energy for ATP synthesis depends on the metabolic status (Fig. 7.2). For example, after a balanced meal, the increase in blood glucose concentration triggers a series of regulatory events that select the carbohydrates as the main class of nutrients to be used as energy source, while fats are stored in the adipose tissue (see Chap. 8). In contrast, during the intervals between the meals or during fasting or low-carbohydrate diets, fatty acids are mobilized from the adipose tissue and become the major source of energy for most of the tissues in the body (see Chap. 9).

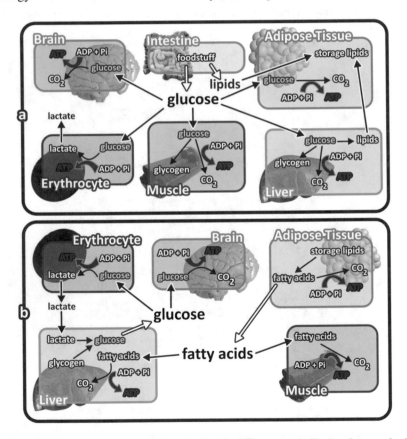

**Fig. 7.2** Overview of the integration of metabolism in different metabolic situations emphasizing the catabolism of different nutrient molecules. (**a**) After a balanced meal, carbohydrate catabolism is the main pathway for ATP synthesis in virtually all the cells in the body. Ingested lipids are stored in the adipose tissue. The excess of glucose is stored as glycogen in muscle and liver cells or converted into lipids in the liver and in the adipose tissue. (**b**) When glucose concentration decreases in the blood, fatty acids are mobilized from the adipose tissue, becoming the main energy source for most of the tissues. Cells lacking mitochondria, as erythrocytes, or isolated from systemic circulation, as those of the central nervous system, are exceptions and maintain glucose catabolism in this situation

## 7.2   Citric Acid Cycle: The Central Pathway for the Oxidation of the Three Classes of Nutrient Molecules

In aerobic metabolism, the products of nutrients' degradation converge to the CAC, a central pathway consisting in eight reactions (Fig. 7.3). In this pathway, the acetyl group of acetyl-CoA resulting from the catabolism of monosaccharides, fatty acids, or amino acids is completely oxidized to $CO_2$ with concomitant reduction of the electron-transporting molecules (NADH and $FADH_2$).

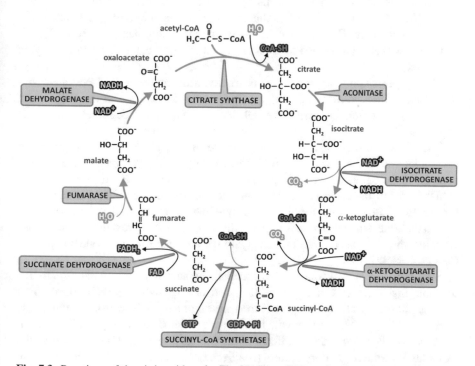

**Fig. 7.3** Reactions of the citric acid cycle. The $NAD^+$ or FAD molecules reduced to NADH or $FADH_2$ are indicated at the respective oxidation reactions, as well as the $CO_2$ released at the decarboxylation reactions. The GTP molecule synthesized by substrate-level phosphorylation is also shown. The names of the enzymes are highlighted in yellow boxes. Remind that succinate dehydrogenase is the Complex II of the electron transport system

## 7.2.1  Citric Acid Cycle Reactions

We start the description of CAC reactions with the condensation of acetyl-CoA and oxaloacetate, which generates the tricarboxylic molecule citrate (Fig. 7.3). The next seven steps, which regenerate oxaloacetate, include four oxidation reactions in which $NAD^+$ or FAD are reduced to NADH or $FADH_2$, whose electrons will be then transferred to $O_2$ in the respiratory chain. Additionally, a GTP or an ATP molecule is directly formed by the mechanism of substrate-level phosphorylation (see Sect. 6.1).

The condensation of the two-carbon acetyl group of acetyl-CoA with the four-carbon molecule oxaloacetate, forming citrate, a six-carbon molecule, is catalyzed by the enzyme citrate synthase. A thioester intermediate, citroyl-CoA, is formed in the active site of the enzyme, and its hydrolysis makes the reaction highly exergonic. The coenzyme A released in this reaction may be recycled in the conversion of pyruvate to acetyl-CoA by pyruvate dehydrogenase complex (see Sect. 7.3).

Citrate is then transformed to isocitrate, in a reaction catalyzed by the enzyme aconitase.

The next reaction is an oxidative decarboxylation catalyzed by isocitrate dehydrogenase, which converts the six-carbon compound isocitrate in the five-carbon molecule α-ketoglutarate, with the release of $CO_2$ and reduction of a $NAD^+$ molecule.

The next step is also an oxidative decarboxylation, the conversion of α-ketoglutarate in the four-carbon molecule succinyl-CoA, catalyzed by the enzyme complex α-ketoglutarate dehydrogenase. In this reaction, $CO_2$ is released, and $NAD^+$ works as the electron acceptor, with the subsequent binding of coenzyme A through a thioester bond forming succinyl-CoA. It is interesting to notice that α-ketoglutarate dehydrogenase is structurally similar to the pyruvate dehydrogenase complex, the enzyme complex that catalyzes the conversion of pyruvate into acetyl-CoA (see Sect. 7.3.1).

Succinyl-CoA is then converted to succinate by succinyl-CoA synthetase. In this reaction, the hydrolysis of the thioester bond of succinyl-CoA with concomitant enzyme phosphorylation is coupled to the transfer of a phosphate group bound to the enzyme to GDP or ADP through the mechanism of substrate-level phosphorylation.

Succinate dehydrogenase oxidizes succinate to fumarate. This enzyme is the Complex II of the respiratory system (see Sect. 6.2), a FAD-linked enzyme through which the electrons from succinate enter in the respiratory chain.

Then, fumarate is hydrated to form malate by the enzyme fumarase.

Finally, oxaloacetate is regenerated by the oxidation of malate catalyzed by malate dehydrogenase, with concomitant reduction of $NAD^+$.

It is important to point out that although the $O_2$ does not participate directly in the CAC, the cycle only operates in aerobic conditions since the oxidized $NAD^+$ and FAD are regenerated in the respiratory chain.

## 7.2.2   Citric Acid Cycle as a Dynamic Pathway

CAC is a pathway fed by molecules other than acetyl-CoA. Likewise, $CO_2$ is not its unique end product. Different four- and five-carbon metabolic intermediates may enter the cycle at different points to be oxidized, as occurs, for instance, with the amino acid glutamate after being converted to α-ketoglutarate by a transamination reaction (for details, see Sect. 7.5). The same occurs with the other amino acids, which are also converted in CAC intermediates.

In addition to receiving metabolic intermediates to be oxidized, CAC has also an anabolic role in metabolism, since many of its intermediates may be withdrawn from the cycle to be used as the precursors of biosynthetic processes (see Box 7.2). Due to its role in both catabolism and anabolism, CAC is considered an amphibolic pathway.

---

**Box 7.2   The Anabolic Role of CAC**

Many CAC intermediates may be used as precursors for biosynthetic pathways (see below figure). Citrate leaves mitochondria yielding acetyl-CoA for the synthesis of fatty acids and other lipids (see Sect. 8.3). α-ketoglutarate and oxaloacetate are the precursors of several amino acids. The synthesis of porphyrins starts with succinyl-CoA. Oxaloacetate generates glucose through the pathway of gluconeogenesis (see Sect. 9.3.1) and is also the precursor of purines and pyrimidines that form the nucleotides. It is important to mention that when CAC intermediates are withdrawn for biosynthetic processes, a replacement mechanism should operate to avoid impairment of the oxidative metabolism. This occurs through replenishing reactions, known as anaplerotic reactions, in which pyruvate or phosphoenolpyruvate (PEP) is carboxylated, generating oxaloacetate or malate (see below figure). The most important of these reactions is the one catalyzed by pyruvate carboxylase (see Sect. 9.3.1). This reaction is highly regulated by acetyl-CoA, so that when this molecule accumulates, indicating that cycle intermediates are not sufficient to perform its complete oxidation, pyruvate is shifted toward the formation of oxaloacetate, increasing the capacity of CAC to oxidize acetyl-CoA. Pyruvate may also be converted to malate by the malic enzyme. The other anaplerotic reaction is that catalyzed by PEP carboxykinase, which generates oxaloacetate from PEP with the transfer of the phosphate from PEP to GDP, forming also a GTP molecule. This reaction occurs in the opposite direction in the gluconeogenesis pathway (see Sect. 9.3.1).

(continued)

**Box 7.2**  (continued)

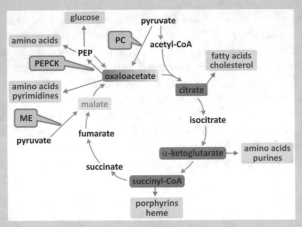

CAC intermediates shown in orange boxes may be used as precursors for the synthesis of fatty acids, cholesterol and steroids, amino acids, purines and pyrimidines, porphyrins, and heme. Anaplerotic reactions (shown by *green arrows*), catalyzed by the enzymes pyruvate carboxylase (PC), phosphoenolpyruvate carboxykinase (PEPCK), or malic enzyme (ME), replenish the cycle, maintaining its oxidative capacity

## 7.2.3   A Historical Overview of CAC Discovery

CAC is also known as the "Krebs cycle" in recognition of the contribution of Sir Hans Krebs to its discovery (see Box 7.3). As stated in Krebs obituary, written by J. R. Quayle, *"his life's work, published in over 360 publications, spanned a whole era of biochemistry, namely, the elucidation of the major part of central intermediary metabolism and its regulation... Perhaps his greatest contribution to biological sciences as a whole was his discovery of the concept of metabolic cycles, how to recognize them, how to establish them unequivocally, and how to delineate and to assess their physiological function."*

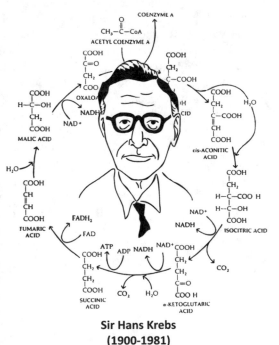

**Sir Hans Krebs**
**(1900-1981)**

**Box 7.3  Sir Hans Krebs and the Discovery of the Citric Acid Cycle**

Sir Hans Krebs was awarded the Nobel Prize in Physiology or Medicine in 1953, for the discovery of CAC. The prize was shared with Fritz Lipmann, who discovered the coenzyme A and its importance for the intermediary metabolism (see Box 7.4). Krebs began the studies that led him to the discovery of CAC working with Otto Warburg (see Box 6.4), who introduced to him the technique of using tissue slices for metabolic studies. This experimental procedure allowed working with intact cells in controlled media with free gas exchange between cells and the medium, an approach only possible before for studies using microorganisms. Using this technique, Krebs could propose for the first time a cycled pathway in the intermediate metabolism, the urea cycle (see Sect. 7.5). In the case of CAC, Krebs based his proposal on several observations carried out during the decade of 1930. In 1935, Albert Szent-Gyorgyi discovered the sequence of reactions from succinate to fumarate to malate to oxaloacetate (see below figure), showing that these dicarboxylic acids present in animal tissues stimulate the $O_2$ consumption. Two years later, Carl Martius and Franz Knoop found the sequence from citrate to α-ketoglutarate to succinate. Krebs integrated these findings adding his own observations about the effect of the addition of tricarboxylic acids to slices of pigeon breast muscle. He found that these acids even in very low concentrations promoted the oxidation of a much higher amount of pyruvate in the muscle slices, suggesting a catalytic effect of these compounds. He also observed that malonate, an inhibitor of succinate dehydrogenase, inhibited the oxidation of pyruvate and that the addition of fumarate to the medium in this condition restored pyruvate consumption (see below figure). Assembling this information, Krebs elegantly showed the cyclic nature of the pathway.

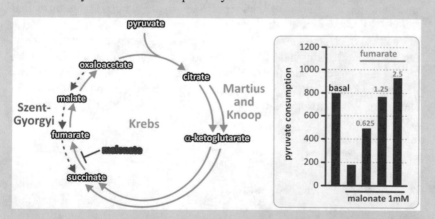

Sequence of reactions described by Szent-Gyorgyi (*purple*), Martius and Knoop (*green*), and Krebs (*blue*). The graph in the right shows the results of the crucial experiment done by Krebs that proved the cycled nature of the reactions for pyruvate oxidation: the inhibition of pyruvate consumption by the addition of malonate (an inhibitor of the enzyme succinate dehydrogenase) was abolished by fumarate addition to the system

When Krebs proposed CAC, in 1937, he thought that citrate was synthesized from oxaloacetate and pyruvate (see Box 7.3). Thus, CAC was seen as a pathway that serves only for carbohydrate oxidation, since pyruvate was known to be the product of carbohydrate metabolization (see Chap. 6). The importance of CAC in the oxidation of other compounds, such as the fatty acids, was only realized after three important discoveries in the middle of the twentieth century: (a) the isolation of coenzyme A (CoA), by Fritz Lipmann, in 1945 (see Box 7.4); (b) the isolation of acetyl-CoA by Feodor Lynen, in 1951; and (c) the subsequent work by Joseph Stern, Severo Ochoa, and Lynen, showing that the reaction between acetyl-CoA and oxaloacetate forming citrate was accompanied by the generation of stoichiometric amounts of sulfhydryl groups, indicating that acetyl-CoA was the molecule that donated the acetyl group to oxaloacetate.

Since acetyl-CoA is the product of the oxidation of pyruvate (and carbohydrates) (see Sect. 7.3) and fatty acids (see Sect. 7.4), as well as being it generated in the metabolization of some amino acids (see Sect. 7.5), one can recognize acetyl-CoA as the convergence point of the oxidative metabolism (see Fig. 7.1).

---

**Box 7.4  Fritz Lipmann and the Discovery of Coenzyme A**

The discovery of coenzyme A led Fritz Lipmann to be awarded the Nobel Prize in Physiology or Medicine in 1953. Lipmann dedicated many years of his scientific career working on the process of ATP synthesis. In 1941, he published a landmark paper in which he introduced the term "energy-rich phosphate bond" and established the basis for the understanding of the substrate-level phosphorylation mechanism of ATP synthesis (see Chap. 6). As a follow-up to these studies, Lipmann got interested in the capacity of some molecules to transfer acetyl groups, which he thought would be an important step in some biosynthetic processes. In the search for this "activated acetate," Lipmann found a dialyzable, heat-stable factor present in all tissues he studied, which could not be replaced by any other known cofactor. He purified and identified this new coenzyme and named it coenzyme A, in which "A" is related to the activation of acetate. Coenzyme A is composed of phosphoadenosine diphosphate linked through a phosphoester bond to pantothenic acid, which is in turn bound through an amide linkage to β-mercaptoethylamine (see below figure).

Chemical structure of the coenzyme A

## 7.2.4   *Regulation of Citric Acid Cycle*

The rate of CAC is basically regulated by the flux through the dehydrogenases—isocitrate dehydrogenase, α-ketoglutarate dehydrogenase, succinate dehydrogenase, and malate dehydrogenase. The activities of the dehydrogenases, in turn, are determined by the mitochondrial ratios $NADH/NAD^+$ and $FADH_2/FAD$, which depend on the ATP/ADP ratio. Therefore, when the concentration of ATP increases, the electron flow through the electron transport chain slows, and the $NADH/NAD^+$ and $FADH_2/FAD$ ratios increase, leading to a decrease in CAC rate. On the other hand, when ATP is used, the increase in ADP levels favors the electron transport, increasing NADH and $FADH_2$ oxidation and then enhancing CAC rate.

## 7.3   Catabolism of Carbohydrates

Human diet contains different carbohydrates, including polysaccharides, such as starch and glycogen; disaccharides, such as sucrose, lactose, and maltose; and monosaccharides, such as glucose and fructose. Polysaccharides and disaccharides are broken in the digestive tract to their respective monosaccharides, which together with the free monosaccharides present in the diet reach the bloodstream and enter the cells. Inside the cells, monosaccharides are converted to a phosphorylated derivative that is degraded in two pyruvate molecules through a metabolic pathway named glycolysis (Fig. 7.4). It is important to remind that in addition to being the pathway for carbohydrate degradation in aerobiosis, glycolysis also affords ATP synthesis in the absence of oxygen (see Chap. 6).

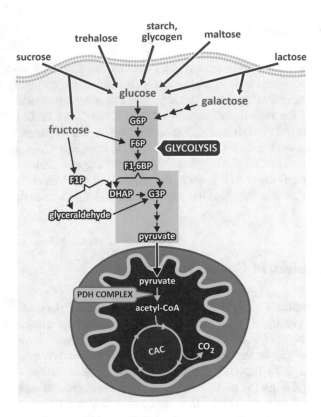

**Fig. 7.4** Dietary carbohydrates are digested and reach the bloodstream mainly as monosaccharides that enter the cells where they are phosphorylated and degraded to pyruvate through the pathway named glycolysis. Pyruvate is transported into the mitochondria where it is converted to acetyl-CoA, which in turn is completely oxidized in the CAC. *G6P* glucose-6-phosphate, *F6P* fructose-6-phosphate, *F1,6BP* fructose-1,6-bisphosphate, *F1P* fructose-1-phosphate, *G3P* glyceraldehyde-3-phosphate, *DHAP* dihydroxyacetone phosphate

### 7.3.1 Carbohydrate Oxidation Reactions

Glycolysis itself is composed of ten reactions, which are presented in detail in Sect. 6.1.3. Thus, in this section we will focus our attention in the conversion of pyruvate to acetyl- CoA, the connecting point between glycolysis and CAC.

Pyruvate molecules, generated in the cytosol as the product of either glycolysis or the degradation of some amino acids, need to be transported to the mitochondrial matrix where it is converted to acetyl-CoA (Fig. 7.4).

The oxidative decarboxylation of pyruvate to form acetyl-CoA is catalyzed by one of the largest cellular multienzyme complexes known, the pyruvate dehydrogenase complex (PDH), with about 50 nm in diameter and a molecular mass of almost 10 MDa (Fig. 7.5a). Eukaryotic PDH is composed of multiple copies of four distinct proteins, three of them presenting enzymatic activity, E1 (pyruvate decarboxylase), E2 (dihydrolipoyl acetyltransferase), E3 (dihydrolipoyl dehydrogenase), and the fourth, E3-binding protein (E3BP), which mediates E3 integration in the complex. Additionally, regulatory kinases and phosphatases are also associated to PDH complex.

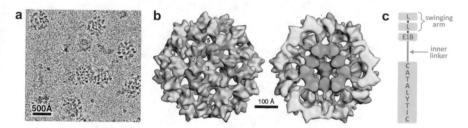

**Fig. 7.5** (**a**) Cryoelectron microscopy image of bovine PDH showing the multienzyme complex with ~500 Å. (**b**) Reconstructed structure of the reconstituted complex and its half section (*right*) showing the dodecahedral core formed by E2 catalytic domains (*green*) and E2 inner linker (*blue*) that connects the core to the E1–E3 outer shell (*yellow*). (**c**) Diagrammatic representation of E2 four structural domains connected by linkers: the catalytic domain, the E1-binding domain (E1B), and the two lipoyl domains (L) that form the "swinging arm." ((**a**) and (**b**): reproduced with permission from Zhou et al. Proc. Natl. Acad. Sci. USA 98:14802–14807, 2001)

Mammalian E1 are tetrameric, composed of two $\alpha$- and two $\beta$-subunits. It is estimated that 20–30 $\alpha_2\beta_2$ E1 tetramers and 6–12 E3 homodimers form PDH complex outer shell (Fig. 7.5b). This outer shell is tightly associated to the enzyme core, which is formed by 60 E2 subunits arranged as a pentagonal dodecahedron of 20 E2 trimers (Fig. 7.5b). E2 core acts as a scaffold for the complex organization.

Human E2 contains four distinct domains connected by linkers (Fig. 7.5c). The C-terminal half comprises the catalytic domain, which self-associates forming the organized core. The N-terminal half is extended, generating an annular gap between the core and the E2E3-outer shell. It contains, besides the E1-binding domain, a "swinging arm" in which two highly flexible lipoyl domains allow an efficient transfer of the reaction intermediates among the active sites. Human E3BP structural organization is very similar to that of E2, and it is suggested that its addition to the core structure replaces part of the E2 subunits generating a final core structure containing up to 48 E2 and about 12 E3BP subunits.

The reaction catalyzed by PDH complex consists of five steps in which the intermediates remain bound to the enzyme components (Fig. 7.6). Firstly, E1 catalyzes the thiamine diphosphate (TPP)-dependent decarboxylation of pyruvate with release of $CO_2$ and formation of a hydroxyethyl derivative bound to TPP. This is followed by the transfer of two electrons and the remaining acetyl group to the oxidized lipoyl group linked to the lipoyl domain of E2, a reaction also catalyzed by E1. The flexible linker that connects the lipoyl domain in E2 allows the movement of the acetyl group to the catalytic site of E2, where its transfer to CoA is catalyzed, generating acetyl-CoA that is released from the enzyme. Then, the lipoyl group, which is reduced at this point, moves to the E3 catalytic site where it is oxidized with the reduction of the enzyme-bound FAD. Finally, the electrons are transferred from E3-bound $FADH_2$ to $NAD^+$, forming NADH. The flexible lipoyl domains allow the reaction intermediates to be channeled through the catalytic sites during the catalytic cycle, in what is called "swinging-arm mechanism" (Fig. 7.6).

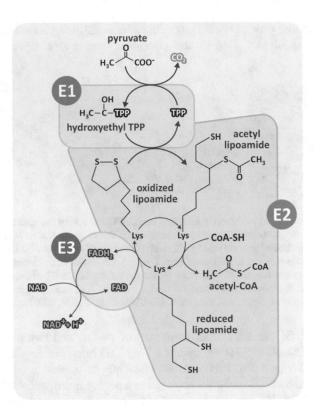

**Fig. 7.6** Schematic representation of the catalytic cycle of PDH complex. Step (1) consists of the TPP-dependent decarboxylation of pyruvate, catalyzed by E1, in which $CO_2$ is released and a hydroxyethyl derivative remains bound to TPP. In the step (2), also catalyzed by E1, two electrons reduce the lipoyl group bound to E2, and the remaining acetyl group is transferred to one of the reduced sulfhydryl groups. In step (3) the acetyl group is transferred to the coenzyme A, forming acetyl-CoA. In step (4) the reduced sulfhydryl transferred the electrons to E3-bound FAD. In step (5) the electrons are transferred from $FADH_2$ to $NAD^+$, generating NADH

## 7.3.2   Regulation of Pyruvate Conversion to Acetyl-CoA

The PDH complex is mainly regulated by phosphorylation and dephosphorylation, catalyzed by pyruvate dehydrogenase kinase (PDK) and pyruvate dehydrogenase phosphatase (PDP), respectively. Both enzymes are bound to the lipoyl domain of E2 in PDH complex.

Three phosphorylation sites were identified in the human enzyme, all of them located in E1: site 1 in Ser264, site 2 in Ser271, and site 3 in Ser203. Phosphorylation leads to enzyme inactivation, being the phosphorylation of site 1 the one that causes the major effect.

Four PDK isoforms with tissue-specific expression have been identified. PDK1 is mostly expressed in heart, PDK2 in most tissues, PDK3 in testis, and PDK4 in heart and skeletal muscles. Two PDP isoforms, PDP1 and PDP2, both requiring $Mg^{2+}$ for their activity, have been identified.

# 7.4 Catabolism of Lipids

Fatty acids are the main source of energy for most of human cells. They are stored in the adipose tissue as triacylglycerol (TAG), a molecule formed by one glycerol esterified to three fatty acid molecules (see Sect. 3.1). TAGs may be obtained from the diet or can be synthesized from carbohydrates and accumulate in adipocytes after being transported from the intestine or from the liver by specialized lipoproteins (see Box 7.5).

**Box 7.5 Lipoproteins and Lipid Transport in the Body**
The transport of lipids in the body fluids needs special requirements due to their chemical properties (see Sect. 3.1). Most lipids are insoluble in aqueous media and would form drops when circulating, causing fat embolisms. Other lipids have amphipathic properties that would affect cellular membrane integrity. These problems can be overcome because lipids circulate associated to lipoprotein complexes. The lipoproteins differ in their size and composition according to their origin (for details on the different types of lipoproteins, see Sect. 3.1.2). Lipids obtained from the diet are incorporated in lipoproteins named chylomicrons, which are released in the lymphatic system to reach the blood. The chylomicrons transport TAGs mainly to the adipose tissue, where TAGs are stored in lipid droplets (see below figure). The excess of carbohydrates is converted to lipids in the liver and in the adipose tissue itself (see Chap. 8). TAGs synthesized in the liver are transported to the adipose tissue associated to the very-low-density lipoprotein (VLDL).

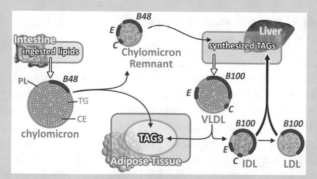

Transport of lipids in the body fluids through different lipoproteins. Lipoproteins contain a core of neutral lipids, mainly triacylglycerols and cholesterol esters (TG and CE, represented as yellow and blue circles, respectively), surrounded by a phospholipid (PL) monolayer with which different proteins associate (represented as the pink structures—apolipoproteins are named with letters, sometimes followed by numbers that correspond to their molecular mass)

In the adipocytes, TAGs accumulate in a large lipid droplet that occupies most of the cell volume (Fig. 7.7a). This organelle buds from the endoplasmic reticulum (ER) after accumulation of lipids in between ER membrane leaflets. Lipid droplets consist in a hydrophobic core containing mainly neutral lipids, as TAGs and cholesterol esters, surrounded by a phospholipid monolayer with associated proteins (Fig. 7.7b). The most abundant protein in the lipid droplets from adipocytes is named perilipin, which, besides its structural function, has a central role in the activation of lipolysis (see next section).

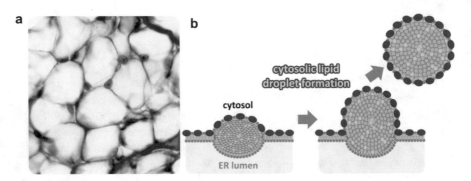

**Fig. 7.7** (**a**) Section of white adipose tissue showing the adipocytes with a large lipid droplet inside. (Figure reproduced with permission of Instituto de Histologia e Biologia do Desenvolvimento, Faculdade de Medicina, Universidade de Lisboa, FMUL) (**b**) Model of lipid droplet formation. Synthesized neutral lipids accumulate between the leaflets of the endoplasmic reticulum membrane until the lipid droplet buds as an independent organelle limited by a phospholipid monolayer with associated proteins among which the most abundant are the perilipins

### 7.4.1   TAG Mobilization and Fatty Acid Transport in the Bloodstream

Hormones, mainly catecholamines, and probably also glucagon, secreted in response of hypoglycemia or other stress situations, trigger a signaling pathway in the adipocytes that culminates with the activation of the protein kinase A (PKA) (see Chaps. 9 and 10). PKA phosphorylates two adipocyte proteins essential for TAG mobilization, perilipin, the major protein on the surface of adipocyte lipid droplets, and the enzyme hormone-sensitive lipase (HSL), which catalyzes the hydrolysis of the ester linkages of the TAG molecules. Phosphorylated perilipin recruits HSL to the surface of the lipid droplet, coordinating the access of HSL to TAG molecules. HSL in turn becomes active upon phosphorylation, converting TAGs to glycerol and fatty acids, which are released in the bloodstream (Fig. 7.8). Fatty acids circulate associated with serum albumin, reaching different tissues in the body where they are used as energy source.

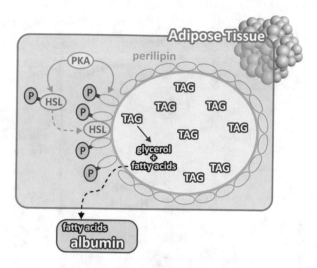

**Fig. 7.8** TAG mobilization in adipocyte. PKA, activated through hormone signaling pathways, phosphorylates the enzyme HSL and the lipid droplet surface protein perilipin. Phosphorylation activates HSL and induces a conformational rearrangement in perilipin that allows access of HSL to TAG molecules inside the lipid droplets. HSL catalyzes the hydrolysis of TAG in glycerol and fatty acids, which are released to the bloodstream. Fatty acids circulate associated with serum albumin. The "P" symbol in a pink circle indicates the phosphate group

## 7.4.2  Activation of Fatty Acids

The oxidation of fatty acids occurs inside the mitochondria. To be transported to the mitochondrial matrix, long-chain fatty acids should be activated in an ATP-dependent reaction catalyzed by the acyl-CoA synthetases, in which fatty acid carboxyl group is linked to the thiol group of the coenzyme A through a thioester bond forming acyl-CoA. In this reaction, ATP is cleaved forming AMP and pyrophosphate. Pyrophosphate is in turn promptly hydrolyzed by the action of the inorganic pyrophosphatase, yielding two phosphate molecules. The AMP is temporally linked to the fatty acyl group, which is then transferred to coenzyme A (Fig. 7.9).

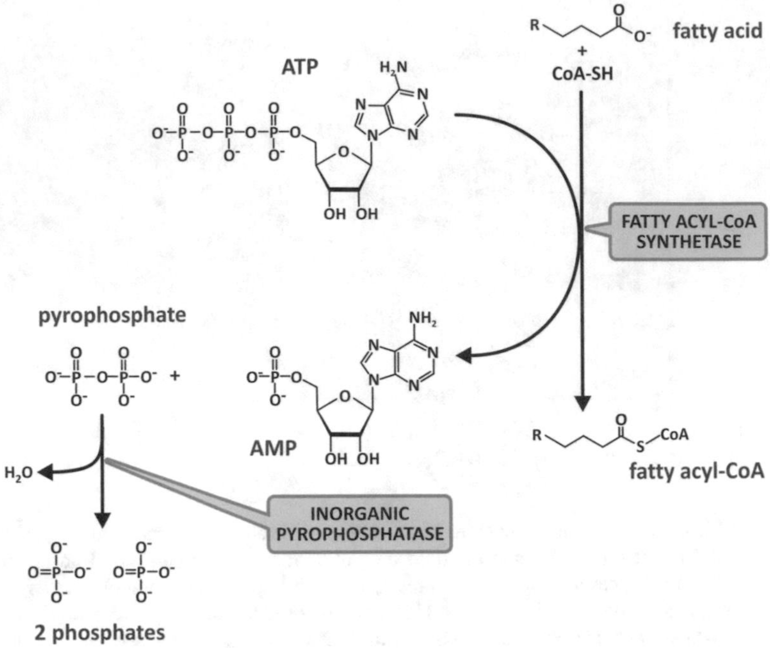

**Fig. 7.9** Activation of fatty acids through the formation of acyl-CoA. Acyl-CoA synthetase links the carboxyl group of the fatty acid molecule to the thiol group of the coenzyme A, in a reaction dependent on ATP hydrolysis in AMP and PPi with an intermediate step in which the fatty acyl group in bound to AMP. PPi is hydrolyzed by the enzyme inorganic pyrophosphatase, driving the entire process toward fatty acyl-CoA formation

Acyl-CoA synthetases consist in a family of isozymes that differ in their subcellular localization and specificity for fatty acids with different chain length. The long-chain acyl-CoA synthetase is associated either to the outer mitochondrial membrane, or peroxisomes, or ER membrane. It acts efficiently for 10- to 20-carbon saturated fatty acids or 16- to 20-carbon unsaturated fatty acids, which are the most common fatty acids used as energy source by the cells. Fatty acids with less than ten carbon atoms cross freely the mitochondrial membrane and become activated in the mitochondrial matrix by the short- or medium-chain acyl-CoA synthetases.

### 7.4.3  Fatty Acid Transport into Mitochondria

The fatty acyl-CoA formed in the cytosol cannot be directly transported into the mitochondrial matrix since the inner mitochondrial membrane is impermeable to coenzyme A. Thus, to be transported, the fatty acyl group is firstly transferred from coenzyme A to a carnitine molecule (Fig. 7.10a). This reaction is catalyzed by an enzyme associated to outer face of the outer mitochondrial membrane named carnitine/palmitoyl transferase I (CPT-I), which generates acyl-carnitine and free coenzyme A in the cytosol (Fig. 7.10b). The acyl-carnitine crosses the inner mitochondrial membrane through the carnitine-acyl-carnitine transporter, a transmembrane protein that exchanges acyl-carnitine with carnitine (Fig. 7.10b). Finally, the acyl group of the acyl-carnitine is transferred to a mitochondrial coenzyme A molecule, in a reaction catalyzed by an enzyme located in the inner side of the inner mitochondrial membrane, the carnitine/palmitoyl transferase II (CPT-II) (Fig. 7.10b).

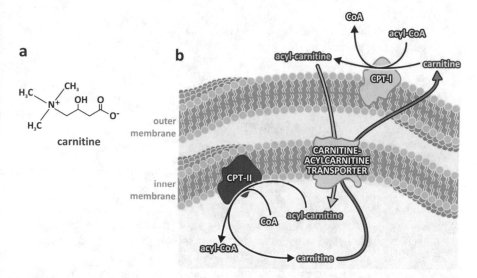

**Fig. 7.10**  (**a**) Carnitine molecular structure. (**b**) Schematic representation of the fatty acyl-CoA transport across the mitochondrial membranes. CPT-I catalyzes the transfer of the acyl group from acyl-CoA to carnitine, generating acyl-carnitine and free coenzyme A in the cytosol. The carnitine-acyl-carnitine transporter in the inner mitochondrial membrane exchanges the acyl-carnitine with free carnitine, generated in the mitochondrial matrix by the action of CPT-II, which catalyzes the transfer of the acyl group of the acyl-carnitine to coenzyme A, regenerating the acyl-CoA in the matrix

### 7.4.4    β-Oxidation: The Pathway for Fatty Acid Degradation

The main pathway of fatty acid oxidation is known as β-oxidation, a process in which two-carbon units are progressively removed from the carboxyl end of the fatty acid molecule. For the oxidation of saturated fatty acids, the process consists in four reactions that generate acetyl-CoA and the acyl-CoA molecule shortened by two carbons, with the concomitant reduction of FAD and NAD$^+$. These reactions are continuously repeated until the acyl-CoA is entirely oxidized to acetyl-CoA.

The experiments carried out by Franz Knoop in the beginning of the 1900s were decisive to the elucidation of the pathway for fatty acid oxidation (Box 7.6). These experiments revealed that the β-carbon atom of the fatty acid molecule is oxidized, leading to the cleavage of the bond between carbons β and α. This made this pathway to be known as β-oxidation. Nowadays the lipid carbons nomenclature has changed (see Sect. 3.1), but the pathway designation remains.

---

**Box 7.6  Franz Knoop and the Discovery of the β-Oxidation of Fatty Acids**

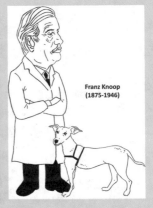

Franz Knoop
(1875-1946)

In 1904, Franz Knoop proposed the basis for the fatty acid oxidation through the conception of a very bright experiment that probably represents the first time in which a tracer was used to follow a compound along its metabolic pathway. Knoop fed dogs with fatty acid molecules of different chain length in which the terminal methyl groups has been linked to a phenyl ring. Then, he observed the phenyl-containing products in the urine of these dogs. He found that, when the ingested fatty acid had an even number of carbon atoms, the final product was always phenylacetate, excreted as phenylaceturic acid, a glycine conjugate (see below figure). On the other hand, when the fatty acid molecules contained an odd number of carbon atoms, they yielded benzoate, excreted as hippuric acid, also a glycine conjugate (see below figure). From these results, Knoop concluded that the fatty acids were degraded by the oxidation of the β-carbon followed by the cleavage of the chemical bond between carbons α and β, in a sequential process that generated molecules of two carbons, which Knoop assumed to be acetate. Now we know that the two-carbon product of fatty acid oxidation is acetyl-CoA.

| diet | product of oxidation | excretion |
|---|---|---|

odd-chain fatty acid        $(n+1)C_2$  benzoic acid            hippuric acid

even-chain fatty acid       $(n+1)C_2$  phenyl-acetic acid      phenylaceturic acid

Representation of the chemical reaction occurring during Knoop's experiment

### 7.4.4.1  Oxidation of Saturated Fatty Acids with Even Number of Carbons Atoms

The four reactions of the β-oxidation of saturated fatty acids are repeated in successive cycles. Each cycle consists in a sequence of a FAD-dependent oxidation, a hydration, a NAD-dependent oxidation, and a thiolysis.

The first reaction is the oxidation of the acyl-CoA to trans-enoyl-CoA with the concomitant reduction of an enzyme-bound FAD molecule (Fig. 7.11). This reaction is catalyzed by four types of acyl-CoA dehydrogenases (ACADs). ACADs present distinct but overlapping fatty acid chain length specificities, being named as very long-, long-, medium-, and small-chain acyl-CoA dehydrogenases (VLCAD, LCAD, MCAD, and SCAD), which show optimum activities for fatty acids of 16, 14, 8, and 4 carbons atoms, respectively. VLCAD is a homodimer bound to the inner mitochondrial membrane, whereas the other three ACADs are homotetramers located in the mitochondrial matrix. The resulting $FADH_2$ bound to the ACADs is reoxidized by the electron transferring flavoprotein (ETF), which in turn transfers the electrons to another flavoprotein, the ETF/ubiquinone oxidoreductase. Finally, the latter enzyme transfers the electrons to the ubiquinone in the electron transport chain (see Sect. 6.2.3).

The three other reactions of β-oxidation are also catalyzed by families of enzymes with chain-length specificities. The reactions are the hydration of the trans-enoyl-CoA to β-hydroxyacyl-CoA, catalyzed by enoyl-CoA hydratases; the oxidation of the β-hydroxyacyl-CoA to β-ketoacyl-CoA, with concurrent NAD reduction, catalyzed by β-hydroxyacyl-CoA dehydrogenases; and the cleavage of the bond between carbons α and β of the β-ketoacyl-CoA with the transfer of the resultant acyl residue shortened in two carbons to coenzyme A, catalyzed by thiolases (Fig. 7.11a).

The enzymes with long-chain specificity form a complex associated to the inner mitochondrial membrane, the trifunctional β-oxidation complex (Fig. 7.11b). This complex channels the substrates from one enzyme to the other so that the intermediates are not released in the mitochondrial matrix. Thus, in the initial stage β-oxidation, the long-chain intermediates are processed by the membrane-associated enzymes (VLCAD and the trifunctional β-oxidation complex), which lose affinity to the acyl-CoA molecules when they reach about 12 to 8 carbons.

The enzymes with preference for medium-chain and short-chain acyl-CoA are soluble mitochondrial matrix proteins that perform the β-oxidation reactions until the acyl-CoA molecules are entirely oxidized to acetyl-CoA, which may finally enter CAC.

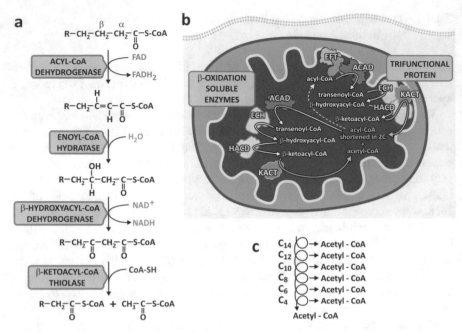

**Fig. 7.11** (**a**) Reactions of β-oxidation. (**b**) Schematic representation of the β-oxidation enzymes, with those with long-chain specificity associated to the inner mitochondrial membrane and those with short-chain specificity, localized in the matrix. Long-chain fatty acyl-CoA are oxidized by the membrane-associated enzymes until being shortened to about 12 to 8 carbons atoms, when they become substrates to the matrix enzymes. (**c**) Scheme representing the sequential β-oxidation reactions resulting in the conversion of a fatty acid of n atoms of carbons in n/2 acetyl-CoA molecules

The oxidation of a fatty acid of 16 carbon atoms results in eight acetyl-CoA, seven FADH$_2$, and seven NADH molecules, after seven cycles of the four β-oxidation reactions. It is important to point out that β-oxidation reactions themselves reduce the electron carriers FAD and NAD, which in turn transfer the electrons to the respiratory chain components. FADH$_2$ associated to ACADs transfers the electrons to ubiquinone via ETF, and NADH is oxidized by the NADH dehydrogenase complex or Complex I of the respiratory chain (see Sect. 6.2.3). Thus, even if acetyl-CoA does not enter CAC, the β-oxidation pathway itself results in ATP synthesis through oxidative phosphorylation.

It is important to mention that β-oxidation of some fatty acids may occur in another organelle, the peroxisome. In this case the pathway is not directly associated to ATP synthesis but to the formation of hydrogen peroxide (H$_2$O$_2$) (see Box 7.7).

**Box 7.7 β-Oxidation in Peroxisomes**

Peroxisomes are cellular organelles (see below figure) whose major function is to break down very-long-chain fatty acids through β-oxidation. Peroxisomes contain acyl-CoA oxidases instead of ACADs. The acyl-CoA oxidase is also a FAD-associated enzyme, but the electrons transferred to FAD via the oxidation of the acyl-CoA molecule directly reduce $O_2$, generating $H_2O_2$. Thus, $O_2$ is consumed but ATP is not produced in these organelles. Since catalase is present in the peroxisomes, $H_2O_2$ is readily converted to $H_2O$. The other steps of peroxisomal β-oxidation are the same as those of mitochondrial β-oxidation, but the peroxisomal enzymes are specific for long-chain and branched-chain acyl-CoAs. Thus, the products of β-oxidation in peroxisomes are short-chain acyl-CoAs (mainly octanoyl-CoA), acetyl-CoA, and NADH. The short-chain acyl-CoA and the acetyl-CoA molecules are then sent to mitochondria through a carnitine transport system also present in peroxisomes.

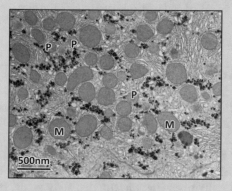

Transmission electron micrograph of a liver thin section showing the peroxisomes (P) and the mitochondria (M). (The image is a courtesy from Prof. Marlene Benchimol, from the Federal University of Rio de Janeiro, Brazil)

The fate of acetyl-CoA molecules generated in β-oxidation depends on the metabolic situation or on the cellular type: they may be completely oxidized in the CAC, may be used in the synthesis of the ketone bodies (see Sect. 7.4.6), or may act as substrate for the synthesis of some amino acids.

### 7.4.4.2 Oxidation of Odd-Chain Fatty Acids

Human diet is mainly composed of fatty acids with an even number of carbon atoms, but fatty acids with odd number of carbons also occur, especially in diets containing vegetables and marine organisms. Odd-chain fatty acids are also oxidized through the β-oxidation pathway. The difference from the oxidation of the even-chain fatty acids is that the last cycle of reactions generates one acetyl-CoA and one propionyl-CoA (a three-carbon molecule), instead of two acetyl-CoA molecules. Propionyl-CoA is then converted to succinyl-CoA, a CAC intermediate, through a sequence of three reactions that occur in the mitochondrial matrix (Fig. 7.12).

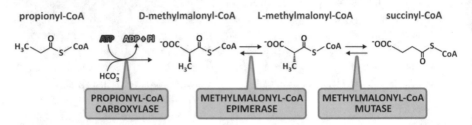

**Fig. 7.12** The propionyl-CoA formed in the last cycle of the β-oxidation of the odd-chain fatty acids is converted to succinyl-CoA by an ATP-dependent carboxylation reaction followed by structural rearrangements

### 7.4.4.3  Oxidation of Unsaturated Fatty Acids

Human diet also contains monounsaturated fatty acids, such as oleic acid, or poly-unsaturated fatty acids, such as linoleic acid and linolenic acid. These molecules are oxidized through β-oxidation pathway until reaching the double bond. At this point, if the double bond is in trans-configuration and occurs between carbons α and β, this intermediate will enter β-oxidation in the enoyl-CoA hydratase reaction. However, in the naturally occurring unsaturated fatty acids, the double bonds are generally in the cis configuration. Thus, an additional enzyme is required: an enoyl-CoA isomerase that converts the cis-double bond in a trans-double bond. The product, a trans-enoyl-CoA, can then proceed in the β-oxidation pathway. Additionally, if the double bond is not at the right position, a reductase is also required.

## 7.4.5  Regulation of Fatty Acid Oxidation

The major regulation point of fatty acid oxidation is the reaction catalyzed by CPTI, which is inhibited by malonyl-CoA. Malonyl-CoA is the product of the carboxylation of acetyl-CoA, a reaction catalyzed by the enzyme acetyl-CoA carboxylase (ACC). ACC reaction is generally activated when blood glucose concentration is high (see Sect. 8.3.1). Thus, when glucose is highly available as energy source, fatty acid oxidation is inhibited, whereas when blood glucose concentration decreases, the intracellular levels of malonyl-CoA diminishes allowing fatty acyl-CoA transport into mitochondrial matrix where they are oxidized.

### 7.4.6 Fatty Acid Conversion to Ketone Bodies

The acetyl-CoA molecules produced in β-oxidation do not always enter CAC; their fate may differ depending on the metabolic situation. As mentioned in the beginning of this chapter (see Sect. 7.1), fatty acids, mobilized from the adipose tissue, become the major source of energy for most of the tissues in the body when the glucose concentration in the blood starts to decrease during the periods in between the meals or in fasting situations. In these situations, the liver plays an essential role in the maintenance of glycemia by synthesizing glucose through a pathway known as gluconeogenesis (see Chap. 9).

In gluconeogenesis, CAC intermediates act as substrates for the synthesis of glucose, and thus, these molecules, especially oxaloacetate, are removed from the cycle (Fig. 7.13a). The low levels of oxaloacetate decrease the capacity of CAC to oxidize acetyl-CoA, which accumulates in the mitochondrial matrix. Acetyl-CoA, even when in excess, continues to be produced to ensure ATP synthesis in the liver (remember that the oxidation steps of β-oxidation generate $FADH_2$ and NADH, allowing ATP synthesis through oxidative phosphorylation without requirement of CAC). Thus, to recycle mitochondrial coenzyme A pool, acetyl-CoA is converted to acetoacetate, β-hydroxybutyrate, and acetone (Fig. 7.13b), molecules known as the ketone bodies, a nomenclature rather inappropriate since these molecules are not insoluble as the word "bodies" would suggest, neither the β-hydroxybutyrate has a ketone function in its molecular structure. In fact, two of the "ketone bodies" are carboxylic acids, making this functional group as important as ketone group in "ketone bodies."

The ketone bodies are released in the bloodstream, and several extrahepatic tissues, especially the brain, kidney, heart, and skeletal muscle, may use these molecules as energy source (Fig. 7.13a). In these tissues, β-hydroxybutyrate and acetoacetate are converted to acetyl-CoA (Fig. 7.13c), which enters CAC being completely oxidized. The use of ketone bodies by the brain only occurs during long periods of fasting, as it will be deeply discussed in Chap. 9.

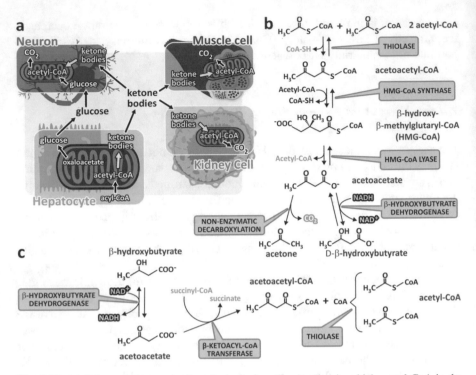

**Fig. 7.13** (**a**) Schematic representation of a hypoglycemia situation in which acetyl-CoA in the liver is converted to ketone bodies, which are released in the bloodstream and used as energy source by several tissues. The *yellow arrow* in the hepatocyte highlights the pathway of ketone bodies synthesis, occurring only in the liver; the *green arrows* highlight the pathway of degradation of ketone bodies, which occurs in different extrahepatic tissues. (**b**) Reactions for the synthesis of the ketone bodies, a pathway restricted to the liver. (**c**) Reactions for the degradation of the ketone bodies, which occur in several extrahepatic tissues, including the skeletal and cardiac muscles, kidneys, and brain. The names of the enzymes are highlighted in *yellow boxes*

## 7.5    Catabolism of Amino Acids

The catabolism of the amino acids varies greatly depending on the metabolic situation, but it is calculated that about 15% of the resting energy expenditure comes from amino acid oxidation in different tissues, in particular the liver, the main site of amino acid metabolization, and muscle, especially in the case of the branched-chain amino acids.

It is important to bear in mind that amino acids cannot be stored in the body, using the word storage in the strict sense of its meaning. Proteins in the body play a number of functions, and even the proteins that occur in very high amounts, such as the muscle contractile proteins, cannot rigorously be seen as an energy storage. However, proteins are constantly being synthesized and degraded, both as a result of the normal process of protein turnover as well as due to a controlled degradation of a specific protein, and it is believed that between 200 and 300 g of protein are formed and destroyed each day in an adult of average size.

Additionally, the protein content in the diet is also a determining factor for their use as energy sources. Thus, the balance between availability of amino acids and their requirement for protein synthesis may result in an excess of free amino acids, which then become available for oxidative degradation.

## 7.5.1   An Overview of the Amino Acid Catabolism

As occurs in carbohydrate and fatty acid catabolism, the products of amino acid degradation converge to CAC (see Fig. 7.1). However, the fact that there are at least 20 different amino acids as well as the presence of nitrogen in their molecular structure makes the catabolism of amino acids more complex than that of glucose and fatty acids.

Before the metabolic use of amino acids, nitrogen should be separated from the amino acid carbon skeleton to be excreted or recycled for the biosynthesis of nitrogenous compounds, such as the nucleotides and other amino acids (Fig. 7.14). Nitrogen is removed from the amino acid molecules in the form of ammonia ($NH_3$) that in solution, at physiological pH, exists as the ammonium ion ($NH_4^+$). This occurs through deamination reactions, which must be precisely regulated since ammonia is very toxic. The liver is the main site of amino acid metabolization, where $NH_4^+$ is converted to urea, the nitrogenous compound that is excreted in humans.

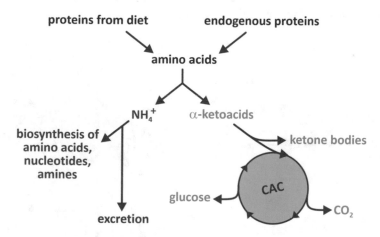

**Fig. 7.14**  Overview of amino acid metabolization in the liver

Distinct degradation pathways to each carbon skeleton generated from different amino acids are necessary, but all amino acids are ultimately converted to pyruvate, acetyl-CoA, or some of the intermediates of CAC, which can be then completely oxidized. In the liver, the products of amino acid degradation may also be converted in glucose or ketone bodies that are released in the bloodstream and used by other tissues (Fig. 7.14).

## 7.5.2  Amino Acid Metabolism in the Liver

The liver is the major site of amino acid metabolization. The pathway that converts $NH_4^+$ in urea, the nitrogenous compound that is excreted in humans, occurs only in the liver, as well as the main metabolic strategy used to separate the nitrogen from the carbon skeleton of the amino acids, which consists in the combination of a transamination reaction followed by a deamination reaction, a process known as transdeamination.

### 7.5.2.1  Oxidative Deamination of the Amino Acids

The evolutionary strategy to cope with amino acid diversity relied on the conversion of all the amino acids into a specific one, Glu, which then undergoes a highly specific and efficient deamination. The oxidative deamination of Glu is, thus, of central importance for the amino acid metabolism and consists in the main deamination reaction in the liver. This reaction generates α-ketoglutarate and $NH_4^+$ and is catalyzed by Glu dehydrogenase (GDH), a hexameric enzyme restricted to liver mitochondria that displays an unusual ability to use both $NAD^+$ and $NADP^+$ as the electron acceptor (Fig. 7.15). The fact that the first steps of the conversion of $NH_4^+$ in urea also occur in the mitochondria of the hepatocytes prevents ammonia dissemination in the body, thus avoiding its toxic effects. Since the GDH uses only Glu, a mechanism for the conversion of all other amino acids into Glu is required. This occurs through transamination reactions.

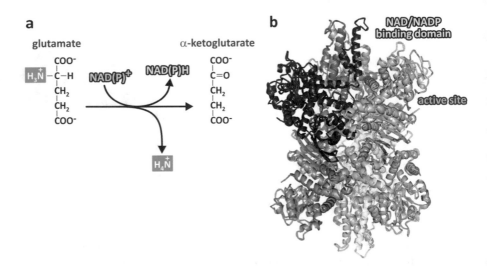

**Fig. 7.15**  (a) Reaction catalyzed by GDH. (b) Structure of the hexameric human GDH with each subunit represented in a different color (PDB 1L1F), showing the NAD/NADP-binding domain and the active site

## 7.5.2.2   Amino Acid Interconversion: The Transamination Reactions

The enzymes transaminases or aminotransferases convert amino acids in their respective α-ketoacids by transferring the amino group of one amino acid to an α-ketoacid, in a reaction dependent on the cofactor pyridoxal phosphate, or vitamin B6, which is covalently bound to the enzyme. This allows the amino acids to be interconverted (Fig. 7.16).

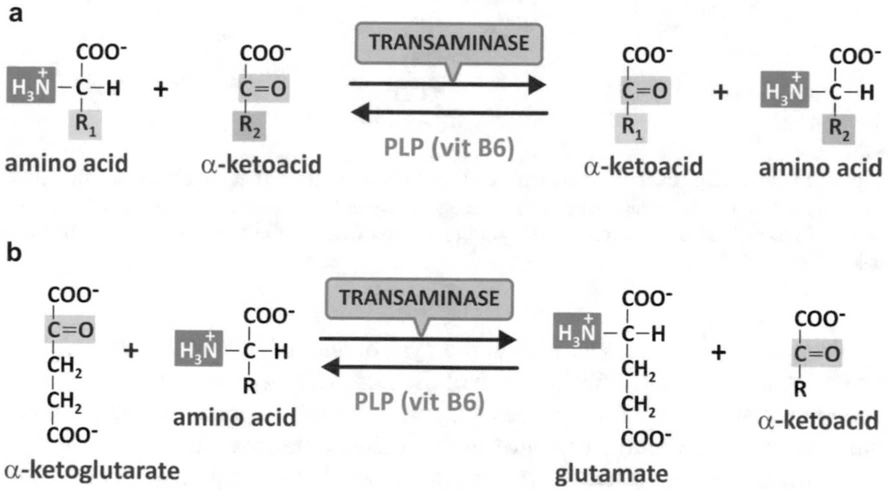

**Fig. 7.16** (**a**) A general scheme of the transamination reaction. Radical groups R1 and R2 are swopped between an amino acid and an α-ketoacid. (**b**) Transamination reaction that makes all the amino acids to be converted into Glu, which uses α-ketoglutarate as the α-ketoacid that accepts the amino group of the amino acid, generating Glu and the α-ketoacid correspondent to the amino acid

The main α-ketoacid that acts as the amino group acceptor is the α-ketoglutarate, which is the product of the GDH reaction and whose transamination generates Glu. Thus, the transamination reactions ultimately convert all the amino acids in Glu, which in turn is deaminated by GDH (Fig. 7.17). Through this strategy, the amino groups of all the amino acids are separated from their carbon skeleton.

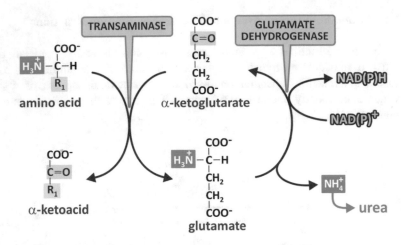

**Fig. 7.17** Coupling of transamination and oxidative deamination reactions in the liver. Transamination of each amino acid with α-ketoglutarate generates Glu, which in turn is the substrate of GDH. This enzyme removes the amino group of Glu in the form of NH$_4$$^+$, regenerating α-ketoglutarate

Transaminases are present in both the cytosol and mitochondria, but their high activities in the cytosol imply that most of the formed Glu is transported to the mitochondrial matrix where GDH is located. It is important to point out that the transaminases are also widely distributed in other tissues besides the liver, where their main function is amino acid interconversion instead of Glu formation, since GDH and urea synthesis are restricted to the liver.

### 7.5.2.3  Pathways for the Metabolism of the Amino Acid Carbon Skeletons

As mentioned in the beginning of this section, the existence of at least 20 different amino acids implies a number of distinct pathways for their metabolization. Here we will give only a general overview of this process. The detailed pathways for the metabolization of each amino acid may be found in more specialized literature.

For some amino acids, a single transamination step directly generates an intermediate of the central metabolic pathway. This is the case of the transamination of Ala, Asp, or Glu that forms directly pyruvate, oxaloacetate, or α-ketoglutarate, respectively. For other amino acids, a complex set of reactions is required. Additionally, some amino acids are converted to other amino acids before removal of the amino group, such as Phe, which is converted to Tyr, and Gln, His, Pro, and Arg, which are converted to Glu.

But regardless whether the metabolization pathway is very simple or complex, the end product is one of the following six intermediates of the central metabolic pathway: pyruvate, acetyl-CoA, or one of the four intermediates of CAC, α-ketoglutarate, succinyl-CoA, fumarate, or oxaloacetate (Fig. 7.18).

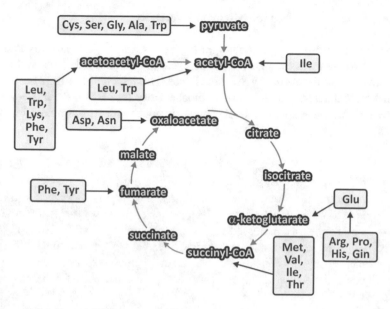

**Fig. 7.18**  Entry of deaminated amino acids into CAC

Depending on the situation, the carbon skeleton generated by transdeamination in the liver, instead of being directly oxidized in hepatocytes, may be converted to glucose or ketone bodies that are released into the bloodstream to be used as energy source by other tissues (see Chap. 9).

The complete oxidation of amino acids' carbon skeletons in CAC depends on the point they enter the cycle. When pyruvate and/or acetyl-CoA are generated, the complete oxidation can proceed directly. On the other hand, metabolites that enter into CAC in positions other than acetyl-CoA increase the amount of CAC intermediates, meaning that they increase the oxidative capacity of the cycle, but this does not result in the actual oxidation of the entering compound to $CO_2$. In this case, the intermediates may be converted to oxaloacetate, which can be converted into phosphoenol-pyruvate (PEP) by the enzyme PEP carboxykinase (see Sect. 9.3.2). PEP is then converted to pyruvate by the glycolytic enzyme pyruvate kinase (see Sect. 6.1.3) and then to acetyl-CoA, which can now be completely oxidized through CAC reactions.

Glucose is synthesized from the amino acids that generate CAC intermediates or pyruvate, which are converted in oxaloacetate that in turn follows the pathway known as gluconeogenesis (see Chap. 9). These amino acids are classified as gluconeogenic amino acids. Amino acids that generate only acetyl-CoA (Leu and Lys) cannot be converted to glucose. They are converted to ketone bodies (see Sect. 7.4.6), being classified as ketogenic amino acids.

### 7.5.2.4 Synthesis of Urea for Nitrogen Excretion

As commented in Box 7.3, in the beginning of this chapter, the pathway that converts $NH_4^+$ to urea was the first metabolic cycle to be described. This landmark in biochemistry was discovered by Sir Hans Krebs. This took place in 1932 but remains an illustrative timeless plan of work in biochemistry (see Box 7.8). Indeed, Krebs may be seen nowadays as one of the most, if not the most, prominent biochemist ever.

---

**Box 7.8 Sir Hans Krebs and the Discovery of the Urea Cycle**

The motivation that led Krebs to study the synthesis of urea came from the experience of working with tissue slices in the period he was at Otto Warburg laboratory. Krebs was impressed by the new possibilities that the use of tissue slices technique opened, allowing approaches that before were only feasible using microorganisms. Since Warburg used this technique to study only degradative metabolic pathways (glycolysis and respiration), Krebs decided to evaluate whether it would also be possible to explore biosynthetic processes, at that time restricted to the use of entire organisms and perfused organs. He chose urea synthesis because it seemed to occur through a simple pathway at a very high rate. Firstly, Krebs developed a simple and reproducible method to measure urea and created a new reaction medium that simulated blood plasma. With this system, Krebs and Kurt Henseleit, a medical student whose M.D. thesis was supervised by Krebs on this subject, designed a working plan that included the quantification of urea synthesis using as substrates the combination of ammonia with different amino acids. These experiments resulted in two main observations with nonobvious connection: (a) the unexpected finding that ornithine together with ammonia yields urea in exceptionally high rates, and (b) the presence, in the liver, of high levels of the enzyme arginase, an enzyme that converts the amino acid Arg into ornithine and urea. The connection between these findings came with the subsequent observation that for each ornithine added to the medium, 20 urea molecules were formed and that the total urea nitrogen was equivalent to the ammonia disappearance.

This led to the crucial proposal that ornithine would act as a catalyst: it was combined to ammonia generating an intermediate in the production of Arg, whose hydrolysis recovered ornithine levels in the system, yielding urea as the final product. The next steps were the search for the intermediates between ornithine and Arg. At the same time, two biochemists independently identified citrulline as a molecule that would fulfil this gap. Indeed, when Krebs tested the effect of citrulline on the urea synthesis from ammonia, it worked in the same way as ornithine. So, Krebs proposed the first metabolic cycle as the pathway of urea synthesis. The finding of additional intermediates later completed the cycle as we know it today (see below figure).

(continued)

**Box 7.8**  (continued)

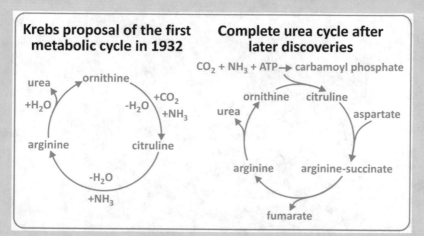

| **Krebs proposal of the first metabolic cycle in 1932** | **Complete urea cycle after later discoveries** |

Krebs proposal of urea cycle (*left*) and the complete pathway after the later discoveries (*right*)

Urea is synthesized almost exclusively in the liver, from which it is released in the bloodstream, reaching the kidneys for excretion in the urine. The molecule of urea contains two nitrogen atoms, one coming from the $NH_4^+$ produced by the reactions catalyzed by GDH or glutaminase within the mitochondria and the other coming from the amino group of the amino acid Asp, generated in the hepatocyte cytosol by transamination of any amino acid with oxaloacetate. In a series of reactions, these two amino groups ultimately become part of an Arg molecule, which after hydrolysis forms urea and ornithine (see figure in Box 7.8). The detailed sequence of reactions that comprise the pathway is represented in Fig. 7.19 and described below.

Firstly, the $NH_4^+$ reacts with $HCO_3^-$ and two molecules of ATP in the mitochondrial matrix, in a reaction catalyzed by the enzyme carbamoyl phosphate synthase I, yielding two ADPs and one Pi and the intermediate carbamoyl phosphate, which enters the cycle by transferring its carbamoyl group to ornithine, forming citrulline and Pi, in a reaction catalyzed by ornithine transcarbamoylase. Citrulline leaves mitochondria and reacts with Asp in an ATP-dependent reaction catalyzed by the enzyme argininosuccinate synthase, which generates argininosuccinate, AMP, and PPi. The enzyme argininosuccinase cleaves the argininosuccinate into fumarate and arginine, which finally is hydrolyzed to urea and ornithine by the enzyme arginase (Fig. 7.19). It is important to point out that the oxaloacetate molecules removed from CAC to be transaminated, generating Asp, are replenished since the reaction of argininosuccinase releases fumarate, which enters CAC to regenerate oxaloacetate.

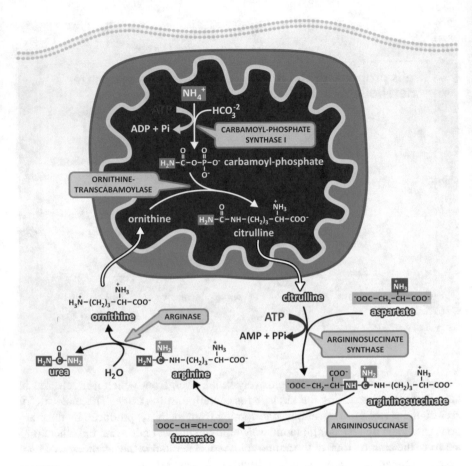

**Fig. 7.19** The urea cycle. Urea is synthesized in the liver through a cyclic pathway with part of the reactions occurring in the mitochondria and part in the cytoplasm. One of the nitrogen atoms of urea molecule comes from $NH_4^+$ produced by the GDH or glutaminase reactions (highlighted with a *blue box*) within the mitochondria. The other comes from the amino group of Asp (highlighted with a *purple box*). The names of the enzymes are highlighted in *yellow boxes*

## 7.5.3 Amino Acid Metabolism in Other Tissues

Although much more information is available about amino acid metabolism in the liver, other tissues also oxidize amino acids, with a special emphasis on the use of branched-chain amino acids by muscle.

Amino acid deamination reactions in different tissues form $NH_4^+$ that is generally combined to Glu yielding Gln, in a reaction catalyzed by the enzyme glutamine synthetase (Fig. 7.20). Gln is then released in the bloodstream, being the main transporter of amino groups in blood. This explains why Gln concentration in blood is much higher than that of other amino acids. Gln is taken up mainly by the liver and the kidney, where the enzyme glutaminase removes the amido group from Gln, generating Glu and $NH_4^+$. In the liver, $NH_4^+$ enters the urea cycle, and Glu undergoes oxidative deamination by GDH (see previous section). In kidney, $NH_4^+$ is directly excreted (Fig. 7.20).

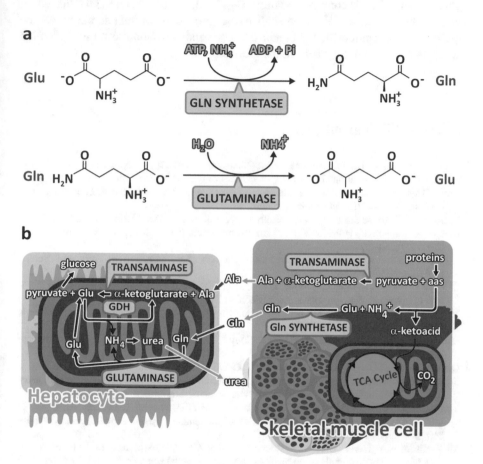

**Fig. 7.20**  (**a**) Reactions catalyzed by the enzyme glutamine synthetase and by the enzyme glutaminase. (**b**) Integration of amino acid metabolism. In extrahepatic tissues (especially skeletal muscle), amino acids from diet, intracellular protein turnover, or proteolysis undergo: (i) transamination with pyruvate generating Ala; or (ii) deamination forming $NH_4^+$ that is combined to Glu yielding Gln. Gln and Ala are released in the bloodstream reaching the liver, where Ala is transaminated with α-ketoglutarate, forming pyruvate and Glu. Glu undergoes oxidative deamination, generating $NH_4^+$ that enters the urea cycle. Pyruvate is converted in glucose or is completely oxidized. Gln is deaminated yielding $NH_4^+$ that enters the urea cycle and Glu that undergoes oxidative deamination. Urea is released in the bloodstream and excreted by the kidneys. The names of the enzymes are highlighted in *yellow boxes*

Amino acids in muscle also undergo transamination with pyruvate, generating Ala. This is especially important in fasting, when the amino acids released from the degradation of contractile proteins are converted to Ala, which in turn is released in the bloodstream reaching the liver, where pyruvate is regenerated (and converted to glucose) and $NH_4^+$ is converted to urea (Fig. 7.20; see also Sect. 9.3.4). Indeed, in fasting, the proportions of amino acids released from the muscle do not reflect the muscle protein composition. Ala and Gln correspond to about 60% of the amino acid released from muscle cells in the bloodstream.

## Selected Bibliography

Carpenter KJ (1994) The life and times of W. O. Atwater (1844-1907). J Nutr 124:S1707–S1714

Ghisla S (2004) β-Oxidation of fatty acids. A century of discovery. Eur J Biochem 271:459–461

Krebs HA (1953) The citric acid cycle. Nobel lecture. http://www.nobelprize.org/nobel_prizes/medicine/laureates/1953/krebs-lecture.html

Krebs HA (1973) The discovery of the ornithine cycle of urea synthesis. Biochem Educ 1:19–23

Kresge N, Simoni RD, Hill RL (2005a) Fritz Lipmann and the discovery of coenzyme a. J Biol Chem 280:164–166

Kresge N, Simoni RD, Hill RL (2005b) Severo Ochoa's contributions to the citric acid cycle. J Biol Chem 280:138–140

Li M, Li C, Allen A, Stanley CA, Smith TJ (2012) The structure and allosteric regulation of mammalian glutamate dehydrogenase. Arch Biochem Biophys 519:69–80

Manchester KL (1998) Albert Szent-Gyorgyi and the unravelling of biological oxidation. Trends Biochem Sci 23:37–40

McAndrew RP, Wang Y, Mohsen A, He M, Vockley J, Kim JP (2008) Structural basis for substrate fatty acyl chain specificity. J Biol Chem 283:9435–9443

Quayle JR (1982) Obituary: Sir Hans Krebs, 1900-1981. J Gen Microbiol 128:2215–2220

Schultz H (2002) Oxidation of fatty acids in eukaryotes. In: Vance DE, Vance JE (eds) Biochemistry of lipids, lipoproteins and membranes. Elsevier Science, Amsterdam, pp 127–150

Vijayakrishnan S, Kelly SM, Gilbert RJ, Callow P, Bhella D, Forsyth T, Lindsay JG, Byron O (2010) Solution structure and characterisation of the human pyruvate dehydrogenase complex core assembly. J Mol Biol 399:71–93

Yu X, Hiromasa Y, Tsen H, Stoops JK, Roche TE, Zhou ZH (2008) Structures of the human pyruvate dehydrogenase complex cores: a highly conserved catalytic center with flexible N-terminal domains. Structure 16:104–114

## Challenging Case 7.1: Into the Wilderness of Metabolic Deregulation

### *Source*

This case is based on the book titled *Into the Wild*, published in 1996 by Jon Krakauer, an international bestseller translated to tens of different languages, and adapted into a film, directed by Sean Penn, in 2007. It tells the journey of Chris McCandless across North America in pursuit of simple living with minimal supplies. McCandless' tragic story also inspired documentaries later on.

*Into the Wild* book cover, reproduced with permission of Pan Macmillan

### *Case Description*

Christopher Johnson McCandless, who also went by the name Alexander Supertramp, was born in California, in February 1968, and died in Stampede Trail, in Alaska, in August 1992. After graduating from Emory University (USA), with very good grades, in 1990, Chris McCandless felt compelled to abandon his possessions, promising career, and family to set off for a cross-country journey, reaching the Denali National Park and Preserve in the beginning of 1992.

Chris McCandless's story was brilliantly recovered in an article written by Jon Krakauer for The New Yorker magazine.[1] On September 6, 1992, the decomposed body of Christopher McCandless was found by hunters inside an abandoned bus. Taped to the door was an almost illegible note on a page torn from a novel by Nikolai Gogol:

> *attention possible visitors.*
> *s.o.s.*
> *i need your help. i am injured, near death, and too weak to hike out of here. i am all alone, this is no joke. in the name of god, please remain to save me. i am out collecting berries close by and shall return this evening. thank you,*
> *chris mccandless*
> *august ?*

According to McCandless's diary, it appeared that he had been dead for 19 days. A document found among his possessions, issued 8 months before, indicated that he was 24 years old and weighed 140 pounds (63.5 kg). His body was sent to autopsy

---

[1] Krakauer. How Chris McCandless died. The New Yorker. https://www.newyorker.com/books/page-turner/how-chris-mccandless-died.

which indicated it weighed 67 pounds (30.4 kg) and lacked visible subcutaneous fat. The probable cause of death, according to the autopsy report, was starvation.

Until now, the actual cause Chris McCandless death is unclear and a matter of debate and investigation. According to Jon Krakauer's research, he died because of an excessive intake of roots and seeds of either wild potato (*Hedysarum alpinum*), which became a staple of his daily diet, or wild sweet pea, *Hedysarum mackenzii*. Wild potato is safe to eat, and Jon Krakauer speculated that McCandless had mistakenly consumed the seeds of wild sweet pea—a toxic plant that is hard to distinguish from wild potato. Jon Krakauer took samples of both plants (roots and seeds) and had them screened for toxic compounds. Indeed, it seemed that their seeds contained an alkaloid, perhaps swainsonine, known to inhibit glycoprotein metabolism in animals, leading to neurological disorders, loss of appetite[2], and altered erythrocytes.[3] In the *Into the Wild* novel, the toxic alkaloid swainsonine was assigned to be the cause of Chris McCandless's death.

However, 17 years after the book was published, Jon Krakauer himself, in his article in The New Yorker,[1] revisited the cause of Chris McCandless's death, which had been challenged by several botanists, researchers, and other professionals who had different hypotheses, not necessarily scientifically tested. At that time, Jon Krakauer stated death was caused not by an alkaloid but, rather, an amino acid—β-N-oxalyl-L-alpha-beta diaminoproprionic acid, referred to as β-ODAP. Indeed, other chemical analyses ran on the same plant samples—roots and seeds—revealed the presence of this amino acid in significant high amounts. β-ODAP is considered a toxin, and it is the cause of neurolathyrism, characterized by lack of strength in or inability to move the lower limbs. Yet, the story of the cause of death of McCandless was not over: 2 years later, Jon Krakauer published another article in the same magazine, The New Yorker, re-revisiting this subject and came to the conclusion that what killed Chris McCandless was indeed an amino acid but L-canavanine instead of β-ODAP.[4]

## Questions

1. Compare the structure of all three compounds suspected of having killed Chris McCandless (swainsonine, β-ODAP, and L-canavanine) to similar molecules used in human metabolism and elaborate a hypothesis to explain why these molecules are capable of interfering in human metabolism.

---

[2] Stegelmeier BL et al (1995) The lesions of locoweed (*Astragalus mollissimus*), swainsonine, and castanospermine in rats. Vet Pathol 32:289–298.

[3] Chui D et al (1997) Alpha-mannosidase-II deficiency results in dyserythropoiesis and unveils an alternate pathway in oligosaccharide biosynthesis. Cell 90:157–167.

[4] Krakauser. How Chris McCandless died: an update. The New Yorker. https://www.newyorker.com/books/page-turner/chris-mccandless-died-update.

2. Would simple starvation not caused by any culpable chemical compound be ruled out as possible cause of death of Chris McCandless?
3. When Chris McCandless' body was found by the hunters, it weighted about 33 kg less than he weighed when he left his house. Based on the fact that the adipose tissue of a 70 kg man stores about 12 kg of triglycerides and the daily energy expenditure in rest is about 2200 kcal, calculate how long Chris McCandless would have survived to starvation.

## 7.6.4 Biochemical Insights

1. There is notorious structural resemblance between all the phytotoxins (i.e., toxins from plants) in this case and the endogenous molecules of human metabolism:

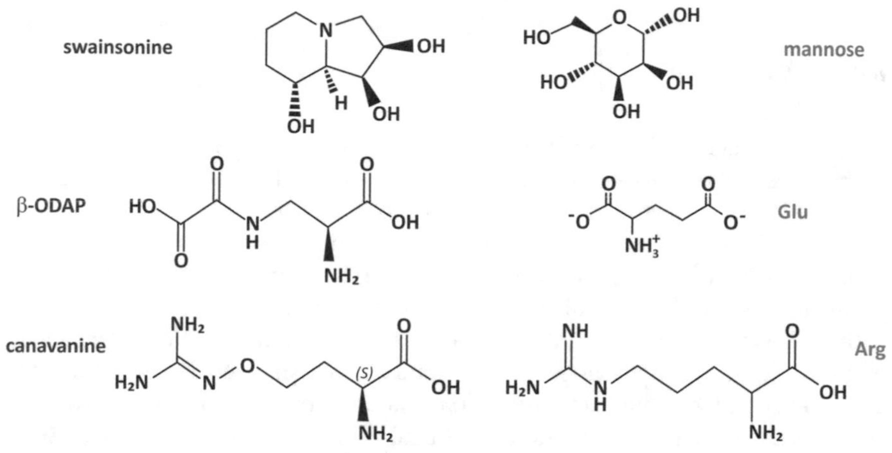

Chemical structures of phytotoxins swainsonine, β-ODAP, L-canavanine, and MCPA and the respective similar endogenous metabolite of human metabolism mannose, Glu, Arg, and carnitine

It is therefore likely that these phytotoxins may interact with molecules that bind their "human counterparts," namely, proteins that bind, catalyze, or transport these molecules, i.e., receptors, enzymes, or transporters. The binding of phytotoxins may therefore hinder or alter the action of the proteins responsible for important functions in the human body. This is the biochemical ground for their noxious action. For swainsonine, for instance, it is known how the toxin inserts in the active center of the Golgi α-mannosidase II (where mannose, the natural substrate of the enzyme, also binds) and the exact interactions it establishes with the amino acid residues of the protein. Swainsonine "sequesters" the active center of the enzyme, which therefore will not be available to process mannose.

2. From the data available, one cannot exclude the hypothesis that death occurred due to simple starvation caused by the lack of enough nutrients in general as well as essential nutrients present in meat and some vegetables in particular. The presence of phytotoxins in some of his improvised food in the wild may have contributed to worsen his condition, but general body weakness followed by starvation as the cause of death may not be ruled out.
3. Chris McCandless weighed 63.5 kg when he left his house, so we can estimate a triglyceride store of 10.9 kg. Since the net energy yielded by the metabolism of lipids is 8.9 kcal/g (see Sect. 7.1), McCandless adipose tissue stored 97,000 kcal, which would allow 44 days of survival if used as the single energy source. However, as it will be discussed in Chap. 9, some tissues need glucose instead of lipids as the main energy source. In starvation, body proteins are used to produce glucose (see Sect. 9.3.2), so usually death by starvation occurs not only by the depletion of the energy stores but also by the weakening of essential muscles such as the diaphragm.

## Final Discussion

Although we address this case from the biochemical perspective, it is impossible not to think that the primary cause of death of Chris McCandless was his reckless attitude and lack of proper planning for his dangerous endeavor. He clearly underestimated the dangers of going alone into the wild without a contingency plan. His dream of natural life in the wild became a terrifying journey into a nightmare of degradation and suffering, which could have been avoided.

In addition to biochemistry, this case also raises important questions in the realm of public health worth discussing. For example, lathyrism is epidemic in Ethiopia. This is because the population relies on the consumption of grass pea (*L. sativus*), a species very tolerant to the prevailing drought in this African region.[5] There is an urgent need to address chemical and biochemical methods to decrease the amount of β-ODAP in the food preparation. Importantly, β-ODAP spontaneously isomerizes to the corresponding alpha-isomer. This isomerization reaction occurs during the preparation of *dal*, a dish prepared from split peas. The α-ODAP isomer seems to be neither acutely nor chronically neurotoxic, probably because the different geometry of the molecule is not enough to bind to glutamate receptors.

Detoxification of grass pea through aqueous leaching of the neurotoxin is an important method of food preparation that reduces the risk of development of lathyrism, as pointed out by Peter S. Spencer and Valerie S. Palmer from Third World Medical Research Foundation, Portland, Oregon, USA.[6] Steeping dehusked seeds in hot water for several hours and boiling the seeds in water remove 70–80% of the

[5] Tekle-Haimanot R et al (1993) Pattern of Lathyrus sativus (grass pea) consumption and beta-N-oxalyl-α-β-diaminoproprionic acid (β-ODAP) content of food samples in the lathyrism endemic region of northwest ethiopia. Nutrition Res 13:1113–1126.
[6] Spencer PS, Palmer VS (2003) Lathyrism: aqueous leaching reduces grass-pea neurotoxicity. Lancet 362:1775–1776.

neurotoxin into the supernatant, which is discarded. Simulated kitchen experiments show that steeping grass pea in a large volume of water for 3 min and decanting the excess water leaches about 30% of the neurotoxin. Water during drought is therefore needed for safe food preparation as well as hydration. Appropriate techniques of food preparation are important to reduce to a minimum the neurotoxic hazards associated with consumption of grass pea. Thus, education programs designed to increase awareness of these simple methods are likely to reduce risk for lathyrism.

*This case had the collaboration of Dr. Tatiana El-Bacha (Instituto de Nutrição Josué de Castro, Universidade Federal do Rio de Janeiro, Brazil).*

## Challenging Case 7.2: A Mystic Man with an Aura of Flowers

### Source

Raul Seixas is one of the greatest icons of Brazilian rock, whose life and career is documented in the film "Raul – O Início, o Fim e o Meio" (Raul— The Beginning, the End and In Between), by Walter Carvalho.[7] The documentary unveils Raul Seixas' many facets, his partnership with Paulo Coelho (currently a worldwide bestselling author), and the testimonies of his five wives and shows how his eccentric personality and very rich discography continue to mobilize legions of fans 30 years after his death. The title of the film is a reference to the final verses of the song Gîtâ, composed together with Paulo Coelho, alluding to one of the sacred texts of Hinduism, the Bhagavad-Gita.[8]

Raul Seixas. (Image reproduced from Wikimedia Commons under a CC-BY license)

### Case Description

Raul Seixas was born in 1945 and died from chronic pancreatitis at age of 44. His discography consists of 21 albums released from 1968 to 1989, the year of his death, besides dozens of posthumous, live, and compilations albums. In the 1970s, Raul Seixas became very popular by the satirical and sarcastic lyrics of his songs, full of references to historical and fictional personalities. At this time, he began a partnership with Paulo Coelho, and under the influence of the controversial English mystic Aleister Crowley, idealized the "Alternative Society," an anarchist community based on the premise of "Do what thou wilt' shall be the whole of the Law," as cited in lyrics of the song "Sociedade Alternativa,"[9] which means "Alternative Society" in Portuguese. The "Alternative Society" project was considered subversive during the Brazilian military dictatorship (1964–1985), and after detention by government repressive agents, Raul Seixas got into self-exiling in the USA. Although composed in 1974, "Sociedade Alternativa" is still a reference in the Raul Seixas' legacy. For example, in 2013, to the delirium of Raul Seixas' fans, the famous American singer

---

[7] https://www.youtube.com/watch?v=Nw7lombOTt0&t=109s.

[8] version in Portuguese: https://www.youtube.com/watch?v=RezuYwS0ngI; and a modified version in English: https://www.youtube.com/watch?v=6-6Li3CKflQ.

[9] https://www.youtube.com/watch?v=isGCtqPdcg0.

Bruce Springsteen opened his participation in the "Rock in Rio," interpreting this song in Portuguese.[10]

Possibly, the eccentric and unquiet personality of Raul Seixas, as illustrated by the verses of the song "Metamorfose Ambulante" ("Ambulant Metamorphosis," reproduced below), pushed him to mysticism and a quest for new approaches in music and esthetics, which ultimately led him to drug addiction, including an overt problem of severe alcoholism.

| "Quero dizer agora o oposto do que eu disse antes<br>Eu prefiro ser essa metamorfose ambulante<br>Do que ter aquela velha opinião formada sobre tudo…" | (I want to say now the opposite of what I said before<br>I rather be this ambulant metamorphosis<br>Than having a settled opinion on everything…) |
| --- | --- |

At the age of 12, Raul Seixas created his first rock'n'roll group named "Raulzito e os Panteras." Even being a teenager, he started drinking and smoking cigarettes compulsively. Later, at the time of the "Alternative Society" project, Paulo Coelho introduced him to a plethora of recreational drugs, as the writer himself tells in an interview reproduced in the film "Raul – O Início, o Fim e o Meio":

> "I introduced Raul to all the drugs. All, all, just all. From cannabis to acid; from hallucinogenic mushrooms' tea to mandrax. I was the bad boy. It was part of my culture."

Curiously, Raul was not a youngster anymore as he was 25 years old at the time.

Despite having experienced many different drugs, Raul Seixas had used constantly cocaine and alcohol. Alcoholism interfered with his personal and professional life, ruining his marriages and being the direct cause for the cancellation or abrupt interruption of several of his live shows. In some of them, Raul barely could stand up so drunk he was. Discredit among producers, colleagues, and the public arose rapidly. In the 1980s, Raul was hospitalized several times, even compulsorily, to attempt treatment of alcohol addiction. However, not only could he not quit the alcoholism but also he began to inhale ethyl ether, aggravating the deterioration of his already fragile condition.

Alcoholism and smoking are the two most significant risk factors for the development of chronic pancreatitis.[11] Pancreas inflammation progressively causes pancreatic fibrosis, leading to the loss of the exocrine and endocrine functions of the organ. In 1979, Seixas' pancreatitis reached such a point that the singer had to have two thirds of his pancreas removed. One of the consequences of chronic pancreatitis is the development of diabetes due to the inability to produce insulin. This condition may become very severe and can ultimately lead to death. Doctors told Raul's family in the year of the surgery that his estimated life span was no longer than 10 years. They were right: Raul Seixas died in 1989. Thousands of desperate fans in grief were present at his burial ceremony.

---

[10] https://www.youtube.com/watch?v=OuF6DiDDxWs.

[11] Sankaran SJ et al (2015) Frequency of progression from acute to chronic pancreatitis and risk factors: a meta-analysis. Gastroenterology 149:1490–1500.

## Questions

1. In the film "Raul – O Início, o Fim e o Meio," both the third and fifth wives of Raul Seixas, Tania Menna Barreto and Lena Coutinho, commented that Raul exhaled a sweet flower scent.

   Tania Menna Barreto: "… he emanated [like a] flower. The entire room was impregnated with flower. It was a very strong flower scent. Almost unbreathable."

   Lena Coutinho: "Raul smelled sweet, which can be seen as a poetic thing. But he had diabetes. He had sweet breath."

   Raul Seixas developed a severe pancreatitis, probably due to alcoholism as well as the use of drugs such as cocaine. Based on the information that the development of pancreatitis leads to an impairment of insulin production/action, stimulating of TAG mobilization in the adipose tissue, hypothesize the cause for Raul Seixas' flower scent.

2. Albert Lester Lehninger (1917–1986), the author of the classical biochemistry textbook *Principles of Biochemistry*, was a North American scientist who made fundamental contributions in the field of bioenergetics. In one of his experiments, published in 1945, Lehninger analyzed the correlation between the formation of the ketone "body" acetoacetate and the oxidation of fatty acids.[12]

   With an experimental design similar to the one used by Sir Hans Krebs in the classic experiment that led to the elucidation of CAC, the cycle that is also known as Krebs cycle (see Box 7.3), Lehninger measured oxygen uptake and the formation of acetoacetate and citrate after addition of octanoate (a fatty acid with eight carbon atoms) or pyruvate in a rat liver preparation (see the results, shown in the following table).

| Substrate | Fumarate | $O_2$ uptake ($\mu$mol) | Change in acetoacetate ($\mu$mol) | Citrate formation ($\mu$mol) |
|-----------|----------|------------------------|-----------------------------------|------------------------------|
| None      | –        | 0.04                   | 0.00                              | 0.15                         |
| None      | +        | 3.97                   | 0.13                              | 1.33                         |
| Octanoate | –        | 9.64                   | 6.38                              | 0.21                         |
| Octanoate | +        | 14.24                  | 2.54                              | 2.61                         |
| Pyruvate  | –        | 7.46                   | 7.05                              | 0.20                         |
| Pyruvate  | +        | 11.56                  | 2.19                              | 3.20                         |

Taking into account that, as in the Krebs' experiment design, all the measurements are performed in the presence of malonate, an inhibitor of succinate dehydrogenase, elaborate on:

   (a) The differences observed by Lehninger in acetoacetate and citrate formation, in the absence and in the presence of fumarate, when octanoate was used as the respiratory substrate (compare the results shown in rows 3 and 4 in the table).

---

[12] Lehninger AL (1945) Fatty acid oxidation and the Krebs tricarboxylic acid cycle. J Biol Chem 161:413–414.

(b) Why, even with CAC inhibited (i.e., without addition of fumarate), oxygen consumption greatly increases when octanoate is added (compare the results shown in rows 3 and 1 in the table).

(c) The increased oxygen uptake when pyruvate was used as a substrate without the addition of fumarate (compare row 5 to row 1 in the table).

3. Existence of an "ethanol metabolism" in humans, fit for moderate intake, may be puzzling at first glance as ethanol has been part of the human diet only since the fabrication of fermented drinks, which occurred very recently on an evolutionary time scale.

(a) What could be the explanation for the occurrence of ethanol metabolism in humans?

(b) Research in scientific literature the main routes of metabolic clearance of ethanol, and elaborate on the alcohol dehydrogenase reaction pathway, which oxidases ethanol to acetaldehyde, using the principles learned in Chap. 4. Correlate your findings to the biochemical principles underlying hangovers.

## Biochemical Insights

1. In adipocytes, storage and mobilization of TAGs are controlled by the antagonic action of insulin or glucagon/adrenaline, respectively. A main player in this control is the hormone-sensitive lipase (HSL), the enzyme that catalyzes the hydrolysis of the ester linkages of the TAG molecules, converting TAGs to glycerol and fatty acids. This enzyme is activated by glucagon- or adrenaline-induced phosphorylation and inhibited by insulin-mediated dephosphorylation. Pancreatitis, and the consequent impairment of insulin action, leads fatty acids to be released into the bloodstream, where they associate with serum albumin and circulate, reaching different tissues in the body to be used as energy source.

In the liver, impairment of insulin action favors glucose synthesis through gluconeogenesis, even with diabetic patients experiencing several periods of high blood glucose concentration. Since oxaloacetate and malate are used as substrates for the synthesis of glucose (see Sect. 9.3.2), the acetyl-CoA molecules produced in hepatic β-oxidation cannot be further oxidized in CAC, accumulating in the mitochondrial matrix, where they are converted to the "ketone bodies" acetoacetate, β-hydroxybutyrate, and acetone (see Fig. 7.13).

Thus, the uncontrolled lipolysis and the consequent high rates of hepatic β-oxidation result in the production of high levels of ketone bodies. Among them, acetoacetate and β-hydroxybutyrate are acidic molecules that may reach concentrations that exceed the blood buffering capacity, leading to a dangerous decrease in blood pH, known as ketoacidosis (see Final Discussion). Acetone, on the other hand, is volatile, being responsible for the characteristic fruity (flower) smell of diabetic patients. As commented in the editorial of one of the 2014 volumes of the periodical Annals of Translational Medicine: "*In 1920*

*(2 years before the discovery of insulin) … the average lifespan of a patient with diabetes was 6 to 12 months as patients often universally succumbed to diabetic ketoacidosis. Hospitals had the distinctive odor of acetonemia as described in The Discovery of Insulin by Michael Bliss: 'It was a sickish sweet smell, like rotten apples, that sometimes pervaded whole rooms or hospital wards.'*[13] Interestingly, the "smell of pear drops" of diabetic patients was used by Oxford University researchers as a principle for developing a breath test to detect type 1 diabetes in children before they start to show symptoms.[14]

2. The experimental design used by Lehninger, i.e., the inhibition of succinate dehydrogenase by malonate, simulates a condition in which CAC intermediates are being used as substrates for gluconeogenesis, as occurs in type 1 diabetes. Thus, in the absence of fumarate addition, acetyl-CoA cannot condense with oxaloacetate, which is unavailable. Addition of fumarate recovers oxaloacetate formation, allowing acetyl-CoA to enter the cycle.

   (a) Octanoate is a fatty acid with eight carbons, which undergoes $\beta$-oxidation in the mitochondrial matrix, generating acetyl-CoA (see Sect. 7.4.4). If fumarate is not added, the inhibition of succinate dehydrogenase impairs oxaloacetate formation, leading to the accumulation of acetyl-CoA, which is converted into acetoacetate. When fumarate is added, oxaloacetate is formed (fumarase converts fumarate to malate, which in turn is converted to oxaloacetate by malate dehydrogenase; see Fig. 7.3), and acetyl-CoA can enter CAC resulting citrate formation. This explains the observed results: oxaloacetate availability (due to the addition of fumarate, in the case of the experiment) allows citrate formation while acetoacetate formation is favored in the absence of fumarate.

   (b) $O_2$ is reduced to $H_2O$ by the Complex IV in the electron transport system (ETS). Thus, $O_2$ consumption is proportional to ETS feeding by electrons coming either from NADH via Complex I or from $FADH_2$ through three different FAD-associated complexes (Complex II, electron-transferring flavoprotein, or glycerol-phosphate dehydrogenase, see Fig. 6.11). In $\beta$-oxidation pathway, octanoate results in four acetyl-CoA, three $FADH_2$, and three NADH molecules. So even if acetyl-CoA does not enter CAC (where NAD molecules are reduced in the reactions of isocitrate dehydrogenase, $\alpha$-ketoglutarate dehydrogenase and malate dehydrogenase, and FAD molecule is reduced in Complex II; see Sect. 7.2.1), the $\beta$-oxidation itself results in $O_2$ consumption, explaining great increase in $O_2$ consumption when octanoate is added even when CAC is inhibited.

---

[13] Unger J (2014) Measuring the sweet smell of success in diabetes management. Ann Transl Med 2:119.

[14] Spencer B New breath test for type 1 diabetes could spot signs in children long before they begin to develop symptoms. The Daily Mail. https://www.dailymail.co.uk/health/article-2849683/New-breath-test-Type-1-diabetes-spot-signs-children-long-begin-develop-symptoms.html. Accessed 26 Nov 2014.

(c) Pyruvate oxidative decarboxylation to acetyl-CoA catalyzed by pyruvate dehydrogenase results in NAD reduction to NADH, which may also transfer electrons to Complex I (see Sect. 7.3.1). So, as in the case of octanoate addition (previous item), pyruvate addition leads to an increase in $O_2$ consumption even when CAC is inhibited.

3. Ethanol can be regarded as an "involuntary nutrient" of natural human diet because fermentation performed by intestinal flora releases ethanol that is easily absorbed. Ethanol is a small molecule soluble both in aqueous media and lipid bilayers, so it diffuses freely in tissues and reaches the blood therefore entering systemic distribution. Exposure to ethanol has probably occurred for a long period, starting from ancestors of humans.

In human liver, ethanol has three main routes of clearance (see next figure): the catalase pathway, in peroxisomes; the cytochrome P450 2E1 (CYP2E1) complex pathway, in microsomes; and the reversible oxidation catalyzed by alcohol dehydrogenase (ADH), in the cytosol. All these routes convert ethanol into ethanal, which is commonly named acetaldehyde. Aldehyde dehydrogenase (ALDH2) oxidizes acetaldehyde to acetate in mitochondria. Ethanol, ethanal, and acetate are small soluble molecules free to diffuse in tissues. Acetaldehyde accumulation, in particular, leads to physiological effects such as cardiac arrhythmias, nausea ("hangover"), anxiety, and the typical facial flushing of drinkers.

The main contributor to ethanol clearance is the ADH pathway, so we will focus on it and overlook the catalase and CYP2E1 pathways.

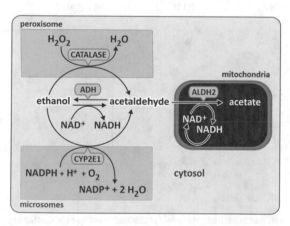

Routes of ethanol clearance in the liver. Ethanol is converted to acetaldehyde by: (i) catalase in peroxisomes, (ii) alcohol dehydrogenase (ADH) in the cytosol, and (iii) cytochrome P450 2E1 (CYP2E1) complex in microsomes. Then, acetaldehyde is oxidized to acetate in mitochondria by the aldehyde dehydrogenase (ALDH2)

A physiologically based model for ethanol and acetaldehyde metabolism in humans has been devised considering the kinetic parameters (KM and Vmax) measured in vitro for liver ADH and ALDH2, as well as the absorption and distribution of the metabolites in the main organs of the human body (stomach, gastrointestinal tract, blood, and muscle/adipose).[15] The goal was to understand the evolution of acetaldehyde levels in blood after ethanol ingestion, which depends heavily on the balance between the rates of formation and consumption of acetaldehyde in the liver.

The kinetic parameters for the formation of acetaldehyde by oxidation of ethanol and clearance of acetaldehyde (either by the action of ADH in reverse sense, forming ethanol, or oxidation to acetoacetate by the action of ALDH2) are in the following table:

| Parameter | Experimental | Used in calculations | Units |
| --- | --- | --- | --- |
| $V_{maxAl}$ | 2.0, 2.4–4.7 | 2.2 | mmol* (min*kg liver)$^{-1}$ |
| $K_{MAl}$ | ~1 | 1 | mM |
| $V_{rev}$ | 11–110 | 60.5 | mmol* (min*kg liver)$^{-1}$ |
| $K_{rev}$ | ~1 | 1 | mM/mM |
| $V_{maxAc}$ | – | 2.7 | mmol* (min*kg liver)$^{-1}$ |
| $K_{MAc}$ | 0.2–3 | 1.6 | µM |

Subscript *Al* refers to alcohol consumption (ADH), subscript *Ac* refers to acetaldehyde consumption by ALDH2, and *rev* refers to the action of ADH in the reverse sense (ethanol formation by ADH). $K_{rev}$ is formally similar to a Michaelis constant with adaptations and $V_{rev}$ is the maximum velocity of the revere reaction. $V_{maxAc}$ was estimated theoretically

The net results are presented in the following figures:

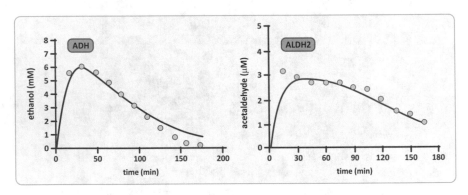

Average ethanol (*left*) and acetaldehyde (*right*) concentration in blood after ingestion of a dose of 96% ethanol (0.25 g/kg body mass) in 10 adult male subjects. Theoretical model prediction based on the kinetic parameters of ADH and ALDH2 are plotted against experimental data. (Figure adapted from the article cited in the footnote 15)

---

[15] Umulis DM et al (2005) A physiologically based model for ethanol and acetaldehyde metabolism in human beings. Alcohol 35:3–12.

Excellent agreement between theory and experimental data in the previous figures supports the idea that physiological outcomes can be accounted for based on foundational biochemistry principles, in this case from enzyme kinetics. The peak of blood ethanol concentration is reached 35–40 min after ingestion, concomitant with the onset of maximal levels of acetaldehyde in blood. Yet, the concentration of the aldehyde is kept at much lower levels than ethanol for the entire duration of the clearance period. Importantly, comparing Vrev to $V_{maxAc}$ and $V_{maxAl}$, it is clear that the reverse reaction of ADH is favored 5–50-fold, being the most important mechanism in keeping the concentration of acetaldehyde low thus minimizing the toxic effects of this metabolite in case of excessive intake of ethanol. Reversibility in the catalysis performed by ADH is a backup plan against the accumulation of acetaldehyde in case ethanol accumulates, as the catalysis of ALDH2 is relatively slow, which hampers fast processing of acetaldehyde through this pathway.

## *Final Discussion*

Ketoacidosis, the acidification of blood pH as a consequence of the high levels of the "ketone bodies" acetoacetate and β-hydroxybutyrate, is a metabolic complication with a high incidence in type 1 diabetic patients (about 14 per 1000 in developed countries), which was invariably fatal until the discovery of insulin in the 1920s.[16] It mostly occurs in uncontrolled type 1 diabetes mellitus as this form of diabetes is associated with the lack of insulin and corresponding elevation of glucagon levels in the blood. In type 2 diabetes, insulin production occurs, but its action is hampered due to insulin resistance. Usually, the amount of insulin produced in type 2 diabetic patients is sufficient to suppress ketogenesis.

Diabetic ketoacidosis consists of the biochemical triad of hyperglycemia, ketonemia, and metabolic acidosis, as shown in the next table. Metabolic acidosis is caused by the high blood concentration of acetoacetate and β-hydroxybutyrate, which overwhelms the bicarbonate buffering system. Hyperventilation, which lowers the blood carbon dioxide levels, is one of the adaptive mechanisms to compensate acidosis. Indeed, in severe diabetic ketoacidosis, breathing becomes rapid with a deep, gasping character.

---

[16] Nyenwe EA, Kitabchi AE (2016) The evolution of diabetic ketoacidosis: an update of its etiology, pathogenesis and management. Metabolism 65:507–521.

| Diagnostic criteria and classification for diabetic ketoacidosis | | | |
|---|---|---|---|
|  | Mild | Moderate | Severe |
| Plasma glucose (mg/dL) | >250 | >250 | >250 |
| Arterial pH | 7.25–7.30 | 7.00–<7.24 | <7.00 |
| Serum bicarbonate (mEq/L) | 15–18 | 10–<15 | <10 |
| Urine ketone | Positive | Positive | Positive |
| Serum ketone | Positive | Positive | Positive |
| Mental status | Alert | Alert/drowsy | Stupor/coma |

Table adapted from the article cited in the footnote 16

Persistent hyperglycemia results from an impairment of glucose uptake, especially by the tissues expressing the insulin-dependent glucose transporter, as the muscle and adipose tissue (see Sect. 8.4.3.1), together with an increase in glucose release by the liver (a process that is normally suppressed by insulin and stimulated by glucagon) both from glycogen via glycogenolysis and through gluconeogenesis (see Sect. 9.4.1.1). Hyperglycemia induces osmotic diuresis, which promotes net loss of multiple minerals and electrolytes such as sodium, potassium, calcium, magnesium, chloride, and phosphate (see next table). The resulted dehydration is severe, with an average total body water shortage of about 6 L.

| Total body deficits of water and electrolytes in diabetic ketoacidosis | |
|---|---|
| Total water (L) | 6 |
| $Na^+$ (mEq/kg) | 7–10 |
| $Cl^-$ (mEq/kg) | 3–5 |
| $K^+$ (mEq/kg) | 3–5 |
| $PO_4$ (mmol/kg) | 5–7 |
| $Mg^{+2}$ (mEq/kg) | 1–2 |
| $Ca^{+2}$ (mEq/kg) | 1–2 |

Table adapted from the article cited in the footnote 16

# Chapter 8
# Metabolic Responses to Hyperglycemia: Regulation and Integration of Metabolism in the Absorptive State

Carbohydrates, lipids, and proteins are the main components of foods and serve as "fuel" molecules that provide energy for the organism. A regular meal is generally composed of 45–65% carbohydrates, 20–35% lipids, and 10–30% protein. After ingestion, these nutrients are broken down into smaller molecules, which are subsequently absorbed and metabolized (see Chap. 7).

The end products of carbohydrate and protein digestion are monosaccharides (mainly glucose) and small peptides and amino acids, respectively, which reach the portal vein after absorption by intestinal cells (Fig. 8.1). Before reaching the systemic circulation, these metabolites pass through the liver, which absorbs and stores from 50 to 75% of the nutrients coming from the digestive tract (see Box 8.1).

Monoacylglycerol and long-chain fatty acids originated from lipid digestion are not soluble in aqueous medium and need specific transport mechanisms to be distributed in the body. They are reconverted to triacylglycerols in the intestinal mucosal cells and then are released into the lymph in the form of lipoproteins named chylomicrons (Fig. 8.1).

This chapter will focus on the metabolic adaptations that take place after the ingestion of nutrients, which allow the components of the diet to be used as energy sources or to be stored as reserves that will be mobilized in the postabsorptive or fasting states. This main topic will be introduced with a discussion on how the cells can sense the variations in blood glucose concentration, how they respond to this, and how this sensing mechanism controls the secretion of insulin. Then, we will discuss in detail the metabolic pathways involved in the biosynthesis of the main energy storage molecules, glycogen and triacylglycerol. Additionally, we will present the mechanism of action of insulin and its effects on different target tissues, giving an overview of the fate of the main metabolites absorbed after a meal in different cell types.

© Springer Nature Switzerland AG 2021                                            413
A. T. Da Poian, M. A. R. B., *Integrative Human Biochemistry*,
https://doi.org/10.1007/978-3-030-48740-9_8

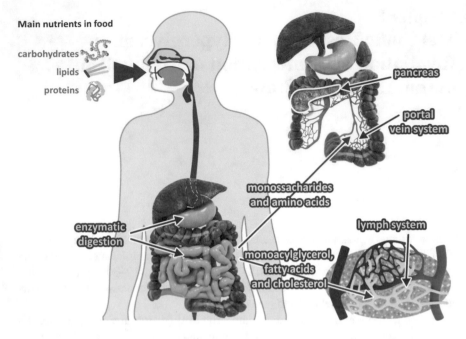

**Fig. 8.1** Absorption of the main nutrients. Digestion of carbohydrates and proteins generates monosaccharides and amino acids that are absorbed and travel through the portal vein system to the liver, entering the general circulation by the way of the hepatic vein. Digestion of lipids generates fatty acids and monoacylglycerol, which are released into the lymph

**Box 8.1  The Hepatic Portal System**

Blood is supplied to the liver by the hepatic portal vein and hepatic arteries. About 75% of the blood entering the liver is venous blood drained from the intestine, pancreas, and spleen, which converges to the portal vein. Thus, everything absorbed by the digestive tract, including nutrients and toxins, as well as the pancreatic secretions and the blood cells and their degradation products released from the spleen, passes through the liver before reaching the systemic circulation. The remaining 25% of the blood supply to the liver is arterial blood coming from the hepatic arteries. The functional units of the liver are the hepatic lobules, a polygonal arrangement of plates of hepatocytes radiating outward a central vein (see below figure). The hepatocytes make contact with the blood in the sinusoids, vascular channels that receive the blood coming from the terminal branches of both the portal vein and hepatic arteries. The sinusoids are lined with highly fenestrated endothelial cells, allowing a considerable amount of plasma to be filtered into the space between the endothelium and hepatocytes. Blood flows through the sinusoids and empties into the central vein of each lobule that coalesces into hepatic veins,

(continued)

**Box 8.1  (continued)**

which leave the liver. Of special importance in the context of this chapter is the fact that the liver is responsible for removing part of glucose from the blood, converting it into glycogen or lipids, before it reaches the systemic circulation.

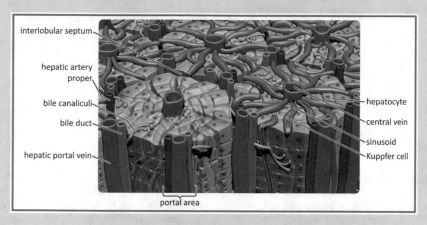

Schematic representation of the hepatic lobules. Portal vein branches and the sinusoids, carrying venous blood, are shown in *blue*, and the hepatic arteries branches are shown in *red*. The bile ducts that transport the bile from the hepatocytes to the gallbladder or the duodenum are shown in *green*

## 8.1   Glucose Sensing by Cells

The increase in blood glucose concentration is the main signal that transmits the information that food was ingested to most cells in human body, directly or indirectly, by regulating the secretion of different hormones, which in turn control the energy metabolism. After a carbohydrate-rich meal, the increase in blood glucose concentration triggers the secretion of the hormone insulin (Fig. 8.2a). The action of this hormone on its target cells as well as the increase in glycemia itself result in a rapid glucose utilization by different tissues, both as energy source and as the precursor for the synthesis of storage molecules.

Glucose metabolism homeostasis depends on rapid responses to changes in glycemia. These responses occur when blood glucose concentration decreases (see Chap. 9) but also when glycemia rises. In both cases, the metabolic adaptations operate until the basal level of blood glucose concentration, which is approximately 5 mM, is reestablished (Fig. 8.2a). This set point is maintained due to the opposite effects of glucose on the rate of pancreatic secretion of the hormones glucagon and insulin, which shows a crossover point of 5 mM glucose (Fig. 8.2b), concentration below which glucagon action predominates, whereas above it insulin effects prevail.

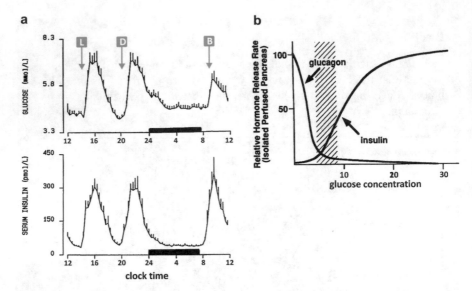

**Fig. 8.2** (**a**) Day profile of plasma glucose and serum insulin of nine human subjects (mean + SE). L, D, and B indicate lunch, dinner, and breakfast time, respectively. The *black bar* represents the sleep period. (Figure reproduced with permission from Biston et al. Hypertension 28:863–871, 1996) (**b**) Hormone secretion profile in isolated perfused pancreas as a function of glucose concentration. The dashed region corresponds to the physiological blood glucose range. (Figure reproduced with permission of the American Society for Clinical Investigation, from Matschinsky et al. J. Clin. Invest. 92:2092–2098, 1993)

Insulin is produced and secreted by the β-cells of pancreatic Langerhans islets. The mechanism involved in secretion induction will be discussed in Sect. 8.4, but a question that may be posed at this point is how can the β-cells sense the increase in blood glucose concentration and promptly respond by increasing insulin secretion, as observed in Fig. 8.2a.

To answer this question, we will first analyze two important steps that precede and determine glucose utilization in cellular metabolism: (a) the transport of glucose across the plasma membrane and (b) its phosphorylation to glucose-6-phosphate.

Glucose enters the cells by facilitated diffusion mediated by membrane transport proteins, the glucose transporters (GLUTs). Fourteen GLUT isoforms are expressed in human cells, which are characterized by distinct kinetic and/or regulatory properties, substrate specificity, cellular localization, as well as tissue-specific expression. The best studied GLUTs are the isoforms 1–4 (Table 8.1), which play a central role in glucose homeostasis by conferring specific properties to glucose uptake depending on the cell type. The other isoforms seem to transport fructose, myoinositol, and urate, in addition to glucose, and are probably also involved in the transport substrates yet to be identified.

**Table 8.1** Properties of the four best studied GLUT isoforms

| Transporter | $K_M$ (mM) | Distribution | Features | Tridimensional structure[a] |
|---|---|---|---|---|
| GLUT-1 | 1–2 | Ubiquitous, erythrocytes | Constitutive glucose transporter | |
| GLUT-2 | 20 | Liver, β-cells, intestine, kidney | Low affinity, high capacity transporter | extracellular medium / cytoplasm |
| GLUT-3 | 1 | Neurons, placenta | High affinity transporter | |
| GLUT-4 | 5 | Adipose tissue, skeletal muscle, heart | Insulin-dependent transporter | |

[a]Structure of the human GLUT3 bound to D-glucose (*orange*) (PDB 4ZW9). It contains 12 segments that cross the plasma membrane, forming a "pore" through which the sugar is transported

In the liver and pancreatic β-cells, the uptake of glucose occurs predominantly through GLUT2, which is characterized by a very high $K_M$ for glucose uptake (~20 mM) and by a very high expression level that gives these cells a high capacity for transport. These properties ensure the fast equalization of extracellular and cytosolic concentrations of glucose after any change in the physiological glycemic levels. Thus, although GLUT proteins are important for the regulation of the glucose uptake in other tissues (see Sect. 8.4.3 for glucose transport in the muscle and adipose tissue), in the case of β-cells, GLUT2 properties are important for balancing the extra and intracellular glucose concentrations.

To complement the answer to our question about how β-cells sense the changes in glycemia, we have also to take into account the next step of glucose utilization by cells. Glucose metabolization starts with its phosphorylation to glucose-6-phosphate, a reaction catalyzed by a family of enzymes named hexokinases (HK). Once glucose is phosphorylated, it cannot leave the cell unless the phosphate group is removed by the action of glucose-6-phosphatase. Since this enzyme is only expressed in liver and kidney cells (see Sect. 9.3.1), glucose phosphorylation in most cells, including β-cells, marks it for metabolization inside the cell.

There are four isoforms of HKs (HK 1–4). The isoform expressed in β-cells is the HK 4, also named glucokinase (GK) since glucose is its preferred substrate. GK is a monomer with low affinity for the glucose, with an $S_{0.5}$ of 8 mM, and cooperativeness with a Hill number of 1.7, giving to the curve that describes its activity as a function of glucose concentration an inflection point of approximately 4 mM glucose (Fig. 8.3). Additionally, unlike other HK isoforms, GK is not inhibited by the reaction product, glucose-6-phosphate. These kinetic properties make it possible that changes in GK activity occur exactly over the range of physiological blood glucose concentrations, as can easily be seen in Fig. 8.3, which compares the activity profile of HK isoforms as a function of glucose concentration.

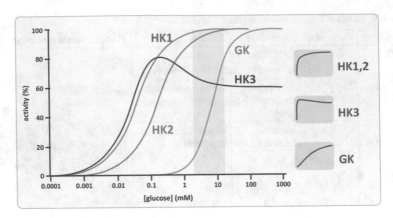

**Fig. 8.3** Comparison of the kinetic properties of HK isoforms. The physiological range of blood glucose concentration is indicated by the *grey section*. The *boxes on the right* show the curves redrawn with a linear scale of glucose concentration to better illustrate the kinetic profile. (Figure adapted from Cardenas et al. Biochim. Biophys. Acta 1401:242–264, 1998, with permission from Elsevier)

HK 1 activity is the same (and maximal) over all glucose concentrations in the physiological range, while for GK activity a great increment in glucose phosphorylation occurs as the glucose concentration increases from 2 to 12 mM. Reminding that extra and intracellular concentrations of glucose are equalized due to the properties of GLUT2 in β-cells, the kinetic features of GK makes it possible that the rate of glucose metabolization in β-cells is determined by blood glucose concentration, making GK the intracellular sensor of glycemia. Nevertheless, mutations in the gene encoding GK cause different pathologies characterized by the impairment in glycemia control (Box 8.2).

**Box 8.2  GK-Related Pathologies**

The role of GK as the glucose sensor in β-cells is strongly supported by the consequences of several mutations in GK gene, which are manifested in at least three distinct pathologies: (a) persistent hyperinsulinemic hypoglycemia (PHHI), caused by mutations that activate the enzyme; (b) permanent neonatal diabetes mellitus (PNDM), which occurs in newborns carrying inactivating mutations in the two alleles of the gene; and (c) maturity-onset diabetes of the young (MODY), occurring due to one defective allele with a mutation that inactivates the enzyme. The β-cell threshold for glucose stimulation of insulin release (GSIR) may be as low as 1.5 mM in PHHI patients because of lowered glucose $S_{0.5}$ (substrate concentration that results in 0.5 $V_{max}$) and/or increase in $k_{cat}$, (the turnover number, the number of times each enzyme site converts

(continued)

**Box 8.2** (continued)

substrate to product per unit time), while it may be increased to 7 mM due to a single inactivating mutation in a MODY patient (see below figure). Therefore, these syndromes can be explained on the basis that glucose concentration threshold for GSIR is determined by the catalytic capacity and/or substrate affinity of β-cell GK, supporting the concept that this enzyme plays a role in glucose sensing. Other pathologies, such has insulinomas, are related to GK and glucose sensing. This matter is addressed in Challenging Case 8.2, in the end of this chapter.

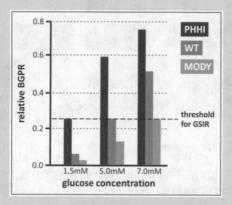

Comparison of the relative glucose phosphorylation rate, BGPR, as a function of glucose concentration for the wild type and two different mutants of GK. The *dashed line* indicates the threshold for glucose stimulation of insulin release, GSIR. (Figure based on data from Matschinsky, Diabetes 51:S394–S404, 2002)

GK seems to play the role of glucose sensor in all GK-containing cells, which include, besides the β-cells, the hepatocytes, the specialized hypothalamic neurons, and the endocrine enterocytes. However, the responses to glucose sensing may be different in each of these cells. In β-cells, the increase in blood glucose concentration stimulates insulin secretion (see Sect. 8.4.2), whereas in hepatic cells, it induces the expression of glycolytic and lipogenic genes. Additionally, glucose sensing in hypothalamic cells participates in the regulation of food ingestion and energy expenditure, impacting on the control of body mass, as it will be discussed in Chap. 11.

Before describing the mechanisms involved in insulin secretion and the direct effects of this hormone on the regulation of metabolism of different target cells, we will present in detail the main metabolic pathways that are activated when glycemia rises.

Figure 8.2a showed us how glucose is rapidly removed from the bloodstream in a period of 2 h after a carbohydrate-rich meal. In this period, the major fates of glucose are: (a) its use as energy source by most, if not all, of the cells in the body (which does not imply that it is the only energy source of the human body); (b) its incorporation into glycogen molecules, in which glucose is stored by polymerization, therefore, without major changes in its molecular structure; and (c) its transformation into fatty acids, which are incorporated in the triacylglycerols, the main energy storage molecule in the human body (Fig. 8.4).

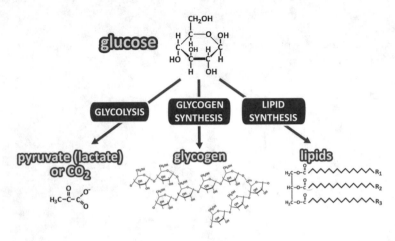

**Fig. 8.4** Possible fates of glucose after its uptake by the cells. The metabolic pathways that may be followed by glucose are shown in the *blue boxes*

The use of glucose as energy source has already been discussed in Chaps. 6 and 7. In this chapter, we will detail the metabolic pathways that convert glucose in the energy reserves, the synthesis of glycogen and the synthesis of lipids. We will also present the pentose-phosphate pathway, an additional pathway for glucose-6-phosphate oxidation that generates NADPH for the reductive biosynthesis and pentose phosphate for nucleotide synthesis.

## 8.2   Biosynthesis of Glycogen

Glycogen is a branched polymer of glucose that forms granular structures corresponding to aggregates of glycogen itself associated with the enzymes involved in its synthesis and degradation as well as their regulatory machinery (Fig. 8.5).

In humans, glycogen is accumulated mainly in the liver and skeletal muscle in response to an increase in blood glucose concentration that generally occurs after a meal. It serves as an immediate source of glucose, which in the liver is released to maintain glycemia (see Sect. 9.2.1), while in muscle it serves as a fuel reserve to ensure ATP synthesis during intense contractile activity (see Sect. 10.4.2). The general pathways for storing and mobilizing glycogen are the same in both tissues, with slight differences in the enzymes involved, especially with respect to the mechanisms of their regulation.

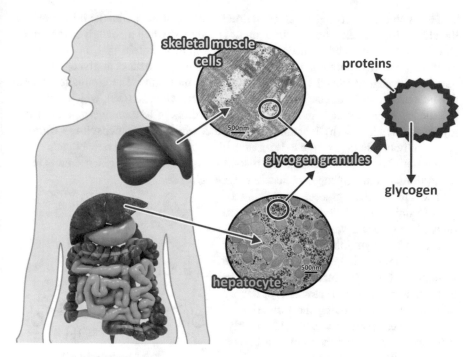

**Fig. 8.5** Liver and muscle glycogen granules. Glycogen, a polymer of glucose molecules, is stored mainly in the liver and in muscles as granules with the glycogen molecule in the center surrounded by proteins and enzymes involved in its metabolism. The electron micrographs show sections of a skeletal muscle cell (*top*) and a hepatocyte (*bottom*) in which glycogen depots can be seen as electron dense granules. (Transmission electron micrograph images: skeletal muscle cell section, reprinted from the archives of Prof. David Ferreira, with the permission of Karin David Ferreira; hepatocyte section, courtesy from Prof. Marlene Benchimol)

In the beginning of the twentieth century, Carl and Gerty Cori discovered and studied glycogen phosphorylase (GP), the enzyme that catalyzes glycogen degradation (see Sect. 9.2.1). Since they demonstrated that the reaction catalyzed by this enzyme was reversible in vitro, the synthesis of glycogen was firstly believed to occur as the reversal of glycogenolysis.

The existence of a different pathway for glycogen synthesis became more evident after the discovery of the cause of McArdle's disease, the most common disorder of glycogen metabolism. This disease is caused by a mutation in the muscle isoform of GP, resulting in a nonfunctioning enzyme and in a complete block of glycogen degradation in muscle. Patients with McArdle's disease have chronic high muscle glycogen levels, showing that glycogen synthesis occurs independently of GP activity.

Both in the liver and muscle, blood glucose is the major precursor for glycogen synthesis, which is mainly regulated by insulin action, as it will be discussed in Sect. 8.4. In the skeletal muscle, the transport of glucose into the cells is also insulin-dependent (GLUT4 transporter).

The synthesis of glycogen can be divided in two distinct parts, the initiation and the elongation phases. The initiation phase is required when a new glycogen molecule is synthesized, and is mediated by glycogenin, a protein that, in addition to catalyzing the incorporation of the first glucose units in the nascent glycogen chain, remains covalently linked to the final glycogen molecule. The elongation phase depends on two different enzymes: (a) the glycogen synthase (GS), which catalyzes the successive addition of α-1,4-linked glucose units to a nonreducing end of the glycogen branch, and (b) the branching enzyme, which has a glucosyl(4 → 6)transferase activity that creates the α-1,6-glucosidic bond that starts a new branch.

For both initiation and elongation reactions, the donor of the glucose units to be incorporated in the glycogen molecule is UDP-glucose.

The role of UDP-glucose as a donor of glucose units for the synthesis of disaccharides and glycogen was discovered by Luis Leloir, an Argentine biochemist. After Leloir's findings, it became clear that the formation of a sugar nucleotide was a general activation step in the reactions of hexose polymerization, including the formation of disaccharides, glycogen, extracellular polysaccharides, and aminohexoses and deoxyhexoses found in some of these polysaccharides. Due to this great scientific

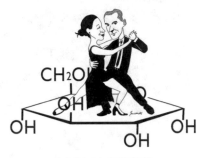

Luis F. Leloir (1906-1987)

contribution that opened the way for the understanding of carbohydrate biosynthesis, Leloir was awarded the Nobel Prize in Chemistry in 1970.

### 8.2.1   Formation of UDP-Glucose

The pathway from blood glucose to UDP-glucose differs in the liver and muscle in the beginning but converges to the same reactions in the end (Fig. 8.6a).

In the liver cells, glucose enters the cells through GLUT2 and is phosphorylated by GK. As already discussed for β-cells (see Sect. 8.1), this makes the synthesis of glucose-6-phosphate dependent on glucose concentration in the blood. Additionally, since the liver cells express the enzyme glucose-6-phosphatase, which removes the phosphate group from glucose-6-phosphate, when intracellular concentration of glucose-6-phosphate becomes high, some of the molecules may be dephosphorylated to glucose, which then leaves the cells. Thus, part of glucose molecules that enter the liver are metabolized in this organ, but the excess returns to the bloodstream to be used by other tissues.

In the muscle cells, the GLUT4 transports glucose into the cytoplasm. This transport is dependent on insulin action, as it will be discussed in detail in Sect. 8.4. Once inside the cell, glucose is phosphorylated to glucose-6-phosphate by HK, which has a high affinity for glucose.

Once glucose-6-phosphate is formed, in both the liver and muscle, the enzyme phosphoglucomutase converts glucose-6-phosphate to glucose-1-phosphate, which is the substrate for UDP-glucose synthesis. Glucose-1-phosphate reacts with UTP to form UDP-glucose, which is the immediate precursor for glycogen synthesis (Fig. 8.6b). This reaction is catalyzed by UDP-glucose pyrophosphorylase, with the formation of pyrophosphate in addition to UDP-glucose. Although the enzyme name, UDP-glucose pyrophosphorylase, refers to the reverse reaction, the synthesis of UDP-glucose is irreversible in physiological conditions. This occurs because pyrophosphate is readily hydrolyzed by the enzyme inorganic phosphatase, in a very exergonic reaction that pushes the glycogen synthesis forward.

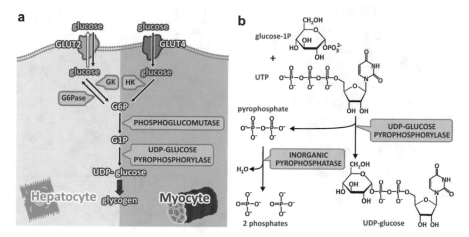

**Fig. 8.6** (**a**) Pathway from blood glucose to UDP-glucose in the liver and muscle cells. Glucose enters the cell through GLUT2 or GLUT4 and is phosphorylated by GK or HK in the liver or muscle cells, respectively. Glucose-6-phosphate (G6P) is converted to glucose-1-phosphate (G1P), which is the substrate for the formation of UDP-glucose in both cell types. UDP-glucose is the substrate for glycogen synthesis. The names of the enzymes are highlighted in *yellow boxes*. (**b**) Reaction for UDP-glucose synthesis catalyzed by UDP-glucose pyrophosphorylase

## 8.2.2 Reactions for the Initiation of Glycogen Synthesis from UDP-Glucose

The initiation of glycogen synthesis depends on a very unusual protein, named glycogenin (Fig. 8.7). This protein occupies the core of glycogen molecule and is also the enzyme that catalyzes the initiation of glycogen synthesis.

Glycogenin is a member of glucosyltransferase superfamily. It is dimeric and shares with other glucosyltransferases the nucleotide-binding fold comprising a four-stranded $\alpha\beta$ domain that is responsible for the majority of the interactions with the UDP moiety of UDP-glucose (Fig. 8.7a). In mammals, there are two tissue-specific isoforms: glycogenin-1, predominantly expressed in muscle, and glycogenin-2, mainly expressed in the liver.

The initial reaction catalyzed by glycogenin is unique, since the protein is the substrate, the catalyst, and the product of the reaction. First, glycogenin catalyzes the glucosylation of its Tyr195 by transferring a glucose unit from UDP-glucose to form a glucose-1-O-tyrosyl linkage (Fig. 8.7b). Then, glycogenin sequentially incorporates α-1,4-linked glucose residues to the nascent glycogen chain starting from the Tyr195-linked glucose and using UDP-glucose as substrate until the chain reaches eight glucose residues, with the chain remaining attached to glycogenin (Fig. 8.7b).

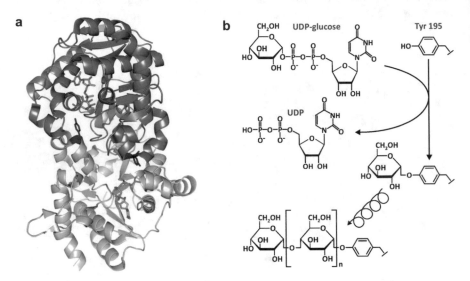

**Fig. 8.7** (**a**) Structure of dimeric human glycogenin-1 (PDB 3T7O), highlighting Tyr195 (*red*), the residue that is glycosylated by the transfer of a glucose unit from UDP-glucose (*orange*), which is shown bound to the nucleotide-binding domain. (**b**) The two chemically distinct reactions that are catalyzed by glycogenin: the initial glucosylation of Tyr195 through the formation of a glucose-O-tyrosyl linkage (*top*) and the subsequent formation of the α-1,4-glucosidic linkages (*bottom*)

## 8.2.3   Reactions for the Elongation of Glycogen Chain

Elongation of glycogen chain depends on two different enzymes, the GS and the branching enzyme. GS catalyzes the formation of an α-1,4-glucosidic bond between a glucose unit from UDP-glucose and a nonreducing end of a glycogen chain (Fig. 8.8a). When the glycogen chain reaches at least 11 residues, it becomes a substrate for the branching enzyme, which catalyzes the transfer of a terminal segment of 7 glucose residues to a hydroxyl group on the position 6 of a glucose residue in the same or in a neighboring chain, creating an α-1,6-linked branch (Fig. 8.8b). Extra glucose residues may be added to the new branch or to the original chain by the action of GS.

Branching enzyme activity is important not only because the branches increase glycogen solubility inside the cell but also because it increases the number of non-reducing ends that are the sites of action of GS and GP, making glycogen molecule more readily metabolized.

Glycogen molecule usually grows until approximately 55,000 glucose residues are incorporated, which gives to the molecule an average molecular mass of $10^7$ kDa and a diameter of 21 nm.

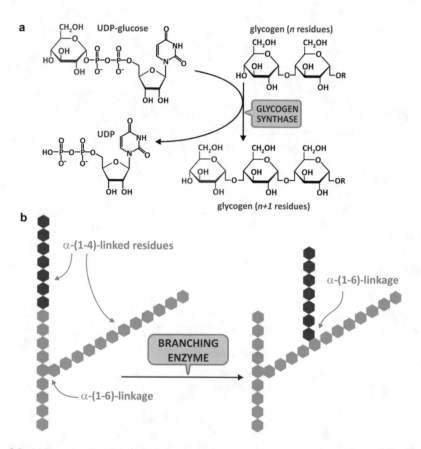

**Fig. 8.8** (**a**) Reaction for glycogen chain elongation, catalyzed by GS. (**b**) Reaction for glycogen ramification, catalyzed by the branching enzyme

## *8.2.4   Regulation of Glycogen Synthesis*

GS is the key enzyme in the regulation of glycogen synthesis. GS activity is inhibited by phosphorylation, which can occur in multiple sites by at least 11 different protein kinases (Box 8.3).

**Box 8.3  The Multisite Phosphorylation in GS Inactivation**

In humans, there are two isoforms of GS with 70% identity: the muscle isoform, which is expressed in most tissues, and the liver isoform, whose expression is tissue-specific. The catalytic site of the enzyme is located at the central region of its primary sequence and presents the highest degree of identity between the two isoforms (80%). The N- and C-terminal ends show lower degree of homology between the isoforms (50 and 46%, respectively) and contain multiple phosphorylation sites: 2 and 2a in the N-terminus and 3a, 3b, 3c, 4, 5, 1a, and 1b in the C-terminus (see below figure). At least 11 different protein kinases are involved in GS phosphorylation, which results in enzyme inactivation: cAMP-dependent protein kinase (PKA), Ca$^{2+}$-calmodulin protein kinase (CaMK), protein kinase C (PKC), phosphorylase kinase (PK), cGMP-dependent protein kinase (PKG), AMP-activated protein kinase (AMPK), ribosomal protein S6 protein kinase II (S6KII), mitogen-associated protein kinase 2 (MAPK2), casein kinase I (CKI), glycogen synthase kinase 3 (GSK3), and casein kinase II (CKII). It is interesting to note that this mechanism of GS regulation involving multiple phosphorylations is very different from that of GP, the enzyme that catalyzes glycogen breakdown, which is activated by phosphorylation in a single serine residue by phosphorylase kinase (see Sect. 9.2.2).

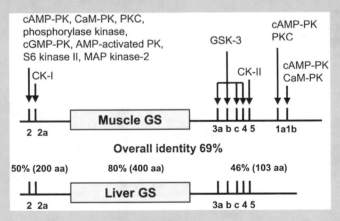

Schematic representation of the primary sequence of the muscle and liver GS isoforms showing the phosphorylation sites and the possible protein kinases involved in phosphorylation. (Figure reproduced from Ferrer et al. FEBS Lett. 546:127–132, 2003, with permission from Elsevier)

Although phosphorylation causes GS inhibition, this effect can be completely overcome by the binding of glucose-6-phosphate, the major allosteric activator of GS (Fig. 8.9).

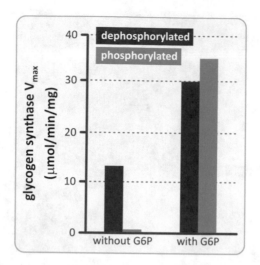

**Fig. 8.9** Effects of phosphorylation and glucose-6-phosphate on GS activity. GS is activated by glucose-6-phosphate (G6P) either when it is phosphorylated (orange) or dephosphorylated (blue), but without G6P, the basal activity is inhibited by phosphorylation. (Figure based on data from Pederson et al. J. Biol. Chem. 275:27753–27761, 2000)

The molecular mechanisms that underlie the switch between the inhibitory effect of phosphorylation and the enzyme activation by glucose-6-phosphate depend on a cluster of six conserved Arg residues located in the C-terminal region of the enzyme (Box 8.4).

**Box 8.4 The Three-State Model for GS Regulation**

The crystal structure of yeast GS revealed that the eukaryotic enzymes exist as tetramers. The location of the helix containing the conserved Arg residues indicates that these residues interact with the phosphate groups of either glucose-6-phosphate or the phosphorylated sites, suggesting a three-state model for the regulation of GS activity (see below figure). The I state, which corresponds to the dephosphorylated form, shows an intermediate basal activity due to its low affinity for glycogen. The T state is the phosphorylated form of the enzyme, in which the interaction between the phosphorylated residues and the conserved Arg locks the active site cleft, leading to enzyme inactivation. The R state is achieved by the binding of glucose-6-phosphate, whose phosphate group interacts with two conserved Arg residues, resulting in the opening of the active site cleft, allowing UDP-glucose and glycogen to bind and react.

(continued)

**Box 8.4** (continued)

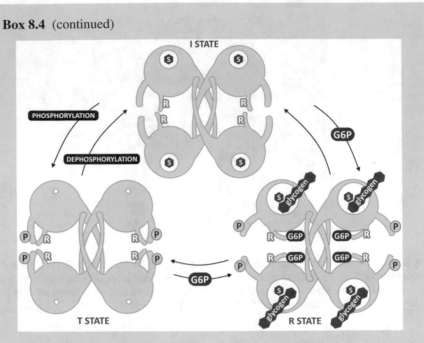

Schematic representation of GS conformational states proposed in Baskaran et al. Proc Natl Acad Sci USA 107:17563–17568, 2010. The regulatory Arg residues (labeled with an R) interact with the phosphate groups of Tyr 668 in the phosphorylated form, locking the enzyme in the T state. Glucose-6-phosphate (G6P) binding frees this constraint, leading the enzyme to the R state, which is fully active. The hexagon labeled with S represents the UDP-glucose molecule

The major protein kinase involved in GS inactivation is the glycogen synthase kinase 3 (GSK3), which catalyzes the sequential incorporation of phosphate groups in adjacent phosphorylation sites of the enzyme. However, evidence accumulated during the 1980s indicated that GSK3 alone cannot phosphorylate and inactivate GS, being necessary a concerted action of more than one protein kinase (Box 8.5).

**Box 8.5  Inactivation of GS by GSK3**
The results of in vitro phosphorylation experiments, performed with the recombinant muscle GS, clearly demonstrated that GSK3 alone cannot phosphorylate and inactivate GS, requiring previous phosphorylation by the casein kinase II (see below figure). Casein kinase II is able to incorporate one phosphate group per subunit of GS (at the phosphorylation site 5, see Box 8.3), but this phosphorylation alone has little effect on the enzyme activity. On the other hand, once phosphorylated by casein kinase II, GS becomes an effective substrate for GSK3, which introduces additional four phosphate groups in the

(continued)

**Box 8.5**  (continued)

enzyme (at the phosphorylation sites 4, 3c, 3b and 3a), resulting in a great decrease of its activity. This occurs because GSK3 needs to bind to a priming phosphate in order to catalyze the phosphorylation of the subsequent phosphorylation site (see the schematic representation in the figure of GSK3 structure). This latter site, when phosphorylated, becomes another site for GSK3 to bind and to act on the next available site. This sequential phosphorylation continues until five phosphate groups are incorporated in the GS C-terminal phosphorylation sites.

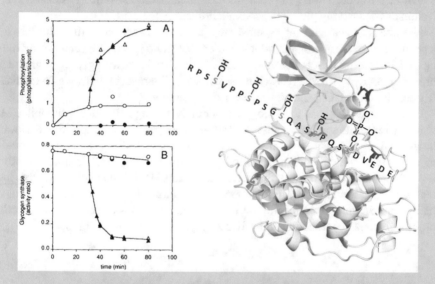

*Left:* Quantification of phosphate groups incorporated (*upper panel*) and the effect on enzyme activity (*lower panel*) after incubation with casein kinase II alone (*open circles*), GSK3 alone (*filled circles*), or GSK3 added after 30 min incubation with casein kinase II (*triangles*). (Figure reproduced from Zhang et al. Arch. Biochem. Biophys. 304:219–225, 1993, with permission from Elsevier) *Right:* GSK3 structure (PDB 1H8F). The positively charged residues Arg96, Arg180, and Lys205 (lateral chains shown in *red*) interact with the phosphate group incorporated in GS (represented by the primary sequence of the 5-3a C-terminal phosphorylation sites) by casein kinase II, allowing the occupation of GSK3 active site (highlighted in *green*) with the GS subsequent phosphorylation. Thus, GS slides allowing the incorporated phosphate to interact with the positively charged GSK3 residues, allowing the next phosphorylation site to occupy the active site

It is interesting to point out that although GSK3 was firstly discovered as acting in insulin-mediated activation of GS, it is now recognized as a constitutively active protein kinase that phosphorylates a wide range of substrates. GSK3 is involved in many cellular processes, such as cell proliferation, neuronal function, oncogenesis, and embryonic development, besides glycogen metabolism and insulin signaling. Interestingly, many of GSK3 substrates, besides GS, need this "priming phosphorylation" in a Ser/Thr residue located +4 residues of the site of GSK3 phosphorylation.

Since GSK3 is a constitutive enzyme, GS is maintained constantly phosphorylated and inactive until GSK3 is inhibited. GSK3 is inhibited by phosphorylation mediated by insulin action (see the detailed mechanism in Sect. 8.4). Thus, when glycemia increases and insulin is secreted, GSK3 becomes inhibited, allowing GS to be dephosphorylated and activated. In addition to mediating the inhibition of GSK3 activity, insulin action also stimulates glycogen synthesis by inducing the dephosphorylation of GS. GS dephosphorylation is catalyzed by the protein phosphatase 1 (PP1), the major protein Ser/Thr phosphatase in eukaryotic cells, which associates to glycogen granules in response to insulin signaling pathway and also acts on GP and on phosphorylase kinase, the enzyme that catalyzes the phosphorylation of GP (see Sect. 9.2.2), leading to their inactivation. Thus, PP1 promotes the simultaneous activation of GS and inactivation of GP, favoring glycogen accumulation inside the cells.

PP1 does not act on all of its substrates in the cell at the same time. Its activity depends on regulatory proteins, which work as substrate specifying subunits, recruiting PP1 to specific targets inside the cell. PP1 is associated with glycogen granules through a family of these target subunits, known as G target subunits, which mediate PP1 activity on GS, GP, and phosphorylase kinase (Fig. 8.10).

In muscle, PP1 binds to glycogen essentially through the target subunit GM, which increases its activity against GS, GP, and phosphorylase kinase (Fig. 8.10). In the liver GL and GC (or PTG) are expressed at the same levels. GL contains a binding site to GP*a*, which inhibits PP1 activity. Glucose binding to GP inhibits its activity and promotes its release from GL, activating PP1, and, consequently, the dephosphorylation and activation of GS. Thus, during hyperglycemic conditions, the increase in glucose concentration inside the hepatocytes directly and indirectly inhibits glycogen degradation and activates glycogen synthesis (Fig. 8.10).

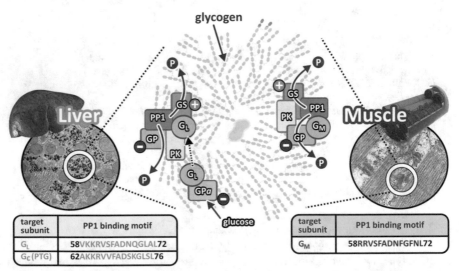

| target subunit | PP1 binding motif |
|---|---|
| G$_L$ | 58VKKRVSFADNQGLAL72 |
| G$_C$(PTG) | 62AKKRVVFADSKGLSL76 |

| target subunit | PP1 binding motif |
|---|---|
| G$_M$ | 58RRVSFADNFGFNL72 |

**Fig. 8.10** The enzymes of glycogen metabolism (GS, GP, and phosphorylase kinase, PK) are associated with glycogen granules in the liver (*left*) or muscle (*right*) cells. In liver cells, the increase in intracellular glucose concentration inactivates GP and induces the dissociation of GL, which binds to glycogen granule and recruits PP1, which, in turn, dephosphorylates GS, activating it, and GP and PK, inactivating them. In muscle, GM recruits PP1, increasing its activity against GS, GP, and PK. PP1-binding motifs in each of the target subunits (as identified in the work by Fong et al. J. Biol. Chem 275:35034–35039, 2000) are represented in the *boxes*

Muscle and liver GS also differ in their intracellular localization. In muscle cells, when intracellular glucose concentration is low, GS accumulates in the nucleus. On the other hand, when insulin is released in the bloodstream and glucose concentration increases inside the cell due to the insulin-dependent glucose uptake by GLUT4, the enzyme is translocated to the cytosol, where it becomes active, both by the direct binding of glucose-6-phosphate and due to its dephosphorylation mediated by insulin. In liver cells, GS moves from a diffuse distribution in the cytosol to the cell periphery in response to elevation of the intracellular glucose concentration, which in this case, reflects directly the increase in glycemia due to GLUT2 and GK properties. As glycogen synthesis proceeds, glycogen deposits grow from the periphery toward the center of the cell and GS maintains its co-localization with glycogen molecules.

## 8.3 Biosynthesis of Lipids

The major energy reserve in humans is constituted of lipids, which are stored as triacylglycerols in lipid droplets inside the adipocytes, the main cellular type in the adipose tissue (Fig. 8.11). Adipose tissue corresponds to 12–20% of the body weight of a non-obese man and may yield more than 100,000 kcal upon degradation of the triacylglycerols.

Additionally, other lipids play equally important roles in human organisms, such as the phospholipids and cholesterol that are constituents of cellular membranes, the steroids that act as hormones, the eicosanoids that are extra- and intracellular messengers, and vitamin K that acts as cofactor of enzymatic reactions, among others.

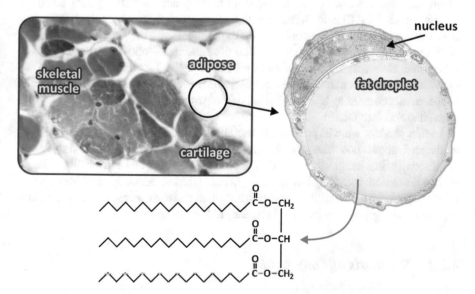

**Fig. 8.11** Histological section showing the adipose tissue. (Figure reprinted with the permission of Instituto de Histologia e Biologia do Desenvolvimento, Faculdade de Medicina, Universidade de Lisboa, FMUL). In detail, an electron microscopy of an adipocyte, showing a lipid droplet inside, which stores triacylglycerol molecules

Lipids have been regarded for a long time as relatively inert components of the body, at least in part due to the fact that the earlier methodologies to study metabolism were developed for aqueous systems. This picture started to change significantly after the pioneering studies of Rudolf Schoenheimer, who revolutionized the concept of metabolic activity by merging atomic physics and physiology through the incorporation of isotopic atoms into specific positions of organic molecules. This approach allows the pathway and the fate of different molecules to be traced, overcoming the limitations of the attachment of detectable chemical groups, as in the experiments previously performed by Franz Knoop (see Sect. 7.4), since the binding of these chemical groups would introduce structural changes in the target molecules and would alter them greatly in comparison to the natural analogs.

Schoenheimer fed mice with deuterated fatty acids and found that ingested lipids interchanged with those stored in the tissues. Even in mice that lost weight, a large proportion of the labeled fatty acids was deposited, indicating that fats were not catabolized directly following absorption, but they were continously stored and mobilized. From these results, Schoenheimer concluded: "The fat tissues are generally regarded as storage for a time of need and have frequently been compared with a dead storage or food cellar for emergencies. On the basis of the new findings, the fat tissues might better be compared with the refrigerator in which excess is stored for the short interval between meals."

Rudolf Schoenheimer (1898-1941)

Schoenheimer's findings provided a new view of metabolism, with the concept of a "dynamic state of body constituents," which, although already noticed by Claude Bernard in the nineteenth century (see Sect. 9.2), was only clearly proved with Schoenheimer's experiments.

Since the isotopic labeling has revealed the rapid metabolic turnover of body fat, it becomes clear that the conversion of carbohydrate to fat would be a major route of glucose metabolism. Indeed, several experiments in which animals were fed with a fat-free diet containing labeled glucose demonstrated the incorporation of glucose-derived atoms into fat.

In this section, we will present the metabolic pathway for the synthesis of fatty acids and show how these molecules are incorporated into triacylglycerols. Additionally, the integration of carbohydrate metabolism and fatty acid synthesis will be discussed, including how glucose is converted in fatty acids and how a shunt of the glycolytic pathway, the pentose-phosphate pathway, is essential for generating the reducing power necessary for the lipid biosynthesis.

### 8.3.1    Synthesis of Fatty Acids

Until the decade of the 1950s, fatty acid synthesis was thought to be the reversal of β-oxidation. This idea was sustained by the fact that, at that time, all the enzymes of β-oxidation had been purified and each reaction had been shown to be reversible. However, now it is known that, although the synthesis of fatty acid involves the sequential addition of acetyl units into a nascent fatty acid chain, this pathway occurs in the cytosol in contrast to the mitochondrial localization of β-oxidation, and is catalyzed by different enzymes, two unusually large multifunctional enzymes named acetyl-CoA carboxylase (ACC) and fatty acid synthase (FAS). Regarding tissue localization, fatty acid synthesis occurs mainly in liver and adipose tissue, as well as in mammary gland during lactation.

### 8.3.1.1   Reactions for Fatty Acid Synthesis

Incorporation of acetyl units into the nascent fatty acid chain requires the previous formation of a three-carbon molecule, malonyl-CoA. This molecule is the substrate for fatty acid synthesis and also the negative regulator of fatty acid oxidation by inhibiting the carnitine acyltransferase 1 and, consequently, the transport of the fatty acyl-CoA into the mitochondria (see Sect. 7.4.3).

Malonyl-CoA is generated by the carboxylation of acetyl-CoA in a reaction catalyzed by the enzyme ACC (Fig. 8.12). ACC is one of the biotin-dependent carboxylases, a group of enzymes that use the same biochemical strategy to overcome the high energy barrier of the carboxylation reaction. These enzymes have a biotin prosthetic group linked to the biotin-carboxyl carrier protein (BCCP) component and two enzymatic activities catalyzed by their biotin carboxylase (BC) and carboxyltransferase (CT) components, performing a two step-reaction in which the enzyme-linked biotin is carboxylated in an ATP-dependent reaction (Box 8.6).

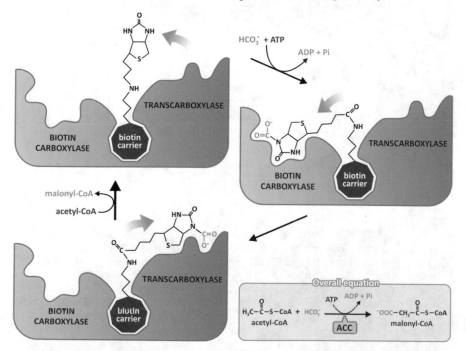

**Fig. 8.12** Reaction catalyzed by ACC. First the BC component of the enzyme catalyzes the ATP-dependent carboxylation of the biotin prosthetic group linked to the BCCP component of the enzyme. Then, the biotin arm moves the carboxybiotin to the CT component of the enzyme, which catalyzes the transfer of the carboxyl group from the biotin to acetyl-CoA, forming malonyl-CoA

**Box 8.6  The Biotin-Dependent Carboxylases**

Biotin-dependent carboxylases include acetyl-CoA carboxylase (ACC), propionyl-CoA carboxylase (PCC), 3-methylcrotonyl-CoA carboxylase (MCC), geranyl-CoA carboxylase, pyruvate carboxylase (PC; see Sect. 9.3.1), and urea carboxylase (UC). They contain three distinct components in their structure: the biotin carboxylase (BC), carboxyltransferase (CT), and biotin-carboxyl carrier protein (BCCP) components, which occur, depending on the organism, as separate subunits (usually in bacteria) or part of a multi-domain protein (in eukaryotes), with some intermediate structural organizations also found in nature. These enzymes contain a biotin prosthetic group covalently linked to a Lys residue of the BCCP component and display two distinct enzymatic activities. First, the BC component catalyzes carboxylation of the biotin cofactor, using bicarbonate as the $CO_2$ donor, and then the CT component catalyzes the carboxyl group transfer from carboxybiotin to the substrate (see below figure).

Biotin linkage to the Lys side chain in BCCP forms a highly flexible 16 Å-length arm composed of eight methylene groups with ten rotatable single bonds, so that a "swinging-arm model" has been suggested as the mechanism for the translocation of biotin from BC to CT active sites. However, structure determination of several biotin-dependent carboxylases revealed that the distance between BC and CT active sites ranges from 55 to 85 Å, suggesting that the translocation of the BCCP domain is also necessary for biotin to visit the BC and CT active sites, in what was called a "swinging-domain model" for the catalytic activity of the biotin-dependent carboxylases.

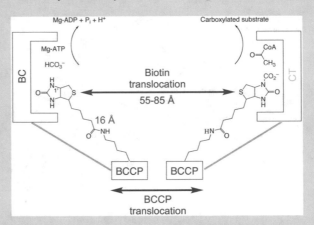

Schematic representation of the biotin-dependent carboxylases-catalyzed reactions. First BC component catalyzes the carboxylation of the biotin prosthetic group linked BCCP component of the enzyme, in a reaction dependent on $Mg^{2+}$-ATP hydrolysis. Then, biotin arm flexibility together with BCCP translocation allows carboxylated biotin to move in the direction of the TC component, which catalyzes the transfer of carboxyl group attached to biotin to the substrate (in this example acetyl-CoA), generating the carboxylated molecule. (Figure reproduced with permission from Tong, Cell. Mol. Life Sci. 70:863–891, 2013)

Human ACC is a ~250 kD protein in which BC and CT activities are separated in two domains of the same polypeptide chain. A third domain of the enzyme, the BCCP component, contains the biotin prosthetic group covalently linked to a Lys residue. The structural organization of the enzyme allows that biotin move from one catalytic site to the other during catalysis (see Box 8.6).

Two human ACC isoforms, presenting 73% identity of amino acid sequences, were identified. They are highly segregated within the cell and play different roles in metabolism (Fig. 8.13). ACC1 is expressed in all tissues but in high levels in the liver, adipose tissue, and lactating mammary gland and produces malonyl-CoA for fatty acid synthesis. ACC2 is highly expressed in the skeletal muscle and heart and is associated with the mitochondrial membrane, where the synthesized malonyl-CoA acts as an inhibitor of carnitine/palmitoyl shuttle system, impairing fatty acid oxidation. The implications of the fate of malonyl-CoA produced by each ACC isoform in the regulation of lipid metabolism will be discussed in the next section.

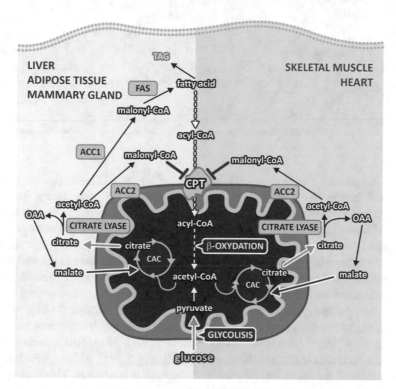

**Fig. 8.13** ACC isoforms and their intracellular localization. ACC1 is highly expressed in the cytosol of liver, adipose tissue, and lactating mammary gland cells, where it converts acetyl-CoA to malonyl-CoA, which is used as substrate for fatty acid synthase (FAS) in the fatty acid synthesis. In contrast, ACC2 is associated with the mitochondrial membrane, and the malonyl-CoA produced acts as an inhibitor of the carnitine/palmitoyl transferase (CPT), impairing the transport of the acyl-CoA to mitochondrial matrix and, thus, its oxidation. The substrate for both ACC isoforms is acetyl-CoA, produced in the cytoplasm from citrate, which leaves the mitochondrial matrix after its accumulation due to the high rate of glucose metabolization. The *dashed lines* indicate inhibited reactions

Malonyl-CoA produced by ACC1 is the immediate substrate for fatty acid synthesis, which has as the main end product palmitate, a 16-carbon saturated fatty acid (see Sect. 3.1). Fatty acids with other chain length and/or containing insaturations are produced by modification of this previously synthesized palmitate.

The overall process of palmitate synthesis is catalyzed by multifunctional enzyme FAS (see Box 8.7) and occurs through sequential condensation of seven two-carbon units derived from malonyl-CoA to a primer acetyl group derived from acetyl-CoA.

---

**Box 8.7  The FAS Isoforms**

Although fatty acid biosynthesis occurs through a conserved set of chemical reactions in all organisms, the structural organization of FAS varies. In animals and fungi, FAS is a huge cytosolic multifunctional enzyme referred as type 1 FAS, whereas in plant and bacteria, the different enzymatic activities for fatty acid synthesis are accomplished by separate polypeptide chains that compose the type 2 FAS system. It is interesting to note that this is similar to what is observed for bacterial ACC, which shows its three components as separate subunits, in contrast to the multi-domain protein of eukaryotes (see Box 8.6). It is believed that the ancestor of type 1 FAS resembled the dissociated type 2 FAS, and gene duplication, loss of function, and gene fusion gave rise to the multienzyme of mammals.

---

The human FAS is a cytosolic homodimer of 270 kDa monomers, highly expressed in the liver, adipose tissue, and mammary glands during lactation, although it is also detected in some level in almost all tissues of the human body. It performs seven enzymatic activities comprised in six domains of the same polypeptide chain. Additionally, a seventh domain, the acyl carrier protein, contains the prosthetic group phosphopantetheine, to which the intermediates are linked during the synthesis reactions. The seven domains are named accordingly to their respective enzymatic activities and occur in the following order from N- to C-terminal ends of the enzyme: $\beta$-ketoacyl synthase (KS); malonyl/acetyl transferase (MAT), which is a bifunctional domain catalyzing acetyl and malonyl transferase reactions; $\beta$-hydroxyacyl dehydratase (DH); enoyl reductase (ER); $\beta$-ketoacyl reductase (KR); acyl carrier protein (ACP); and thioesterase (TE) (Fig. 8.14a). Although each monomer contains all the activities, only the dimeric form is functional. This is explained by the fact that in the dimer, the monomers form an intertwined, X-shaped, head-to-head homodimer, in a way that a complete cycle of reactions depends on activities located at distinct monomers (Fig. 8.14b, c).

Palmitate synthesis starts with an acetyl and a malonyl groups attached to the enzyme through thioester linkages. The acetyl group comes from acetyl-CoA and is

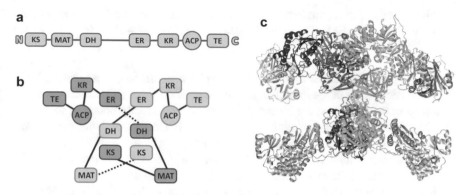

**Fig. 8.14** (**a**) Primary structure of FAS showing the position of the seven domains from the N- to the C-terminus of the sequence: β-ketoacyl synthase (KS), malonyl/acetyl transferase (MAT), β-hydroxyacyl dehydratase (DH), enoyl reductase (ER), β-ketoacyl reductase (KR), acyl carrier protein (ACP), and thioesterase (TE). (**b**) Schematic representation of the X-shaped dimeric form of FAS. Each monomer is shown in a *different color*. (**c**) Crystallographic structure of FAS (natively purified from pigs). The solved structure covers five catalytic domains (the ACP and TE domains remained unresolved). One subunit is colored by different shades of *purple pink* and the other of *blue green*. Two additional nonenzymatic domains were identified in the crystal structure: a pseudoketo-reductase (ΨKR, colored in *yellow*) and a pseudo-methyltransferase (ΨME, colored in *beige*) that is probably a remnant of an ancestral methyltransferase domain maintained in some related enzymes. (Figure adapted from Liu et al. Int. J. Biochem. Mol. Biol. 1:69–89, 2010, using the PDB file 2VZ8)

transferred to a thiol group of a Cys residue of the KS portion of the enzyme by the acetyl transferase activity of FAS MAT domain. The malonyl group is derived from malonyl-CoA and is transferred to the thiol group at the end of the phosphopantetheine cofactor in the ACP domain of the enzyme by the malonyl transferase activity of FAS MAT domain (Fig. 8.15a).

A decarboxylative condensation reaction between acetyl and malonyl groups, with the elimination of $CO_2$, forms an acetoacetyl group bound to ACP thiol group. This step is catalyzed by FAS KS activity and is energetically favored by the decarboxylation of the malonyl group (Fig. 8.15a).

The acetoacetyl-ACP is reduced to β-hydroxybutyryl-ACP by the FAS KR activity, with concomitant oxidation of NADPH to $NADP^+$. β-hydroxybutyryl-ACP is dehydrated, generating a double bond that forms the trans-butenoyl-ACP, in a reaction catalyzed by the FAS DH activity. Then, the double bond is reduced by the FAS ER activity, in a reaction coupled to oxidation of another NADPH molecule to $NADP^+$, which generates the saturated butyryl-ACP (Fig. 8.15a). Note that these three last steps correspond to the reverse of β-oxidation first reactions (see Sect. 7.4.4), except that the electrons come from NADPH instead of being transferred to $NAD^+$ or FAD.

After the first cycle of synthesis, the butyryl group bound to ACP is transferred to the thiol group of the Cys residue of KS (Fig. 8.15a), without dissociating from the enzyme. Then, another malonyl-CoA is transferred to the unoccupied thiol group of the phosphopantetheine in the ACP, starting the second round of the fatty acid synthesis, which gives rise to a six-carbon intermediate by its condensation with the butyryl group attached to the thiol group of KS, to which the subsequent reactions are repeated.

After seven cycles of condensation and reduction, the palmitoyl group is released from ACP by the FAS TE activity, generating free palmitate in the cytosol (Fig. 8.15b), which may be incorporated into triacylglycerol or phospholipid molecules.

The long and flexible structure of the phosphopantetheine group bound to ACP allows the delivering of the intermediates to the active sites of all catalytic domains. Throughout the synthesis, the intermediates remain covalently bound as thioesters to a thiol group in the enzyme, either in KS or in ACP. The first evidence for this was obtained in the decade of the 1950s, in experiments in which after the incubation of an enzyme preparation from pigeon liver with isotopically labeled acetate, the radioactivity was found associated only to fatty acids with 16 carbons and not uniformly distributed among intermediates of different chain lengths (Fig. 8.15c).

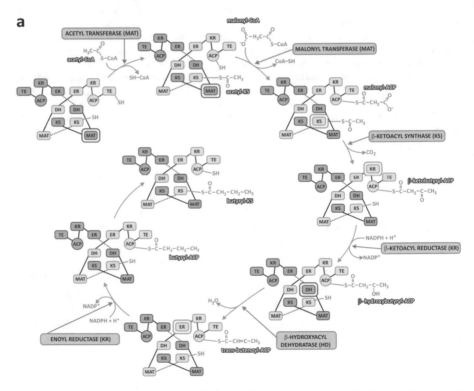

**Fig. 8.15** (**a**) Schematic representation of fatty acid synthesis, starting with the transfer of acetyl and malonyl groups from acetyl-CoA and malonyl-CoA molecules to the SH groups of KS and ACP domains, respectively. KS activity condenses the acetyl and malonyl groups, generating β-ketobutyryl bound to ACP, which is sequentially reduced, dehydrated, and reduced, by the action of the KR, DH, and ER activities, respectively. In the reduction reactions, the electron donor molecule is NADPH. These series of reactions generate a butyryl group bound to ACP, which is transferred to the SH group of KS, allowing the onset of the second round of fatty acid synthesis by making the ACP free to receive another malonyl group from malonyl-CoA. The FAS enzymatic activities operating in each step are shown in *yellow boxes*, with the respective position highlighted in the dimer structure. (**b**) After seven cycles starting with malonyl transfer to ACP, palmitate is released by the TE activity. (**c**) Distribution of $^{14}$C-long-chain fatty acid synthesized in pigeon liver preparation incubated with $^{14}$C-acetate after separation by a paper chromatography system. (Figure reproduced from Porter & Tietz, BBA 25:41–50, 1957, with permission from Elsevier)

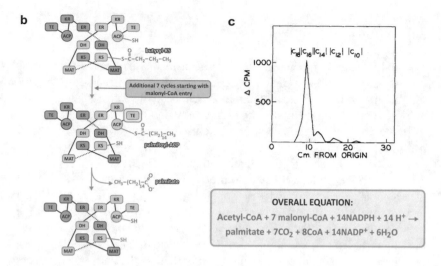

**Fig. 8.15**   (continued)

The three-dimensional structure of FAS dimer shows that the whole structure can be divided into two portions: the condensing portion, containing KS and MAT domains, and the modifying portion, containing DH, ER, and KR domains (see Fig. 8.14c).

The biosynthesis of fatty acid, as well as other biosynthetic processes, takes place entirely in the cytosol, in opposition to most of the oxidation reactions that occur in the mitochondrial matrix. The use of NADPH instead of NADH as the electron donor in the fatty acid synthesis is very important in this context, since glycolysis, which occurs simultaneously to fatty acid synthesis in the cytosol, requires a low NADH/NAD$^+$ ratio to allow the oxidation glyceraldehyde-3-phosphate to 1,3-biphosphoglycerate, a NAD$^+$-dependent reaction catalyzed by glyceraldehyde-3-phosphate dehydrogenase (see Sect. 6.1). Thus, cytosolic high NADPH/NADP$^+$ and low NADH/NAD$^+$ ratios enable the oxidative reaction of glycolysis to occur at the same time and cellular compartment that the reductive fatty acid synthesis. It is interesting to note that the phosphate group in NADPH/ NADP$^+$ does not interfere with the redox reaction itself. It only grants specificity for enzyme recognition (see Sect. 5.5).

Acetyl-CoA, in addition to being the immediate substrate for fatty acid synthesis (both as substrate for ACC, generating malonyl-CoA, and as the donor of the first acetyl unit transferred to FAS KS), is also the precursor for the cholesterol synthesis (Box 8.8). The precursor of acetyl-CoA molecules used for these biosynthetic processes is mainly the ingested carbohydrate, as it will be discussed in the next topic.

### Box 8.8  Cholesterol Biosynthesis

Cholesterol is a 27-carbon steroid with key roles in the metabolism. It is an essential structural component of animal cell membranes, modulating their fluidity and permeability (see Sect. 3.1), and is also the precursor of the steroid hormones, bile acids, and vitamin D. Cholesterol synthesis is very complex (see below figure) and may be didactically divided in three parts.

In the first phase, three acetyl units are used to synthesize the isopentenyl pyrophosphate, which act as the building block of cholesterol synthesis. This phase includes the key step in the regulation of the process, the formation of 3-hydroxyl-3-metylglutaryl-CoA from acetyl-CoA and acetoacetyl-CoA. In the second phase, six molecules of isopentenyl pyrophosphate are condensed to form squalene, a 30-carbon isoprenoid. In the third phase, squalene cyclizes to form a tetracyclic structure, which rearranges to form lanosterol. Lanosterol is then converted to cholesterol through 19 reactions, catalyzed by enzymes associated with endoplasmic reticulum membrane, including some members of the cytochrome P450 superfamily (see Box 8.9).

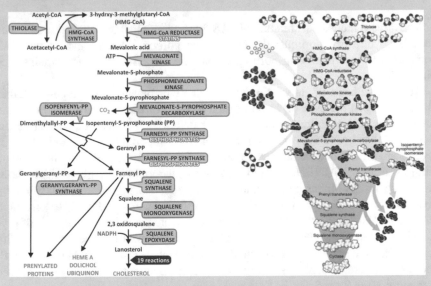

Cholesterol biosynthesis with the molecular transformations shown in the right and the reactions depicted in left. The names of the enzymes are shown in *yellow boxes*, with some inhibitors indicated. (Figure reproduced from Goodsell DS. The machinery of life. Springer, New York, 2009)

### 8.3.1.2 Origin of the Acetyl-CoA for the Fatty Acid Synthesis

Fatty acid synthesis occurs mainly in liver and adipose tissue due to the high expression of both ACC1 and FAS in hepatic and adipose cells. In both tissues, the use of glucose is greatly increased as glycemia rises, either by kinetic properties of GLUT2/GK in the liver (see Sect. 8.1) or by the insulin-dependent glucose uptake in adipose cells (see Sect. 8.4.3).

As glycolysis proceeds and pyruvate is completely oxidized in CAC, the resulting NADH and $FADH_2$ feed the electron transport chain, whose activity culminates with ATP synthesis through oxidative phosphorylation (see Chap. 6). When the ATP/ADP ratio is high, ATP synthesis decreases due to the lack ADP, resulting in the accumulation of reduced electron carriers (NADH and $FADH_2$) in mitochondrial matrix. This leads to the inhibition of CAC dehydrogenases, which depend on the oxidized electron carriers ($NAD^+$ and FAD) in order to work. As a consequence, citrate accumulates, leaving mitochondria through the citrate transporter (Fig. 8.16).

In the cytosol, citrate is broken into acetyl-CoA and oxaloacetate, through an ATP-dependent reaction catalyzed by citrate lyase. The acetyl-CoA molecule is then used as a substrate for the fatty acid synthesis.

Oxaloacetate does not directly return to mitochondria due to the absence of a transporter in the mitochondrial membrane. It is firstly converted by the cytosolic malate dehydrogenase to malate, which enters the mitochondria, where it regenerates oxaloacetate. Malate dehydrogenase reaction is also important to reoxidize the NADH produced in glycolysis (in the reaction catalyzed by glyceraldehyde dehydrogenase; see Sect. 6.1.3), allowing the maintenance of a high $NAD^+$/NADH ratio in the cytosol, necessary for glycolytic pathway to proceed (Fig. 8.16).

Alternatively, malate may be converted to pyruvate, by the malic enzyme, in a reaction that uses $NADP^+$ as the electron acceptor, generating part of the NADPH necessary for fatty acid synthesis (Fig. 8.16).

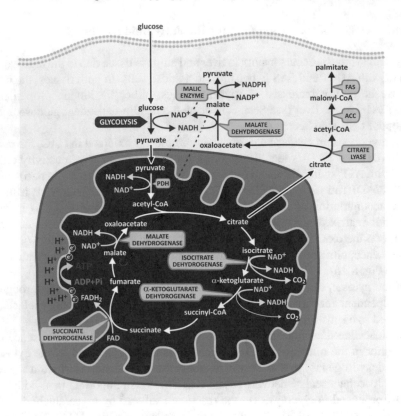

**Fig. 8.16** Schematic representation of the transformation of glucose in fatty acids in the liver and adipose tissue. Complete oxidation of glucose results in NADH and FADH$_2$ that feed the electron transport chain, whose activity generates ATP. When the ATP/ADP ratio is high, ATP synthesis decreases due to the lack ADP, resulting in the accumulation of NADH and FADH$_2$ in mitochondrial matrix, inhibiting CAC dehydrogenases, leading to citrate accumulation. Citrate leaves mitochondria, being broken to acetyl-CoA and oxaloacetate by citrate lyase in the cytosol. Acetyl-CoA is used in the fatty acid synthesis. Oxaloacetate is converted to malate, with concomitant oxidation of cytosolic NADH to NAD$^+$, favoring glycolysis, which requires NAD$^+$ for the reaction of glyceraldehyde-3-phosphate dehydrogenase. Malate may enter the mitochondria, where it regenerates oxaloacetate, or may be converted to pyruvate, by the malic enzyme, in a reaction that generates part of the NADPH necessary to the fatty acid synthesis. The names of the enzymes are highlighted in *yellow boxes*

### 8.3.1.3   Origin of NADPH for the Fatty Acid Synthesis

Most of the NADPH used in fatty acid synthesis is provided by the pentose-phosphate pathway, an alternative pathway that oxidizes glucose-6-phosphate (see also Box 8.9).

The first evidence of an oxidative pathway that converted glucose-6-phosphate in a pentose phosphate was demonstrated in yeast, in the decade of the 1930s, by pioneering studies of Otto Warburg, who also had shown that this pathway was dependent on a coenzyme different from that required in glycolysis (named by Warburg at that time as TPN, from triphosphopyridine nucleotide, which is now known as NADPH). Further work by Frank Dickens proved that this pathway occurred in many animal tissues and proceeded independently of the glycolytic route, since high concentrations of glycolytic inhibitors, such as iodoacetamide, had no effects on its activity.

The pentose-phosphate pathway may be divided in two phases: (a) an oxidative phase, in which glucose-6-phosphate is converted to a pentose-phosphate, with two NADP+-dependent oxidation steps, the last one being coupled to a decarboxylation reaction, and (b) a non-oxidative phase, in which the carbon skeletons of the pentose phosphates are rearranged to generate intermediates of glycolytic pathway, fructose-6-phosphate and glyceraldehyde-3-phosphate. A schematic representation of these two phases is shown in Fig. 8.17.

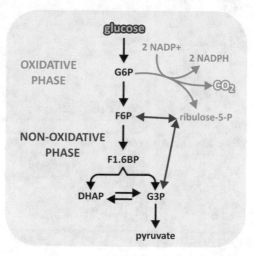

**Fig. 8.17** Schematic representation of the two phases of pentose-phosphate pathway and their relationships with glycolysis

**Box 8.9 Other Roles of the Pentose-Phosphate Pathway**

In addition to providing the reducing power in the form of NADPH for the biosynthetic processes, the pentose-phosphate pathway plays other important roles in human metabolism. In cells with a rapid rate of division, such as bone marrow, skin, and intestinal mucosa cells, glucose-6-phosphate follows the oxidative phase of pentose-phosphate pathway, generating ribose-5-phosphate, which is used in the synthesis of nucleotides and nucleic acids as well as of NADH, $FADH_2$, and coenzyme A (see figure in the top). Another important role of the NADPH produced in the pentose-phosphate pathway is to prevent oxidative damage. This is the case of the erythrocytes, which are directly exposed to oxygen and are highly dependent on glutathione reduction as a protective mechanism against oxidative stress (see figure in the middle). Glutathione is a tripeptide composed of Glu, Cys, and Gly that occurs in a high intracellular concentration (~5 mM). The sulfhydryl group of the Cys residue acts as a reducing agent, which is oxidized with the formation of a disulfide bridge that links two glutathione molecules (represented as GSSG), preventing oxidative damage to cell molecules. Glutathione is converted back to its reduced form (GSH) by the enzyme glutathione reductase, in a reaction dependent on NADPH. Thus, to maintain glutathione in the reduced form, erythrocytes are highly dependent on the pentose-phosphate pathway to

(continued)

**Box 8.9** (continued)

generate                NADPH. Additionally, the metabolization of a number of drugs and other toxic chemicals, as well as endogenous compounds such as lipids and steroidal hormones, is accomplished by a family of enzymes that uses NADPH produced in the pentose-phosphate pathway as the electron donor. These enzymes are heme-containing monooxygenases known as cytochromes P450 (abbreviated as CYP). CYPs catalyze the hydroxylation of organic substrates using an oxygen atom from $O_2$ and electrons from two NADPH molecules (see figure in the bottom), in a reaction that involves the iron atom of the enzyme heme group (see complete cycle in Fig. 2.6). In the human genome, 57 genes codifying cytochrome P450 enzymes are found. Most of them are expressed in the liver, where CYPs play crucial roles in drug metabolization and detoxification, but these enzymes are also found in extrahepatic tissues where their main function is processing endogenous substrates such as steroids and vitamins. Most CYPs are located on the endoplasmic reticulum membranes, but mitochondrial isoforms are also found, especially those involved in the steroidal hormones processing in the adrenals.

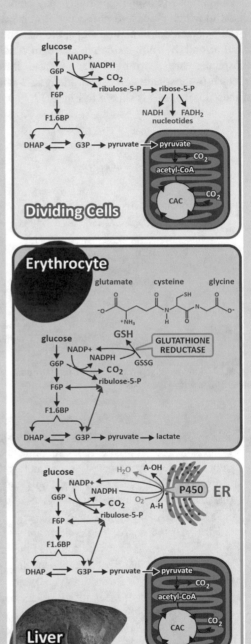

Roles of pentose-phosphate pathway in providing ribose-5-phosphate for nucleotide synthesis in dividing cells (*top*); NADPH for the glutathione reductase reaction that protects erythrocytes against oxidative damage (*middle*); and the cytochrome P450 reaction cycle responsible for drug and endogenous substrates metabolization in the liver (*bottom*)

The reactions of the pentose-phosphate pathway are detailed in Fig. 8.18.

The oxidative phase starts with the oxidation of glucose-6-phosphate to 6-phosphogluconolactone, in a reaction catalyzed by glucose-6-phosphate dehydrogenase, in which $NADP^+$ is the electron acceptor, generating NADPH. Then, a hydrolysis reaction forms 6-phosphogluconate that undergoes an $NADP^+$-dependent oxidation and decarboxylation, catalyzed 6-phosphogluconate dehydrogenase, which forms ribulose-5-phosphate and the second molecule of NADPH. Ribulose-5-phosphate may be converted to its isomer ribose-5-phosphate by the enzyme phosphopentose isomerase or may be epimerized to xilulose-5-phosphate, in a reaction

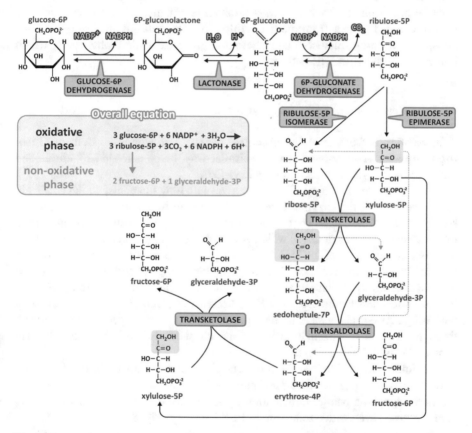

**Fig. 8.18** Reactions of the pentose-phosphate pathway. In the oxidative phase, $NADP^+$-dependent oxidation of glucose-6-phosphate generates NADPH and 6-phosphogluconolactone, which is hydrated to 6-phosphogluconate, which in turn undergoes an oxidative decarboxylation generating $CO_2$, NADPH, and ribulose-5-phosphate. Ribulose-5-phosphate may be isomerized or epimerized to ribose-5-phosphate or xilulose-5-phosphate, respectively, which follow the non-oxidative phase in which two- or three-carbon fragments are interchanged to finally form fructose-6-phosphate and glyceraldehyde-3-phosphate. The reaction stoichiometry is shown in the *box*. The names of the enzymes are shown in *yellow boxes*

catalyzed by pentose-5-phosphate epimerase. Both ribose-5-phosphate and xilulose-5-phosphate follow the non-oxidative phase of the pentose-phosphate pathway. It is important to note that if ribose-5-phosphate is required, for example, for nucleotide synthesis, the pentose-phosphate pathway ends at this point.

The non-oxidative phase comprises two types of reactions: (a) the transfer of a two-carbon fragment from a ketose donor to an aldose acceptor, catalyzed by the enzyme transketolase, and (b) the transfer of a three-carbon fragment also from a ketose donor to an aldose acceptor, catalyzed by the enzyme transaldolase. Transketolase first transfers a two-carbon fragment of xylulose-5-phosphate to ribose-5-phosphate, forming the seven-carbon product sedoheptulose-7-phosphate and glyceraldehyde-3-phosphate. Then, transaldolase transfers a three-carbon fragment from sedoheptulose-7-phosphate to glyceraldehyde-3-phosphate, forming fructose-6-phosphate and erythrose-4-phosphate. Finally, transketolase transfers the two-carbon fragment from another xylulose-5-phosphate this time to erythrose-4-phosphate, forming fructose-6-phosphate and glyceraldehyde-3-phosphate, which may return to the glycolytic pathway.

### 8.3.1.4  Regulation of Fatty Acid Synthesis

ACC reaction is the main point of control of fatty acid synthesis. Both ACC iso-forms are allosterically regulated in a similar way, being activated by citrate and inhibited by long-chain saturated fatty acyl-CoA.

In the presence of citrate, ACC polymerizes into long filaments, which consist in the active form of the enzyme. Citrate is, indeed, an important signal for nutrient storage. The increase in its concentration in the cytosol is a result of the inhibition of the dehydrogenases of CAC, which occurs due to a high ATP/ADP ratio that, in turn, is a consequence of the high availability of glucose (see previous section). In addition to activating ACC and, thus, favoring the storage of carbons from ingested carbohydrate as lipids, citrate also acts as an allosteric inhibitor of the glycolytic enzyme phosphofructokinase 1 (PFK1), decreasing glycolytic flow and allowing part of glucose-6-phosphate to be incorporated into glycogen or metabolized through the pentose-phosphate pathway, which yields NADPH for the fatty acid synthesis (Fig. 8.19; see also the previous section).

It is important to note that the products of the non-oxidative phase of pentose-phosphate pathway, fructose-6-phosphate and glyceraldehyde-3-phosphate, may return to glycolytic pathway, maintaining the carbon flow in the direction of fatty acid synthesis, even with a reduction in PFK1 activity (Fig. 8.19).

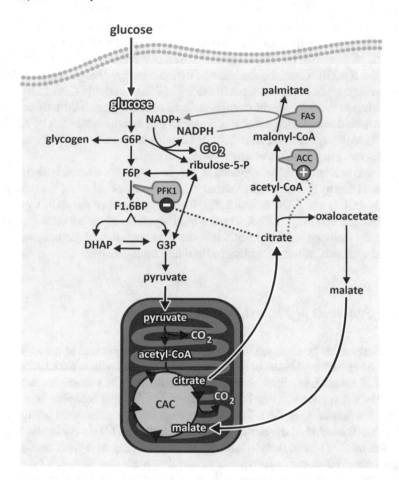

**Fig. 8.19** Role of cytosolic citrate in the regulation of metabolic pathways. As a consequence of the high ATP/ADP ratio due to nutrient availability, NADH and FADH₂ accumulate in mitochondrial matrix, inhibiting dehydrogenases of CAC and leading to citrate accumulation. Citrate leaves mitochondria where it acts as the allosteric activator of ACC, inducing its polymerization to the active polymeric form. Additionally, citrate acts as an allosteric inhibitor of PFK1, decreasing glycolytic flow and allowing glucose-6-phosphate to be used in glycogen synthesis or to be oxidized in the pentose-phosphate pathway, generating NADPH for fatty acid synthesis

ACC is also regulated by phosphorylation/dephosphorylation. Phosphorylation is either triggered by hormones (mediated by PKA, see Sects. 9.4.1 and 10.5.2 for details on this signaling pathway) or by the changes in cellular energy charge (mediated by AMP-activated protein kinase, AMPK, see Sect. 10.4.1 for detail on this kinase), resulting in ACC dissociation into monomers and, consequently, its loss of activity. Insulin mediates ACC dephosphorylation, allowing citrate-induced polymerization.

PKA phosphorylates ACC1 isoform at Ser1200. Thus, when the effects of glucagon (in hypoglycemia; see Chap. 9) or adrenaline (during intense physical activity or stress; see Chap. 10) predominate, ACC1 is phosphorylated and, consequently, inhibited, impairing fatty acid synthesis due to the lack of malonyl-CoA.

AMPK phosphorylates ACC1 isoform at Ser80 and Ser81 and ACC2 isoform at Ser219 and Ser220. Thus, the decrease in cellular energy charge, which indicates that nutrients are not in excess, leads to the inhibition of malonyl-CoA formation by ACC1, whose fate is fatty acid storage, as well as the malonyl-CoA formation by ACC2, whose decrease in cytosolic concentration stops the inhibition of carnitine/palmitoyl shuttle system (see Sect. 7.4.3), allowing the transport of fatty acyl-CoA molecules into mitochondria where they are oxidized.

Fatty acid synthesis is also regulated at the gene expression level, both by dietary components and by hormones. A diet rich in carbohydrates induces the transcription of ACC1, ACC2, and FAS. Additionally, insulin upregulates the ACC1 promoter, while glucagon downregulates it. Thus, a high ingestion of carbohydrates, directly or indirectly (through insulin action), induces the expression of the enzymes of the fatty acid synthesis pathway, leading to lipid accumulation.


## 8.3.2    Synthesis of Triacylglycerols

Most of fatty acids synthesized in the organism are incorporated into triacylglycerols, which are fatty acyl esters of glycerol (see Sect. 3.1). To form the triacylglycerol molecule, fatty acids are firstly esterified to coenzyme A in a reaction catalyzed by fatty acyl-CoA synthetase (Fig. 8.20), the same reaction that activates fatty acids to undergo β-oxidation (see Sect. 7.4.4). Thus, regulatory mechanisms are important to avoid the oxidation of the newly synthesized fatty acids. This occurs through the inhibition of CPTI by malonyl-CoA, which impairs fatty acyl-CoA transport into the mitochondrial matrix, where β-oxidation takes place (see Sect. 8.3.1).

The glycerol moiety of the triacylglycerol molecule comes from glycerol-3-phosphate, which is formed from dihydroxyacetone phosphate, an intermediate of glycolysis, in an NADH-dependent reaction catalyzed by glycerol-3-phosphate dehydrogenase. Liver cells also have the enzyme glycerol kinase, which catalyzes the formation of glycerol-3-phosphate from glycerol (Fig. 8.20).

The two free hydroxyl groups of glycerol-3-phosphate are firstly acylated by two molecules of fatty acyl-CoA, in a reaction catalyzed by the acyl transferase, yielding diacylglycerol-3-phosphate, also called phosphatidic acid. The phosphate group of phosphatidic acid is then hydrolyzed by phosphatidic acid phosphatase to form a 1,2-diacylglycerol, which is esterified with a third fatty acyl-CoA to form the triacylglycerol molecule.

When synthesized in adipocytes, triacylglycerols are directly stored in the lipid droplets inside these cells (Fig. 8.20).

Triacylglycerols synthesized in the liver must be transported to adipose tissue to be stored, which is not a trivial task due to their extremely low solubility in the aqueous blood environment. To overcome this barrier, they are incorporated into the very-low-density lipoprotein (VLDL), which travels through the bloodstream from the liver to the adipose tissue, where it delivers part of their lipid content, becoming a lipoprotein of intermediate density (IDL) (Fig. 8.20). IDL returns to the liver or is converted to LDL (see details about lipoproteins' structure in Sect. 3.1.2).

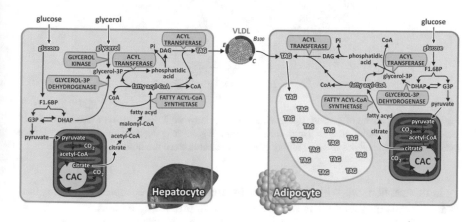

**Fig. 8.20** Schematic representation of triacylglycerol synthesis in the liver and adipose tissue. Fatty acids are synthesized in both tissues from the excess of glucose. To be incorporated into the triacylglycerol molecule, the main human energy store, fatty acids are firstly esterified to coenzyme A by fatty acyl-CoA synthetase. Part of the dihydroxyacetone-phosphate (DHAP) molecules formed in glycolysis may be used to generate glycerol-3-phosphate (glycerol-3P), in a reaction catalyzed by the glycerol-3P dehydrogenase. In the liver, glycerol-3P may also be formed by the phosphorylation of glycerol by glycerol kinase. The acyl groups of two molecules of fatty acyl-CoA are transferred to each of the free hydroxyl groups of a glycerol-3P molecule, generating the phosphatidic acid. Then, the phosphate group of the phosphatidic acid molecule is hydrolyzed, and a fatty acyl group of another fatty acyl-CoA is transferred to the resulting hydroxyl group of the diacylglycerol (DAG) molecule formed, generating a triacylglycerol (TAG) molecule. In adipose tissue, TAGs are directly stored in lipid droplets (LD). TAG molecules synthesized in the liver are incorporated in VLDL to be transported to the adipose tissue. After delivering TAGs in adipose tissue, VLDL becomes an IDL, which returns to the liver

## 8.4 Hormonal Responses to Hyperglycemia: Role of Insulin

The main hormonal response to hyperglycemia is insulin secretion, whose immediate global effect is to reduce blood glucose concentration. The loss of insulin-mediated effects results in an incapacity of controlling glycemia, which characterizes the diseases known as diabetes mellitus.

The role of insulin in reducing glycemia can be illustrated by comparing the result of the oral glucose tolerance test between nondiabetic and diabetic subjects (Fig. 8.21). While in nondiabetic subjects the glycemia promptly recovers the basal levels due to insulin action, diabetic subjects experience high glycemia during a long period after glucose ingestion.

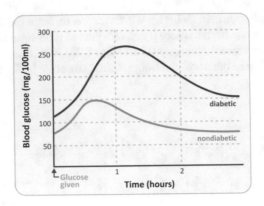

**Fig. 8.21** Comparison of glucose tolerance test results between a diabetic and a nondiabetic subjects. In this test, a solution with a high concentration of glucose is ingested and blood samples are taken afterward to determine how quickly glucose is cleared from the blood

In this section, we will discuss how insulin, through a complex signaling cascade, controls the metabolism in different tissues, promoting a rapid use of glucose that results in the decrease of its concentration in the bloodstream. The main metabolic responses to insulin will be described in detail, including (a) glucose uptake in muscle and adipose tissue; (b) glucose utilization as energy source by all the cells in the organism; (c) glucose storage as glycogen in the muscle and liver; (d) glucose conversion into storage lipids in liver and adipose tissue; and (e) inhibition of glucose production through gluconeogenesis and glycogenolysis in the liver. It is important to bear in mind that, in addition to these metabolic effects, insulin controls several other biological processes, including growth and differentiation, which will not be the focus of this chapter.

## 8.4.1   Discovery of Insulin

Until the beginning of twentieth century, insulin deficiency, which is known as type 1 diabetes, was a devastating disease. The impact of insulin discovery can be realized by reading the introduction of the Michael Bliss's book, in which he states that the discovery of insulin was "one of the most dramatic events in the history of the treatment of disease ... and those who watched the first starved, sometimes comatose, diabetic receive insulin and return to life saw one of the genuine miracles of modern medicine, ... the closest approach of the resurrection of the body... the elixir of life for millions of people around the world" (Bliss M. *The discovery of insulin*. University of Chicago Press, 1982).

The history of insulin discovery is indeed remarkable. In the 1880s, Josef von Mering and Oskar Minkowski showed that the total pancreatectomy in experimental animals, usually dogs, resulted in the development of acute hyperglycemia and glycosuria, leading the dogs to die from severe diabetes. This raised the possibility that glucose metabolism was promoted by an "internal secretion" produced by the pancreas (in opposition to the digestive enzymes secreted by the pancreas into the gut—the "external secretion").

During more than 30 years, several researchers have tried to isolate this substance, which has already been named insulin due to the speculation that it was produced by the Langerhans islets, but all the efforts were unsuccessful probably due to its proteolytic digestion by the pancreatic enzymes. The turning point in the history occurred when Frederick G. Banting, a young surgeon, intuited, while studying the pancreatic function to prepare a lecture, that the ligation of pancreatic ducts would allow the isolation of pancreatic internal secretion free of proteolysis.

In 1921, Banting, together with the medical student Charles H. Best, with the support of John J. K. Macleod, a prominent Canadian scientist in the field of carbohydrate metabolism, performed decisive experiments that definitely demonstrated that a substance secreted by the pancreatic islets was able to drop plasma glucose levels of diabetic dogs.

James Bertram Collip (1892-1965)
Charles Herbert Best (1899-1978)
Frederick Grant Banting (1891-1941)

In just 1 year, this substance was (a) isolated and tested in dogs; (b) further purified by James Bertram Collip, who joined the group to make the preparation suitable to be injected in humans; and (c) successfully used in the first patient, Leonard Thompson, a 14-year-old boy, who recovered from a very severe diabetic state. In the same year, the preparation was used to treat other diabetic volunteers.

In 1923, only 2 years after Banting idealized the first experiments, Banting and Macleod were awarded the Nobel Prize in Physiology or Medicine, in a controversial ceremony in which Banting announced that he would share his half of the award with Best, which led Macleod to announce that he would share his half with Collip.

## 8.4.2   Mechanisms of Insulin Action

Insulin is a small protein whose mature form is composed of two polypeptide chains (A and B) linked by disulfide bonds (Fig. 8.22; see also Sect. 3.3.1). It is synthesized as a single polypeptide called preproinsulin. Its N-terminal sequence, which consists in a signal peptide, is cleaved as the polypeptide is translocated into ER. This proteolytic cleavage induces the formation of three disulfide bonds, yielding the proinsulin, which is transported to the trans-Golgi network. The action of prohormone convertases, endopeptidases that hydrolyzes a proinsulin internal segment named C peptide, as well as the carboxypeptidase E, generates the mature insulin.

Insulin is packaged inside intracellular granules that are secreted through exocytosis together with the C peptide upon stimulation. Insulin secretion is dependent on β-cells' energetic metabolism. As discussed in Sect. 8.1, glucose utilization by β-cells occurs in parallel with the increase in glycemia due to GLUT2 and glucokinase high $K_M$ values. Thus, the enhancement in the blood glucose concentration is followed by an increase in the ATP/ADP ratio inside β-cells (Fig. 8.22). The high intracellular concentration of ATP inhibits the ATP-sensitive K⁺ channels in β-cell plasma membrane, resulting in K⁺ retention inside the cell. The elevation of intracellular K⁺ concentration leads to membrane depolarization, which favors the opening of the voltage-dependent Ca²⁺ channels, allowing the increase in the cytosolic concentration of Ca²⁺, which activates the secretory pathway, leading to insulin release from β-cells (Fig. 8.22). It is important to have in mind that the preexistence of intracellular vesicles loaded with insulin allows a very rapid insulin secretion upon stimulation.

The mechanism of insulin action on its target cells is complex and initiates with insulin binding to its receptor on cell surface. Insulin receptor is expressed in virtually all the cells, but differences in the components of the intracellular signaling pathway as well as the presence or not of the target enzymes or their specific isoforms result in distinct effects of insulin in each tissue.

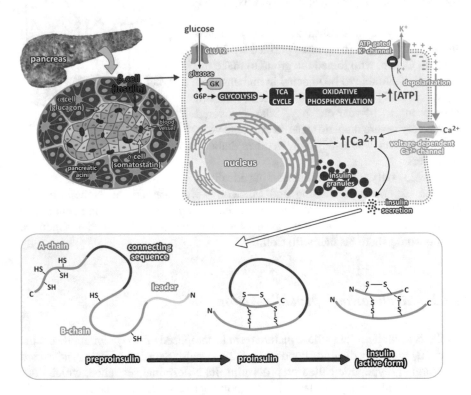

**Fig. 8.22** Schematic representation of a Langerhans islet, showing β-cell localization (*left*) and the mechanism of insulin secretion in detail (*right*). The increase in ATP levels in β-cells due to the increase in glucose uptake and metabolization inhibits the ATP-sensitive K⁺ channels, leading to membrane depolarization and Ca²⁺ release from ER and uptake from the extracellular medium. The high cytosolic Ca²⁺ concentration stimulates the secretory pathway resulting in insulin release from β-cell. Mature insulin is formed by proteolytic processing of the preproinsulin (*bottom*). First, preproinsulin is converted in proinsulin by the cleavage of its N-terminal signal peptide, which leads the formation of three disulfide bonds. Then, mature insulin is formed by removal of the C peptide segment through the action of the prohormone convertases

The insulin receptor is a tetrameric protein composed of two extracellular α-subunits, which contain the insulin-binding site, and two transmembrane β-subunits, which display an intrinsic tyrosine kinase activity (Fig. 8.23). Insulin binding to α-subunit promotes a conformational change that is transmitted to the β-subunit, causing its autophosphorylation and further activation. The phosphorylated β-subunit recruits and phosphorylates a family of proteins known as insulin receptor substrates (IRS), which connect insulin receptor activation with the downstream signaling pathways (see Box 8.10), regulating a broad array of physiological functions.

Although very complex, insulin-mediated effects on cellular function may be summarized in a simplified double pathway: (a) the pathway mediated by phosphatidylinositide-3-kinase (PI3K), which controls metabolism, and (b) the pathway mediated by mitogen-activated protein kinases (MAPK), which regulates gene expression and cell proliferation (Fig. 8.23). The pathway-mediated MAPK will not be discussed in this book.

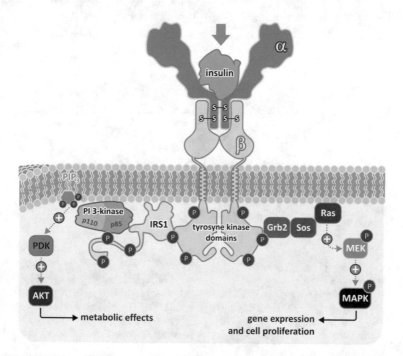

**Fig. 8.23** Overview of the insulin signaling pathway. Insulin receptor is composed of α-subunits, which contain the insulin-binding site, and β-subunits, which display tyrosine kinase activity responsible for receptor autophosphorylation. Phosphorylated β-subunit recruits and phosphorylates insulin receptor substrates (IRS), which mediate two main pathways in the target cells, one that results in regulation of gene expression and other that modifies the activity of metabolic enzymes. The metabolic effects are mediated by recruitment of p85, the regulatory subunit of phosphatidylinositide-3-kinase (PI3K), which binds and activates the PI3K catalytic p110 subunit. Active PI3K phosphorylates the plasma membrane phosphatidylinositol 4,5-bisphosphate (PIP$_2$) at the position 3 of the inositol ring, generating phosphatidylinositol 3,4,5-trisphosphate (PIP$_3$), which mediates many metabolic effects of insulin action. The alterations in gene expression are mainly mediated by the activation of the mitogen-activated protein kinases (MAPK). The Grb2 protein associates to the phosphorylated receptor and recruits the protein Sos, which in turn activates Ras that activates Raf. Raf is a protein kinase that phosphorylates and activates another protein kinase, MEK, which, in turn, phosphorylates and activates MAPK

### Box 8.10  IRS Proteins, Adapter Proteins in Insulin Signaling

IRS proteins do not have enzymatic activity but play a key role in connecting insulin receptor activation to the downstream enzymes that regulate a number of cellular functions. IRS proteins are characterized by three major structural elements that enable them to transmit insulin signal. The N-terminal end folds in two conserved domains: a "pleckstrin homology" domain (PH), which contains conserved basic residues that allow interaction with anionic phospholipids, recruiting the protein to the cellular membrane, and a phosphotyrosine binding domain (PTB), through which the protein interacts with the phosphorylated insulin receptor (see below figure). The C-terminus contains several Tyr phosphorylation motifs that when phosphorylated are binding sites to enzymes or proteins containing another type of protein-protein interaction domain, the "Src homology 2" (SH2) domains. SH2 proteins, in turn, frequently possess SH3 domains that recognize phosphorylated motifs in other intracellular proteins, leading to further downstream signal transduction.

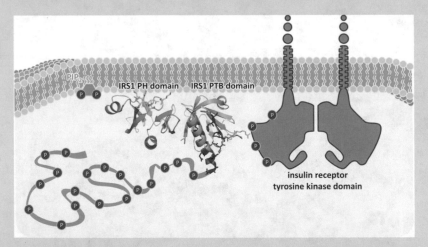

Schematic representation of the insulin receptor and its interaction with IRS proteins. IRS is represented by a combination of the crystal structure of its pleckstrin homology-phosphotyrosine binding (PH-PTB) targeting region (PDB 1QQG) and a schematic representation of its unstructured C-terminus containing the Tyr phosphorylation motifs

The ability of PH domain to interact with phospholipids not only facilitates IRS proteins to bind insulin receptor but also favors the recruitment of phosphatidylinositide-3-kinase (PI3K), a SH2 enzyme, to the plasma membrane, where its substrate, phosphatidylinositol 4,5-bisphosphate ($PIP_2$), is located (see Sect. 8.4.3). Additionally, the C-terminal end also contains Ser/Thr-rich regions that may be recognized by kinases that act as negative regulators of insulin action (see Sect. 1.3.1).

## 8.4.3 Effects of Insulin on Energy Metabolism

Insulin effects on metabolism are mainly mediated by PI3K, an enzyme composed of two subunits, a regulatory subunit p85 and a catalytic subunit p110. When p85 binds to the Tyr-phosphorylated IRS, it recruits and activates the p110 catalytic subunit, which phosphorylates phosphatidylinositol 4,5-bisphosphate ($PIP_2$) at the position 3 of the inositol ring, generating phosphatidylinositol 3,4,5-trisphosphate ($PIP_3$) (Fig. 8.23). $PIP_3$ recruits several Ser/Thr kinases to the plasma membrane, including 3-phosphoinisitide-dependent protein kinase (PDK). At the plasma membrane, PDK phosphorylates and activates another Ser/Thr kinase, AKT (also known as protein kinase B, PKB), which, in turn, phosphorylates several target proteins, such as GSK3, leading to the activation of glycogen synthesis (see Sect. 8.2.4), and AS160, leading to the increase in glucose uptake through GLUT4 in muscle and adipose tissue (see next section).

### 8.4.3.1 Effects of Insulin on Glucose Uptake by Muscle and Adipose Tissue

In muscle and adipose tissue, glucose is transported across the plasma membrane by GLUT4. This GLUT isoform is not constitutively present on the cell surface but remains sequestered in intracellular vesicles until the insulin signaling pathway is activated (Fig. 8.24). Thus, while blood glucose concentration is maintained around the basal levels, muscle and adipose tissue are not able to uptake glucose from the bloodstream and their main metabolic substrates are the fatty acids. However, after a carbohydrate-rich meal, when blood glucose rises and insulin is secreted, this picture changes completely. Insulin binding to its receptor in muscle and adipose tissue cells triggers the sequence of events described above, leading to AKT (or PKB) activation. One of the AKT substrates is the protein AS160 (AKT substrate 160), which negatively regulates the migration and fusion of intracellular vesicles to plasma membrane. AS160 is inactivated by insulin-mediated phosphorylation, thus allowing the translocation of the GLUT4-containing vesicles to the cell surface (Fig. 8.24).

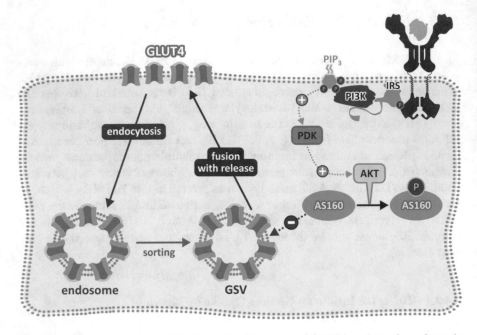

**Fig. 8.24** Schematic diagram of insulin-mediated exposure of GLUT4 on the surface of muscle and adipose tissue cells. Insulin signaling cascade (described in Fig. 8.23) results in the activation of PDK, which phosphorylates AKT, leading to its activation. AKT catalyzes the phosphorylation and inactivation of AS160, which negatively regulates GLUT4 translocation to cell surface

### 8.4.3.2    Effects of Insulin on Metabolic Pathways

Insulin signaling transmitted through the activation of PI3K alters the activity of several enzymes in different metabolic pathways.

Glycogen is stored due to the simultaneous activation of its synthesis and inhibition of its degradation. This occurs through the combined effects of the phosphorylation of GSK3 by AKT, which results in its inhibition, and the activation of PP1. PP1 is bound to glycogen granules, where it catalyzes the dephosphorylation of GS, promoting its activation, and also acts on GP and on phosphorylase kinase, leading to their inactivation (see details in Sect. 8.2.4). Thus, GS is maintained active through its dephosphorylation by PP1 together with the inhibition of its phosphorylation by GSK3, leading to glycogen synthesis both in the liver and muscle (Fig. 8.25).

Glycolysis is stimulated in the liver due to the dephosphorylation of bifunctional enzyme phosphofructokinase-2/fructose-2,6-bisphosphatase (PFK-2/F2,6-BPase; see Sect. 9.4.1 for details), which results in the activation of the PFK2 and in the inhibition of the F2,6BPase activities, leading to a great increase in the concentration of fructose-2,6-bisphostate that in turn activates PFK1 and inhibits fructose-1,6-bisphosphatase, favoring glycolysis and stopping gluconeogenesis (Fig. 8.25).

Fatty acid and triacylglycerol synthesis is stimulated by insulin, whose signaling pathway leads to ACC dephosphorylation, allowing its citrate-induced polymerization and activation both in the liver and adipose tissue (Fig. 8.25). Gene expression regulation by insulin also favors lipid biosynthesis since it induces the expression of ACC gene.

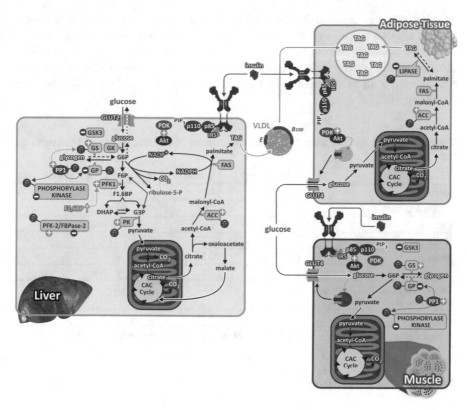

**Fig. 8.25** Metabolic interplay in hyperglycemia, showing the enzymes and the metabolic pathways regulated in each cell

## 8.5   Metabolic Interplay in Response to Hyperglycemia

Responses to hyperglycemia start with liver and β-cells sensing the higher levels of blood glucose through GLUT2/GK sensor (see Sect. 8.1). β-cells respond to the increase in glycemia by secreting insulin, which acts on tissues and organs stimulating further glucose utilization and storage.

In liver cells, blood glucose is internalized by GLUT2 and converted to glucose-6-phosphate, which has at least four direct fates: (a) conversion to glycogen (see Sect. 8.2); (b) degradation through glycolysis (see Sect. 7.4); (c) metabolization through the pentose-phosphate pathway (see Sect. 8.3); and (d) reconversion to glucose by glucose-6-phosphatase. Insulin action on hepatocytes results in an increase in hepatic accumulation of glycogen and in a stimulus to glycolysis with the subsequent transformation of glucose in fatty acids, which are then incorporated into triacylglycerol molecules. Triacylglycerols travel in the bloodstream after their incorporation to VLDL to reach the adipose tissue, where they are stored (Fig. 8.25).

GLUT4 is the major glucose transporter in muscle and adipose tissue. Thus, the insulin-dependent translocation of GLUT4 to the cell surface results in a great enhancement in glucose uptake by these cells. Insulin action also results in a rapid switch to glucose usage as the major metabolic substrate in these cells. Since muscle and adipose tissues together represent more than 60% of the body weight and contribute to almost one third of the basal metabolic rate of the organism (Table 8.2), their metabolic activity in this situation causes a rapid removal of glucose from the bloodstream with its storage as glycogen in muscle and triacylglycerol in adipose tissue (Fig. 8.25).

**Table 8.2** Weight distribution and estimated contribution of different organs and tissues to the basal metabolic rate[a]

| Tissue | Weight (kg) | % body weight | % basal metabolic rate |
|---|---|---|---|
| Muscle | 28 | 40 | 22 |
| Adipose tissue | 15 | 21.4 | 4 |
| Liver | 1.8 | 2.6 | 21 |
| Brain | 1.4 | 2.0 | 20 |
| Heart | 0.33 | 0.5 | 9 |
| Kidney | 0.31 | 0.4 | 8 |
| Other (skin, intestine, bones, glandules, etc.) | 23.16 | 33.1 | 16 |
| Total | 70 | 100 | 100 |

[a]Adapted from Elia. Nutr. Res. Rev. 4:3–31, 1991, with permission

# Selected Bibliography

Baskaran S, Roach PJ, DePaoli-Roach AA, Hurley TD (2010) Structural basis for glucose-6-phosphate activation of glycogen synthase. Proc Natl Acad Sci U S A 107:17563–17568

Cárdenas ML, Cornish-Bowden A, Ureta T (1998) Evolution and regulatory role of the hexokinases. Biochim Biophys Acta 1401:242–264

Chaikuad A, Froese DS, Berridge G, von Delft F, Oppermann U, Yue WW (2011) Conformational plasticity of glycogenin and its maltosaccharide substrate during glycogen biogenesis. Proc Natl Acad Sci U S A 108:21028–21033

Ferrer JC, Favre C, Gomis RR, Fernández-Novell JM, García-Rocha M, de la Iglesia N, Cid E, Guinovart JJ (2003) Control of glycogen deposition. FEBS Lett 546:127–132

Guo S (2014) Insulin signaling, resistance, and the metabolic syndrome: insights from mouse models into disease mechanisms. J Endocrinol 220:T1–T23

Leloir LF (1970) Two decades of research on the biosynthesis of saccharides. Nobel lecture. http://www.nobelprize.org/nobel_prizes/chemistry/laureates/1970/leloir-lecture.html

Liu H, Liu J-Y, Wu X, Zhang J-T (2010) Biochemistry, molecular biology, and pharmacology of fatty acid synthase, an emerging therapeutic target and diagnosis/prognosis marker. Int J Biochem Mol Biol 1:69–89

Matschinsky FM (2002) Regulation of pancreatic beta-cell glucokinase: from basics to therapeutics. Diabetes 51(Suppl 3):S394–S404

Roth J, Qureshi S, Whitford I, Vranic M, Kahn CR, Fantus IG, Dirks JH (2012) Insulin's discovery: new insights on its ninetieth birthday. Diabetes Metab Res Rev 28:293–304

Sun L, Zeng X, Yan C, Sun X, Gong X, Rao Y, Yan N (2012) Crystal structure of a bacterial homologue of glucose transporters GLUT1-4. Nature 490:361–366

Thorens B, Mueckler M (2010) Glucose transporters in the 21st century. Am J Physiol Endocrinol Metab 298:E141–E145

Tong L (2013) Structure and function of biotin-dependent carboxylases. Cell Mol Life Sci 70:863–891

## Challenging Case 8.1: Beyond the Enigmatic Smile—The Occult Hyperlipidemia of Mona Lisa

### *Source*

The "Mona Lisa," also referred to as "La Gioconda," was painted by Leonardo Da Vinci between 1503 and 1506. It is a masterpiece of the Italian Renaissance, considered the most famous painting in the world. The painting is believed to be the portrait of the Italian noblewoman Lisa Maria de Gherardini, the wife of the Florentine silk merchant Francesco del Giocondo. Mona Lisa's enigmatic expression and the mysterious nature of the painting have interested innumerous scholars, but only in 2004, the painting called medical attention. In an article published in the periodical Medical Archeology, a team of medical doctors suggested that a skin lesion noted on Mona Lisa's left upper eyelid and the swelling seen on her right-hand dorsum were evidence of a lipid disorder.[1]

Mona Lisa, by Leonardo Da Vinci, 77 × 53 cm oil on poplar panel, which is on permanent display at the Louvre Museum in Paris since 1797; public domain

### *Case Description*

When one appreciates an art piece, one cannot wonder everything the author intended to transmit with the work. When one dissects a Leonardo Da Vinci's work, the possible "hidden messages" increase in orders of magnitude.

Leonardo Da Vinci was a quite unique person, who, although having no formal academic training, acted as a master in areas as distinct as painting and engineering, music and anatomy, as well as architecture and botany, being considered a "universal genius." His innovative techniques for painting together with his detailed knowledge of anatomy, as well as his interest in the way humans register emotion in facial expression and gesture, were probably determinants for him to produce what is now considered the most popular portrait ever made: Mona Lisa (or La Gioconda). There are countless examples of the fascination this painting has ignited over the years. For instance, almost 450 years after this masterpiece was created, Nat King Cole, a

[1] Dequeker J et al (2004) Xanthelasma and Lipoma in Leonardo da Vinci's Mona Lisa (1503–1506). Med Archaelogy 6:505–506.

singer from the USA, immortalized it in a song that became a classic, titled "Mona Lisa."[2]:

"(...)
*For that mona lisa strangeness in your smile?*
*Do you smile to tempt a lover, mona lisa?*
*Or is this your way to hide a broken heart?*
(...)
*Are you warm, are you real, mona lisa?*
*Or just a cold and lonely lovely work of art?"*

The mysteries surrounding the Mona Lisa painting are many, including whether the model was indeed Lisa de Gherardini; what was the meaning of her enigmatic expression; and why the model is positioned in front of an imaginary landscape, which is painted in a way that almost fuses with the human figure, representing the Da Vinci's idea of a direct connection between nature, or the macrocosmos, and the microcosmos of the human body.

All these aspects have been extensively explored by scholars in different areas, but only recently, two cutaneous abnormalities highlighted by Da Vinci in the painting (see figure) called the attention of medical doctors. This happened in 2004, when a group at the Leuven University Hospitals published an article on this topic, stating that "*a careful clinical examination of the famous painting Mona Lisa ... reveals a yellow irregular leather-like spot at the inner end of the left upper eyelid and a softy bumpy well-defined swelling at the dorsum of the right hand beneath the index finger about 3 cm long.*"[1] Based on these observations, they postulated that the yellow spot constituted a xanthelasma and that the swelling on the hand was suggestive of a subcutaneous lipoma and hypothesized that Lisa suffered from familial hyperlipidemia. They also suggested that the dyslipidemia, a risk factor for ischemic heart disease in middle age, would have been the indirect cause of the premature death of Lisa de Gherardini (she is thought to have died at the age of 37).

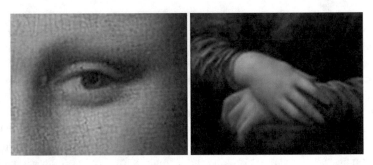

Details of Mona Lisa portrait showing evidence of hyperlipidemia: the yellowish irregular spot in the upper corner of the left orbital cavity, suggestive of xanthelasma (*left*), and the rounded relief plate on the back of the right hand, suggestive of subcutaneous lipoma (*right*)

---

[2] "Mona Lisa" (1950) written by Ray Evans and Jay Livingston for the film Captain Carey, USA. Paramount Pictures. The song won the Oscar for Best Original Song in 1950.

It is unlikely that those physical features, suggestive of a lipid metabolic disorder, had been intentionally represented by Da Vinci to reveal that Lisa de Gherardini suffered from dyslipidemia. Neither the physical signs nor the symptoms of this type of disease had been described at that time. Probably they just reflect a precise and detailed representation of the model's appearance. In any case, Da Vinci registered, probably for the first time, the occurrence of dyslipidemia already in the sixteenth century.

After the publication of the article by the medical doctors from Leuven, other medical hypotheses had been raised to explain Mona Lisa's physical features. One example is a Letter to the Editor of the Mayo Clinic Proceedings in which the authors discredit the hypothesis of Lisa de Gherardini's death by genetic dyslipidemia, suggesting that her hyperlipidemia would be a secondary symptom of hypothyroidism.[3] They argued the painting revealed not only the already identified xanthelasma and lipoma but also the high forehead, thinned and coarse hair, absent eyebrows, overall yellowish hue of the skin, as well as presence of a diffuse neck enlargement such as a goiter.

## Questions

1. The Renaissance, the cultural movement that took place in the historical moment in which Da Vinci and his Mona Lisa lived, impacted profoundly European life, affecting not only art, architecture, and science, but also changing human behavior. In the book *Feast: The History of Grand Eating*[4], Roy Strong describes the extravagances of the nobility's eating habits during the Renaissance, usually used to show power and wealth. The banquets always contained lots of spices and sugar, which had been recently brought from the Americas, and recipes made with white flour, which at that time was more expensive than meat. Considering this historical context, discuss how Mona Lisa, a young woman of the Florentine nobility, may have developed hyperlipidemia.

2. Synthesized TAGs can be incorporated into lipoproteins to be either secreted from the cell or stored in intracellular organelles, the lipid droplets (LDs), which are responsible for their local storage. Considering these two fates for the newly synthesized lipids, hypothesize another possible lipid disorder, besides hyperlipidemia, that Mona Lisa may have developed.

3. The thyroid hormones regulate a broad range of physiological processes including the body energy expenditure through the increase in the metabolic rate and thermogenesis (see Sect. 11.2.2). Considering that lipid metabolism, especially cholesterol and lipoprotein synthesis, is also regulated by thyroid hormones, discuss the hypothesis of hypothyroidism as the cause of Mona Lisa's hyperlipidemia.

---

[3] Mehra MR, Campbell HR (2018) The Mona Lisa decrypted: allure of an imperfect reality. Mayo Clin Proc 93:1325–1326.

[4] Strong R (2013) Feast: the history of grand eating. Harcourt, 368 pages. ISBN-13: 978-0151007585.

## *Biochemical Insights*

1. Hyperlipidemia is a disorder in lipid metabolism characterized by an increase in the concentration of circulating lipids. It may occur as a consequence of a high dietary intake of lipids or a result of an upregulation of the de novo lipid synthesis, which occurs mainly in the liver and adipose tissue under a positive energy balance condition (see Sect. 8.3). The carbohydrates consumed in the diet, such as sucrose, starch, and lactose, are broken down during the digestive process generating glucose, the molecule responsible for insulin secretion by pancreatic β-cells. Insulin is primarily responsible for inducing translocation of GLUT4 to the membrane of muscle and adipose tissue cells, making them able to internalize glucose (see Sect. 8.4.3.1). In a condition in which more glucose is internalized than the energy demand of the cell, glucose will be converted to glycogen in the liver and muscle or triacylglycerol (TAG) in the liver and adipose tissue.

   Glycogen is a highly hydrated molecule, being associated with three or four parts of water per part of glycogen itself. This makes the cellular capacity of glycogen storage very limited, reaching about 100 g in the liver and 400 g in the entire muscle tissue. On the other hand, lipid synthesis and storage may persist until blood glucose concentration is maintained above the basal levels. In adipose tissue, lipids are produced and stored directly, increasing the tissue volume. In the hepatic tissue, when lipogenesis increases, TAGs are synthesized and incorporated in very-low-density lipoproteins (VLDL) to be secreted. VLDL circulates and delivers these lipids to other organs, especially the adipose tissue, where they will be stored. The increased levels of circulating lipoproteins characterize the condition named as hyperlipidemia.

2. In the liver, the synthesis of both lipoproteins and lipid droplets occurs through similar pathways. The acyl-CoA molecules, coming either from the absorption of circulating lipids or from the de novo biosynthesis, accumulate between the leaflets of the endoplasmic reticulum (ER) membrane bilayer. These lipids are esterified by enzymes present in the ER membrane: either the diacylglycerol acyltransferases (DGAT1 and DGAT2), which incorporate the acyl-CoA molecules in diacylglycerols, forming TAGs; or the acylCoA:cholesterol-O-acyltransferases (ACAT1 and ACAT2), which esterify cholesterol using acyl-CoA. The accumulation of TAGs and cholesterol esters between ER membranes' leaflets leads to the formation of oil lenses that bud toward the cytoplasm, carrying one of the ER leaflets.[5,6] This gives rise to the lipid droplets (LDs), intracellular organelles delimited by a phospholipid monolayer. Alternatively, budding may occur toward the ER lumen, generating lipoproteins, which are directed to the Golgi and secreted. The budding toward ER seems to

---

[5] Bhatt-Wessel B et al (2018) Role of DGAT enzymes in triacylglycerol metabolism. Arch Biochem Biophys 655:1–11.

[6] Yen CLE (2008) DGAT enzymes and triacylglycerol biosynthesis. J Lipid Res 49:2283–2301.

be driven by the interaction with the apoprotein B100, after its lipidation by the microsomal triacylglycerol transfer protein (MTP) (see below figure).

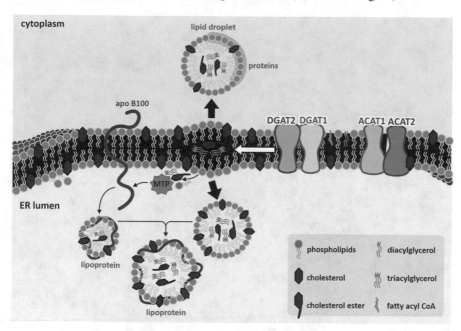

Hypothetical model illustrating the formation of LD and lipoproteins. Triacylglycerol and cholesterol esters synthesized by DGAT and ACAT, respectively, accumulate in between ER membrane leaflets, leading to the formation of LDs, when budding occurs toward cytoplasm, or lipoproteins, when budding occurs toward ER lumen, directly or via apoB100 lipidation by MTP. (Figure adapted from the article cited in the footnote 6)

The increase in the number or size of LDs characterizes a severe pathology named hepatic steatosis. When more than 5% of hepatocytes present this alteration, the condition is known as nonalcoholic fatty liver disease (NAFLD).[7] This condition may progress to the nonalcoholic steatohepatitis (NASH), a more severe disease that involves severe liver dysfunction and the induction type 1 collagen production by the stellate cells. The formation of this excess of fibrous connective tissue in the organ induces the replacement of a large amount of the hepatocytes by this scar tissue, characterizing what is known as cirrhosis (see figure). The development of hepatic cirrhosis can further aggravate and trigger hepatocellular carcinoma, a major cause of liver transplantation.

---

[7] Cohen JC et al (2011) Human fatty liver disease: old questions and new insights. Science 332:1519–1523.

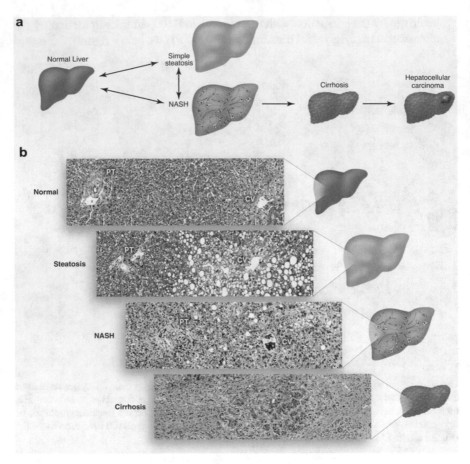

The spectrum of nonalcoholic fatty liver disease (NAFLD). (**a**) Schematic representation of the liver aspect in the different stages NAFLD progression. TAG accumulation in the hepatocytes' lipid droplets causes steatosis. Development of inflammation, cell death, and fibrosis causes NASH, which can progress to cirrhosis. (**b**) Histological sections of normal liver, steatosis, NASH, and cirrhosis. Blue staining indicates collagen fibers. (Figure reproduced with permission from article cited in the footnote 7)

Obesity and type 2 diabetes are highly correlated to the onset of hepatic steatosis. Patients with type 2 diabetes show reduced sensitivity to insulin. Therefore, although type 2 diabetic patients are able to secrete insulin, glucose internalization by their muscle and adipose tissue is impaired, as it occurs mainly through the insulin-dependent GLUT4, leading to a persistent hyperglycemic state. The liver, on the other hand, expresses GLUT2, which internalizes glucose independent of insulin action. In a condition of a persistent hyperglycemia, glucose continuously enters the hepatic tissue, being converted to lipids. Thus, similar to hyperlipidemia caused by the increased production of circulating lipoproteins, hepatic steatosis may be developed as a consequence of a high caloric diet. Therefore, although imperceptible to Da Vinci's eyes, we cannot rule out that Mona Lisa also developed hepatic steatosis.

3. The thyroid hormones, thyroxin (T4) and triiodothyronine (T3), regulate metabolism by inducing gene expression (see Sect. 11.2.2). They bind to nuclear receptors (the thyroid hormone receptors, THR), which act as a hormone-activated transcription factors, regulating gene expression through the binding to specific hormone-responsive elements in DNA (see Fig. 11.7). By activating lipases, thyroid hormones reduce triacylglycerol (TAG) content of low-density lipoproteins (LDL). Additionally, triiodothyronine (T3) increases the expression of LDL receptors.[8] Thus, the low levels of thyroid hormones occurring in hypothyroidism may then increase the amount of circulating lipids by either maintaining TAG-rich lipoproteins or by limiting the ability of tissues to internalize LDL.

## Final Discussion

As we discussed in the question 2, the metabolic alterations induced by obesity may lead to hepatic steatosis that may progress to a spectrum of fatty liver complications known as nonalcoholic fatty liver disease. However, currently the main causes of fatty liver diseases are the alcoholism and infection by hepatitis C virus (HCV).

Both alcoholism and HCV infection cause chronic changes in liver metabolism. The development of hepatic steatosis in both cases is associated with the modulation of the expression of key enzymes involved in the lipid metabolism via two distinct families of transcription factors: the sterol regulatory element-binding proteins (SREBP) and the peroxisome proliferator-activated receptors (PPARα).

SREBPs are transcription factors located at the nuclear envelope and ER membrane. When activated, they migrate to the nucleus where they bind to the sterol regulatory elements (SREs) in DNA, inducing the expression of enzymes involved in the synthesis of fatty acids and cholesterol.

PPARα is a nuclear receptor that acts as ligand-activated transcriptional factor when activated via dimerization with retinoid X receptors (RXR). It is major regulator of lipid metabolism in the liver, controlling the expression of genes associated with oxidation, transport, and exportation of fatty acids.

---

[8] Duntas LH, Brenta G (2018) A renewed focus on the association between thyroid hormones and lipid metabolism. Front Endocrinol (Lausanne) 9:511.

**Alcoholic Fatty Liver (AFL)**

Chronic high exposure to alcohol (more precisely ethanol) affects hepatic metabolism mainly due to an increase in the NADH/NAD+ ratio. In the liver, ethanol is metabolized in the cytoplasm by the enzyme alcohol dehydrogenase (ADH), generating acetaldehyde, which is be converted to acetate by mitochondrial aldehyde dehydrogenase (ALDH2) (see also Challenging Case 7.2).

The products of these two steps are involved in the activation of lipid synthesis in hepatocytes. Acetate is a direct substrate for lipid synthesis through its conversion to acetyl-CoA by acetyl-CoA synthetases (ACSS1 and ACSS2). Acetaldehyde induces the expression of SREBP while inhibiting PPAR alpha expression, ultimately activating fatty acid synthesis and inhibiting its oxidation, respectively.

Additionally, both ADH and ALDH2 reactions use NAD+ as an electron acceptor, producing NADH. Reducing the availability of the oxidized form of NAD impairs other NAD+-dependent reactions such as the citric acid cycle and β-oxidation, reducing ATP production capacity by these pathways, and leading to lipid accumulation in the hepatocytes.

**HCV Infection**

Viruses are obligate intracellular parasites that depend on host cell metabolic pathways for successful infection and replication. HCV is a member of the Flaviviridae family, a group of viruses that particularly alter host energy metabolism, especially lipid metabolism, to favor their replication.

HCV is able to inhibit the microsomal triacylglycerol transfer protein (MTP), responsible for ApoB lipidation (see question 2), resulting in secretion of a VLDL particle with lower TAG content, while increasing the amount of TAG in cytoplasmic LDs, favoring the development of hepatic steatosis.

HCV also controls other points of host lipid metabolism. For instance, the viral core protein associates with the LD surface, favoring the biogenesis of this organelle. This event is essential for replication since LD acts as a platform for the assembly of the newly synthesized viral particle.

Interestingly, it is well described that HCV circulates in the body in the form of lipoviroparticles. These particles are formed because HCV nucleocapsids (the core protein associated with the viral genome) remain associated with LD and fuse to VLDL. Therefore, when VLDL is secreted, it carries an HCV nucleocapsid inside, using the mechanism of VLDL internalization to disseminate the virus.

*This case had the collaboration of Dr. Lorena O. Fernandes-Siqueira (Instituto de Bioquímica Médica Leopoldo de Meis, Universidade Federal do Rio de Janeiro, Brazil).*

## Challenging Case 8.2: Dolores Bothered to Kill Her Dying Husband

### Source

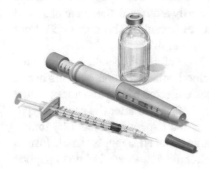

Dolores Miller, a practical nurse, was convicted of murdering her 10th husband, who was dying after a not totally successful brain surgery. The murder weapon was unusual: insulin! Overdose of this hormone drug impacts severely on metabolism and may be lethal. Its use as a weapon may not be so surprising after all. This case is based on real events described in *Insulin Murders: True Life Cases*.[9]

Insulin syringe and pen. (Image reproduced from Blausen.com staff (2014). Medical gallery of Blausen Medical 2014. WikiJournal of Medicine 1(2))

### Case Description

Dolores Miller's 10th husband was Erroll, also known as Roy. Thirteen days after getting married, Roy was admitted to hospital for an incurable brain tumor. The previous 2 months were quite agitated for him due to the onset of a personality change related to fits. He left a secure job and his family after withdrawing his life savings from the bank and before meeting Dolores, a practical nurse, through a lonely hearts club. They lost no time in getting married.

Roy underwent brain surgery, but the surgeons found it impossible to remove the entire tumor, and, according to their expert opinion, he was expected to almost certainly live less than 6 months. Recovery from surgery was uneventful despite Roy's inability to speak until the sixth postoperative day, when he became unconscious and had to be resuscitated. This was unexpected and happened while he was being taken for an X-ray brain scan. Blood collected for routine analysis showed severe hypoglycemia. Doctors immediately decided for glucose administration. First, bolus doses were given intravenously; then constant intravenous infusion was provided. In all, a massive 260 g of glucose were required over a 9–11 h period before glycemia raised to "normal" levels. Nonetheless, the outcome was tragic: Roy did not recover consciousness, remained comatose, and died 4 days later. Autopsy revealed the persistent brain tumor and no evidence of pancreas tumor or other apparent cause for his death. Doctors had suspected hypoglycemia would have been

---

[9] Mark V, Richmond C (2007) Insulin murders. True life cases. Chap. 14. The Royal Society of Medicine Press; ISBN 978-1-85315-760-0.

caused by insulin someone had injected into Roy and the autopsy aggravated suspicion. His recent wife, being a nurse, was "flagged" by the medical team since the first day of coma. During visits, she was never left alone in his room. However, on the day Roy died, during a visit Dolores collapsed over the intravenous infusion lines that were trailed across his bed while the nurse was adjusting his medication. As she flung herself over Roy, Dolores shouted that he was dying but the nurse noticed nothing different about him. However, shortly afterward, the nurse observed on the cardiac monitor the heartbeat had become irregular. The nurse could not reverse this situation and Roy died shortly after this episode.

The most recent blood samples, collected just before his death, had revealed glycemia near zero. The same happened with blood collected soon after his death. Blood retrieved from blood samples that had been collected before and after the first hypoglycemic attack, but curiously not the ones before and after the second hypoglycemic attack, was sent for analysis for insulin and C-peptide content by immunoassay. The results were consistent with insulin-induced hypoglycemia. Insulin was present in high concentration in the blood after the first attack, but no C-peptide was detected.

Dolores confessed to the police to giving her husband insulin intravenously but not with the intention to kill but solely to have him transferred to the intensive care ward. The criminal investigation found that Dolores had, prior to Roy's death, withdrawn all of her new husband's savings from the bank and contacted several undertakers to find out how quickly a cremation could be arranged. A stolen gun, lists of members of the lonely hearts club and several types of medicines and syringes, but no insulin, were found in her car.

The immediate cause of death was never ascertained. A second administration of insulin, with a sudden decrease of blood glucose levels, followed by an equally sudden release of adrenaline, with severe impact in his heart given his ill condition? It was plausible but not demonstrated. Nevertheless, a jury in Franklin County Court, Missouri, USA, in February 1984, was convinced that Dolores was guilty of murder. She was sentenced to life imprisonment.

## Questions

1. Why wasn't glucagon used to counteract hypoglycemia when trying to recover normal blood glucose levels?
2. What was the importance of determining C-peptide in the blood samples?
3. The autopsy did not reveal the existence of pancreatic tumors. What is the relevance of this finding?
4. What physical chemical properties of insulin can be manipulated in order to produce insulin variants having fast or prolonged action?

## Biochemical Insights

1. Glycemia homeostasis in the human body is heavily dependent on the insulin–glucagon balance (see Fig. 8.2b). Glucagon is produced and released by pancreatic alpha cells in response to hypoglycemia (see Sect. 9.4.1) the same way insulin is produced and released by pancreatic beta cells in response to hyperglycemia (see Sect. 8.4.2). The coordinated action of both hormones contributes to stable glucose levels. At first glance it could seem that administration of glucagon would counteract the action of excess insulin, but this is illusory.

   Glucagon's main effect on metabolism is the degradation of liver glycogen and stimulation of gluconeogenesis (see Sect. 9.4.1.1). However, the amount of liver glycogen is typically about 100 g, which is much lower than needed to keep normal blood glucose levels in victims of insulin overdose in the days after intake. In a case thoroughly described in the literature, a patient required nearly 2.5 kg total glucose administered for 120 h at an infusion rate of 3–6 mg/kg/h.[10] Moreover, the pancreas of patients will produce glucagon, which in principle is already impacting on liver glycogenolysis. Exogenous glucagon would no doubt accelerate glycogenolysis and gluconeogenesis, but it would not be enough. Even in a case in which suicide was attempted and exogenous sugars were taken in food and drinks soon after insulin injection to revert the intended outcome of insulin overdose, admission to hospital was needed with infusion of glucose for many hours and administration of glucose at hypoglyccmic peaks.[11] In conclusion, external administration of D-glucose (sometimes called "dextrose" in older medical literature) or even concentrated saccharose (common sugar) is simpler, faster, and more cost-effective than glucagon-based therapies.

   One simple additional action of glucose administration to improve the clinical situation of insulin-poisoned patients is to excise the insulin injection site.[12]

2. C-peptide is released in blood together with insulin from pancreatic β-cells (see Sect. 8.4.2). In contrast, exogenous insulin is C-peptide-free. A high concentration of insulin concomitant with a low concentration of C-peptide is thus indicative of exogenous insulin administration. In this case, hyperinsulinemia-derived hypoglycemia can be assigned to human action, not natural causes.

3. Pancreatic tumors may in some cases cause overproduction and release of insulin. Insulinoma, for instance, is a rare tumor of the pancreas associated with β-cells, which results in frequent hyperinsulinemia-derived hypoglycemia.

---

[10] Matsumura M et al (2000) Electrolyte disorders following massive insulin overdose in a patient with type 2 diabetes. Intern Med 39:55–57.

[11] Tofade TS, Liles EA (2004) Intentional overdose with insulin glargine and insulin aspart. Pharmacotherapy 24:1412–1418.

[12] Campbell IW, Ratcliffe JG (1982) Suicidal insulin overdose managed by excision of insulin injection site. Br Med J 285:408–409.

Insulinomas are generally small (≥90% are ≤2 cm), usually not multiple (90%), and only 5–15% are malignant. They almost invariably occur only in the pancreas, distributed equally in the pancreatic head, body, and tail. Serum levels of insulin and C-peptide after a 72 h fast are used to diagnose insulinoma.

Interestingly, the biochemical mechanism behind insulinoma-associated hyperinsulinemia is quite straightforward. It all resorts to the kinetic parameters of glucose transporters (GLUT; see Table 8.1) and glucose-phosphorylating enzymes (see Fig. 8.3) in β-cells. In insulinoma patients, β-cells contain GLUT1, while GLUT2 is absent or expressed in low concentrations.[13] The former has $K_M$ of 1–2 mM, and the latter has $K_M$ much higher, 20 mM. Importantly, GLUT2 is expressed in tissues that express glucokinase (GK in Fig. 8.3). There is a correlation between GLUT isoforms and glucose affinity of the hexokinases that are co-expressed in the same cell.[14] Hexokinase 1 (HK1 in Fig. 8.3) has been detected is some insulinoma patients.[15]

The combined action of GLUT2 and GK, both having high $K_M$ and $V_{Max}$ (see next figure), ensures that glucose internalization is proportionally responsive to high glycemia and, concomitantly, glucose is rapidly processed via glycolysis, leading to ATP production, which in turn triggers the cascade of events that result in insulin release (see Fig. 8.22). In the low end of the glycemia range (<5 mM), much lower than the $K_M$ of GLUT2 (20 mM) and GK (8 mM), the activity of both is very low, which hampers insulin release and thus prevents insulin-induced aggravation of hypoglycemia. In contrast, combining GLUT1 with other hexokinases, with $K_M$ of 1–2 mM and <1 mM, respectively, leads to high activity in glucose uptake and metabolization in β cells, even in conditions of hypoglycemia. Insulin is thus excreted to blood further aggravating hypoglycemia.

[13] Boden G et al (1994) Glucose transporter proteins in human insulinoma. Ann Intern Med 121:109–112

[14] Wu L et al (1998) Different functional domains of GLUT2 glucose transporter are required for glucose affinity and substrate specificity. Endocrinology 139:4205–4212.

[15] Henquin JC et al (2015) Human insulinomas show distinct patterns of insulin secretion in vitro. Diabetes 2015(64):3543–3553.

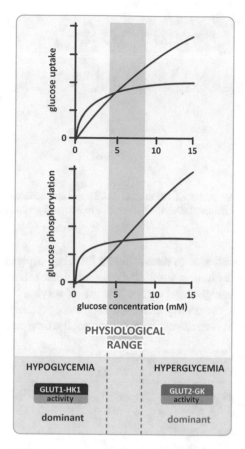

Activities of GLUT1 and HK1 (*blue* curves) or GLUT2 and GK (*red* curves) as a function of glucose concentration. When GLUT2-GK combination is replaced by GLUT1–HK1, there is lower activity in insulin release in hyperglycemia by β-cells and significant activity in insulin release in hypoglycemic conditions. Insulin release is proportional to ATP levels in the cells, which are governed by intracellular levels of phosphorylated glucose

4. Insulin is usually injected forming subcutaneous depots. Release of insulin from these depots depends on its solubility, which is determined by (1) the abundance of adipose cells and irrigation of the microenvironment around the depot, and (2) the hydrophobicity/polarity of the insulin molecules being used in therapeutics. Insulin in high concentration forms hexamers and the hexamer aggregate. Disaggregation and dissociation of the hexamers cause insulin to be available in the monomeric form, which is the one interacting with cellular receptors. Therefore, insulin forms that are more soluble in aqueous environment are faster-acting. The human insulin analogue known as "aspart," for instance, produced by substituting an aspartic acid residue for a proline residue at position 28 of the B-chain (see below figure) dissociates more rapidly and is a fast-acting form of insulin.

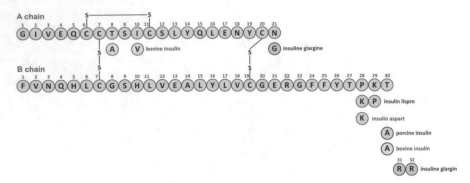

Amino acid sequence of the two chains of human insulin and synthetic variants aspart, lispro, and glargine. Bovine and porcine insulin variations relative to human insulin amino acid sequence are also represented

Insulinemia variation with time differs a lot when aspart insulin is used compared to unmodified human insulin[16] (see next figure). Serum concentration of insulin aspart peaks higher and earlier. The use of varying concentrations of zinc, present in insulin hexamers in pancreas, and cationic proteins such as protamine in insulin formulations is another strategy to control the time course of insulin action.

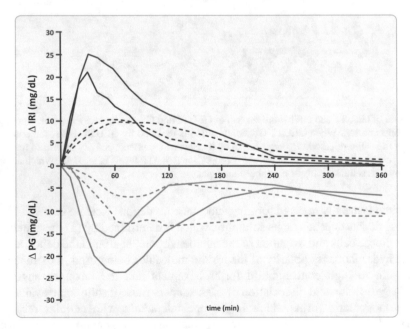

Average variations in immunoreactive insulin (IRI; insulin detected in plasma) and plasma glucose (PG) after 0.025 U/kg body mass (*purple* and *green*) or 0.050 U/kg body mass (*magenta* and *orange*) insulin administration. *Solid lines*, insulin aspart; *dashed lines*, unmodified human insulin. (Data collected from the article cited in the footnote 16)

---

[16] Kaku K et al (2000) Pharmacokinetics and pharmacodynamics of insulin aspart, a rapid-acting analog of human insulin, in healthy Japanese volunteers. Diabetes Res Clin Pract 49:119–126.

Insulins are classified as rapid-acting, short-acting, intermediate-acting, or long-acting. The differences are related to the molecular structure as mentioned above. Most diabetic patients require a combination of a shorter acting insulin to cover increased glucose levels after a meal and longer-acting insulin to maintain a basal level. This is usually accomplished by giving twice-daily doses of intermediate-acting insulin (NPH) combined with a rapid- or short-acting insulin before breakfast and the evening meal. Alternatively, a dose of long-acting insulin is administered in the evening, and doses of rapid-acting or short-acting insulin are given before each meal. Insulin requirements change because of stress, infection, illness, emotional disturbances, exact site of injection, and pregnancy.

## *Final Discussion*

### Insulin as Weapon

Insulin does not come to mind when imagining weapons. However, insulin has been used for murder and suicide, *albeit* not frequently. Its malicious use is almost totally restricted to people in direct contact with the hormone drug: diabetic patients, their caregivers, and medical staff. The first recorded case of an attempted suicide using insulin dates from 1927. Thirty years later, the first use as murder weapon was recorded. *"Insulin is not a very good weapon,"* said Vincent Marks, a world expert on insulin, who assisted some insulin-related trials, to BBC in 2007, *"It is easy to kill babies and old people with insulin, but not adults."* Vincent Marks co-authored with journalist Caroline Richmond the book titled *Insulin Murders: True Life Cases* from which Dolores Miller's true story was taken. In fact, medical literature reviews show that only a small fraction of insulin overdoses results in death. Nevertheless, frequently recovery is not complete. Neurological impairments caused by prolonged hypoglycemia persist.[17,18] In the case that most attracted the attention of the media, a socialite woman was so brain-damaged that she survived in a vegetative state for more than 25 years. His husband went to trial twice and convicted once but later acquitted with the intervention of Professor Marks' expertise.

---

[17] Arem R, Zoghbi W (1985) Insulin overdose in eight patients: insulin pharmacokinetics and review of the literature. Medicine 64:323–332.
[18] "Mona Lisa" (1950) written by Ray Evans and Jay Livingston for the film Captain Carey, USA. Paramount Pictures. The song won the Oscar for Best Original Song in 1950.

**Insulin Concentration: Activity vs. Molar Concentration**

Insulin started being used as therapeutic long before its molecular structure or composition were revealed by Dorothy Hodgkin (see Fig. 2.10). The concentration of insulin could not be expressed in mol per volume because the molar mass was not known. But it could not be precisely purified either so the mass per volume concentration was little help for pharmacological purposes. Thus, insulin concentration was first referred, in 1922, to its effect: a unit of insulin was the amount required to induce hypoglycemia in a rabbit. This is obviously too qualitative as it depends on the body mass and diet of the rabbit among other factors. The unit was later redefined as the amount of insulin required to reduce the concentration of blood glucose in a fasting rabbit to 2.5 mmol/L. Once the chemical structure and mass of insulin were known, the unit of insulin was defined by the mass of pure crystalline insulin required to obtain this effect. The World Health Organization established in 1987 a standard for human insulin with a potency of 26,000 U/g. By means of quantitative amino acid analysis of the human insulin standard, it has been found that 1 mole corresponds to 166.8 million U or that 1 U = 6.00 mmol. Usually the conversion factor is written in the form 1 mU/L = 6.00 pmol/L. There is no factual reason to continue expressing insulin concentration in U as there is a precise conversion for molar concentration, but tradition and cultural factors are hard to overcome. The same happens in biochemical molecular nomenclature (e.g., acetaldehyde is still more frequently used than ethanol).

**Insulin Sources Over Time**

The initial insulin preparations were derived from bovine and porcine sources and contained many impurities. Potency was heterogeneous among different extracts, and uncontrolled side effects were frequent. Over time, the preparation and manufacturing processes improved. Purity improved and the pharmaceutical use became safer and more controlled. Only beef, beef–pork, and pork insulin preparations were available until 1986, when recombinant human insulin started being used. Human insulin was produced by either substituting alanine for threonine on the porcine insulin sequence ("human" semisynthetic insulin) or using genetically altered bacteria ("biosynthetic" human insulin). Today, recombinant human insulin is produced from either genetically altered bacteria (*Escherichia coli*) or yeast (*Saccharomyces cerevisiae*) and has the same amino acid chain sequence as human insulin. Insulin preparations went from 300 to 10,000 parts per million (ppm) of impurities to 10 ppm of impurities. The first insulin analog product, "insulin lispro," was introduced in 1996. Insulin analogs are synthetically derived preparations based on the human insulin structure with a slightly modified amino acid sequence resulting in altered pharmacokinetics, as mentioned above. Currently, only recombinant human insulin and insulin analog preparations are available on the market. Beef and beef–pork insulin preparations were phased out in 2003, and pork insulin production ceased in 2005.

## Beyond Hyperinsulinemia-Induced Hypoglycemia

Several electrolyte disorders, such as hypokalemia, hypophosphatemia, and hypo-magnesemia are observed following insulin overdose.[10,16,17] Pulmonary edema, hypertensive crisis, and respiratory insufficiency have also been described. In patients with hypophosphatemia, depletion of both ATP and 2,3-diphosphoglycerate increases the affinity of hemoglobin for oxygen, which results in decreased delivery of oxygen to tissues, as discussed in [10]. This effect may potentiate cerebral damage by hypoglycemia. Hypomagnesemia is recognized to cause cardiac arrhythmias and potentiates the arrhythmic effect of hypokalemia. Serum electrolytes including phosphorus and magnesium should be measured in cases of severe insulin overdose complicated neuromuscular, central nervous, or cardiovascular disorders.[10] It is worth noting that in the case study presented above, Roy died after suspicious activity of his wife that was followed by cardiac alterations.

# Chapter 9
# Regulation and Integration of Metabolism During Hypoglycemia

In humans, blood glucose is the main reporter of fed and fasting states. Its concentration directly regulates the secretion of hormones, including glucagon and insulin by the pancreas and glucocorticoids by the adrenal cortex. These hormones, in a coordinated action, regulate the energy metabolism in different organs, allowing the blood concentration of glucose to be maintained within a narrow range by a precise system balancing glucose production by the liver and its utilization by peripheral tissues.

The basis for this major role of glucose in controlling human metabolism may be discussed in the context of human evolution. About four million years ago, carbohydrates were important components of the diet of primates and prehuman ancestors, which possibly favored brain and reproductive tissues to develop a specific requirement for glucose as their primary fuel. The subsequent periods of human evolution were dominated by severe Ice Ages that selected the hominids who developed hunting and fishing abilities and consumed high-protein and low-carbohydrate diets. This led to metabolic adaptations to protect the brain and embryonic tissues from the low-glucose availability, resulting in the increased efficiency of hepatic glucose production and the decrease in peripheral glucose utilization. Additionally, the alternate periods of food scarcity and abundance selected metabolic mechanisms to increase deposition of energy reserves during periods of plenty for subsequent use when food was not available, favoring lipid accumulation and the development of the adipose tissue. However, the advent of agriculture after the last Ice Age greatly modified the quantity of carbohydrates consumed by humans, and the Industrial Revolution in the nineteenth century dramatically changed the quality of the carbohydrate ingested (see Box 9.1). These events together are related to an increase of postprandial (i.e., after a meal) glycemia and insulinemia, probably contributing to the predisposition of the modern diseases known as metabolic syndrome (see Chap. 11).

© Springer Nature Switzerland AG 2021
A. T. Da Poian, M. A. R. B., *Integrative Human Biochemistry*,
https://doi.org/10.1007/978-3-030-48740-9_9

Nowadays, in Western cultures, humans have three main meals per day, and the amount of carbohydrates in a regular meal is about 50–60%. After the digestion and absorption of the carbohydrates, glucose reaches the bloodstream, and glycemia rapidly rises from the basal value of 4 mM to more than 10 mM (see Fig. 8.2). In Chap. 8, we discussed the biochemical mechanisms that explain why plasma glucose concentration sharply falls down between the first and the second hour after a meal.

In this chapter, we will turn our attention to the subsequent hours, which are characterized by a slow decrease in glycemia. We will focus on the metabolic pathways and the mechanisms involved in the maintenance of blood glucose concentration when carbohydrates are not ingested, which include the periods in between meals as well as low-carbohydrate diets or prolonged starvation. Additionally, the metabolic interrelationships that take place in the different organs and the hormonal regulation in this situation will also be discussed.

---

**Box 9.1 The Evolution of Human Diet**

Paleolithic and current Western diets show profound differences in composition, especially regarding carbohydrate content and quality (see below figure). Several studies suggest that two major dietary shifts occurred during human evolution: (a) in the Neolithic (~10,000 years ago), when agriculture emerged, leading to an increase in the consumption of domesticated cereals and the adoption of carbohydrate-rich diet and (b) after the Industrial Revolution (~1850), with the introduction of industrially processed flour and sugar. This led to the hypothesis, proposed in a classical article published by Eaton and Konner in 1985, that predisposition of the modern diseases known as metabolic syndrome resulted from an evolutionary discordance between the adaptations established in the Paleolithic era (2.6 million to 12,000 years ago) and the way of life in the industrialized world. Based on this, the authors proposed the "Paleolithic diet" as a reference for human nutrition. However, this view is being questioned nowadays, especially because (a) it implies that human genetic background has not changed since Paleolithic, although humans continued evolving in the Neolithic period, with genetic changes directly related to diet variations, such as on the genes that code for amylase (enzyme that degrades starch) production, and (b) it does not take into account the non-genomic form of inheritance, such as the epigenetic regulation of gene expression, which alters fitness in short-term environmental shifts, such as during in utero development.

The dietary impacts during human evolution can be exemplified with one interesting study that analyzed the genetic diversity of oral microbiota of calcified dental plaque obtained from prehistoric European human skeletons, including the remains of the last hunter-gatherers in Poland and the earliest farming culture in Europe (the Linear Pottery Culture, LBK), as well as late

(continued)

**Box 9.1** (continued)

Neolithic (Bell Beaker Culture), early and later Bronze Age, and medieval rural and urban populations (see in the figure an example of phylum frequencies obtained comparing a specific genome region). This study revealed that oral microbiota became markedly less diverse in the modern times, with the dominance of the potentially cariogenic bacteria, like *S. mutans* (see below figure).

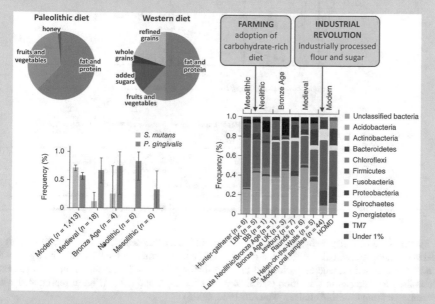

Carbohydrate consumption by man's ancestors corresponds to about 35% of the total energy intake, which comes mainly from fruits and vegetables, while 50% daily energy in a Western diet is obtained from carbohydrate, of which about 15% comes from sugar added to food during processing or consumption and about 20% comes from refined cereals. (Figure based on data from Eaton SB. Proc. Nutr. Soc. 65:1–6, 2006) The changes in carbohydrate content and quality in human diet impact in the diversity of oral microbiota. (Figure reproduced by permission from Macmillan Publishers Ltd.: Adler et al., Nat. Genet. 45:450–456, 2013)

The maintenance of glycemia may be explained by two distinct phenomena: (i) the decrease in glucose utilization by different tissues and (2) the increase in its production and release into the bloodstream by the liver and kidney (Fig. 9.1). When plasma glucose concentration is above the basal levels, insulin secretion is stimulated, and the action of this hormone allows for rapid glucose utilization by all the tissues (see Chap. 8). However, as glycemia decreases, glucagon secretion will predominate, leading to a complete change in metabolism, which is characterized by a decrease in glucose utilization, especially by the muscle and adipose tissues, and a continuous release of glucose in the bloodstream by the liver and, in smaller quantities, by the kidney cortex, as we will discuss in the next sections.

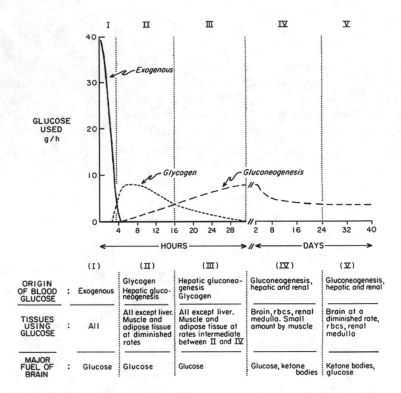

**Fig. 9.1** Classical figure by Dr. George Cahill, in which he describes five metabolic stages between the postprandial state and the near-steady state of prolonged starvation. The figure was constructed based on studies performed in the 1960s by Dr. Cahill and his group with patients submitted to therapeutic starvation (see Box 9.2). *rbcs* red blood cells. (Figure reproduced with permission from Cahill. Ann. Rev. Nutr. 26:1–22, 2006)

## 9.1   Overview of Metabolism During Fasting: Exemplifying with Studies on Therapeutic Starvation

To introduce the issue of glycemia control, we will take advantage of some studies carried out between the 1950s and 1960s, when therapeutic starvation was used as a strategy to treat obesity (Box 9.2). It is important to note that this extreme situation was chosen as an example for clarity and taking into account that the adaptations and the metabolic pathways that will be discussed along this chapter also occur in more common situations, such as low-carbohydrate diets, overnight fasting, or even during the periods in between meals.

**Box 9.2 Therapeutic Starvation**

The studies used here to exemplify the metabolic adaptations to hypoglycemia have been performed by the group of Dr. George F. Cahill (1927–2012) during the 1960s. As pointed out by Richard W. Hanson in a retrospective article on Dr. Cahill's contributions to the understanding of human metabolism, the fact that he was not a biochemist but a physician–scientist made his approach to research integrative and not reductionist in nature. This integrative view allows him to make a crucial discovery regarding brain metabolism during fasting, which explained how humans can survive for more than 60 days without food: ketone bodies, molecules derived from fatty acid metabolism (see Sect. 7.4.6), supply most of the energy requirements of the brain during fasting (see details in Sect. 9.3.4). This finding "resulted in a total reappraisal of the hierarchy of fuels used by different tissues of humans," as stated by Dr. Oliver E. Owen, who worked with Dr. Cahill on these classical experiments. In the 1950s and 1960s, therapeutic starvation of obese subjects was in vogue. The studies described here were performed in the Peter Bent Brigham Hospital, which has a National Institutes of Health-supported clinical research center where the patients were housed and continuously observed during experimental protocols. The patients spent 5–6 weeks fasting with total withdrawal of calories, when the daily intake consisted of one multivitamin capsule, water, and salt replacement. Dr. Owen tells in one of his articles that when someone asked him why he chose a 6-week period for starvation, he answered citing St. Matthew 4:2: "Jesus fasted forty days and forty nights and afterward he hungered." During treatment, the patients volunteered for blood and urine collections for measurements of the plasma concentrations of different metabolites. Furthermore, some of them underwent catheterization to determine the consumption or the production of metabolites by specific organs by measuring the arterial–venous differences of these substances. Some of the data originated from these studies will be used in this chapter to discuss several aspects of human adaptation to fasting.

The variation of blood concentration of different metabolites during the therapeutic starvation of those patients will be used as the starting point for our discussion on the metabolic adaptations to hypoglycemia (Fig. 9.2).

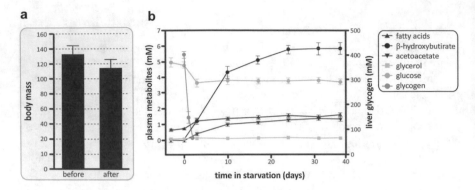

**Fig. 9.2** (**a**) Average body mass of a group of 11 obese patients before and after they were subjected to therapeutic starvation at the Clinical Center of the Peter Bent Brigham Hospital. (**b**) Concentrations of different metabolites in the plasma and glycogen in the liver during fasting. (The figures were prepared using data from Owen et al. J. Clin. Invest. 48:574–583, 1969)

As seen in Fig. 9.2, the concentration of fatty acids in the blood, after a slight increase, remains constant during all the period of fasting due to the equilibrium between its use as energy source by many tissues and its mobilization from triacylglycerols stored in the adipose tissue. Indeed, the content of triacylglycerol stored in the human body may provide energy for approximately 2 months of fasting (Table 9.1).

**Table 9.1** Human nutrient stores (typical of a 70 kg man)

| Molecule | Weight (g) | Energetic value (kcal) | Period as single energy source (days) |
| --- | --- | --- | --- |
| Triacylglycerol (adipose tissue) | 9000 to 15,000 | ~ 108,000 | 60 |
| Glycogen (liver) | 90 | 360 | 0.2 |
| Glycogen (muscle) | 250 | 1000 | 0.55 |
| Glucose (blood and body fluids) | 20 | 80 | 0.044 |
| Proteins (mainly muscle)[a] | 8000 | 32,000 | 17.8 |

[a]It should be stressed that most of the muscular proteins are not readily available for mobilization

As ketone bodies are the main product of fatty acid oxidation in the liver (see Sect. 7.4.6), it is expected that the concentration of these metabolites increases along fasting, as observed in Fig. 9.2. It is possible to note two different phases in the profile of the blood concentration of ketone bodies. In the first 10 days of fasting, there is a constant increase in their concentration, which then gradually reaches a plateau after 25 days. This second phase may be explained by the increase in ketone bodies' consumption and/or excretion by the organism. In fact, brain adaptation to the use of ketone bodies as its main source of energy contributes to their

removal from the bloodstream as well as decreasing whole body glucose requirements (for more details, see Fig. 9.16).

A remarkable observation regarding the profile of plasma metabolites during
fasting is that glucose concentration, after a small decrease in the first 3 days, is
maintained constant during all the period of starvation. From this observation,
immediate questions arise: Why is it necessary to maintain glycemia? Why not use
only fatty acids as the energy source during fasting since they are, quantitatively, the
major fuel reserve in the body (see Table 9.1)? The answer lies in the fact that some
cells depend on glucose as their exclusive or preferential source of energy. This is
the case of cells that lack mitochondria, as erythrocytes and the cells from crystalline
lens, which depend on the anaerobic metabolism (glucose fermentation, see Sect.
6.1) to survive. It is also the case of the cells from the nervous system and the embryonic tissues, which are isolated from the systemic circulation by blood barriers that
do not allow the uptake of fatty acids since these molecules are bound to albumin,
their major form of transport in the bloodstream (Box 9.3). Therefore, it is essential
that glucose is constantly produced and released in the bloodstream during the periods in which it is not ingested. The metabolic pathways involved in the maintenance
of glycemia will be discussed in the next sections.

---

**Box 9.3  Glucose Transport Through the Blood–Brain Barrier (BBB)**
BBB is a highly selective permeability barrier that restricts the passage of
most substances from the circulating blood to the central nervous system
(CNS) fluids (see also Sect. 3.3.4.1). It is formed by three cellular components
(see below figure): (a) the brain capillary endothelial cells; (b) the astrocytic
end-feet, which cover the vessel wall maintaining the endothelial barrier; and
(c) the pericytes. The brain capillary endothelial cells are connected by tight
junctions that make the paracellular transport of substances through BBB
negligible under physiological conditions. Additionally, these cells express a
number of drug efflux transporters, such as the glycoprotein P (Pgp) and several members of the multidrug resistance (MDR) protein family, which
remove or prevent the entry of drugs and other substances in the CNS. Glucose
transport through BBB is mediated by GLUT1 (see Sect. 8.1). It is important
to remind that GLUTs are facilitated-diffusion transporters, which means that
glucose cannot be transported against a gradient from bloodstream to
CNS. Additionally, GLUT1 $K_M$ for glucose transport is about 1–2 mM. Thus,
the human brain cannot be supplied with glucose when its blood concentration is lower than this reference interval (normoglycemia is about 5 mM), so
that the only strategy to protect brain cells from starvation is to prevent
hypoglycemia.

(continued)

**Box 9.3** (continued)

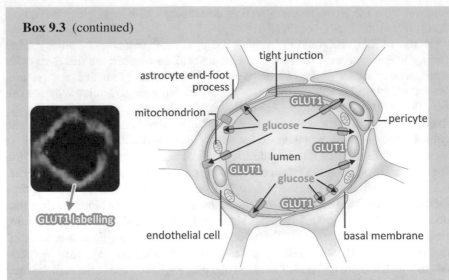

Micrograph of a brain microvessel section showing GLUT1 labeled with a green fluorescent probe (*left*) and the schematic representation of the BBB with its three cellular components (*right*), highlighting the glucose transport through GLUT1 (represented as *green rectangles*). (Figure reproduced by permission from Macmillan Publishers Ltd.: Löscher and Potschka. Nature Rev. Neurosci 6:591–602, 2005)

## 9.2   Glycogen Degradation in the Liver

The crucial role of liver metabolism in producing glucose to maintain glycemia started to be elucidated in the middle of the nineteenth century, with the pioneering studies developed by Claude Bernard (Box 9.4). In 1853, he showed that the liver was able to release glucose in the bloodstream even when carbohydrate was not present in the diet. This finding completely changed the current idea about animal nutrition. It contradicted the accepted concept that animals always decompose complex substances obtained from food and for the first time suggested that organism functions would be maintained by a metabolic interplay among different tissues. Some years later, Claude Bernard isolated from the liver a substance that he named "la matière glycogène" (the substance that generates glucose), the glycogen.

   At that time, it was not yet clear that glucose was released from the liver by a combination of two processes: (1) the degradation of glycogen and (2) the synthesis of glucose from noncarbohydrate precursors. Glycogen may be seen as a transient storage of the ingested sugar that is mobilized when necessary, while different non-glycidic molecules may be converted to glucose by a pathway named gluconeogenesis. Both pathways will be discussed in the next sections of this chapter.

**Box 9.4  Claude Bernard: The Founder of Experimental Medicine**

Maybe the most important accomplishment of Claude Bernard was to introduce the scientific methodology into medicine and physiology, establishing the basic rules of experimentation in the life sciences. This can be noticed in one of his letters to his friend Mme. Raffalovich: "The scientist must have imagination, but he must master this imagination and coldly probe the unknown. However, if he lets himself be carried away by his imagination, he will be overcome by vertigo and, like Faust and others, fall into the chasm of magic and succumb to phantoms of the mind." Early in his career, he aimed to follow the fate of the sugar absorbed from the food in the animal body. The hypothesis at that time was that the ingested sugar was burned in the lungs, passing the liver through the hepatic portal system (see Box 8.1) without any processing to reach the bloodstream. To

Claude Bernard (1813-1878)

confirm this hypothesis (and also to discard that the sugar was not destroyed in the liver, as it seems that he suspected), Claude Bernard measured the amount of sugar in the portal and hepatic veins of a dog fed with sweet milk. He would be convinced when he found a large amount of glucose in the blood that had passed through the liver. However, he was not, as he pointed out on his book "*Introduction à l'étude de la medicine expérimentale*": "More than one researcher would have stopped here and would have thought that any control experiment was useless. But I performed a control experiment because I am convinced that in physiology you should always doubt even if the doubt doesn't seem to be permitted." The control was a similar experiment in which the dog was fed only with meat. For his surprise, he found a large amount of sugar in the hepatic vein, which led him to state: "I don't understand anything anymore." This unexpected finding led Claude Bernard to isolate glycogen some years later, but his contribution to the understanding of animal metabolism did not stop there. The observation that liver releases glucose into the blood led him to establish the concept of "internal secretion." Additionally, Claude Bernard was the first to express the idea that the animals have an inner milieu, different from the external environment, whose constancy would be a requirement for life maintenance (the basis from which the principle of homeostasis emerged), as he stated: "*La fixité du milieu intérieur est la condition d'une vie libre et indépendent*" (The stability of the internal environment is the condition for a free and independent life).

In many tissues, especially in the liver and muscles, glycogen is the storage form of glucose, being observed as dense granules at the microscope (see Fig. 8.5). Liver glycogen content may correspond to about 10% of wet weight of this organ in well-fed humans. In the muscles, glycogen content can account for 1–2% of their wet weight, but since the muscles occupy a much larger area of the body than the liver, total muscle glycogen content is twice as high as that of liver.

Carl Ferdinand Cori (1896-1984)
Gerty Theresa Cori (1896-1957)

The metabolic pathway for glycogen degradation, known as glycogenolysis, was almost entirely elucidated by Carl and Gerty Cori, a couple of scientists whose great contribution to carbohydrate metabolism led them to be awarded the Nobel Prize in Physiology or Medicine in 1947. They discovered (a) the nature of glycogen degradation reaction, a phosphorolysis reaction; (b) the enzyme that catalyzes it, glycogen phosphorylase; and (c) its product, glucose-1-phosphate.

### 9.2.1   Reactions of Glycogen Degradation

Glycogen is a highly branched polymer of glucose containing α1,6-glycoside linkages in the branch points, with 10–12 glucose residues linked by α1,4-glycoside linkages between each branch (see Sect. 3.2.1). Its degradation depends on two enzymes, the glycogen phosphorylase (GP) and the debranching enzyme.

GP catalyzes the phosphorolysis of the α1,4 glycoside linkage at a terminal glucose unit from the nonreducing ends of the molecule, yielding glucose-1-phosphate (Fig. 9.3). In their first studies on glycogen degradation, the Coris showed that the reaction catalyzed by GP was reversible in vitro. However, now it is clear that phosphorolysis is greatly favored in vivo due to the very high intracellular ratio [Pi]/[glucose-1-phosphate], and glycogen synthesis has to occur through another pathway (see Sect. 8.2).

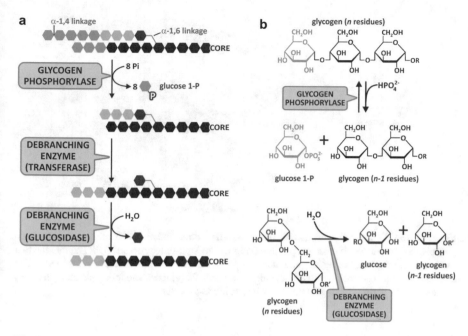

**Fig. 9.3** (**a**) Schematic representation of glycogen degradation. (**b**) Reactions catalyzed by glycogen phosphorylase (*top*) and the glucosidase activity of the debranching enzyme (*bottom*)

The branched structure of glycogen enables its rapid degradation, since about 50% of glucose units are in the outer branches. Glucose units are removed sequentially until 4 units before the branch point, where the enzyme loses its activity, leaving what is known as a limit dextrin (a molecule with short branches of four glucose units in length).

Further removal of the glucose units from a glycogen molecule depends on the activity of the debranching enzyme. This enzyme has two different activities in the same polypeptide chain. The first is a glucosyltransferase activity, in which a trisaccharide unit at the end of a branch is transferred through an α1,4-linkage to a non-reducing end of another chain, resulting in an extended chain susceptible to GP action (Fig. 9.3). The second activity is as an amylo-1,6-glucosidase, in which the α1,6-glycoside linkage in the branch points is hydrolyzed to form glucose. Due to this activity, about 7% of the glucose residues in glycogen are released as glucose and not as glucose-1-phosphate.

Glucose-1-phosphate produced during glycogenolysis may be converted to glucose-6-phosphate by the action of the enzyme phosphoglucomutase.

In the liver, glucose-6-phosphate may be dephosphorylated through the action of glucose-6-phosphatase (G6Pase), allowing glucose units to be removed from glycogen and released in the bloodstream. G6Pase is an integral protein of the endoplasmic reticulum membrane with the active site on the luminal side of this compartment (Fig. 9.4). It is expressed only in the liver and kidney. Because muscle cells lack this enzyme, the glucose units removed from muscular glycogen are only metabolized in muscle tissue itself.

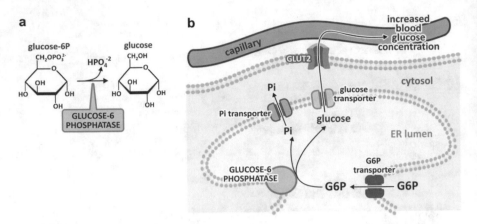

**Fig. 9.4** (**a**) Glucose-6-phosphatase reaction. (**b**) Intracellular localization of glucose-6-phosphatase. The enzyme is an integral protein in the endoplasmic reticulum (ER) membrane that catalyzes the dephosphorylation of glucose-6-phosphate in the lumen of this organelle. The enzyme substrate (glucose-6-phosphate, G6P) and products (glucose and inorganic phosphate, Pi) are transported across ER membrane through specific transporters

In their first physiological studies on carbohydrate metabolism, the Coris proposed that the lactate produced in muscle from glycogen degradation (and subsequent glycolysis) reached the liver through the bloodstream, being then reconverted to glycogen, in a cycle that became known as the "Cori cycle." However, although the idea of cycling metabolites between tissues is a very important concept, the "Cori cycle," exactly as it was proposed, does not occur physiologically. Glucose produced in the liver from muscle lactate, instead of being converted to glycogen, is released into the bloodstream to be used by the brain, as well as by erythrocytes and other fermentation-dependent cells. Additionally, it is very unlikely that glucose produced in the liver goes to the muscles, since its internalization in muscle cells depends on the insulin action, which is not operating in this situation (see Sect. 8.4). In fact, we can construct a more complex picture of glycogen/glucose–lactate cycling in different organs, as presented in Fig. 9.5.

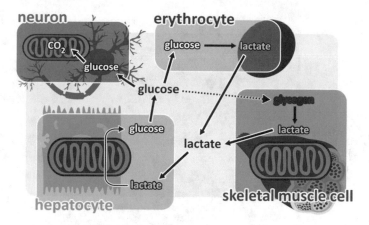

**Fig. 9.5** Glycogen/glucose–lactate cycling: glycogen degradation in the muscles forms lactate, which is released in the bloodstream and enters the liver, where it is converted to glucose by gluconeogenesis. Glucose released from the liver into the bloodstream is taken up by the glucose-dependent cells, such as those from the brain and the erythrocytes. Uptake of glucose by skeletal muscle (forming the so called "Cori cycle") is hypothetical because the physiological conditions that favor lactate production by muscle cells do not favor glucose uptake by these cells (see text)

## 9.2.2  Regulation of Glycogen Degradation in the Liver

Liver glycogen content largely varies in response to food intake (Fig. 9.6). Glycogen accumulates rapidly after carbohydrate ingestion, and then it is gradually mobilized to generate free glucose in between the meals.

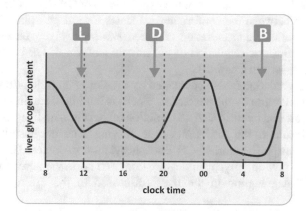

**Fig. 9.6**  Variation in the liver glycogen content after each meal

Glycogen degradation is mainly controlled by regulating GP activity. This enzyme is a dimer that exists in two different conformations, one more active, named phosphorylase $a$, and one much less active, named phosphorylase $b$ (Fig. 9.7). These two forms are interconvertible by phosphorylation/dephosphorylation of a serine residue (Ser14), induced by hormone action (see Sect. 9.3.3).

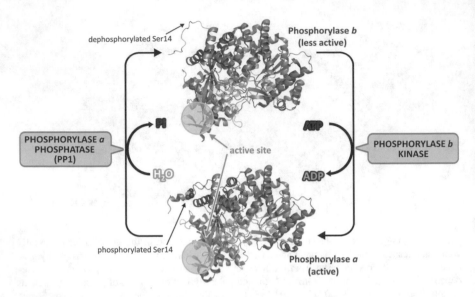

**Fig. 9.7** Regulation of glycogen phosphorylase activity by phosphorylation. In the liver, glucagon action triggers the phosphorylation of the less active form of GP, named phosphorylase *b* (PDB 1FC0), in its Ser14, which promotes a structural change to the active form, phosphorylase *a* (PDB 1FA9). The N-terminal segment, which contains the phosphorylation site (Ser14), converts from a completely disordered conformation to a well-ordered structure, leading to several structural transitions in the active site (highlight by the *green circle*). To facilitate the visualization of enzyme conformational changes, the monomer structures are shown in the figure, although the enzyme exists as homodimers

The major allosteric modulator of the liver isoform of GP is glucose, which shifts the equilibrium between the conformational states toward phosphorylase *b*. Thus, when the intracellular concentration of glucose is high, glycogen degradation is inhibited. The allosteric regulation of the muscle isoform of GP is more complex and will be discussed in Chap. 10.

Glycogen degradation is also controlled by hormones. In a situation of hypoglycemia, insulin secretion by the pancreatic β-cells is inhibited, leading to an increase in glucagon secretion by the α-cells. The action of glucagon on liver tissue (detailed in Sect. 9.4) results in the activation of the enzyme phosphorylase kinase, which catalyzes the phosphorylation of GP. Phosphorylation maintains GP in the active form (Fig. 9.7), favoring glycogen degradation. The hormone adrenaline also controls glycogen degradation in the liver (discussed in the exercise situation in Chap. 10).

It is important to note that, in humans, glycogen stored in the liver lasts between 12 and 24 h during fasting (see Fig. 9.2). Therefore, the contribution of liver glycogenolysis to the control of glycemia is limited, and gluconeogenesis, the other glucose-producing pathway, is required to maintain blood glucose concentration.

## 9.3   Gluconeogenesis

Gluconeogenesis is the synthesis of glucose (and other carbohydrates) from non-glycidic compounds. This pathway occurs mainly in the liver but also in the kidney cortex.

For many years it was thought that gluconeogenesis occurred as a reversal of the glycolytic pathway. However, some of the glycolysis reactions are highly exergonic (Fig. 9.8), which makes very unlike that they could be reversed within the cells. This energy barrier impairs that three reactions of the glycolytic pathway occur reversibly: the reactions catalyzed by (a) pyruvate kinase; (b) phosphofructokinase; and (c) hexokinase. For these reactions to occur, glycolytic enzymes should be bypassed.

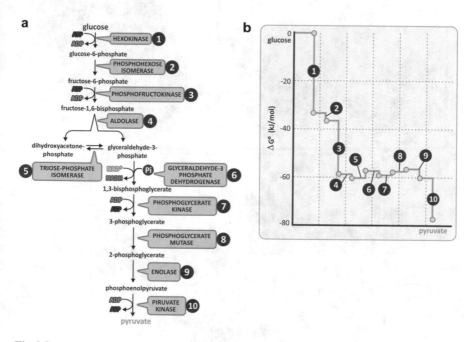

**Fig. 9.8**   (**a**) Glycolysis reactions (for more details, see Chap. 6). (**b**) Free energy variation between each step of glycolysis. The numbers correspond to the reactions indicated in (**a**)

### 9.3.1   Gluconeogenesis Reactions

The first bypass is the conversion of pyruvate in PEP, which requires two reactions involving two enzymes: pyruvate carboxylase (PC) and PEP carboxykinase (PEPCK).

PC is a mitochondrial enzyme that catalyzes the biotin-dependent carboxylation of pyruvate to produce oxaloacetate. Human PC is active in the tetrameric form and contains three functional domains in the same polypeptide chain: biotin–carboxyl carrier protein (BCCP), biotin carboxylase (BC), and carboxyltransferase (CT) domains (Fig. 9.9; see also Box 8.6 for more information on biotin-dependent carboxylases). Additionally, there is a fourth central structural domain that contains the binding site of the allosteric activator acetyl-CoA. Biotin is covalently linked to a Lys residue in the BCCP domain. The reaction occurs in two steps. First, the BC domain catalyzes the carboxylation of biotin in a reaction that uses bicarbonate as a substrate and requires the hydrolysis of one ATP molecule, generating ADP and Pi (Fig. 9.9). Then, the carboxyl group is transferred from carboxybiotin to pyruvate to form oxaloacetate in a reaction catalyzed by the CT domain of the enzyme. In PC tetrameric form, the domains are arranged in such a way that carboxybiotin is transferred from BC domain to the neighboring CT domain on opposing polypeptide chains, explaining why the enzyme is only active as a tetramer (Fig. 9.9).

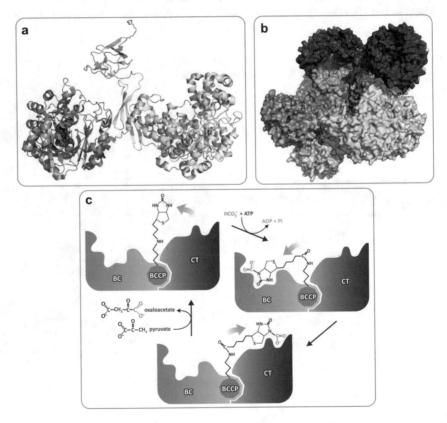

**Fig. 9.9** (**a**) Structure of the *Staphylococcus aureus* PC monomer (PDB 3BG5), highlighting the three functional domains: BC (*olive green*) with ATP bound (*red*), CT (*yellow*) with pyruvate bound (*cyan*), and BCCT (*green*) with the biotin group bound (with carbon skeleton in *green sticks*) and the allosteric domain (*lime*). (**b**) Tetrameric organization of PC subunits. Two monomers are highlighted: one monomer with the domains colored in green/yellow shades as in (**a**) and the other colored in red/purple shades. The localization of biotin group allows its movement from BC active site in the same subunit (green/yellow monomer) to the CT active site in the other subunit (red/purple monomer). ATP (*red*) and pyruvate (*cyan*) are shown in their binding sites in the BC and CT domains, respectively. (**c**) Reactions catalyzed by PC: first bicarbonate is used to carboxylate the enzyme-linked biotin in an ATP hydrolysis-dependent reaction catalyzed by the BC domain of one subunit; then the carboxyl group is transferred to pyruvate generating oxaloacetate in a reaction catalyzed by the CT domain in the adjacent subunit

It is important to point out that in addition to its crucial role in gluconeogenesis, the PC reaction plays an important anaplerotic function, replenishing oxaloacetate that is withdrawn from CAC in many metabolic situations (see Sect. 7.3).

The formation of PEP from oxaloacetate is catalyzed by PEP carboxykinase (PEPCK), in a reaction that requires the transfer of a phosphoryl group from GTP (Fig. 9.10).

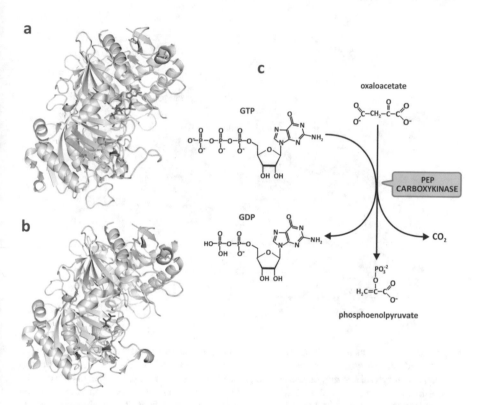

**Fig. 9.10** Structure of human cytosolic PEPCK: (**a**) complexed with a non-hydrolyzable GTP analogue (PDB 1KHE) and (**b**) complexed with PEP (PDB 1KHF). (**c**) Conversion of oxaloacetate in PEP, using GTP as the phosphate donor, as occurs for the reactions catalyzed by PEPCK from animals. Differently, bacterial, fungal, and plant PEPCKs use ATP as the phosphate donor

In human cells, PEPCK is distributed between mitochondria and the cytosol, and the isoform used will depend indirectly on the $[NADH]/[NAD^+]$ in the cytosol. The reason for this is that the reaction catalyzed by glyceraldehyde-3-phosphate dehydrogenase requires NADH to occur in the reverse direction of glycolysis, from 3-phosphoglycerate to glyceraldehyde-3-phosphate (Fig. 9.11).

When $[NADH]/[NAD^+]$ is very low, NADH equivalents are transported from the mitochondrial matrix to the cytosol by the malate–oxaloacetate shuttle. Oxaloacetate is converted to malate by the mitochondrial malate dehydrogenase, with NADH oxidation to $NAD^+$ into the mitochondria. Malate crosses the mitochondrial membranes and reaches the cytosol where it is reconverted into oxaloacetate through the action of the cytosolic malate dehydrogenase, with NADH formation in the cytosol. Therefore, in this situation, cytosolic PEPCK is used to generate PEP, while mitochondrial PEPCK is preferred when cytosolic $[NADH]/[NAD^+]$ allows glyceraldehyde dehydrogenase reaction to occur in the direction of glyceraldehyde-3-phosphate formation (Fig. 9.11).

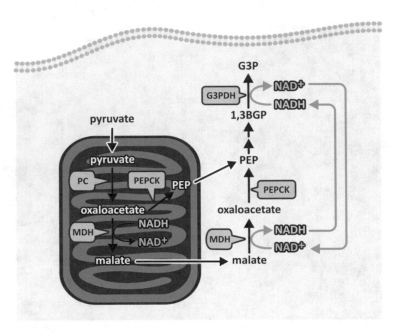

**Fig. 9.11** Intracellular localization of the first gluconeogenesis bypass. PC is a mitochondrial enzyme, and thus pyruvate is converted to oxaloacetate inside the mitochondria. PEPCK is distributed equally in cytosol and mitochondria, so oxaloacetate may be converted to PEP in both cellular compartments. When cytosolic $[NADH]/[NAD^+]$ ratio allows the reaction catalyzed by glyceraldehyde-3-phosphate dehydrogenase (G3PDH) to occur from 1,3-bisphosphoglycerate (1,3-BPG) to glyceraldehyde-3-phosphate (G3P), the mitochondrial isoform may be used. When the formation of additional NADH is required, oxaloacetate is converted to malate in the mitochondrial matrix by malate dehydrogenase (MDH) with NADH oxidation. Malate is transported to the cytoplasm where it is converted to oxaloacetate by the cytosolic MDH, generating NADH for the G3PDH reaction

The second bypass is the conversion of fructose-1,6-bisphosphate to fructose-6-phosphate. In this reaction, the phosphoryl group associated to carbon 1 of fructose-1,6-bisphosphate is removed by hydrolysis through the action of the enzyme fructose-1,6-bisphosphatase.

Finally, the third bypass is the conversion of glucose-6-phosphate to glucose. As discussed in the previous section, in the liver as well as in the kidneys, the enzyme glucose-6-phosphatase catalyzes the dephosphorylation of glucose-6-phosphate generating glucose, which can be released into the bloodstream (see Fig. 9.4).

## 9.3.2 Precursors for the Synthesis of Glucose

At this point it is already possible to visualize how different precursors can enter the gluconeogenesis pathway. As detailed below, the main precursors are lactate, amino acids, and glycerol. Additionally, propionyl-CoA formed in the oxidation of odd-chain fatty acids can also contribute to glucose synthesis, although in a much lower scale due to the low availability of odd-chain fatty acids in human metabolism.

Several studies in the beginning of the twentieth century showed that lactate could be converted to glucose (and glycogen) in liver tissue. At that time, the interest was focused on the synthesis of glycogen in the body, and less attention was given to the reactions involved in the conversion of lactate to glucose. Now we know that lactate conversion to glucose (gluconeogenesis) and glucose conversion to glycogen (glycogenogenesis) are processes that do not occur simultaneously (see Sect. 9.4.1.1). Probably as a result of this, there were many controversial observations at that time.

Lactate enters the gluconeogenesis pathway through its conversion to pyruvate in a reaction catalyzed by the enzyme lactate dehydrogenase (Fig. 9.12). In this reaction, cytosolic $NAD^+$ is reduced to NADH, increasing [NADH]/[$NAD^+$] in the cytosol and favoring PEP formation inside the mitochondria (see previous section).

Evidence that amino acids can act as substrates for glucose synthesis came from the early studies on glycogen formation by Claude Bernard in the nineteenth century, in which he fed dogs with a meal consisting solely of meat and found an increase of glucose released by the liver (see Box 9.4).

To enter gluconeogenesis, the amino group of the amino acids should be removed by transamination and/or deamination. After this metabolization, 18 from the 20 more common amino acids generate α-ketoglutarate, succinyl-CoA, fumarate, oxaloacetate, or pyruvate (see Sect. 7.5). The CAC intermediates converge to malate through the action of the enzymes of the cycle. Malate leaves mitochondria and is converted to oxaloacetate, which then forms PEP in the cytosol (Fig. 9.12).

The product of the metabolization of leucine or lysine is acetyl-CoA, which cannot be used to synthesize glucose since its two carbon atoms are completely oxidized to $CO_2$ in CAC, thus impairing the net accumulation of carbons to be incorporated in the newly synthesized glucose molecule. This is also the reason why fatty acids cannot be transformed in glucose in animals.

Finally, glycerol, which is converted to glycerol phosphate after being transported into the liver, is converted to dihydroxyacetone phosphate by glycerol-phosphate dehydrogenase, entering gluconeogenesis at this point (Fig. 9.12).

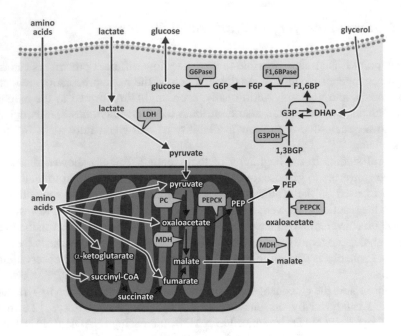

**Fig. 9.12** Entry of precursors in the gluconeogenesis pathway. Lactate is converted to pyruvate by lactate dehydrogenase (LDH). Amino acids have their amino groups removed by transamination and/or deamination generating CAC intermediates that converge to malate, which is converted to oxaloacetate and then PEP in the cytosol. Glycerol is converted to glycerol-phosphate and then to dihydroxyacetone-phosphate (DHAP), entering gluconeogenesis. *F1,6-BP* fructose-1,6-bisphosphate, *F1,6-BPase* fructose-1,6-bisphosphatase, *F6P* fructose-6-phosphate, *G6P* glucose-6-phosphate, *G6Pase* glucose-6-phosphatase, and other abbreviations are the same as in Fig. 9.11. The names of the enzymes are highlighted in *yellow boxes*

### 9.3.3　Regulation of Gluconeogenesis

Glycolysis and gluconeogenesis share many reactions, and both pathways are exergonic in the intracellular conditions. Thus, these pathways must be reciprocally regulated to allow one of them to predominate over the other in each specific situation. As a matter of fact, in the liver, the rates of glycolysis and gluconeogenesis are adjusted to maintain the blood glucose concentration stable.

The main point of the reciprocal regulation of glycolysis and gluconeogenesis is the interconversion between fructose-6-phosphate and fructose-1,6-bisphosphate by the enzymes phosphofructokinase-1 (PFK-1) in glycolysis and fructose-1,6-bisphosphatase (F1,6BPase) in gluconeogenesis. This point is controlled by hormonal action with a key role of the molecule fructose-2,6-bisphosphate (Fig. 9.13).

Fructose-2,6-bisphosphate in submicromolar concentrations simultaneously activates PFK-1 and inhibits F1,6BPase. This molecule is synthesized through the phosphorylation of fructose-6-phosphate in a reaction similar to that catalyzed by

PFK-1 but with the transfer of the phosphoryl group of ATP to the carbon 2 instead of the carbon 1 of fructose. The enzyme that catalyzes this reaction was named phosphofructokinase-2 (PFK-2) to be distinguished from the classic PFK-1. Fructose-2,6-bisphosphate is hydrolyzed to fructose-6-phosphate by the enzyme fructose-2,6-bisphosphatase (F2,6BPase).

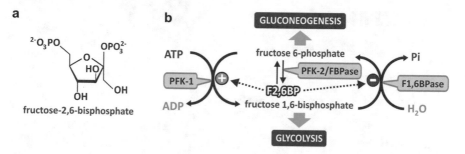

**Fig. 9.13** (**a**) Fructose-2,6-bisphosphate structure. (**b**) Synthesis and degradation of fructose-2,6-bisphosphate and its effect on the activities of PFK-1 and F1,6BPase in the liver

It is very interesting to note that these two antagonic activities, PFK-2 and F2,6BPase, are present in the same polypeptide chain, an example of a bifunctional enzyme (Box 9.5).

The bifunctional enzyme presents a regulatory domain in its N-terminal end, which contains a serine residue (Ser32) that can be phosphorylated or dephosphorylated in response to the action of glucagon (see Sect. 9.4.1) or insulin (see Sect. 8.4), respectively. When Ser32 is phosphorylated, the enzyme undergoes a conformational change that favors the F2,6BPase activity. In contrast, when it is dephosphorylated, the enzyme activity is turned to PFK-2.

---

**Box 9.5 Evolution of the Bifunctional Enzyme PFK-2/F2,6BPase**

It is believed that the bifunctional enzyme PFK-2/F2,6BPase resulted from the fusion of genes encoding different enzymes. This seems to have occurred very early during evolution in a common ancestor of all eukaryotes. In all taxonomic groups, the enzyme has a central catalytic core with the PFK-2 and F2,6BPase domains *in tandem* and extensions in N- and C-terminals [see examples in (**a**)]. In some cases, one of the activities is damaged by deletions or insertions. The kinase and the phosphatase domains are structurally similar to one-domain kinases or phosphatases (see below figure).

(continued)

**Box 9.5** (continued)

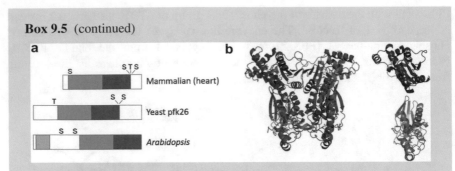

(**a**) Schematic representation of domain localization of PFK-2 and F2,6BPase primary sequence in different organisms. (**b**) Comparison of the structure of the dimeric human bifunctional enzyme with those of gluconate kinase, in *red*, and PhoE phosphatase, in *blue*. (Figure reproduced with permission from Michels and Rigden. IUBMB Life 58:133–141, 2006)

Gluconeogenesis is also regulated at the first bypass level. Carboxylation of pyruvate to oxaloacetate is completely dependent on the presence of acetyl-CoA, which acts as a specific activator of PC by its reversible binding to the allosteric domain of the enzyme (see PC structure in Fig. 9.9). This is an interesting example of how the information of a physiological situation can be transmitted locally to a specific cellular compartment by means of an allosteric modulator. In hypoglycemia, triacylglycerols stored in the adipose tissue are mobilized, generating glycerol and fatty acids (see Sect. 7.4.1). The increased availability of fatty acids in the bloodstream allows them to be used as energy source for many tissues, including the liver. β-oxidation of fatty acids, which occurs within the mitochondria (see Sect. 7.4.4), increases the concentration of acetyl-CoA in the mitochondrial matrix, where PC is located. Thus, the mobilization of triacylglycerols as a response to hypoglycemia results in the generation of an essential activator of an important enzyme for the synthesis of glucose, a crucial pathway in this situation.

The activity of PEPCK is regulated only at the transcriptional level. The gene that encodes the cytosolic form of PEPCK in the liver contains several hormone response elements, including response units for glucocorticoid, cyclic AMP, insulin, thyroid hormones and retinoic acid. Through different mechanisms of actions, glucocorticoids and glucagon enhance the transcription of PEPCK gene (see Sect. 9.4), while insulin represses its basal and hormone-induced expression.

### 9.3.4  Dynamic Utilization of Gluconeogenesis Precursors

Daily glucose requirement for the adult human organism is around 120 g. In a fasting situation, in which no glucose is ingested, the contribution of each of the gluconeogenesis precursors is shown in Table 9.2.

**Table 9.2** Contribution of each gluconeogenesis precursors to glucose production[a]

| Precursor | Amount of glucose produced daily (g) | |
| --- | --- | --- |
| | 1 day starvation | 5 weeks starvation |
| Lactate | 39 | 39 |
| Glycerol | 19 | 19 |
| Amino acids | 60 | 16 |

[a]Data from Newsholme EA, Leech AR. Biochemistry for the medical sciences. Chap. 14, pp. 541. John Wiley & Sons, 1994

Taking into account that each precursor is originated in a specific tissue, the blood concentration of glucose can be maintained due to the interplay among different tissues of the organism (Fig. 9.14). Glycerol is generated by the hydrolysis of triacylglycerols in the adipose tissue. Proteolysis of the contractile proteins in muscle cells generates amino acids that undergo transamination, mainly with pyruvate or α-ketoglutarate within muscle cells, forming alanine and glutamine. Lactate is continuously produced by the fermentation-dependent cells such as the erythrocytes.

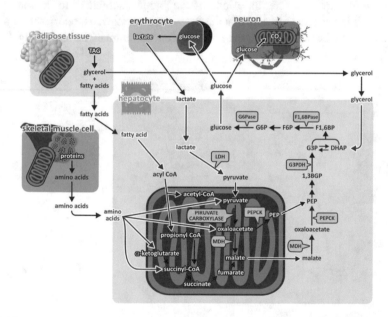

**Fig. 9.14** Metabolite interplay among the organs. Hydrolysis of triacylglycerols in the adipose tissue generates glycerol and fatty acids that are released in the bloodstream. Even-chain fatty acids form acetyl-CoA, while each molecule of odd-chain fatty acid forms one molecule of propionyl-CoA besides acetyl-CoA. Propionyl-CoA may enter gluconeogenesis after its conversion to succinyl-CoA and then malate through CAC reactions. Glycerol enters the liver cells where it is converted to DHAP and then to glucose. Proteolysis of the contractile proteins in muscle cells generates amino acids that are released in the bloodstream entering the liver, where they undergo transamination/deamination generating pyruvate or CAC intermediates that ultimately form malate to enter the gluconeogenesis pathway. Lactate is continuously produced by and released from the fermentative cells such as the erythrocytes, reaching the liver where it is converted to pyruvate, which enters the gluconeogenesis pathway. Glucose produced is mainly used by the cells from central nervous systems and by the fermentation-dependent cells. Abbreviations are the same as used in Fig. 9.12

It is important to note that this situation changes if glucose is not ingested for longer periods. If protein mobilization was maintained at the same rate as in the beginning of fasting, after about 1 month without food ingestion, half of the body proteins would be consumed. This can be calculated from the data shown in Tables 9.1 and 9.2 (for these calculations, the stoichiometry of glucose synthesis from amino acids must be taken into account: to produce 1 g of glucose, about 2 g of amino acids are necessary).

However, the human organism is adapted to survive much longer periods of starvation. Even in the therapeutic starvation for obesity treatment as presented in the beginning of this chapter, the patients spent more than 1 month without any food ingestion. Indeed, the studies performed with these patients made possible to understand this adaptation. Catheterization of brain vessels demonstrated that two thirds of brain fuel consumption in long starvation corresponded to the metabolization of β-hydroxybutyrate and acetoacetate, markedly diminishing the need of glucose production and, consequently, muscle proteolysis to provide gluconeogenic precursors. In fact, the decrease in protein degradation during fasting can be inferred by observing the profile of nitrogen excretion of one of the patients submitted to the therapeutic starvation. It is interesting to note that the decrease in nitrogen excretion shows a clear correlation with the increase in ketone bodies' concentration in the blood (Fig. 9.15a).

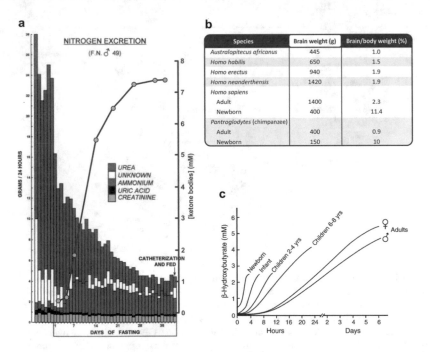

**Fig. 9.15** (**a**) Excretion of different nitrogenated compounds by a human subject during starvation (figure reproduced from Owen, Biochem. Mol. Biol. Educ. 33:246–251, 2005) and its correlation to the concentration of ketone bodies (β-hydroxybutyrate and acetoacetate) in the blood. (**b**) Encephalization during human evolution. (Figure reproduced from Cunnane & Crawford, Comp. Biochem. Physiol. 136:17–26, 2003, with permission from Elsevier) (**c**) Age-dependence production of β-hydroxybutyrate in response to fasting. (Figure reproduced with permission from Cahill. Ann. Rev. Nutr. 26:1–22, 2006)

To better understand this adaptation, it is important to have some information regarding the brain uptake of ketone bodies. Ketone bodies are transported across the blood–brain barrier through the monocarboxylate transporters (MCTs), especially MCT1, which is highly expressed in endothelial cells that form blood–brain barrier vessels. MCT1 transports a wide range of short-chain monocarboxylates, including lactate, pyruvate, acetoacetate, and β-hydroxybutyrate. The $K_M$ values for these substrates are in the range of 5–10 mM. Thus, the increase in ketone bodies' concentration in the blood during fasting greatly favors the transport of these molecules across the blood–brain barrier. As seen in Fig. 9.15a, blood concentration of β-hydroxybutyrate reaches the range of the $K_M$ for MCT1 after the first days of fasting, in such a way that this molecule becomes increasingly available to brain cells as fasting proceeds. Additionally, prolonged ketonemia, as it occurs in starvation or low-carbohydrate diets, induces MCT1 gene expression, also contributing to the increase of MCT1-mediated ketone bodies' transport to the cells of the central nervous system. Thus, along the first week of fasting, the requirement of glucose as the energy supply for the central nervous system cells greatly decreases as ketone bodies become available for these cells. Since brain metabolism accounts for most of the use of glucose in the body (approximately 100 g of 120 g necessary daily), it is easy to imagine that gluconeogenesis rate can be considerably reduced as the blood concentration of ketone bodies increases.

The use of ketone bodies by the brain seems also to have been very important during human evolution, which was characterized by a remarkable increase of brain weight (Fig. 9.15b). The cost of encephalization is the increase in energy demands. The brain of a modern human adult corresponds to 2.3% of its body weight but accounts for 23% of energy consumption of the organism. In children, the energy demand for the brain is even greater (approximately 75% of the total organism energy demand).

The fact that the brain/body ratio is similar between newborn humans and chimpanzees (11.4 and 10, respectively; see Fig. 9.15b) suggests that primates have, in general, the potential to have large brains. During development, however, the brain/body ratio in chimpanzees becomes less than a half of that of humans (0.9 and 2.3, respectively). An interesting observation that may explain this fact is that, among the primates, only humans accumulate fat just after birth (human babies are usually fatty). This probably enables babies to produce ketone bodies for brain use during development. As a matter of fact, human newborn metabolism is essentially ketotic, since the larger the brain/body ratio is, the more rapidly ketosis develops (Fig. 9.15c).

# 9.4   Hormonal Responses to Hypoglycemia

One of the major metabolic adaptations to hypoglycemia is the production of glucose by the liver. As discussed throughout the chapter, this is possible by the combination of hepatic glycogen degradation (in the first hours) and glucose production from non-glycidic precursors. Additionally, mobilization of triacylglycerol in the adipose tissue occurs simultaneously, ensuring the energy supplies for most of the tissues.

The simultaneous activation of these pathways is mainly regulated by the action of glucagon and glucocorticoids. The secretion of these hormones is enhanced by the decrease in blood glucose concentration, and, through their action, the information about the hypoglycemia situation is transmitted to the different cells and organs.

In this section, the signaling pathways as well as the effects of glucagon and glucocorticoids on their main target cells will be detailed.

## 9.4.1   Glucagon: Mechanism of Action and Effects on Energy Metabolism

Glucagon is a peptide hormone of 29 amino acid residues, secreted by the pancreatic α-cells, which compose the Langerhans islets together with β-cells that secrete insulin (see Sect. 8.4) and γ-cells that secrete somatostatin (Fig. 9.16).

Glucagon was discovered in the decade of the 1920s as a hyperglycemic factor produced by the pancreas. This finding was correlated to glycogen degradation in the liver, a subject extensively studied at that time (as discussed in the beginning of this chapter), giving glucagon its original name of "hyperglycemic–glycogenolytic factor." During the twentieth century, the structure of the gene that encodes glucagon and the complex processing of its product as well as the glucagon signaling pathway and its effects on the target cells have been elucidated.

At low levels of glucose, a set of ion channels on α-cell membrane generates action potentials that activate voltage-dependent L-type $Ca^{2+}$ channels, leading to intracellular $Ca^{2+}$ waves that induce the exocytosis of glucagon granules as well as the expression of glucagon gene (Fig. 9.16). Membrane repolarization by $K^+$ flowing through A-type channels triggers oscillatory $Ca^{2+}$ signals, so that glucagon secretion follows a pulsatile pattern. Although it seems clear that glucagon is released constitutively from α-cells, some factors, such as catecholamines and amino acids (mainly arginine), can act as positive regulators of its secretion by increasing the amplitude of the electric pulses in α-cells. However, the main control of blood concentration of glucagon occurs through the inhibition of its secretion. Glucose and insulin are the major negative regulators of glucagon secretion, so that the profile of blood glucagon concentration is the mirror image of that of glycemia. Therefore, glucagon transmits to the different organs in the body the information of hypoglycemia, leading to appropriate responses of specific tissues that allow the organism to deal with this situation.

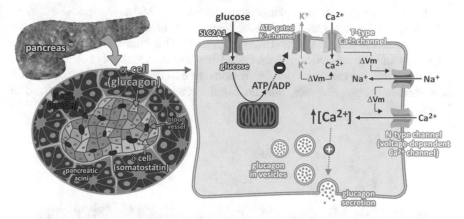

**Fig. 9.16** Representation of a Langerhans islet, showing the α-cell and the mechanism of glucagon secretion in detail. Secretion of glucagon is controlled by a set of ion channels in α-cells that generate action potentials of $Na^+$ and $Ca^{2+}$. At low levels of glucose, the activity of the ATP-sensitive $K^+$ channels renders a membrane potential that stimulates T-type $Ca^{2+}$ channels to open, leading to membrane depolarization that, in turn, activates $Na^+$ and N-type $Ca^{2+}$ channels. Intracellular $Ca^{2+}$ waves caused by $Ca^{2+}$ entry through N-type $Ca^{2+}$ channels induce the exocytosis of glucagon granules. $K^+$ flow through A-type channels mediate membrane repolarization, so that this oscillatory electrical activity results in a pulsatile pattern of glucagon secretion. An increase in glucose consumption by α-cells rises the intracellular ATP/ADP ratio, blocking the ATP-sensitive $K^+$ channels. This causes membrane depolarization and decrease in $Ca^{2+}$ influx, inhibiting glucagon secretion

Glucagon, like other hydrophilic hormones, acts on its target cells by binding to a receptor on the cell surface. Glucagon receptor is not uniformly distributed among the different tissues: it is expressed in high levels in the liver, kidneys, and pancreas and has also been detected in the adipose tissue and in the heart, but it is absent in the skeletal muscle cells.

Glucagon receptor belongs to the superfamily of G protein-coupled receptors (see Fig. 5.19 and also Sect. 10.5.1.1). The main effects of glucagon on its target tissues are mediated by the increase of the intracellular levels of cyclic AMP (cAMP), through the classic mechanism of action of G protein-coupled receptors (Fig. 9.17). G proteins are composed of three subunits, α, β, and γ. α-Subunit binds GTP and catalyzes its hydrolysis, producing GDP, which remains bound to the protein maintaining it in an inactive state, associated to β- and γ-subunits. When glucagon binds to its receptor, it undergoes a conformational change that is transmitted to G protein, leading to the replacement of GDP by GTP. When bound to GTP, the α-subunit dissociates from βγ-subunits and moves freely on the inner side of cellular membrane until it reaches the enzyme adenylate cyclase. This enzyme is activated by G protein α-subunit and catalyzes the conversion of ATP in cAMP, leading to an increase in the intracellular concentration of this molecule. The affinity of the α-subunit to adenylate cyclase decreases after GTP hydrolysis, leading the α-subunit to dissociate from adenylate cyclase and reassociate to βγ-subunits.

The increase in the intracellular concentrations of cAMP as a response to glucagon binding to its receptor promotes the activation of an important enzyme, the
cAMP-dependent protein kinase (PKA) (Fig. 9.17). This kinase is composed of two
regulatory and two catalytic subunits. cAMP binds to the regulatory subunits, which
dissociate from the catalytic ones that become free to catalyze the phosphorylation
of several important enzymes, thus modulating their activities (as summarized in
Fig. 9.17).

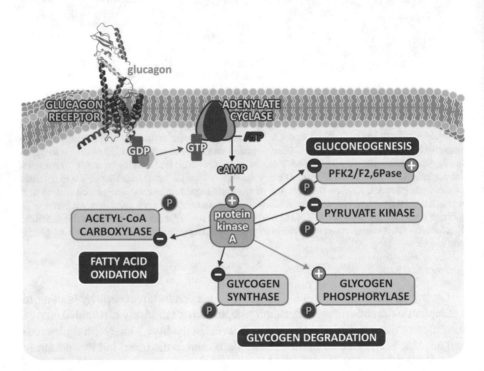

**Fig. 9.17** Mechanism of glucagon action, showing its effects on enzyme activities in the liver.
Glucagon (structure shown in *green*, PDB 1GCN) binds to its G protein-coupled receptor (PDB
4ERS and 4L6R), promoting the replacement of GDP by GTP in the G protein α-subunit, which
dissociates from βγ-subunits and associates to adenylate cyclase, leading to an increase in the
intracellular concentration of cAMP. This second messenger binds to the regulatory subunits (not
represented in this simplified figure) of the protein kinase A (PKA), leading to the release of the
active catalytic subunits. Active PKA catalyzes the phosphorylation of several enzymes (represented by the "P" in the *pink circle*), changing their activities as indicated by *green* "+" for activation, and the *red* "−" for inhibition. The pathways that are activated due to these changes in the
enzyme activities are shown in the *blue boxes*

### 9.4.1.1   Effects of Glucagon on Liver Metabolism

Liver metabolism is drastically affected by glucagon. Hepatic glycogenolysis is activated, while glycogen synthesis is inhibited. Simultaneously, gluconeogenesis is activated and glycolysis inhibited. As a result, glucose is produced and released in the bloodstream. Additionally, fatty acid synthesis is inhibited (see Sect. 8.3), allowing the incoming fatty acids to undergo β-oxidation, ensuring ATP generation. The coordinated control of these different metabolic pathways is possible because key enzymes of each of these pathways have their activities modulated by phosphorylation promoted directly or indirectly by PKA (Fig. 9.18). Additionally, the increase in the intracellular concentration of cAMP can also regulate gene expression, since several genes are present in their promoter region a cAMP-responsive element (CRE; see Fig. 5.20).

The activation of GP, the key enzyme of glycogen degradation, occurs through the phosphorylation of its Ser14, catalyzed by phosphorylase kinase (PK; see Sect. 9.2.2). This enzyme, in turn, is also activated by phosphorylation, in this case catalyzed by PKA. Thus, PKA activation in response to glucagon binding to the liver cells results in the phosphorylation of GP, which becomes active and promotes glycogen degradation (Fig. 9.18).

The signaling pathway triggered by glucagon not only stimulates glycogen degradation but also impairs its synthesis. This is possible because the activity of the key enzyme in glycogen synthesis, GS, is also regulated by PKA-mediated phosphorylation (for details on glycogen synthesis regulation, see Sect. 8.2.4). However, the phosphorylation, in this case, causes the inhibition of the enzyme activity (Fig. 9.18). PKA can directly phosphorylate GS or can phosphorylate another protein kinase, GSK3, which also phosphorylates GS. Thus, the hyperphosphorylation of GS impairs glycogen synthesis (see Box 8.3).

As discussed in Sect. 9.3.3, fructose-2,6-bisphosphate is a key molecule in the reciprocal regulation of glycolysis/gluconeogenesis. This compound is formed/degraded by the action of the bifunctional enzyme PFK-2/F2,6BPase, which is also a substrate for PKA. PKA-mediated phosphorylation of Ser32 of the bifunctional enzyme stimulates the F2,6BPase and inhibits the PFK-2 activities, leading to fructose-2,6-bisphosphate degradation and, therefore, to activation of gluconeogenesis and inhibition of glycolysis (Fig. 9.18).

The uptake and the ability of the liver to use amino acids are also increased in response to glucagon. This seems to occur mainly at the transcriptional level, due to the induction of the expression of the genes encoding the hepatic alanine transporter and different transaminases, such as alanine aminotransferase, aspartate aminotransferase, and tyrosine aminotransferase. This control enables the more efficient use of the amino acids as precursors of glucose synthesis through the gluconeogenesis pathway.

Lipid metabolism in the liver is also regulated by PKA-dependent phosphorylation. The enzyme that has its activity regulated is ACC, a key enzyme in the fatty acid synthesis pathway (see Sect. 8.3). The phosphorylated ACC is inactive, impairing fatty acid synthesis. Additionally, malonyl-CoA, the product of the reaction catalyzed by ACC and an important intermediate of fatty acid synthesis, is a potent inhibitor of the transport of acyl-CoA molecules across the mitochondria membranes. Since malonyl-CoA is not formed due to glucagon-mediated inhibition of ACC, liver β-oxidation can proceed during the hypoglycemia situation.

### 9.4.1.2   Effects of Glucagon on the Adipose Tissue

An important effect of glucagon on the adipose tissue is the activation of lipolysis, which ensures the increase of fatty acid concentration in the blood, making these molecules available to different tissues as the main energetic supply. The phosphorylation of the adipocyte enzyme hormone-sensitive lipase by PKA activates it, increasing the hydrolysis of triacylglycerols to glycerol and fatty acids, which are then released in the bloodstream (Fig. 9.18; see also Sect. 7.4.1). Glycerol may be used as substrate for gluconeogenesis, whereas fatty acids undergo β-oxidation in different tissues.

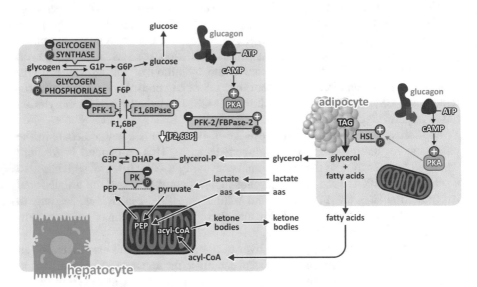

**Fig. 9.18** Integration of the effects of glucagon on the liver and adipose tissue, showing the enzymes and the metabolic pathways regulated in each cell. The P in the *pink circle* represents the phosphate group introduced in the enzyme by PKA; + and − indicate the activated or inhibited states, respectively. The *dashed lines* represent inhibited reactions. *PFK-1* phosphofructokinase-1, *F1,6BPase* fructose-1,6-bisphosphatase, *PFK-2/F2,6BPase* phosphofructokinase-2/fructose-2,6-bisphosphatase, *F2,6BP* fructose-2,6-bisphosphate, *HSL* hormone-sensitive lipase. Other abbreviations are the same as used in Figs. 9.11, 9.12, and 9.14

## 9.4.2   Glucocorticoids: Mechanism of Action and Effects on Energy Metabolism

Glucocorticoids are steroid hormones synthesized from cholesterol in the cortex of adrenal glands (Fig. 9.19). In humans, the main glucocorticoids produced are cortisol (80–90%) and corticosterone.

The synthesis of the glucocorticoids is stimulated by the adrenocorticotropic hormone (ACTH), a peptide hormone produced in the anterior pituitary gland. The release of ACTH by the pituitary is, in turn, stimulated by another hormone, the corticotropin-releasing hormone (CRH). This hormone is produced in the hypothalamus and reaches the pituitary through a portal system (Fig. 9.19). Thus, the synthesis and secretion of glucocorticoids by the adrenal glands are under control of the hypothalamic–pituitary–adrenal axis, in a classical example of the integration of nervous and endocrine systems. Furthermore, there is an efficient feedback control of glucocorticoid production, in which the release of both ACTH and CRH is suppressed when circulating glucocorticoid levels reach a specific threshold.

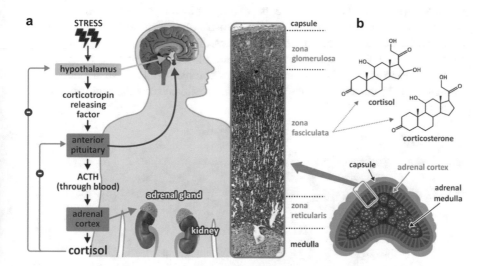

**Fig. 9.19** (**a**) Hypothalamic–pituitary–adrenal axis and glucocorticoid secretion. (**b**) Representation of the adrenal gland, showing the two distinct parts: the adrenal cortex and the adrenal medulla. The three zones of the cortex, the zona glomerulosa, the zona fasciculata, and the zona reticularis, are shown in a histological image (reprinted with the permission of Instituto de Histologia e Biologia do Desenvolvimento, Faculdade de Medicina, Universidade de Lisboa, FMUL). The natural human glucocorticoids, cortisol and corticosterone (chemical structures shown), are synthesized in the zona fasciculata and in the zona reticularis

Glucocorticoid secretion may be regulated in two levels. The first level is constitutive and follows the circadian rhythm, leading to a peak in cortisol concentration in the blood in the beginning of the morning. This pattern is adjusted by individual habits through the light/dark cycles. The second level corresponds to a response to virtually all types of physical or mental stress (Fig. 9.19), in which hypoglycemia can be included.

Due to their hydrophobic nature, the glucocorticoids cross the cellular membrane and bind to intracellular receptors. The hormone–receptor complex migrates to the nucleus, where it interacts with specific regions of the genome, inducing or repressing the expression of several genes.

Glucocorticoid receptor belongs to the superfamily of the nuclear hormone receptors, which also includes the receptors for mineralocorticoids, androgen, progesterone, thyroid hormones, retinoic acid, and retinol. The structure of all these receptors shows three distinct domains: (a) a N-terminal domain that interacts with the DNA and/or with other transcriptional factors; (b) a central domain with two zinc finger motifs responsible for the recognition and binding to specific DNA sequences; and (c) a C-terminal domain that binds to the hormone.

When not bound to the hormone, the glucocorticoid receptor associates to a multiprotein complex, which includes some of the heat shock proteins (Hsp). The interaction with the glucocorticoid results in receptor dissociation from this complex and in its hyperphosphorylation, which exposes nuclear localization sequences that allow the hormone–receptor complex, if in the cytosol, to migrate to the nucleus. In the nucleus, the hormone–receptor complex forms dimers, and the DNA-binding segments become exposed, enabling the regulation of gene expression (Fig. 9.20). The sequences to which the glucocorticoid-receptor complex binds are small palindromic sequences of 15 nucleotides known as the glucocorticoid responsive elements.

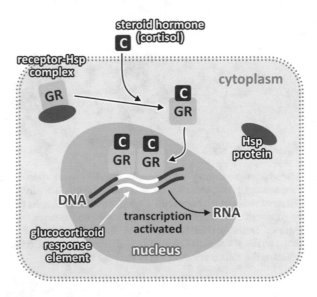

**Fig. 9.20** Mechanism of action of the glucocorticoids. Intracellular binding of glucocorticoid to its receptor promotes receptor dissociation from its inhibitory multiprotein complex (*Hsp* heat shock proteins) and its migration to the nucleus where the hormone–receptor complex binds to segments in DNA containing the glucocorticoid response elements, leading to gene expression

### 9.4.2.1 Effects of Glucocorticoids on Muscle Metabolism

The regulation of protein and amino acid metabolisms by glucocorticoids is among the first actions of these hormones to be characterized. Glucocorticoids seem to have a crucial role in the mobilization of muscular proteins in hypoglycemia, yielding amino acids for liver and kidney gluconeogenesis. Indeed, several studies have shown that the glucocorticoids stimulate protein degradation. The mechanisms of this regulation are still not completely understood, but glucocorticoid-induced proteolysis seems to be restricted to the muscles.

Measurements of the arteriovenous differences in the muscles during fasting revealed that a net release of amino acids from the muscles occurred and, more interestingly, that 60% of them were alanine and glutamine (Table 9.3), although these two amino acids correspond to approximately 10% of the amino acids that compose muscle proteins. Therefore, it seems clear that the different amino acids produced by muscle proteolysis should be preferentially converted into Ala and Gln before being released from muscle cells.

**Table 9.3** Amino acid released from skeletal muscle during fasting in man[a]

| Amino acid | Arteriovenous difference (μmol/L) | Percentage of total |
|---|---|---|
| Alanine | −70 | 30 |
| Glutamine | −70 | 30 |
| Glycine | −24 | 10 |
| Lysine | −20 | 9 |
| Proline | −16 | 7 |
| Threonine | −10 | 4 |
| Histidine | −10 | 4 |
| Leucine | −10 | 4 |
| Valine | −8 | 3 |
| Arginine | −5 | 2 |
| Phenylalanine | −5 | 2 |
| Tyrosine | −4 | 2 |
| Methionine | −4 | 2 |
| Isoleucine | −4 | 2 |
| Cysteine | +10 | – |
| Serine | +10 | – |

[a]Data from Felig. Annu. Rev. Biochem. 44:933–955, 1975

At this point, the role of glucocorticoids is very important. The genes encoding the different transaminases have glucocorticoid-responsive elements and, therefore, have their expression induced by these hormones. The increase in transaminase synthesis and consequently in their activities allows the interconversion of different amino acids into Ala and Glu (see Sect. 7.5), with the subsequent amination of Glu to generate Gln (Fig. 9.21). Ala and Gln released from the muscles into the bloodstream during fasting or low-carbohydrate diets are used as precursors for gluconeo-genesis. Quantitatively, Ala is the most important amino acid that is used as a gluconeogenic precursor in the liver, whereas Gln is preferentially used by the kidney (Fig. 9.21).

### 9.4.2.2  Effects of Glucocorticoids on Liver Metabolism

The induction of the expression of the genes encoding the transaminases also occurs in the liver as a result of glucocorticoid action. This facilitates transamination of the available amino acids allowing them to enter gluconeogenesis (Fig. 9.21). Additionally, the genes of different gluconeogenic enzymes contain glucocorticoid-responsive elements and are under the control of these hormones, so that glucocorticoids induce the synthesis of PEPCK, F1,6BPase, and glucose-6-phosphatase, whose concentrations greatly increase in the liver cells. Thus, glucocorticoid action on liver cells enhances the hepatic capacity to perform gluconeogenesis.

However, it is important to take into account that glucocorticoid action alone is not sufficient to induce gluconeogenesis, since it is not only the presence of a given enzyme in high concentrations that will ensure its activation. Depending on the enzyme, changes in its phosphorylation state or the presence of an allosteric modulator is also necessary to allow the enzyme to operate. But since glucagon and glucocorticoids are both secreted as a response to hypoglycemia, in the case of gluconeogenesis, additive effects of these hormones occur, ensuring the activation of this metabolic pathway.

Moreover, some effects of glucocorticoids represent another type of metabolic control. This is the case, for example, of their action on the genes encoding ACC (a key enzyme in the synthesis of fatty acid) and GS (a key enzyme in the synthesis of glycogen). Glucocorticoids induce the expression of both enzymes, but the effect of glucagon, which is operating simultaneously, leads to the phosphorylation and inactivation of these enzymes. However, when glycemia increases (e.g., after a meal), glucagon action stops, and these enzymes, in high concentration due to glucocorticoids effect, become active. Thus, the longer the period in hypoglycemia, the higher the concentrations of these enzymes. This will ensure the storage of the nutrient after a meal. Therefore, glucocorticoids prepare the organism to become more efficient in using the nutrients after a period of scarcity. Certainly, this role of glucocorticoids was very important for human beings in the past. However, today, when food is easily available in vast areas of the globe, this adaptation to starvation may contribute to the alarming increase in obesity in the world.

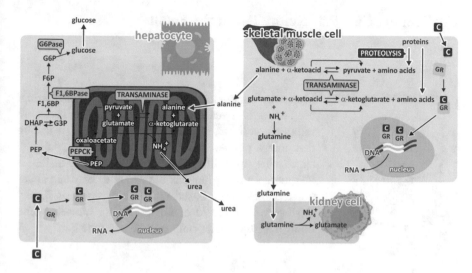

**Fig. 9.21** Integration of the effects of glucocorticoids in different organs showing the regulation of metabolic pathways in each cell type. The *red square* with a C represents cortisol; the *yellow squares* represent the glucocorticoid receptor; the enzymes in *yellow boxes* are the ones whose syntheses are induced by cortisol

## Selected Bibliography

Bhattacharya I, Boje KM (2004) GHB (gamma-hydroxybutyrate) carrier-mediated transport across the blood-brain barrier. J Pharmacol Exp Ther 311:92–98

Cahill GF Jr (2006) Fuel metabolism in starvation. Annu Rev. Nutr 26:1–22

Cori CF, Cori GT (1947) Polysaccharide phosphorylase. Nobel lecture. http://www.nobelprize.org/nobel_prizes/medicine/laureates/1947/cori-gt-lecture.html

Jitrapakdee S, St Maurice M, Rayment I, Cleland WW, Wallace JC, Attwood PV (2008) Structure, mechanism and regulation of pyruvate carboxylase. Biochem J 413:369–387

Michels PA, Rigden DJ (2006) Evolutionary analysis of fructose 2,6-bisphosphate metabolism. IUBMB Life 58:133–141

Owen OE (2005) Ketone bodies as a fuel for brain during starvation. Biochem Mol Biol Educ 33:246–251

Owen OE, Felig P, Morgan AP, Wahren J, Cahill GF Jr (1969) Liver and kidney metabolism during prolonged starvation. J Clin Invest 48:574–583

Owen OE, Morgan AP, Kemp HG, Sullivan JM, Herrera MG, Cahill GF Jr (1967) Brain metabolism during fasting. J Clin Invest 46:1589–1595

Quesada I, Tudurí E, Ripoll C, Nadal A (2008) Physiology of the pancreatic alpha-cell and glucagon secretion: role in glucose homeostasis and diabetes. J Endocrinol 199:5–19

Rath VL, Ammirati M, LeMotte PK, Fennell KF, Mansour MN, Danley DE, Hynes TR, Schulte GK, Wasilko DJ, Pandit J (2000) Activation of human liver glycogen phosphorylase by alteration of the secondary structure and packing of the catalytic core. Mol Cell 6:139–148

Rose AJ, Vegiopoulos A, Herzig S (2010) Role of glucocorticoids and the glucocorticoid receptor in metabolism: insights from genetic manipulations. J Steroid Biochem Mol Biol 122:10–20

van Schaftingen E, Gerin I (2002) The glucose-6-phosphatase system. Biochem J 362:513–532

Young FG (1957) Claude Bernard and the discovery of glycogen. A century of retrospect. Br Med J 1:1431–1437

## Challenging Case 9.1: Kenny Faced the Fa(s)t

### *Source*

Metabolic adaptation to fasting is certainly interesting, but fasting is much more than biochemistry. There are religious, cultural, and aesthetic approaches to fasting that many people find attractive. There are political approaches too, with hunger strikes. Voluntary fasting is ubiquitous among humans but rarely to extreme situations. This case is based on the biography of Kenny Saylors who directed a movie starring himself while doing a 55-day water-only fast: "Facing the Fat."

"Facing the Fat" DVD cover

### *Case Description*

Kenny Saylors directed a movie about his own story while fasting for 55 days. The movie is like a reality show, in which details of Kenny's life during the prolonged fast period were filmed and shared with public. Kenny's motivation for the fast was a strong will to lose "weight" (body mass, to be more precise) and "detox the body." He was determined to fast for 40 days first, a duration inspired by the same events mentioned in the Holy Bible, as Dr. Owen (see Box 9.2), but then continued for extra 15 days to try to beat the Guinness World Record. During this period, Kenny made regular visits to a doctor who requested analyses of biochemical parameters to monitor the impact of fast on his health. He advised Kenny from the beginning about "biochemical and physiological changes in your body" (sic) and the need to drink a lot of water taking electrolytes on a daily basis.

In 1994, at the age of 18 years old, Kenny's body mass was 160 pounds (72.6 kg), and he was in good health. He explains in the movie that a former girlfriend broke his heart in 2006, which caused an eating disorder and triggered overweight ending in obesity. In 1 year, he gained 80 pounds (36.3 kg). Two years later, at the age of 32, he went to the supermarket to buy a large supply of water with added electrolytes but no carbohydrates or other nutrients (zero calories!) and started strict fasting. His body mass was 315.2 pounds (142.9 kg) and body mass index (BMI) was 47.8 kg m$^{-2}$. Estimated percentage of lipids in body mass was 53%.

For the 55-day period, Kenny was monitored for body mass, BMI, estimated % lipid mass, and ketonuria through colorimetric strips. In total, he drank 220 L of water (4 L/day) and urinated 366 times (6–7 per day, 0.6 L on average each time). Put together in a table, the mass and BMI data available from the movie is:

| Day fast | Mass/lbs | Mass/kg | BMI/kg m$^{-2}$ | % lipids |
|----------|----------|---------|-----------------|----------|
| *Before* (age 18) | 160 | 72.6 | – | – |
| 0 | 315.2 | 142.9 | 47.8 | 53 |
| 7 | 305.2 | 138.4 | – | 58 |
| 15 | 301.6 | 136.8 | – | 54 |
| 40 | 274.4 | 124.5 | 41.6 | 50 |
| 55 | 271 | 122.9 | – | 46 |

Starting from the end of day 1, Kenny had headaches that remained for days. At day 3, Kenny had a troubled night, feeling anxiety, discomfort, and sudoresis. Stomach was growling. Hunger peaked some days after starting but attenuated afterward. In the first week, Kenny made a colon cleansing to clear his intestines from resident remaining feces. At day 10, against the recommendation of his doctor, Kenny exercised. He got extenuated and vomited. His doctor had recommended he abstain from any strenuous activity while fasting as consequences can be serious.

Later, at day 50, in one of the scenes of the movie, Kenny visits a cardiologist who had been a body builder. During their conversation, the doctor highlights the dangers of infant obesity, a one-way ticket to diabetes, and the consequences of obesity on heart health. In addition, he recommends caution in interpreting body mass loss data as not all mass loss pertains to lipids. Water loss first and muscle degradation in a second phase account for body mass loss too. Muscle loss in turn leads to metabolic rate decrease and limitations in exercising, which in turn favor body mass conservation. "Weight" loss ends up being a very slow process in fasting.

After his 55-day water-only fast, Kenny initiated a juice diet before ingesting solid food meals. He had lost 20 kg total, from which estimated 19 kg were fat. In the movie Kenny admits he did not lose as much "weight" as he hoped for.

The Guinness Book world record was never awarded. Fasting is dangerous and should be thoroughly followed by medical surveillance; it is not a sport, performance, or a hobby. There is no Guinness award for period spent without eating because no one should be stimulated to starve to death in pursuit of a world record.

## Questions

1. Calculate the BMI values missing in table above for "before" (18 years old, in 1994) and days 7, 15, and 55. Was Kenny overweight or obese in 1994? And after the 55-day fasting?
2. What is ketosis and acidosis? How are they related in Kenny's case?
3. How can ketosis be monitored in urine?
4. Why was Kenny recommended to drink water frequently? Why water with added electrolytes?
5. When is muscle degradation expected to occur in prolonged fast?
6. Calculate the rate of body mass ("weight") loss for Kevin on a daily basis during the whole fasting period, and interpret the results using biochemical reasoning.

## Biochemical Insight

1. When starting the diet, Kenny's BMI was $47.8$ kg $m^{-2}$, and body mass was $142.9$ kg, which means that Kenny height is $h = \sqrt{m}\, / \, \mathrm{BMI} = 1.72\mathrm{m}$. Therefore, BMI for days 0, 7, 15, and 55 are 24.2, 46.3, 45.8, and 41.1 kg $m^{-2}$, respectively. According to the World Health Organization, WHO, BMI between 18.5 and 25 kg $m^{-2}$ is a statistical indicator of having a healthy condition, while BMI between 25 and 30 kg $m^{-2}$ means, statistically speaking, overweight.[1] BMI $> 30$ kg $m^{-2}$ is considered a statistical indicator of obesity. Kenny evolved from a healthy condition in 1994 to very severe obesity (BMI $> 40$ kg $m^{-2}$) and was still severely obese after 55 days fasting.

2. Ketosis is a metabolic state characterized by a rise in the blood concentration of ketone bodies, which results from the increase in the synthesis of these molecules in the liver due to an accumulation of acetyl-CoA (see Sect. 7.4.6). Acetyl-CoA reacts with itself having ketone bodies as end products. When glycemia drops and liver glycogen stores get depleted, ketogenesis (production of ketone bodies) accelerates, and ketone bodies' concentration in plasma may go over 0.5 mM, the typical hallmark of ketosis. Gluconeogenesis favors ketosis as oxaloacetate is used as precursor for glucose synthesis, thus not being available to react with acetyl-CoA and "feed" CAC. Long-term ketosis may result from long-term fasting (see Fig. 9.2) or persistent diets low on carbohydrates.

   Acidosis is a state characterized by a drop in the pH of the blood plasma. Acidosis is a delicate condition due to the Bohr effect (see Sect. 3.3.3), i.e., the effect protonation and deprotonation of amino acid residues has on protein structure and function. Blood plasma is tightly buffered (see Sect. 2.1 and Fig. 2.8): the onset of acidosis is pH drop below 7.35, while alkalosis (the alkaline counterpoint of acidosis) starts at pH 7.45 and on. It is believed that the pH range admissible for the plasma of mammals is 6.8–7.8. Acidosis has two main sets of causes: respiratory and metabolic. Respiratory acidosis is caused by increased partial pressure of carbon dioxide in blood (see also Challenging Case 2.1). Metabolic acidosis is caused by the increased production of acid metabolites, such as lactic acid (fermentation) or the acidic ketone bodies. When acidosis is caused by acidic ketone bodies (i.e., acetoacetate and β-hydroxybutyrate but not acetone), it is called ketoacidosis. The common causes of ketoacidosis are diabetes, prolonged alcoholism, and starvation or fasting such as in Kenny's case.

3. In alkaline medium, acetone and acetoacetate react with sodium nitroprusside forming a colored complex. In urine, acetone is barely present because it is a volatile molecule and is excreted through lungs and tongue, leading to a fruity scented breath (see also Challenging Case 7.2) and a harsh sensation in the month. Most detection strips to be used in urine are based on a local alkaline salt and sodium nitroprusside. The intensity of the magenta color formed in this strip

---

[1] BMI Classification (2006) Global database on body mass index. World Health Organization. http://apps.who.int/bmi/index.jsp?introPage=intro_3.html.

is proportional to the concentration of acetoacetate in urine. β-hydroxybutyrate is not detected because it is not a ketose and does not react with the iron.

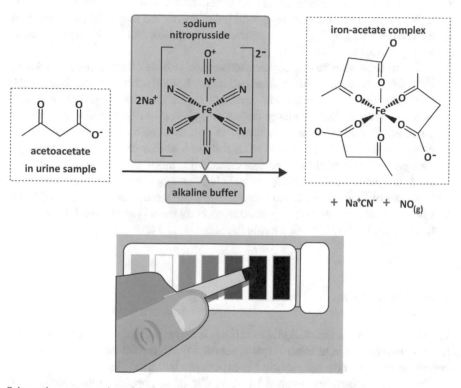

Schematic representation of the biochemical test to detect acetoacetate in urine samples

4. To diminish the risk of acidosis associated with ketosis, fast clearance of ketone bodies from plasma is desirable. Drinking water frequently helps elimination of ketone bodies through the kidneys in urine. As drinking unusually high amounts of water may lead to electrolyte imbalance due to continuous removal of salts in urine, it is desirable that drinking water has ionic concentrations similar to physiological conditions in human plasma so that water intake is not accompanied by ionic imbalance. Isotonic water used in sports has carbohydrates and high concentration of electrolytes to compensate for losses after exercise. Drinking water in Kenny's case had a different motivation. Isotonic water used in sports would not be totally adequate. Regular drinking water with added electrolytes was used instead.

5. Figure 9.1 shows that liver glycogen is no longer the main source of plasma glucose after 16 h fasting and has no contribution at all after 28 h. Amino acids are used as precursors for gluconeogenesis in compensation for the limitation of liver glucose. Muscle proteins are degraded, and the resulting amino acids are converted mainly to Gln and Ala, which are released into the blood (Fig. 7.20) and will serve to feed gluconeogenesis in the liver. Therefore, muscle mass

degradation is concomitant with lipid conversion to ketose bodies in prolonged fasting adding a doubt about the efficacy of starvation to consume body fat.

6. Using the data in the table presented in the description of this case, it is calculated that Kevin lost, on average, 642 g/day during the first week, 200 g/day on the second week, 492 g/day between days 15 and 40, and 320 g/day on the following 2 weeks.

In the beginning fasting appears to be a rewarding effort. Kenny lost 4.5 kg in 7 days. Notwithstanding, the estimated lipid fraction of the body mass increased. Body mass reduction occurs at higher rate first as glycogen is consumed. Glycogen has an energy conversion yield lower that lipids measured on a mass-basis and releases three times its own mass of hydration water. Muscle degradation for the use of amino acids follows a similar pattern. As much as 5 kg "weight" change might not be associated with any fat loss.[2] At later periods, lipids are intensively used. Lipids have high energy conversion yield (meaning less mass of lipids is needed to generate the same amount of ATP) and no hydration water mass to release, so reduction on body mass is moderate. Diet-induced thermogenesis may also play a role (see Sect. 11.2.1.2).

## *Final Discussion*

Fasting is for many a temptation. Sometimes it seems a highly committed sacrifice to look thinner; while at other times it seems a purifying attitude. Kenny Saylors combined both with the will to turn his experience into a movie.

The second motivation evoked by Kenny is quite popular among individuals without scientific background: "detox" the body. "Detox" is an extremely ambiguous term bringing the idea of purification to our intuitive perception. However, there is not a scientific link between "detox" and prolonged starvation. On the contrary, rapid mobilization of lipids from the adipose stores may free into the circulation hydrophobic toxins that were trapped in lipid agglomerates.[3] So, ironically, processes to accelerate lipid mobilization from the adipose tissue may, in fact, help "toxify" the body.

Overall, one learns from Kenny Saylor's story that prolonged fast is a dangerous and worthless strategy to lose body mass. Moreover, a quick search through the Internet reveals that Kenny regained all body mass lost in his unwise endeavor soon after finishing the 55-day water fasting. Not only he did lose less "weight" than he thought; he could not maintain his final "weight." His sacrifice was worthless in the end.

---

[2] Diaz EO et al (1992) Metabolic response to experimental overfeeding in lean and overweight healthy volunteers. Am J Clin Nutr 56:641–655.

[3] Brown RH (2019) Mobilization of environmental toxicants following bariatric surgery. Obesity 27:1865–1873.

## Challenging Case 9.2: The Lychee Paradox: Deliciously Sugary, Dangerously Hypoglycemiant

### *Sources*

This case is based on a series of reports of outbreaks of acute hypoglycemic encephalopathy, occurring especially among children, during lychee fruit ripening season, in India. Since 1995, seasonal outbreaks of a fatal neurological illness have been reported in Muzaffarpur, the largest lychee cultivation region in India. The cause of the disease was unraveled more than 20 years later: a hypoglycemiant toxin present in lychee fruit itself.

Lychee fruit; image reproduced from Wikimedia Commons, under a CC-LY licence

### *Description*

In June 2019, several deaths caused by an acute encephalitis syndrome were reported in the city of Muzaffarpur, state of Bihar, India.[4, 5] The disease, which caused seizures, altered mental state and death in more than a third of cases, is locally known as *chamki bukhar*. This dramatic situation could have been avoided since the cause of the disease, as well as the way to prevent it, had been revealed 2 years earlier, with a good repercussion in the media.[6,7]

In 2017, a joint investigation by the National Center for Disease Control of India government and the Centers for Disease Control and Prevention (CDC), in Atlanta, USA, published an article[8] solving the "mystery of the deadly outbreaks," as named by The New York Times. The researchers hypothesized that the cause of the unexplained neurological illness would be a toxin present in the lychee (or litchi) fruit. They based this hypothesis on the fact that the outbreaks occurred in the same region, annually and during summer, coinciding with the lychee season. During this period the fruit is widely consumed by the children, who also usually skip dinner. The authors also made a parallel between *chamki bukhar* and a similar hypoglyce-

---

[4] https://edition.cnn.com/2019/06/13/health/encephalitis-india-outbreak-deaths-lychee-intl/index.html.

[5] https://www.theguardian.com/world/2019/jun/13/at-least-31-children-in-india-killed-by-toxin-in-lychees.

[6] https://www.bbc.com/news/world-asia-india-38831240.

[7] https://www.nytimes.com/2017/01/31/world/asia/lychee-litchi-india-outbreak.html.

mic encephalopathy occurring in children in the Caribbean region, which had been known for decades to be caused by a toxin present in the fruit of the ackee plant, a member of Sapindaceae family, as lychee is.

The researchers proved that lychees were the culprit of children deaths after a very careful study, in which they collected specimens (blood, cerebrospinal fluid, and urine) from 390 patients admitted between May and July 2014 to referral hospitals in Muzaffarpur and tested them for the presence of a number of pathogens as well as non-infectious agents, such as pesticides, toxic metals, and other toxins.[8] No infectious agents or pesticides were found. On the other hand, they found two modified amino acids shown to cause the hypoglycemia induced by ackee fruit: L(methylenecyclopropanyl)alanine, known as hypoglycin A (HGA), and methylenecyclopropylglycine (MCPG). Analyses of the lychee cultivated in the region also revealed the presence of both substances in high concentrations.

Additionally, the authors clearly indicated how to prevent the disease: *"To prevent illness and reduce mortality in the region, we recommended minimising litchi consumption, ensuring receipt of an evening meal and implementing rapid glucose correction for suspected illness."* Indeed, after the article was published, health officials advised the parents to limit lychee consumption by the children and to be sure they had an evening meal. In two seasons, the number of reported cases per year dropped from hundreds to less than 50.

## Questions

1. HGA is a protoxin, i.e., it is not a toxin itself, but it is physiologically transformed into methylene cyclopropyl acetic acid (MCPA), which was shown to inhibit β-oxidation. In the study performed in 2017, the researchers found that 90% of the case-patients had abnormal plasma acylcarnitine profiles, consistent with a severe inhibition of fatty acid oxidation.[8] Observe MCPA chemical structure and propose a hypothesis to explain its effect on β-oxidation.

2. Discuss a general mechanism through which the inhibition of β-oxidation would cause hypoglycemia.

3. In an in vivo study published in 2019, the authors infused MCPA and MCPG into mice,[9] showing a 50% reduction in hepatic glucose production (see below figure).

[8] Shrivastava A (2017) Association of acute toxic encephalopathy with litchi consumption in an outbreak in Muzaffarpur, India, 2014: a case-control study. Lancet Glob Health 5:e458–e466.

[9] Qiu Y et al (2018) *In vivo* studies on the mechanism of methylenecyclopropylacetic acid and methylenecyclopropylglycine-induced hypoglycaemia. Biochem J 475:1063–1074.

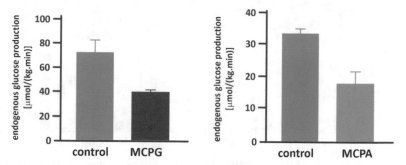

Endogenous glucose production assessed 4 h after mice have been treated with vehicle (saline) or MCPG (left panel) or with vehicle or MCPA (right panel). (Figure adapted from the article cited in the footnote 9, with permission)

These results indicate that not only the use of glucose by the different tissues is increased but also that hepatic gluconeogenesis is being directly inhibited. This observation led the authors to hypothesize that the substances promote a decrease in the activity of the enzyme pyruvate carboxylase (PC). Discuss the possible mechanisms by which MCPA and MCPG would affect PC activity.

## Biochemical Insights

1. The chemical structures of HGA, MCPA, and MCPG resemble that of carnitine (see below figure). This suggests that these molecules would inhibit the carnitine acyl transferase, impairing not only the transport of fatty acids into mitochondria but also limiting the availability of coenzyme A, impairing mitochondrial fatty acid metabolization in β-oxidation (see Sect. 7.4.3).

Chemical structures of HGA, MCPA, MCPG, and carnitine

Differently to the long-chain fatty acids, fatty acids with up to 10 carbons atoms permeate the inner mitochondrial membrane in the non-esterified form, being activated to their CoA-derivatives in the mitochondrial matrix (see Sect. 7.4.2). Thus, if the inhibition of β-oxidation by MCPA and MCPG occurs through the inhibition of the carnitine shuttle, it is expected that these substances did not affect the oxidation of medium-chain and short-chain fatty acids. Indeed, infusion of MCPG into mice resulted in an accumulation of hepatic long-chain fatty acids together with a decrease in the levels of short-chain fatty acids (see below figure).

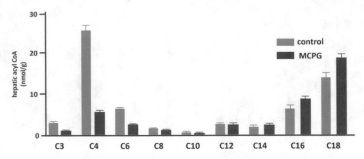

Hepatic content of acyl CoA with different chain lengths assessed 4 h after mice have been treated with vehicle (saline solution) or MCPG. (Figure adapted from the article cited in the footnote 9, with permission)

2. As illustrated in the simplified representation of energetic metabolism in the next figure, ATP is generated from the metabolization of fatty acid or glucose (or other saccharides). Amino acids may also be used to generate ATP, but this largely depends on correlations with glucose metabolism.

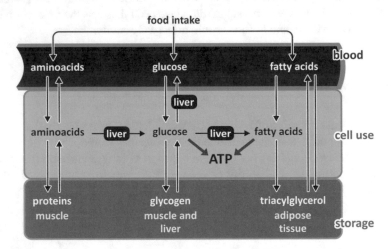

Simplified representation of energetic metabolism. Metabolization of glucose, fatty acids, and, to a lesser extent, amino acids, generates ATP necessary for the cellular functions to be performed. When in excess, glucose may be converted to glycogen or fatty acids to be stored. Fatty acids, obtained from the diet or synthesized from glucose, may be stored in the adipose tissue as triacylglycerols. Maintenance of blood glucose concentration is essential since some cellular types, such as those from the nervous system, are highly dependent on this nutrient. Glucose may be synthesized from amino acids and other precursors in the liver through the gluconeogenesis pathway, being released in the bloodstream, ensuring glycemia control. When ATP production from fatty acids is impaired or severely diminished, glucose becomes the main source for ATP synthesis, which limits its availability to keep glycemia above safe threshold levels

With the reduction of fatty acid metabolism caused by MCPA, one would expect an increase in the utilization of glucose by tissues that usually use fatty acids as the main energy source in a fasting situation. This would result in a decrease in glucose concentration in the blood (hypoglycemia), a condition quite stressful for the brain, which is largely dependent on blood glucose to function: brain metabolism accounts for approximately 85% of the glucose needed for the organism (100 g of 120 g necessary daily; see Sect. 9.3.4). The end result may be cellular death in brain tissues and hence encephalopathy.

3. The inhibition of β-oxidation by MCPA and MCPG would decrease the levels of hepatic acetyl-CoA. Indeed, hepatic acetyl-CoA was reduced by 84% in animals infused with MCPA (see below figure).

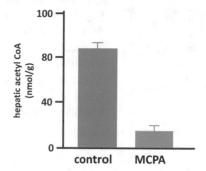

Hepatic acetyl-CoA content assessed 4 h after mice have been treated with vehicle (saline solution) or MCPA. (Figure adapted from the article cited in the footnote 9, with permission)

PC is a mitochondrial enzyme that catalyzes the biotin-dependent carboxylation of pyruvate to produce oxaloacetate, the first reaction necessary for bypassing the glycolytic reaction of pyruvate kinase in gluconeogenesis (see Sect. 9.3).

Since the classical work of Merton Utter, in 1965, acetyl-CoA was shown to be an allosteric activator of PC, required for the first partial reaction (the ATP-dependent carboxylation of biotin; see Sect. 9.3.1). Thus, the drastic decrease in the hepatic acetyl-CoA content promoted by MCPA and MCPG is expected to strongly impact in PC activity, impairing glucose production.

## Final Discussion

### Lychee Breeding and Lychee Overeating

Ingestion of large amounts of lychee by children, when not accompanied by cooked meals, is dangerous. Avoiding lychee poisoning is achievable raising awareness and improving education. However, biotechnology may help. Some lychee trees are genetically determined hyper-producers of HGA, while others are not. Judicious selection of genotypes associated with low contents of hypoglycemic amino acids is both possible and desirable, as discussed in a review article published in 2018.[10] The amount of hypoglycemic amino acids present in the arils of ten lychee coming from "super-toxic" trees is enough to be considered toxic.

---

[10] Kumar S et al (2018) Lychee-associated hypoglycaemic encephalopathy: a new disease of children described in India. Proc Natl Acad Sci India Sect B Biol Sci:1–7.

## Challenging Case 9.3: The Bizarre Death of Elaine Robinson

### Source

Severe hypoglycemia is a life-threatening condition. Combining several hypoglycemia-stimulating factors together is potentially dangerous. The death of Elaine Robinson proves it. Simultaneous intoxication with alcohol and insulin in a social occasion put an end to her life at the age of 43. This case is based on real events described in the book: *Insulin Murders: True Life Cases.*[11]

Insulin syringe. (Image reproduced from Needpix.com)

### Case Description

Elaine Robinson died in April 2000 in a drug abuse scenario, among alcohol and… Insulin! She was born in Port Talbot, Wales, UK, in 1957, ironically the same year the first case on insulin overdose murder ended in conviction in an assize court in Leeds, UK.

Elaine married for the third time when she was 37, with Kevin Johnson, a heavy drinker and drug abuser. At the time she died, with 43 years old, Elaine used to drink 5 L of cider a day, or more. On the afternoon before Elaine died, she and Kevin met two friends, Jacky Jones and Steven O'Keefe, to drink cider at Kevin's house. Steven left at 5:00 p.m., the time another friend, Chris Andrews, arrived. By 7:15 p.m., Steven returned, Chris had left, and the other three friends were asleep. They woke up soon after that and started drinking cider together. At 10:30 p.m., by the time Elaine decided to go to bed, she was acting "too drunk" to walk and had to be carried upstairs by Steven and Kevin. Such an extreme effect after drinking on Elaine had never been noticed before by her friends.

Next morning, Elaine did not respond to Kevin when he tried to wake her up, although he was loud enough to wake Jacky and Steven, who were asleep in the next bedroom. A medical emergency team was called, and an ambulance arrived at 8:13 a.m. The paramedics could not find signs of life in Elaine but tried to revive her. After 20 min she was taken to Neath Hospital. Dr. Emma Meredith noted that Elaine's pupils were dilated and did not respond to light. She had no pulse or audible heartbeat. Nevertheless, she was given glucose, adrenaline and naloxone,

---

[11] Mark V, Richmond C (2007) Insulin murders. True life cases. Chap. 14. The Royal Society of Medicine Press. ISBN 978-1-85315-760-0.

considered an antidote for heroin poisoning, just in case hypoglycemia or narcotic poisoning was responsible for her condition. Maybe because Elaine's death was both sudden and unexplained, Dr. Meredith collected arterial blood from the Elaine's thigh before declaring her death at 9:00 a.m.

The clinical pathology laboratory of the hospital determined the blood glucose concentration in the sample collected by Dr. Meredith: 0.2 mmol/L. Wisely the rest of the serum was kept for other tests, if needed.

Kevin and Jacky did not need medical care and were fully conscious when interviewed by the police surgeon, Dr. Ruth Frager, later that day. They described how Chris had arrived at the party and accepted a drink of cider. Soon after having arrived, Chris took a vial of insulin and offered to inject each of the friends with it. It would act as a "pick you up" or resemble the "speedy" rush of amphetamines, he said. Injection of insulin took place in their thighs, 1 h before he left. Afterward, all three friends continued drinking cider. They could not remember eating during that period. Kevin told Dr. Frager that after the injection he felt drowsy, shaky, and sweaty. Jacky mentioned feeling "sick and shaky" 3 h after being injected, which occurred probably near 6:00 p.m.

Dr. Stephen Leadbeater, a Forensic Pathologist, carried out a *postmortem* examination on Elaine's body the day after she died. His report proclaimed a middle-aged woman, 1.67 m tall, weighing 45.1 kg. No significant physical disorders for her death could be found. Her liver weighed 1,280 kg, and microscopic imaging revealed typical signs of alcohol abuse without progression to permanent cirrhosis yet. Blood alcohol concentration in the legs was modest: 40 mg/100 mL, below typical drunk driving limits. The blood was analyzed again 3 months later on the fourth of July by Dr. Andrew McKinnon. There was no alcohol detected, but he found traces of cannabis and a benzodiazepine-type tranquilizer, though not enough to have killed her. Dr. Leadbeater also collected a sample from the vitreous humor, in Elaine's eyes. β-hydroxybutyrate was present at 200 μmol/L.

In the course of investigation, blood serum was also analyzed 5 days after her death by Dr. Rhys John, a clinical biochemist of the Hormone Reference Laboratory in Cardiff. Her blood contained 57.7 mU/L ($\approx$346 pmol/L) of insulin but no C-peptide. Dr. John also analyzed the contents of a bottle the police had seized from Kevin's house and labeled Human Mixtard and confirmed it contained the expected concentration of around 100 units/mL.

Based on evidence collected during the investigation, Chris Andrews was sentenced to 2 years in jail on 24 October 2001 for the death of Elaine Robinson.

## Questions

1. Elaine stopped drinking at least 12 h before Dr. Stephen Leadbeater found a modest alcoholemia of 40 mg/100 mL in her blood. Between the time she stopped drinking and the certified time of death, at 9:00 a.m., alcoholemia should have dropped at a rate of 15 mg/100 mL/h. So, what was the estimated alcohol-

emia when she has carried to bed? Is it unexpected given Elaine's condition? Is it reasonable to assume it to be the cause of her death?

2. In case you ruled out the possibility of alcohol poisoning as the cause of death, is hypoglycemia a reasonable hypothesis? Why?

3. Dr. Emma Meredith collected a blood sample in which the glucose concentration was nil, in practice. Because the exact time of death was not known, this does not necessarily mean that glycemia had been very low while Elaine was still alive. *Postmortem* blood glucose measurements are known to be unreliable. However, it is reasonable to assume that the time of death was not long before the certified time of death, at 9:00 a.m., and thus there was severe hypoglycemia while Elaine was still alive. Do you agree? Why?

4. The test on the vitreous humor for β-hydroxybutyrate helped to determine the cause of death and convict Chris Andrews. Why?

## Biochemical Insight

1. A timeline is useful to have a clear perspective on the course of events that ended in the tragic death of Elaine Robinson.

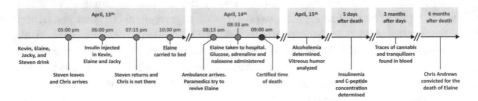

Considering the certified time of death, Elaine had stopped drinking for 10.5 h. The estimated alcoholemia in that moment was $40 + 10.5 \times 15$ mg/100 mL, or ≈197.5 mg/100 mL. It may have been a bit more considering she may have stopped drinking some time before going to bed. This value is high for usual standards, well above drunk driving limits in many countries, but not so excessive for a chronic alcoholic and certainly not fatal, even considering Elaine's severe underweight (body mass index, BMI, of 16.1 kg/m$^2$, for a "normal" interval of 20–25 kg/m$^2$). It should be recalled that Elaine's liver had normal weight and was not extremely damaged.

2. The hypothesis of alcohol poisoning being ruled out, it is reasonable to set the hypothesis that insulin-induced hypoglycemia may have been the cause of death. 57.7 mU/L insulinemia would be considered normal in a fed state individual, but Elaine had not been eating and had nearly nil glycemia. Endogenous insulin in this condition is expected to be absent in the blood circulation. Because Dr. John found no C-peptide in Elaine's plasma, there was solid evidence that insulin was exogenous.

3. In addition to the absence of C-peptide in Elaine's blood, it is also reasonable to assume that glycemia was very low while she was still alive. The exact time of death is not known, but the fact that two paramedics and a hospital doctor had attempted to revive her suggests that either she was in a state of "suspended animation" due to hypoglycemic coma, sometimes wrongly certified as death, or that she was not long dead when the ambulance arrived. At this point, it is important to recall that in humans *rigor mortis* is expected 4 h *postmortem*, approximately. In conclusion, it is reasonable to consider that Elaine was not long dead when the ambulance arrived, and so the 0.2 mmol/L glycemia gains relevance because it points toward hypoglycemia being the cause of death.

4. β-hydroxybutyrate is a ketone body. It was present in the vitreous humor at low concentration (200 μmol/L). The ketone body concentrations in tissues vary little with *postmortem* interval.[12] Ethanol-induced hypoglycemia leads to ketosis due to lipolysis (see Sects. 7.4.6 and 9.3.4, and Challenging Case 7.3). On the contrary, insulin inhibits lipolysis and therefore counteracts ketosis. Alcoholic coma leading to death is expected to lead to a minimal concentration of 531 μmol/L of ketone bodies in blood, which in principle would correspond to a similar concentration of β-hydroxybutyrate in vitreous humor. Thus, it is reasonable to conclude that ethanol-induced hypoglycemia was not by itself the cause of death despite having contributed to the fatal outcome.

## Final Discussion

Vincent Marks, author of *Insulin Murders: True Life Cases* and expert in insulin forensics makes a compelling humanistic description of Elaine's case based on his experience (quotation): "*Strange at is may seem, insulin has been used, probably rarely, by drug abusers for the side effects it regularly produces, and this must have been what motivated Chris Andrews to offer Elaine and her friends 'a fix' with insulin. Diabetic patients who have to take insulin consider these effects undesirable, unpleasant and to be avoided if at all possible. The symptoms are accompanied by an adrenaline surge resembling, in some respects, that produced by the intravenous injection of amphetamines. It was presumably with this in mind rather than any evil intent that Chris offered Elaine and her friends some of the insulin—which he had obtained from a diabetic patient for whom it had been legitimately prescribed. Insulin had been freely available from regular pharmacies without prescription until only a year before Elaine died. It had been put on the prescription-only list because of widespread abuse by bodybuilders, who believed it helped them build up muscles (…). On the basis of my report, the Crown Prosecution Service felt sufficiently confident top charge Chris Andrews with the manslaughter of Elaine*

---

[12] Felby S et al (2008) The postmortem distribution of ketone bodies between blood, vitreous humor, spinal fluid, and urine. Forensic Sci Med Pathol 4:100–107.

*Robinson, to which charge he pleaded guilty on 24th October 2001 at Swansea Crown Court. Three other charges of causing a noxious thing to be taken with intent to endanger life (of Elaine, Kevin and Jacky) were not pursued but ordered to lie on the file. He was sentenced to 2 years in jail. Despite having pleaded guilty, Chris later appealed against his conviction and sentence. His appeal was based on the contention that insulin is not a noxious substance and that Elaine had been in obvious agreement to the injection. The appeal was heard at the Court of Appeal in London on 21st November 2001. It failed. The argument in the Court of Appeal boiled down to the exact nature and wording of the charges against Chris Andrews rather than the nature or heinousness of his crime. Clearly, he did not intend to kill Elaine but did so by behaving recklessly. My opinion is that the question of Chris's guilt and appeal outcome would have been different had insulin been an over-the-counter drug, which it had been until a year earlier. He would probably have been acquitted of any crime."*

# Chapter 10
# Regulation and Integration of Metabolism During Physical Activity

The survival of our ancestors was highly dependent on hunting, gathering, and fighting, behaviors that demanded intense physical activity. A sedentary lifestyle in that environment would certainly result in the elimination of the individuals. This situation imposed a selective pressure directed to adaptations of human physiology toward physical activity, resulting in the development of a very efficient locomotor system, in which the skeletal muscles correspond to about 40% of the body mass and account for a great proportion of the average energy consumption of the organism. The present-day sedentarism is dissonant with the human genetic background selected to favor a physically active lifestyle and probably consists in one of the main causes of the increasing incidence of modern chronic diseases, such as hypertension, obesity, and insulin resistance.

Muscles are tissues specialized in producing force and movement due to an amazing ability to convert the chemical energy of ATP hydrolysis to mechanical work. This energy interconversion is performed by an array of proteins that forms a very organized structure inside the cells. The muscle mechanical activity may change very fast so that the energy sources and the metabolic pathways used to maintain cellular functions need to be finely regulated.

In this chapter, we will discuss the metabolic adaptations to physical exercise, with special attention to the metabolism of the skeletal muscle cells. We will start with a brief review of the structure of the contractile apparatus and the mechanism of muscle contraction, followed by the description of the energy sources and the metabolic pathways involved in muscle activity during physical exercise. We finish the chapter with the mechanism of action and the main metabolic effects of adrenaline, the major hormone secreted during exercise.

© Springer Nature Switzerland AG 2021                                527
A. T. Da Poian, M. A. R. B., *Integrative Human Biochemistry*,
https://doi.org/10.1007/978-3-030-48740-9_10

## 10.1   Muscle Contraction

Muscles are used either for locomotion or for the movements associated to the functions of the internal organs and are classified in three groups according to their histological organization, which is related to the type of movement they generate (Fig. 10.1). The skeletal muscles can be contracted voluntarily allowing the body to move and to maintain the posture, while rhythmic involuntary muscle contractions, such as heart contraction or peristalsis and other autonomous motilities, are performed by cardiac and smooth muscles, respectively.

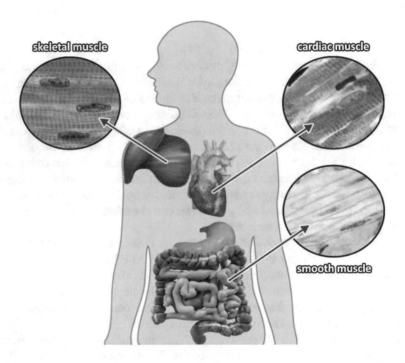

**Fig. 10.1** The three types of muscle tissues. The details show histological images of skeletal, cardiac, and smooth muscles. (Histological images reprinted with the permission of Instituto de Histologia e Biologia do Desenvolvimento, Faculdade de Medicina, Universidade de Lisboa, FMUL)

### 10.1.1   Structural Organization of the Contractile Apparatus

In this chapter, we will focus our attention in skeletal muscles. This tissue is composed of parallel bundles of large multinucleated cells, called muscle fibers. The fibers have 10–100 μm diameter and sometimes can extend over the full length of the muscle, reaching several centimeters. Most of the intracellular volume of the muscle fiber is occupied by 2 μm thick myofibrils formed by the contractile array of proteins (Fig. 10.2a). Observed at the optical microscope, the fiber presents a typical pattern that alternates light and dark bands caused by a regular arrangement of molecules of different densities (Fig. 10.2b).

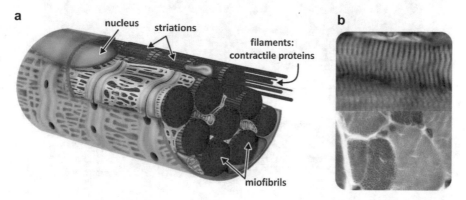

**Fig. 10.2**  (**a**) Schematic representation of a muscle fiber with the nucleus (*blue*) in the periphery of the cell and myofibrils formed by the contractile proteins. (**b**) Histological image of longitudinal (*top*) and transversal (*bottom*) sections of skeletal muscle tissue showing the striated pattern. (Histological images reprinted with the permission of Instituto de Histologia e Biologia do Desenvolvimento, Faculdade de Medicina, Universidade de Lisboa, FMUL)

The organization of the contractile proteins in the myofibril explains the striated pattern observed at the optical microscope: they are arranged in an ordered structure forming thin and thick filaments, clearly seen by electron microscopy, as firstly observed by Hugh Huxley in the 1950s (Fig. 10.3). The different regions were named according to their characteristics. The dense regions are called A (from anisotropic) bands, while the less dense regions are named I (from isotropic) bands. A dark line in the medium point of each I band is also observed and is designated the as Z line (from the German word *Zwischenscheibe*, which means "the disk in between"). A denser line is observed in the middle of the A band, called M line (from the German word *Mittelscheibe*, which means "the disk in the middle"). In the resting muscle, the central region of the A band shows a lighter area, which is called H zone (from the German word *heller*, which means "brighter"). The I band is a region containing only thin filaments, while the A band contains both type of filaments. In the resting muscle, the thin filaments do not reach the center of the A band, explaining the H zone, which contains only thick filaments (Fig. 10.3).

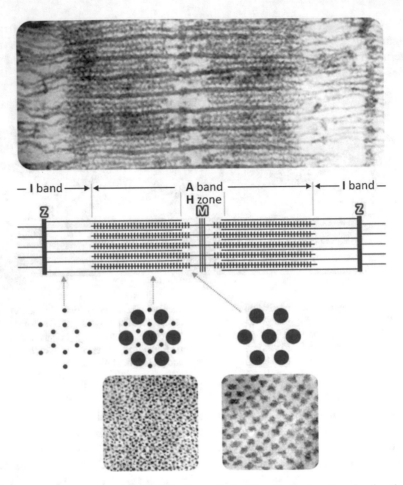

**Fig. 10.3** Electron micrographs of a skeletal muscle (*top*) longitudinal section showing the thin and the thick filaments and (*bottom*) transverse sections of A band and H zone. (Figures reproduced with permission from Huxley, J Biophys Biochem Cytol 3:631–648, 1957) The schematic representation shows filaments' organization in the region in between two Z lines and transverse sections of the I band (*left*), the A band (*medium*), and the H zone (*right*). The region of 2–3 μm in between two Z lines is called sarcomere, which is the contractile unit of the myofibril

The striated pattern occurs in skeletal and cardiac muscles, which are called striated muscles, while the smooth muscles do not present striations.

The thin filaments are inserted in the Z line and are composed of three proteins: actin, tropomyosin, and troponin. The thick filaments are formed by a protein named myosin (see Box 10.1). The region comprised between the Z lines is called sarcomere and corresponds to the contractile unit of the myofibril. It is important to mention that the components of muscle cells are usually designated by a specific nomenclature having the prefix *sarco* (from the Greek root meaning "flesh"): the plasma membrane is known as sarcolemma, the cytoplasm is known as sarcoplasm, and the endoplasmic reticulum is called sarcoplasmic reticulum.

## Box 10.1 Isolation of Myosin

Since the middle of the nineteenth century, it was known that the disruption of muscle cells resulted in the precipitation of an insoluble material in a much higher amount than that observed for homogenates of other tissues.

Still in the nineteenth century, this insoluble material was called "myosin" by Wilhelm Kuehne, but only in the 1940s, with the studies performed mainly by Albert Szent-Gyorgyi, this term was specifically used to name the proteic component that could be solubilized from the insoluble material by treatment with a high ionic strength solution (0.6 M KCl, an ionic strength much higher than the physiological 0.15 M).

Another proteic component could be solubilized from the remaining insoluble material of muscle homogenate by lowering the ionic strength below the physiological range, and this protein was named "actin" (due to its ability to activate the ATPase activity of myosin, as it will be explained in the next section). It is important to note that both myosin and actin are insoluble at the physiological ionic strength, being organized in the filaments seen by electron microscopy.

Albert Szent-Gyorgyi
(1893-1986)

### 10.1.1.1 The Main Proteic Components of the Contractile Apparatus

Myosin represents up to 65% of the total protein that constitute the myofibrils. A landmark finding regarding the role of myosin in contraction came from the work of W. A. Engelhardt and M. N. Lyubimova, who demonstrated that it displayed enzymatic activity of hydrolyzing ATP. It was also shown that the volume of myosin in its insoluble state (at ionic strength lower than 0.6 M; see Box 10.1) contracted after addition of ATP, leading to the supposition that the hydrolysis of ATP by myosin would be the driving force for muscle contraction. This is indeed the basis for the chemical–mechanical energy conversion during muscle contraction, as we will discuss in the next section.

The first electron microscopy images of isolated myosin filaments showed a very peculiar morphology. They present typical projections regularly spaced from each other along the extension of the filament, except for the central part, which is known as the "bare zone" (Fig. 10.4a). The projections are the contact points with the thin filaments.

The morphology of the thick filament can be explained by the structure of myosin units. Each myosin unit in the thick filament shows a golf club shape, with two well-defined regions: a globular "head" from which a "tail" of about 150 nm long extends (Fig. 10.4b). It is a hexameric protein containing two identical heavy chains (with ~220 kDa each) and four light chains (each with ~20 kDa). The head can be separated from the whole myosin by brief digestion with proteases. The generated fragment, containing the N-terminal end of the heavy chains associated with the light chains, is called subfragment 1 (S1) and comprises the ATPase activity and the actin-binding site. The tail is composed of the C-terminal region of the two heavy chains intertwined to form a coiled coil.

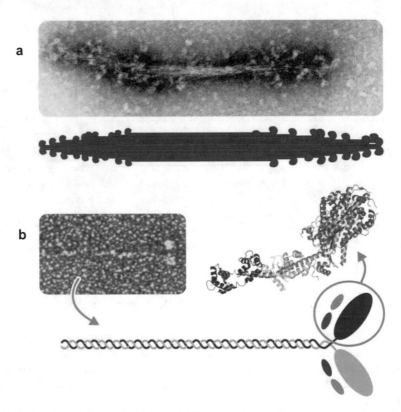

**Fig. 10.4** Structure of the myosin filament. (**a**) Electron micrograph of a reconstituted myosin filament and its schematic representation. (Micrograph reproduced with permission from Huxley, Science 164:1356–1366, 1969) (**b**) Electron micrograph of an isolated myosin molecule and a schematic representation of its structure, which is composed of two heavy chains (*purple* and *pink*) associated to four light chains (*green* and *red*). The three-dimensional structure of the head, or S1 subfragment, is also shown, with the light chains highlighted in *green* and *red* (PDB 1DFL). (Micrograph reproduced from Slayter and Lowey, Proc Natl Acad Sci USA 58:1611–1618, 1967)

Actin is the major component of the thin filament and corresponds to about 20–25% of the muscle proteins. Actin is a protein of 42 kDa that is called G-actin (in a reference of its globular structure) in its monomeric form. To form the thin filaments, actin monomers polymerize in a helical structure, in which it is named F-actin (Fig. 10.5).

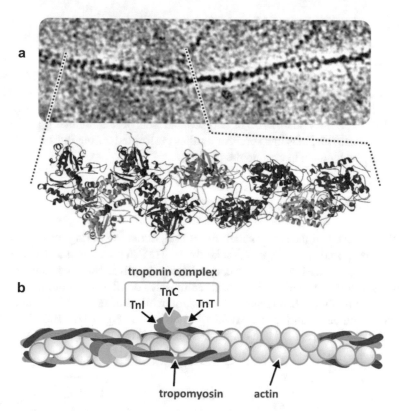

**Fig. 10.5** Structure of the thin filament and its components. (**a**) Section of an electron micrograph from the article that firstly described the organization of an actin filament (image reproduced from Hanson and Lowy, J. Mol Biol 6:46–60, 1963, with permission from Elsevier), with the correspondent molecular structure determined by X-ray crystallography showing 13 actin monomers arranged on six left-handed turns repeated every 36 nm (figure reproduced with permission from Geeves and Holmes, Ann Rev. Biochem 68:687–728, 1999). (**b**) Schematic representation of an actin filament with tropomyosin and troponin bound

Other proteins are associated to the thin filament. Among them, the most important are the tropomyosin and troponin, both involved in the regulation of contraction. Tropomyosin is a dimeric protein of 65 kDa that associates to actin as twisted α-helices that interact in a tail-to-head manner forming long rods along the thin filament (Fig. 10.5b). Troponin is a globular protein of 78 kDa composed of three subunits, Tn-I, Tn-C, and Tn-T. Tn-I (I from inhibiting) binds actin fixing the tropomyosin-troponin complex on the actin surface and blocking myosin binding to thin filament. Tn-C (C from calcium binding) binds calcium ions ($Ca^{2+}$), the main regulator of contraction (see Sect. 10.1.3). Tn-T (T from tropomyosin binding) promotes the association of the two other subunits of troponin and the binding of them to tropomyosin (Fig. 10.5b).

## 10.1.2    Mechanism of Muscle Contraction

After the discovery that the contractile apparatus was constituted mainly by two proteins, actin and myosin, many hypotheses emerged to explain the contraction, most of them suggesting that muscle shortening during contraction was due to changes in protein structure leading to a more packed folding or coiling of the filament. However, the visualization of two separate sets of longitudinal filaments that overlap in certain regions in a series of studies performed during the 1950s led to a completely new model to explain the process.

### 10.1.2.1    The Sliding Filaments Model

The theory accepted today to explain myofibril shortening during muscle contraction was proposed in 1954, independently by H. E. Huxley and J. Hanson, and A. F. Huxley and R. M. Niedergierke, both studies published simultaneously. These scientists observed that either in contracted muscles or after their complete stretching, the length of the A band remained constant, while the length of the I band or the H zone varied according to the extent of contraction (Fig. 10.6). Based on these observations, they proposed that contraction occurred through the sliding of the filaments along each other.

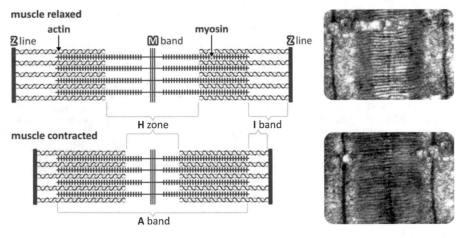

**Fig. 10.6** Schematic representation of the sliding of the filaments showing the relaxed and the contracted pattern with the corresponding electron micrographs. (Micrographs reproduced with permission from Huxley. J Biophys Biochem Cytol 3:631–648, 1957)

One important finding that contributed to the elucidation of the mechanism that allows the filaments to slide was the observation that when the thin filaments were incubated with myosin S1, these subfragments form crossbridges with the thin filament in two orientations: perpendicular to the filament and with an inclination of approximately 45°. When inclined, all myosin heads show the same orientation on one side of the Z line but the opposite orientation on the other side of the Z line, which means that on each side of the Z line, the myosin heads point to opposite directions (Fig. 10.7a). This information could be correlated to the fact that during contraction the thin filaments on each side of the Z line are pulled to the center of the sarcomere.

The current knowledge on the molecular mechanism that leads to the sliding of the filaments is based on the model proposed by R. W. Lymn and E. Taylor, known as the Lymn–Taylor actomyosin ATPase cycle, according to which contraction occurs as myosin heads bind and detach repeatedly to the thin filaments with concomitant ATP hydrolysis, each cycle with the binding occurring in a position closer to the Z line.

The sequence of events that are repeated in each cycle can be summarized as follows, starting with ATP binding to the myosin head for convenience (Fig. 10.7b):

1. Myosin hydrolyzes ATP to ADP and Pi, which remain tightly bound to the protein. The hydrolysis induces a structural change in the myosin head, which points to the thin filament in a perpendicular orientation.
2. Myosin heads bind to the actin molecule in a pre-force-generating state.
3. Binding of the myosin head to actin induces Pi to be released, leading to a conformational change that makes the myosin head to be at a 45° orientation in relation to the filaments. This conformational change allows myosin to perform work.
4. ADP dissociates from myosin being replaced by ATP, whose binding makes myosin heads detach from the actin.

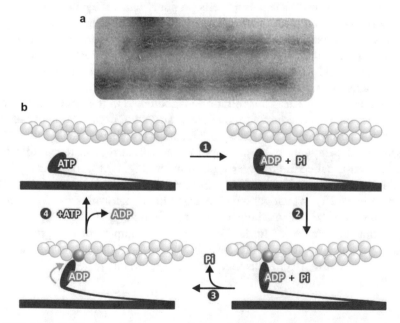

**Fig. 10.7** (a) Actin filaments "decorated" with myosin heads, which show an arrowhead appearance (Micrograph reproduced with permission from Huxley, Science 164:1356–1366, 1969). (b) Schematic representation of actomyosin cycle: (1) myosin head hydrolyzes ATP to ADP and Pi; (2) myosin heads bind to the actin molecule in a perpendicular orientation, in a pre-force-generating state; (3) Pi is released, leading to a conformational change that makes myosin head to be at a 45° orientation in relation to the filaments, allowing myosin to perform work; (4) ADP is replaced by ATP leading to myosin dissociation from actin

Thus, as the cycle is repeated, the thick filaments move along the thin filaments in the direction of the Z line, so that the H zone becomes shorter and the Z lines get nearer to each other.

It is important to note that if ATP is not available, myosin keeps strongly attached to the thin filaments, in a condition known as *rigor*, as occurs after death, when all the body muscles become rigid, which is called *rigor mortis*. In living muscle cells, on the other hand, there is always an excess of ATP due to a constant cycle of hydrolysis and resynthesis, and thus myosin heads remain bound to actin only for short periods of time.

## 10.1.3    Regulation of Muscle Contraction

Knowing that ATP is constantly hydrolyzed and resynthesized within the cells, so that its levels remain almost unchanged, one would ask how muscle contraction can be triggered exactly when it is required and how it is stopped when a specific task has already been performed.

The answer resides in the fact that the process is controlled by the central nervous system. This means that although the contraction is sustained by ATP hydrolysis, the ATP levels inside the muscle cells do not regulate the process extensively. Conversely, ATP concentrations are maintained high by different metabolic pathways that will be discussed in Sect. 10.2. Thus, it becomes clear that an additional player has to be called in to translate the brain signal to a biochemical response. This role is played by $Ca^{2+}$, whose concentration transiently increases in the sarcoplasm inducing the start of the contraction.

To understand how $Ca^{2+}$ concentration is modulated in the sarcoplasm to control contraction, we need firstly to look at the structural organization of the muscle fibers. One particularity of the skeletal muscle cell is that its sarcolemma invaginates perpendicularly to the length of the cell at regular repeat distances, forming what is called the transverse (T) tubules (Fig. 10.8). Inside the cell, the myofibrils are surrounded by a system of membranous vesicles called sarcoplasmic reticulum (SR), which contains enlarged areas, known as the SR terminal cisternae, where $Ca^{2+}$ is stored in a high concentration ($\sim 10^{-3}$ M). The SR terminal cisternae are connected to the T-tubules by a complex of two proteins, the dihydropyridine receptor (DHPR, inserted in the T-tubule membrane) and the ryanodine receptor (RyR, inserted in the terminal cisternae membrane) (Fig. 10.8).

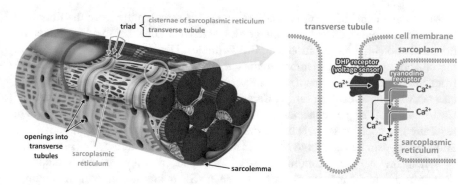

**Fig. 10.8** Schematic diagram of a skeletal muscle fiber section showing the sarcolemma invaginations into the T-tubules and the SR with its terminal cisternae. Dihydropyridine receptor (DHPR), a voltage-dependent $Ca^{2+}$ channel inserted in the T-tubule membrane, interacts with ryanodine receptor (RyR), also a $Ca^{2+}$ channel, inserted in SR terminal cisternae membrane. Propagation of the action potential over the T-tubules activates DHPR, which in turn induces RyR opening, leading to $Ca^{2+}$ release from SR lumen to the cytosol

A nerve impulse induces the opening of ion channels in the sarcolemma leading to an inflow of $Na^+$ into the cell. This causes membrane depolarization due to the dissipation of the membrane potential maintained by the $Na^+/K^+$-ATPase (see Box 10.2).

---

**Box 10.2  $Na^+/K^+$-ATPase and the Maintenance of Cellular Membrane Potential**

In living cells, the distribution of ions inside and outside the plasma membrane is asymmetric, resulting in an electrical voltage between the two sides of the membrane, which is called membrane potential. Membrane potential is regulated by the combined action of: (1) ion channels, which depolarize the membrane, and (2) ion pumps, which actively exchange ions across the membrane, restoring polarization. Membrane potential mainly arises from the exchange of $Na^+$ and $K^+$ through the activity of the $Na^+/K^+$-ATPase. This enzyme is an integral protein in the plasma membrane (see figure) that pumps 3 $Na^+$ out in exchange of 2 $K^+$ in, at the expense of ATP hydrolysis. Its activity accounts for a great part of cellular energy expenditure, being estimated that it is responsible for from 1/3 to 2/3 of ATP hydrolysis in the cells.

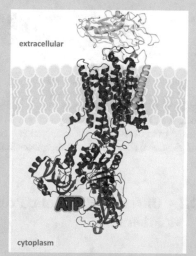

Crystal structure of $Na^+/K^+$-ATPase (PDB 3A3Y), with its three subunits, the α (catalytic, in *red*), β (in *green*), and regulatory (in *blue*). ATP is drawn in its binding site

Membrane depolarization propagates along the sarcolemma from the fiber surface to the T-tubules. DHPR is a voltage-dependent $Ca^{2+}$ channel that is activated by the action potential propagation over the T-tubules. DHPR interaction with RyR in the closely apposed SR membrane causes RyR opening, leading to $Ca^{2+}$ release from SR lumen to the cytosol. The increase of cytosolic $Ca^{2+}$ concentration itself also activates RyR, causing further $Ca^{2+}$ release, in a process known as $Ca^{2+}$-induced $Ca^{2+}$ release. At higher concentrations in the sarcoplasm, $Ca^{2+}$ binds to troponin, resulting in a conformational change that induces the displacement of tropomyosin, allowing myosin heads to bind actin and starting the contraction cycle (Fig. 10.9).

When the nervous stimulus ceases, $Ca^{2+}$ concentration in the sarcoplasm decreases due to the activity of $Ca^{2+}$-ATPase, an SR membrane enzyme that pumps $Ca^{2+}$ from the cytosol to the SR lumen. Thus, in resting muscle, the concentration of $Ca^{2+}$ in the sarcoplasm is maintained very low, in the range of $10^{-7}$ to $10^{-8}$ M, since almost all the intracellular $Ca^{2+}$ is stored inside the SR. In this situation, the complex troponin–tropomyosin is bound to actin in a way that prevents myosin binding (Fig. 10.9). Thus, even in high ATP concentrations, its hydrolysis by myosin occurs in a very slow rate and the muscle remains relaxed.

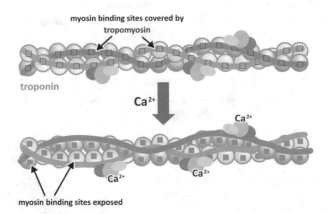

**Fig. 10.9** Model for the $Ca^{2+}$ regulation of contraction. At low concentration of $Ca^{2+}$ in the sarcoplasm, the complex troponin–tropomyosin blocks the myosin binding sites in actin. When $Ca^{2+}$ is released from the SR and its concentration increases in the sarcoplasm, it binds troponin inducing a conformational change that ultimately leads tropomyosin to move away from the myosin binding sites

## 10.2    Different Metabolic Profiles of the Skeletal Muscle Fibers

Skeletal muscles are used to perform very different kinds of activities. Some of these activities require that muscle cells work in their maximal capacity, such as when an elite athlete runs a 100 m sprint, but also in more usual situations, as when you have to quickly run to take a bus that just started to leave the bus stop. Other activities demand muscle work for a long time, such as in running a marathon but also in prolonged walks, riding a bicycle, or cleaning the house. The different metabolic demands required in these diverse activities can be achieved due to the existence of distinct types of muscle fibers, characterized by specific metabolic adaptations that include the type of nutrient metabolized and the metabolic pathways used to synthesize ATP (Fig. 10.10).

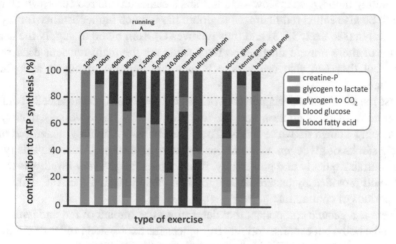

**Fig. 10.10** Contribution of distinct metabolites to ATP synthesis during different types of exercise. The transfer of the phosphate group of phosphocreatine to ADP is the fastest way to regenerate ATP in muscle cells (see next section for details), but the content of phosphocreatine is limited and sustains only short-duration exercises. Muscle glycogen is the main energy source used for ATP synthesis during short- or medium-duration exercises. Depending on the type of exercise, muscle glycogen may be used anaerobically (for instance, in short-distance runs), generating lactate as the end product, or aerobically (as in long-distance runs), being oxidized to $CO_2$. Aerobic metabolism becomes gradually more important in long duration exercises, with also an increasing requirement of fatty acid oxidation as glycogen is depleted. The use of blood glucose by muscle during exercise is almost irrelevant, especially because its transport into the muscle cells is dependent on insulin (see Sect. 8.4). This guarantees glucose availability for the cells that use this nutrient preferentially or exclusively, such as brain cells or erythrocytes, respectively. Ball games, such as soccer or tennis, may be long in duration, but they consist in short and intense runs alternating with resting periods, resulting in a profile closer to that of short distance running. (Data from Newsholme and Leech, Functional biochemistry in health and disease, chap. 13. pg 291, 2010)

From the physiological point of view, there are two major types of muscle fibers: the slow-and the fast-twitch fibers. They are classified according to the expression profile of myosin heavy chain isoforms, which correlate with their physiological role and their biochemical adaptations. Slow-twitch fibers are also known as type 1, while fast-twitch fibers are called type 2 fibers, as they predominantly express type 1 or type 2 myosin heavy chains, respectively. There is an association between fiber type and mitochondrial content, with type 2 fibers tending to have the lowest and type 1 fibers the highest mitochondria content.

Slow-twitch fibers are adapted to prolonged work and are very resistant to fatigue, although they provide relatively low force upon contraction. Their metabolism is mainly dependent on oxidative phosphorylation and thus requires an adequate supply of $O_2$. These cells are very rich in mitochondria and myoglobin and are irrigated by a large amount of blood vessels. The cytochromes in the mitochondria and the high content of myoglobin inside the cells, as well as the hemoglobin in the surrounding blood, give to slow-twitch fibers a characteristic red color, which made them to be also called red fibers. Myoglobin has a much higher affinity for $O_2$ than hemoglobin (see Sect. 3.3.3) and thus receives $O_2$ from blood to supply the aerobic activity of these muscle fibers. The fatty acids are the main nutrient used by red fibers, but they can also use ketone bodies and degrade glycogen and glucose aerobically.

Fast-twitch fibers are adapted to work at low levels of $O_2$. These cells contain a low number of mitochondria and are less supplied by blood vessels than type 1 fibers being known as white fibers. Their metabolism is mainly anaerobic, using muscle stocks of glycogen as the major metabolic substrate. Although this type of fiber contracts quickly and powerfully, it undergoes fatigue very rapidly as the low ATP yield provided by anaerobic metabolism cannot sustain the ATP demand for long periods of contraction.

There is a genetic component that determines the amount of red and white muscles in the body, but exercise training influences the expression profile of contractile proteins resulting in changes in the proportion of fiber types in the muscles.

## 10.3    Overview of ATP Synthesis in the Muscle Cells

ATP consumption in muscle cells may increase 100-fold from resting to vigorous activity. During intense contractile activity, ATP is hydrolyzed to ADP and Pi mainly as a result of three ATPase activities: (a) the ATPase activity of myosin head, which is directly involved in the sliding of the filaments and thus in the contraction; (b) the $Na^+/K^+$-ATPase activity that maintains the $Na^+/K^+$ gradient across the sarcolemma and T-tubules, allowing the membrane potential to be restored; and (c) the SR $Ca^{2+}$-ATPase activity, responsible for pumping $Ca^{2+}$ against the concentration gradient from the sarcoplasm into the lumen of SR. The high rate of ATP hydrolysis during contraction demands that ATP levels are continuously restored within the muscle cells.

ATP is supplied to muscle cells through different pathways depending on the type of the fiber (Fig. 10.11). In type 1 fibers, oxidative phosphorylation is the major mechanism of ATP synthesis, with the fatty acids being the main metabolic substrate used. In contrast, in type 2 fibers, most of the ATP synthesized comes from the substrate-level phosphorylation in glycolysis, being the muscle glycogen the main source of glucose-6-phosphate for glycolysis and lactate the major end product of this pathway.

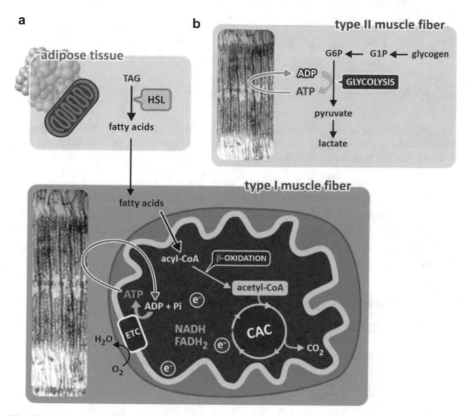

**Fig. 10.11** Main metabolic pathways for ATP synthesis that sustain contraction in muscle fibers. (**a**) Type 1 fibers use fatty acids as the main metabolic substrate. The fatty acids are mobilized by the hydrolysis triacylglycerols (TAGs) coming mainly from the adipose tissue adjacent to muscles, in a reaction catalyzed by the enzyme hormone-sensitive lipase (HSL). Fatty acids are oxidized through the β-oxidation pathway followed by complete oxidation of the resultant acetyl-CoA in CAC. The electrons transferred to the coenzymes NADH and FADH$_2$ are then transported in the electron transport chain, ultimately reducing O$_2$ to H$_2$O. Electron transport promotes the formation of an electrochemical gradient that is the driving force for ATP synthesis. (**b**) Type 2 fibers use muscle glycogen as the main metabolic substrate. Glycogen degradation forms glucose-1-phosphate (G1P), which is converted in glucose-6-phosphate (G6P), which in turn enters glycolysis. ATP is formed by substrate-level phosphorylation in glycolysis. In low availability of O$_2$, the product of glycolysis, pyruvate, is reduced to lactate

Additionally, skeletal muscle contains another mechanism to transiently and rapidly generate ATP: a large amount of phosphocreatine (10–30 mM). The enzyme creatine kinase catalyzes the transfer of phosphate group of phosphocreatine to ADP, regenerating the ATP hydrolyzed in contraction (Fig. 10.12). The resulted creatine can be phosphorylated again during recovery, when the ATP levels increase in the cells. The same enzyme, creatine kinase, catalyzes the reverse reaction using ATP to phosphorylate creatine, restoring the levels of phosphocreatine in muscle cells.

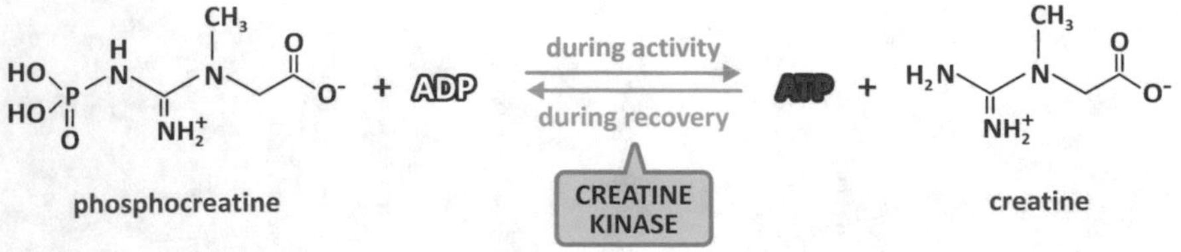

**Fig. 10.12** Creatine kinase catalyzes the reversible reaction of phosphate transfer from phosphocreatine to ADP, generating ATP and creatine during muscle contraction, or from ATP to creatine, forming phosphocreatine and ADP, during recovery

The use of phosphocreatine to phosphorylate ADP is an example of substrate-level phosphorylation, which is also the mechanism of ATP synthesis in glycolysis (see Sect. 6.1.2). This is possible as the $\Delta G^0$ value of phosphocreatine hydrolysis is higher than that of ATP, the same that occurs with phosphoenolpyruvate and 1,3-bisphosphoglycerate (Table 10.1).

**Table 10.1** $\Delta G^0$ values of phosphate hydrolysis of some compounds

| Compound | $\Delta G^0$ (kJ mol$^{-1}$) |
|---|---|
| Phosphoenolpyruvate | −61.9 |
| 1,3-Bisphosphoglycerate | −49.4 |
| Phosphocreatine | −43.1 |
| ATP → ADP + Pi | −30.5 |

## 10.4    Muscle Cell Metabolism During Physical Activity

Due to the extremely high demand for ATP imposed by the contractile activity during intense exercise, a rapid metabolic adaptation to regenerate ATP is required in muscle cells, in which the available substrates are driven to the catabolic pathways. In this section we will discuss the main metabolic steps that are regulated in muscle cell to maintain ATP concentration within the adequate ranges required for cellular functions.

### 10.4.1    Role of the Cellular Energy Charge in the Muscle Cell Metabolism

In the decade of the 1960s, Daniel E. Atkinson proposed that the regeneration and the expenditure of ATP would be regulated by the cellular energy balance itself. This concept has proven to be true for all the cells in the body, at least as a first level of metabolic control. To formalize this idea, Atkinson developed a parameter to describe the cellular energy status based on the relative concentrations of the adenine nucleotides within the total cellular pool, in a given moment or situation. He termed this parameter as the "energy charge of the adenylate system," whose value represents half of the average number of "anhydride-bound phosphates" per adenosine moiety (Box 10.3). A value of energy charge of 0 means that only AMP is present in the cell, whereas if all the adenine nucleotides were in the form of ATP, the cellular energy charge would be 1. In most cells, the energy charge value ranges from 0.8 to 0.95.

---

**Box 10.3  Atkinson's Concept of Cellular Energy Charge**

Daniel E. Atkinson proposed that the energy stored into the cell in the form of adenine nucleotides, referred by him as the adenylate system (AMP + ADP + ATP), resembles an electrochemical storage cell in its ability to accept, store, and supply energy. Based on this view, the adenylate system is fully discharged when all adenylate is in the form of AMP and fully charged when only ATP is present, with the number of anhydride-bound phosphates per adenosine moiety varying from 0 to 2. To normalize, i.e., to have a parameter varying from 0 to 1, he divided the number of anhydride bonds per adenosine nucleotide by 2. Since ATP contains two of anhydride-bound phosphate groups and ADP contains one, Atkinson defined the cellular "energy charge" as the actual concentrations of ATP + ½ ADP relative to the total adenylate system:

$$\frac{[ATP]+1/2[ADP]}{[ATP]+[ADP]+[AMP]}$$

---

It is important to note that during intense contraction, AMP accumulates as a result of the combination of ATP hydrolysis (which generates ADP and Pi) and the reaction catalyzed by the enzyme adenylate kinase (which converts two molecules of ADP in one ATP and one AMP). Therefore, each ATP is ultimately converted in one AMP and two Pi molecules (Fig. 10.13a). Considering the reaction catalyzed by adenylate kinase at equilibrium, Atkinson represented the variation of the concentrations of AMP, ADP, and ATP as a function of the energy charge (Fig. 10.13b). Observing this plot, it becomes clear that AMP is a very sensitive indicator of metabolic status, since its concentration varies in a much greater amplitude when compared to ADP concentration variation (during exercise, AMP concentration may rise more than 100-fold, while no more than a tenfold increase is observed for ADP concentration).

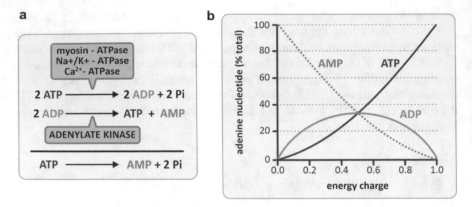

**Fig. 10.13** (**a**) AMP production in muscle cells. ATP is hydrolyzed to ADP and Pi by the ATPase activities of myosin, Na$^+$/K$^+$-ATPase, and Ca$^{2+}$-ATPase. The resulting ADP may be converted to ATP and AMP by the action of the enzyme adenylate kinase. (**b**) Considering the reaction catalyzed by adenylate kinase at equilibrium (using the calculated equilibrium constant of 0.8), the concentrations of the adenine nucleotides in the cell can be represented as a function of the cellular energy charge. (Figure reproduced from Oakhill et al. Trends Endocrinol Metab 23:125–132, 2012, with permission from Elsevier)

AMP is indeed an important activator of the pathways for ATP synthesis in muscle cells. It regulates muscle metabolism acting as an allosteric modulator of many enzymes, such as activating the muscle isoform of the glycogen phosphorylase (GP) and the glycolytic enzyme phosphofructokinase (PFK) and inhibiting fructose-1,6-bisphosphatase (see Sect. 10.4.2 for details).

In addition to acting through its direct binding to several enzymes in different metabolic pathways, another regulatory role has been attributed to AMP: it is the activator of an important regulatory enzyme, the AMP-activated protein kinase (AMPK; not to be confused with the cyclic AMP-dependent protein kinase, PKA, but an enzyme activated by the 5′-AMP).

### 10.4.1.1   The AMP-Activated Protein Kinase: A Cellular Energy Sensor

AMPK is a heterotrimeric protein composed of one catalytic subunit (α) and two regulatory subunits (β and γ) (Fig. 10.14a). In the γ-subunit, there are four adenine nucleotide-binding sites (sites 1–4). Site 2 seems to be unoccupied in mammalian enzymes, and site 4 has a non-exchangeable AMP molecule bound. This indicates that only sites 1 and 3 are involved in AMPK regulation, exchanging ATP for ADP or AMP as the cellular energy charge varies. The N-terminal end of the β-subunit is myristoylated, being this modification important for nucleotide binding to γ-subunit.

AMPK activity is regulated mainly through phosphorylation/dephosphorylation of the Thr172 in the catalytic subunit (Fig. 10.14b). The main protein kinases involved in AMPK phosphorylation are the LKB1 complex and the Ca$^{2+}$/calmodulin-dependent kinase kinase-β (CaMKKβ). Phosphorylation of Thr172 results in an increase of more than 100-fold in AMPK activity. On the other hand, its dephosphorylation catalyzed by phosphatases, such as the protein phosphatase 2C (PP2C), leads to AMPK inactivation.

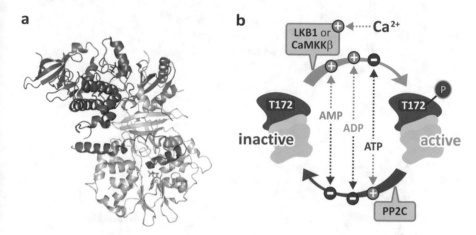

**Fig. 10.14** (**a**) Structure of mammalian AMPK. The structure represents a composition of the available structures of the rat α-subunit (*blue*), with the phosphorylated Thr shown in *red*; the human β-subunit (*light yellow*); and the rat γ-subunit (*lilac*), with an AMP molecule bound shown in *orange* (PDB 4CFH). (**b**) Schematic representation of AMPK regulation. The increase in ADP/AMP concentration triggers Thr172 phosphorylation by the upstream kinases LKB1 and CaMKKβ and simultaneously inhibits Thr172 dephosphorylation by phosphoprotein phosphatase 2C (PP2C). ATP antagonizes this ADP/AMP effect. AMP acts directly as an allosteric activator of AMPK

Binding of ADP or AMP to site 3 of the γ-subunit facilitates AMPK phosphorylation and inhibits its dephosphorylation, maintaining the enzyme in its active state, while binding of ATP to this site causes the opposite effect. This is an interesting example of a posttranslational modification that is modulated allosterically, since binding of the adenine nucleotide to γ-subunit makes the enzyme a better or a worse substrate to the kinases or phosphatases that will introduce or remove the phosphate group in Thr172.

Additionally, when AMPK is phosphorylated, binding of AMP to the site 1 of the γ-subunit further activates the enzyme, although this direct allosteric activation results in only two- to five-fold stimulation of AMPK activity.

It is interesting that AMPK activation/inhibition is not controlled by extracellular signals (e.g., hormones), as occurs with most of the regulatory kinases, but by the intracellular status. Therefore, AMPK can be seen as an adenylate charge regulated protein kinase, since it detects and reacts to changes in the adenine nucleotide ratio.

Although it is unquestionable that protein phosphorylation by AMPK is an important mechanism of control of the energy metabolism in response to changes in cellular energy charge, restoring cellular ATP levels by switching off the anabolic pathways and switching on the catabolic pathways, evidence suggest that the main physiological activator of this enzyme seems to be ADP rather than AMP (see Box 10.4). In muscle cells, the main metabolic pathways regulated by AMPK are the β-oxidation of fatty acids and the glucose uptake via its transport by GLUT4 (see next section for details).

**Box 10.4  ADP and the AMPK Activation**

Studies that measured the nucleotide binding affinity to AMPK revealed that the binding constants ($K_d$) of AMP and ADP to the site 3 of AMPK γ subunit are 80 and 50 µM, respectively. Although during exercise AMP concentration rises dramatically (up to more than 100-fold over the resting levels), the highest AMP concentration reached (about 10 µM) is much lower than the $K_d$ value. On the other hand, ADP concentration, although showing a much more modest increase (from 36 to 200 µM), overcomes the $K_d$ value by up to five-fold. These observations led Bruce E. Kemp and colleagues to propose that is ADP, rather than AMP, that plays the dominant role in activating AMPK. These authors also showed that upon exercise AMPK activation clearly correlates with the increase in ADP concentration (see below figure), further supporting this new view. This makes the historical name of this kinase as well as the extensive discussion in the literature regarding the role of AMP in this regulation an open question.

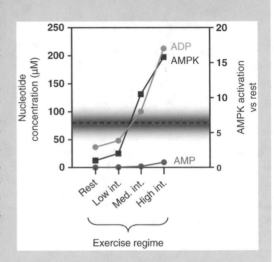

AMPK activation and the concentrations AMP and ADP in human muscle cells after exercises of different intensities. The *black dashed line* indicates the measured $K_d$ for AMP binding at site 3 with the standard deviation indicated by a red zone. (Figure reproduced from Oakhill et al. Trends Endocrinol Metab 23:125–132, 2012, with permission from Elsevier)

## 10.4.2   Metabolic Pathways for ATP Synthesis in the Skeletal Muscle

### 10.4.2.1   Fatty Acid Oxidation in Skeletal Muscle

Fatty acids constitute the preferential class of nutrients used by muscle cells. These molecules are taken up by muscle fibers from blood or mobilized from TAG accumulated either inside the myocytes themselves or in the adipocytes dispersed between or along the fibers (Fig. 10.15).

Albumin-bound fatty acids circulate in the blood after mobilization from the adipose tissue, being the main source of fatty acids for muscle metabolism. Fatty acid can also be obtained from blood through the circulating VLDL. This lipoprotein is the main carrier of TAG in the blood in the postabsorptive state (see Sect. 8.3.2).

VLDL-associated TAG is hydrolyzed by lipoprotein lipase (LPL), an enzyme located at the luminal side of endothelium cells of the capillary bed of muscles, making the resulting fatty acids available to skeletal muscle during exercise (Fig. 10.15).

Muscle cells contain a certain amount of intracellular stores of TAG, especially in type 1 fibers, representing a potential energy source for muscle metabolism during aerobic exercise. Intramyocellular TAGs are probably formed through the re-esterification of the excess of fatty acids that are taken up from blood. Nevertheless, although an increasing number of evidence support the importance of intramyocellular TAG mobilization during muscle activity, it is still difficult to unequivocally distinguish between intracellular and intercellular TAGs (those located in the adipocytes associated with the muscle cells), which should also have a role in supplying muscle cells of fatty acids during exercise. The hydrolysis of intramyocellular TAGs is attributed to the activity of two lipases, the adipose triglyceride lipase (ATGL), which catalyzes the hydrolysis of TAG to diacylglycerol (DAG), and the muscle isoform of the hormone-sensitive lipase (HSL), which exhibits a higher specificity for DAG than TAG (Fig. 10.15).

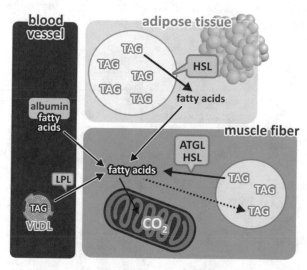

**Fig. 10.15** Sources of fatty acids for muscle metabolism. Fatty acids mobilized from the adipose tissue circulate in association with albumin and are delivered to the muscle cells. VLDL transports the de novo synthesized TAG from the liver to the peripheral tissues, including muscles. VLDL-associated TAGs are hydrolyzed by the enzyme lipoprotein lipase (LPL) at the surface of the endothelium cells of the vessels that irrigate the muscles, and the resulting fatty acids are transported into the fibers. The enzyme adipose triglyceride lipase (ATGL) hydrolyses the TAG molecules stored inside the myocytes, generating DAG, which in turn is further hydrolyzed by the muscle isoform of the hormone-sensitive lipase (HSL). Finally, the TAGs stored in the adipocytes associated to the fibers are also mobilized yielding fatty acids to the muscle cell metabolism

The reactions of fatty acid oxidation are described in detail in Sect. 7.4. Briefly, fatty acids are firstly activated in the cytosol by esterification with coenzyme A (CoA) in an ATP-dependent reaction catalyzed by the acyl-CoA synthetase (ACS). The resulting acyl-CoA molecules are transported into mitochondria where they undergo β-oxidation, generating acetyl-CoA, FADH$_2$, and NADH (Fig. 10.16). To be transported into mitochondria, acyl-CoA molecules are firstly converted to their acyl-carnitine derivatives that are then translocated across the mitochondrial membrane and reconverted to acyl-CoA molecules in the matrix.

The main regulation site of fatty acid oxidation is the transfer of the acyl group from CoA to carnitine, a reaction catalyzed by the carnitine/palmitoyl transferase I (CPT-I). This enzyme is located at the outer mitochondrial membrane and is strongly inhibited by malonyl-CoA. Therefore, a decrease in malonyl-CoA concentration in sarcoplasm increases the transport of the fatty acids into mitochondria, favoring their oxidation (see Fig. 8.13).

Malonyl-CoA is produced by the carboxylation of acetyl-CoA in the cytosol (see also Sect. 8.3.1). In muscle cells, this reaction is catalyzed by the isoform 2 of the enzyme acetyl-CoA carboxylase (ACC2), which is associated to the outer mitochondrial membrane (Fig. 10.16). ACC2 is inhibited by the phosphorylation of its Ser219 and Ser220, catalyzed by AMPK. Therefore, the activation of AMPK during contraction (see previous section) causes the decrease in the concentration of malonyl-CoA in sarcoplasm due to ACC2 inhibition, resulting in the activation of fatty acid β-oxidation (Fig. 10.16).

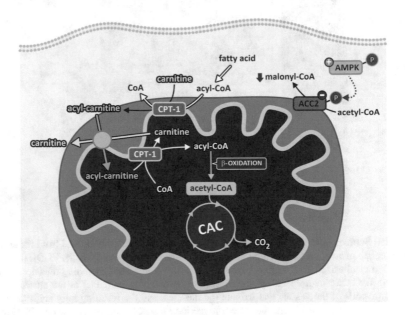

**Fig. 10.16** Regulation of fatty acid oxidation in muscle cells. β-oxidation of fatty acids is activated during contraction due to the decrease in the levels of malonyl-CoA, a potent inhibitor of the transport of the acyl-CoA into the mitochondrial matrix. Activated AMPK phosphorylates the isoform 2 of the acetyl-CoA carboxylase (ACC2), inhibiting the conversion of acetyl-CoA in malonyl-CoA. To be transported into the mitochondrial matrix, the acyl group of acyl-CoA is firstly transferred to carnitine, in a reaction catalyzed by the carnitine/palmitoyl transferase I (CPT-I). The acyl-carnitine is then transported across the inner mitochondrial membrane through the carnitine-acyl-carnitine transporter. In the matrix, the acyl group is transferred to coenzyme A by carnitine/palmitoyl transferase II (CPT-II), regenerating the acyl-CoA, which undergoes β-oxidation. Since malonyl-CoA is a potent inhibitor of the CPT-I, the decrease in its concentration allows the transport and the subsequent oxidation of the acyl-CoA

### 10.4.2.2 Insulin-Independent Glucose Uptake in Skeletal Muscle

Glucose uptake in muscle cells occurs mainly through the isoform 4 of the glucose transporters (GLUT4). In the absence of specific stimuli, GLUT4 is sequestered in intracellular vesicles, restricting the use of blood glucose by the muscle cells. Insulin, the hormone secreted when the concentration of glucose in the blood increases, is the major signal that induces the exposure of GLUT4 on the cell surface, leading to a robust increase in glucose uptake by GLUT4-containing cells (see Sect. 8.4). In the muscle cells, contraction also regulates the migration of GLUT4-containing vesicles to the plasma membrane, allowing an increase in glucose uptake from blood during intense exercise even in the absence of insulin signaling.

The mechanisms by which contraction stimulates the glucose uptake in muscles are not completely understood, but several evidence support that AMPK is involved. The decrease in the cellular energy charge due to the intense ATP hydrolysis during contraction activates AMPK, and this can be correlated to GLUT4 exposure on the cell surface. The signaling pathway that links AMPK activity to GLUT4 translocation has not been elucidated yet, but it seems to involve the phosphorylation of AS160 (AKT substrate 160; see Sect. 8.4) (Fig. 10.17).

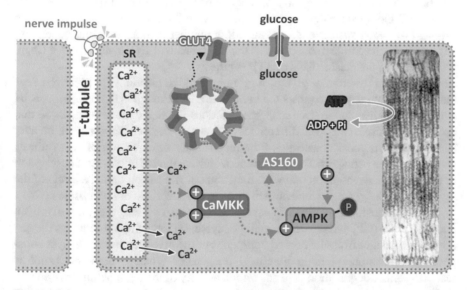

**Fig. 10.17** Increase in glucose transport into the muscle cells during exercise. Intense contraction leads to a high rate of ATP hydrolysis, increasing the concentration of ADP (and AMP due to adenylate kinase activity). ADP/AMP binding to AMPK facilitates its phosphorylation and activation. Simultaneously, T-tubule membrane depolarization caused by nerve impulses induces the release of $Ca^{2+}$ from SR, increasing its concentration in the sarcoplasm and resulting in the activation of the CaMKK, which phosphorylates AMPK. Probably through phosphorylation of AS160 by AMPK, the GLUT4 vesicles migrate to the cell surface, leading to an increase in glucose uptake

As mentioned in Sect. 10.1.3, upon stimulation by a nerve impulse, the membranes of the transverse tubules depolarize causing an increase in cytoplasmic $Ca^{2+}$ concentration due to the release of this ion from the SR stores. $Ca^{2+}$ activates the $Ca^{2+}$/calmodulin-dependent protein kinase kinase (CaMKK), one of the upstream kinases that phosphorylate and activate AMPK (see Sect. 10.4.1.1). Therefore, during exercise, the increase in the glucose uptake by the muscle cells occurs through a coordinated action of ADP/AMP and $Ca^{2+}$, which work simultaneously on the activation of AMPK: while the increase in ADP/AMP concentration makes AMPK susceptible to the action of the CaMKK, the elevation of $Ca^{2+}$ concentration in the sarcoplasm makes this kinase active to phosphorylate and activate AMPK (Fig. 10.17).

### 10.4.2.3   Glycogen Degradation in Skeletal Muscle

Glycogen content in muscle cells corresponds to 1–2% of the net weight of muscles. It consists of an important energy store especially because it can be used either aerobically or anaerobically, although in this latter case, it can sustain vigorous activity only for a short period of time.

The existence of an energy source that can be used independently of the amount of $O_2$ available is especially important for the type 2 muscle fibers as they are less irrigated by blood vessels and contain a low number of mitochondria, depending largely on the anaerobic metabolism as the mechanism of ATP synthesis (see Chap. 6 for review of the mechanisms of ATP synthesis).

The detailed structure of glycogen granules and the reactions for glycogen degradation have already been presented in Sect. 9.2.1. Although in that chapter special attention was given to the liver cells, the reactions per se are the same as those that occur in the muscle cells. The differences between glycogen metabolism in the liver and muscle cells consist basically of the mechanisms of regulation of the synthesis and degradation pathways. Thus, in this section we will give only a brief description of the glycogen degradation pathway, focusing our discussion on the aspects of the regulation of the glycogen metabolism that are characteristic of the muscle cells.

Glycogen degradation depends on the activity of two enzymes, the glycogen phosphorylase (GP) and the debranching enzyme. GP sequentially removes the terminal glucose unit from the nonreducing ends of the glycogen molecule by a phosphorolysis reaction, yielding glucose-1-phosphate. When a branched point is reached, further degradation depends on the activity of the debranching enzyme.

The major site of control of glycogen degradation is the regulation of GP activity. This enzyme exists in two interconvertible conformational states, called GP$a$, the catalytically active form, and GP$b$, the less active form (see Sect. 9.2.2). In resting muscle, the predominant form is the GP$b$, which is converted to the active form by adrenaline-mediated phosphorylation, the main hormonal control that acts on muscle cells. The action of adrenaline on muscle cells (detailed in Sect. 10.5) results in the activation of the enzyme phosphorylase kinase, which catalyzes the phosphorylation of GP. Phosphorylation maintains GP in the active form, favoring glycogen degradation (Fig. 10.18).

In addition to the regulation by phosphorylation, two allosteric modulators activate glycogen degradation in muscle cells: $Ca^{2+}$ and AMP.

As described in Sect. 10.1, the nervous stimulus to contraction induces $Ca^{2+}$ release from SR to the sarcoplasm, where the concentration of this ion increases greatly. $Ca^{2+}$ binds to the phosphorylase kinase, activating this enzyme and leading to the phosphorylation of GP to its active form (Fig. 10.18). This coordinates the first intracellular signal that induces contraction ($Ca^{2+}$ release) to the mobilization of glycogen as an energy source for the process.

The other allosteric activator of glycogen degradation is AMP, which acts directly on GP, favoring glucose-1-phosphate release from the active site of the enzyme and speeding the GP reaction (Fig. 10.18). In resting muscle, ATP, which is in higher concentrations, replaces AMP in the allosteric site, inactivating the enzyme and inhibiting glycogen degradation.

The direct end product of glycogen degradation is glucose-1-phosphate (together with a small amount of glucose that is the product of the glucosidase activity of the debranching enzyme; see Sect. 9.2.1). Glucose-1-phosphate is then converted to glucose-6-phosphate by the action of the enzyme phosphoglucomutase (Fig. 10.18). It is important to remind that muscle cells lack the enzyme glucose-6-phophatase, an enzyme of the gluconeogenesis pathway whose expression is restricted to the liver and kidneys. Thus, glucose-6-phosphate in muscles enters glycolysis, generating ATP to support muscle contraction.

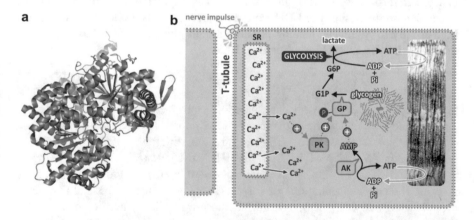

**Fig. 10.18** (a) Structure of the human muscle GP (PDB 1Z8D) showing the phosphorylated Ser14 (*pink*) and the AMP (*yellow*) in its binding site. (b) Activation of the glycogen degradation in contracting muscle. Intense contraction leads to a high rate of ATP hydrolysis, increasing the concentration of ADP, which may be converted to AMP and ATP by the enzyme adenylate kinase (AK). AMP binds to GP and facilitates glucose-1-phosphate (G1P) release from the active site, speeding the reaction of glycogen phosphorolysis. Simultaneously, T-tubule membrane depolarization caused by nerve impulses induces the release of $Ca^{2+}$ from SR, increasing its concentration in the sarcoplasm and resulting in the activation of the phosphorylase kinase (PK), which phosphorylates and activates GP. G1P resulted from the glycogen degradation is converted to glucose-6-phosphate (G6P) by the phosphoglucomutase (PGM). G6P enters glycolysis, which in anaerobiosis generates lactate as the end product

Glycogen stores mobilized during muscle activity are replenished in muscle cells after carbohydrate ingestion. The increase in blood glucose after a carbohydrate-rich meal induces insulin secretion, whose action on muscle cells promotes glucose uptake through GLUT4 and its conversion to glycogen through the activation of the enzyme glycogen synthase (GS). The detailed reactions and regulatory aspects that take place in this situation are described in Sects. 8.2 and 8.4.

### 10.4.2.4   Glycolysis in Skeletal Muscle

The glycolytic pathway in the muscle cells is also activated during physical activity. This occurs mainly by hormonal regulation, through adrenaline-induced phosphorylation of the bifunctional enzyme phosphofructokinase-2/fructose-2,6-bisphosphatase (PFK-2/F2,6BPase; see Box 9.5). PFK-2/F2,6BPase expressed in muscle cells is different from the hepatic isoform, with the phosphorylation resulting in the activation of its PFK-2 activity, leading to an increase in the concentration of fructose-2,6-bisphosphate (see next section). Fructose-2,6-bisphosphate strongly activates the glycolytic enzyme phosphofructokinase-1 (PFK-1), increasing the metabolic flux through glycolysis.

Once glycolysis proceeds rapidly and if there is not enough $O_2$ available, pyruvate is converted to lactate to allow NADH produced in glycolysis to be reoxidized (see Sect. 6.1.2 for details). The conversion of pyruvate to lactate is catalyzed by lactate dehydrogenase (LDH), an enzyme that is expressed as different isoforms depending on the tissue. The LDH isoform expressed in the skeletal muscles has a high affinity for pyruvate, making possible a high glycolytic flow during contraction, especially in anaerobiosis (see Box 10.5).

---

**Box 10.5  LDH Isoforms**

LDH is a tetrameric enzyme that can be formed by a combination of two types of polypeptide chains: the M (from muscle) and the H (from heart) chains (see below figure). Thus, there are five possible different isoforms of LDH (MMMM, MMMH, MMHH, MHHH, HHHH). The M subunits confer to the enzyme a lower $K_M$ for pyruvate, favoring the reduction of pyruvate to lactate even when the concentration of pyruvate is low. This gives to skeletal muscle cells a high capacity of performing lactic fermentation. The presence of H subunits favors the oxidation of lactate to pyruvate. Therefore, cells expressing the H chain-containing isoforms can use the lactate as a metabolic substrate, converting it to pyruvate, which in turn can be oxidized, as occurs in the heart tissue, for example. This is also important in liver cells, where lactate is converted to pyruvate to enter gluconeogenesis (see Sect. 9.3.2). The close similarity of the structures of the M and H subunits suggests that the different $K_M$ observed for each isoforms results from variations in charge surface distribution on the active site.

---

(continued)

**Box 10.5** (continued)

The differences between LDH isoforms make this enzyme a useful tool for diagnosis. The detection of H isoform in the plasma can be used for diagnosis of heart infarction (see Sect. 3.3.4.1).

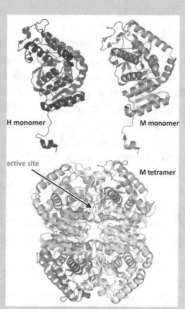

H monomer                    M monomer

active site                    M tetramer

Structures of LDH subunits M and H (*top*, PDB 1I10 and 1I0Z, respectively) and the tetrameric isoform of the enzyme containing only M subunits (*bottom*)

# 10.5  Hormonal Regulation During Physical Activity: Role of Adrenaline

Adrenaline (also known as epinephrine) is the major hormone secreted when the organism is confronted with different stimuli processed in the central nervous system as indicative of an acute stressful situation. It prepares the organism to perform an intense physical activity to deal with dangerous situations—the "fight-or-flight" response. Additionally, the physical activity itself promotes adrenaline secretion in a way dependent on the duration or the intensity of the exercise.

Adrenaline acts on almost all the tissues in the body, triggering many physiological and metabolic responses that prepare the organism for action. It promotes the dilatation of bronchioles, which increases the $O_2$ uptake, and a wide-range effect on the circulatory system, with the increase in the heart rate and blood pressure, and changes in blood flow patterns, leading to a decrease in the peripheral circulation and a reduction in the digestive system activity. These effects guarantee $O_2$ delivery to different organs, especially to the brain, allowing an increase in alertness. Nutrient availability is also tightly controlled by adrenaline, with the activation of glucose production by the liver and the mobilization of TAG in the adipose tissue. Finally, adrenaline prepares the skeletal muscle for contraction, with the activation of ATP-generating pathways, either through the anaerobic use of muscle glycogen or through the aerobic use of fatty acids.

In this section, we will focus on the metabolic effects of adrenaline on the muscle, liver, and adipose tissues.

### 10.5.1  Molecular Mechanisms of Adrenaline Action

Adrenaline belongs to a group of substances known as catecholamines. It is synthe-sized by the adrenal gland, from which its name is derived. The adrenal glands are localized at the top of the kidneys and can be divided in two regions, the cortex, which secretes steroid hormones such as the glucocorticoids (see Sect. 9.4.2), and the medulla, where adrenaline is produced, more specifically in the chromaffin cells (Fig. 10.19a). The pathway for the adrenaline synthesis starts with the amino acid tyrosine and consists of four enzymatic steps, detailed in Fig. 10.19b.

Adrenaline secretion is triggered by a direct stimulus from the sympathetic ner-vous system that propagates through preganglionic nerve fibers reaching the adrenal gland (Fig. 10.19a). The concentration of adrenaline in the blood may increase more than 50-fold upon stimulation of the adrenal gland, but since the half-life of this hormone is too short, about 2 min, the effects of adrenaline on the body may be seen as acute and short-term responses to stress, which is compatible with an evolution-ary adaptation to deal with "fight-or-flight" situations.

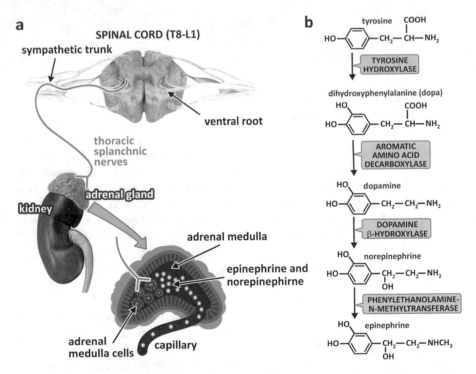

**Fig. 10.19**  (a) Representation of the adrenal gland, showing two distinct parts: the cortex and the medulla. The medulla chromaffin cells produce and secrete adrenaline and noradrenaline. The stimulus for hormone secretion comes from the sympathetic nervous system and is transmitted to the adrenal gland through thoracic nerve fibers. (b) Metabolic pathway for adrenaline synthesis: the enzyme tyrosine hydroxylase converts the amino acid Tyr to L-dopa, which is decarboxylated by the enzyme aromatic amino acid decarboxylase, generating dopamine. The enzyme dopamine β-hydroxylase transforms dopamine in noradrenaline, which is converted to adrenaline by the action of the phenylethanolamine-N-methyl transferase

### 10.5.1.1    Cellular Receptors for Adrenaline

The action of adrenaline on its target tissues depends on the binding of the hormone to receptors present on the cell surface. There are different types of receptors for adrenaline divided in two main classes, the $\alpha$- and the $\beta$-adrenergic receptors, which are in turn subdivided in $\alpha 1$ and $\alpha 2$, and $\beta 1$, $\beta 2$, and $\beta 3$ subtypes, respectively.

The different, and sometime antagonic, effects of adrenaline on each tissue, such as the relaxation of the smooth muscles in the airways that increases the respiratory rate and the contraction of the muscles of the arterioles that causes vasoconstriction, can be explained by a tissue-specific expression of the different types of receptors. The different responses occur because each receptor subtype is coupled to distinct signaling systems, whose activation upon hormone binding triggers distinct intracellular responses (Fig. 10.20).

All the adrenergic receptors belong to the superfamily of G protein-coupled receptors (see also Sect. 9.4.1). They contain seven transmembrane helices and are bound to the G protein, a trimeric protein composed of $\alpha$-, $\beta$-, and $\gamma$-subunits, which associates to the internal face of the plasma membrane (Fig. 10.20). The G protein $\alpha$-subunit binds GDP, which maintains the protein in its inactive form, associated to $\beta$- and $\gamma$-subunits. Upon hormone binding to the receptor, GDP is replaced by GTP, causing the $\alpha$-subunit to dissociate from $\beta\gamma$-subunits and to move on plasma membrane until reaching a target enzyme, which may be the adenylate cyclase or the phospholipase C, depending on the type of G protein. The $\alpha$-subunit has an intrinsic GTPase activity that terminates the signaling pathway through the conversion of the bound GTP in GDP, leading its reassociation to $\beta\gamma$-subunits.

The three subtypes of $\beta$-adrenergic receptors are linked to $G_s$ ("s" from stimulatory G protein), which is a G protein type that activates the adenylate cyclase. The signaling pathway mediated by $G_s$-coupled receptors is the same as that involved in the glucagon mechanism of action, detailed in Sect. 9.4.1. Briefly, upon activation by the $G_s$ subunit, adenylate cyclase catalyzes the conversion of ATP in cyclic AMP (cAMP), leading to an increase in the intracellular concentration of this molecule (Fig. 10.20). cAMP promotes the activation of the cAMP-dependent protein kinase (PKA), which phosphorylates several enzymes, modulating their activities.

The $\beta 2$-adrenergic receptors mediate the main effects of adrenaline on energy metabolism, which will be the focus of the next sections.

The $\alpha 1$-adrenergic receptors are mainly expressed in the smooth muscles, causing vasoconstriction and the decrease of the gastrointestinal tract motility. They are coupled to the type $G_q$ of the G proteins, which activates the phospholipase C, an enzyme that hydrolyzes the phosphatidylinositol in the plasma membrane generating diacylglycerol (DAG) and inositol triphosphate ($IP_3$). DAG activates the protein kinase C (PKC), which mediates the smooth muscle contraction through the phosphorylation of proteins and ion channels. $IP_3$ binds to SR inducing $Ca^{+2}$ release from the SR stores to the cytosol, stimulating muscle contraction (Fig. 10.20).

The $\alpha 2$ receptors are coupled to the type $G_i$, a G protein that inhibits the adenylate cyclase, leading to opposite effects of those mediated by $G_s$-coupled receptors.

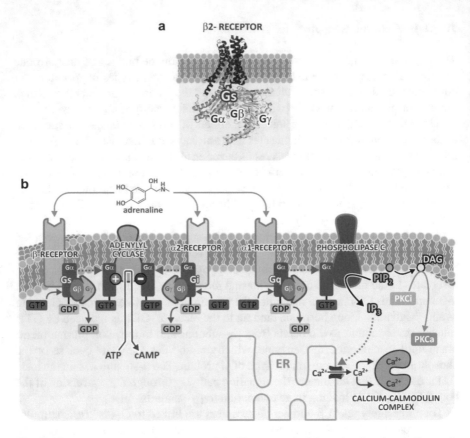

**Fig. 10.20** (**a**) Crystallographic structure of the β2-receptor (*red*) bound to the trimeric G$_s$ protein (PDB 3SN6), with the α, β, and γ subunits highlighted in different colors (*green*, *light green*, and *blue*, respectively). (**b**) Schematic representation of adrenaline signaling pathways through the different types of adrenergic receptors showing the different G protein α-subunits: G$_s$ activates adenylate cyclase, while G$_i$ inhibits this enzyme; G$_q$ activates phospholipase C. β-adrenergic receptors are bound to G$_s$ protein. When adrenaline binds to the receptor, the GDP bound to G protein is replaced by GTP and the G protein α-subunit moves on the membrane surface until it reaches adenylate cyclase, activating this enzyme, which converts ATP into cAMP. cAMP activates PKA leading to the phosphorylation of different targets in the cell. α2 receptor is bound to G$_i$, which inhibits the adenylate cyclase, leading to opposite effects of those mediated by G$_s$-coupled receptors. α1-adrenergic receptor is bound to G$_q$ protein. When adrenaline binds to the receptor, the GDP bound to G protein is replaced by GTP and the G protein α-subunit moves on the membrane surface until it reaches the phospholipase C, which hydrolyzes the phosphatidylinositol (PIP$_2$) in DAG and IP$_3$. DAG activates PKC, and IP$_3$ induces the release of Ca$^{2+}$ from the SR stores to the cytoplasm

## 10.5.2    *Effects of Adrenaline on Energy Metabolism*

### 10.5.2.1    Effects of Adrenaline on the Adipose Tissue Metabolism

The main effect of adrenaline on the adipose tissue is the activation of lipolysis, with the release of fatty acids in the bloodstream, leading to an increase of their availability to be used as an energy source. Lipolysis is activated through the

PKA-mediated phosphorylation of two adipocyte proteins, the perilipin present on the surface of the lipid droplets and the hormone-sensitive lipase (HSL), which catalyzes the hydrolysis of the ester linkages of the TAG molecules. Phosphorylation activates the HSL and promotes a conformational change in perilipin that allows HSL recruitment to the surface of the lipid droplet, where it gains access to the TAG molecules (Fig. 10.21; see also Sect. 7.4.1 for details).

Adrenaline also interferes with the activity of the lipoprotein lipase (LPL), an enzyme involved in the transfer of lipids from the lipoproteins, especially from chylomicrons and VLDL, to the adipocyte, where they are stored after ingestion or de novo synthesis. The effects of adrenaline on LPL involve the inhibition of the translation of its mRNA, ultimately leading to a decrease in the lipid uptake by the adipose tissue. This makes the lipids available for use by the muscle cells.

### 10.5.2.2 Effects of Adrenaline on the Liver Metabolism

The main hepatic metabolic response upon adrenaline binding is the increase in glucose release in the bloodstream. This occurs both through the degradation of the liver glycogen and through the gluconeogenic pathway. Additionally, fatty acid synthesis is inhibited, allowing the incoming fatty acids to undergo β-oxidation.

Figure 10.21 provides a schematic overview of the adrenaline effect on liver metabolism. Phosphorylation by PKA directly or indirectly modulates the activity of key hepatic enzymes, resulting in a coordinated regulation of different metabolic pathways, namely, (a) the activation of GP and the inhibition of GS, leading to glycogenolysis; (b) the activation of the F2,6BPase activity of the bifunctional enzyme, leading to the decrease of fructose-2,6-bisphosphate concentration and the consequent activation of F1,6BPase and inhibition of PFK-1, which results in the occurrence of gluconeogenesis over glycolysis; and (c) the inhibition of ACC1, which decreases malonyl-CoA concentration, leading to the inhibition of fatty acid synthesis and the activation of the acyl-CoA transport into the mitochondrial matrix where they undergo β-oxidation. A more detailed description of the regulation of each of these key enzymes can be found in Sect. 9.4.1, since the effects of adrenaline on the liver metabolism are the same as those induced by glucagon (both β-adrenergic and glucagon receptors are $G_s$-coupled receptors).

### 10.5.2.3 Effects of Adrenaline on Muscle Metabolism

In the muscle cells, adrenaline-mediated signaling pathway results in glycogen degradation and in a strong activation of glycolysis, adaptations that are especially important for the anaerobic metabolism of the type 2 fibers.

The regulation of glycogen metabolism is similar to that occurring in the liver cells. PKA phosphorylates the phosphorylase kinase, which in turn phosphorylates and activates GP. Simultaneously, PKA-mediated phosphorylation inactivates GS. This results in an intense glycogen degradation, yielding glucose-6-phosphate to the glycolytic pathway (Fig. 10.21).

Glycolysis is activated by the increase in the concentration fructose-2,6-bisphosphate, a potent activator of the glycolytic key enzyme PFK-1. The isoform of the bifunctional enzyme expressed in muscle cells differs from the liver isoform in its regulatory phosphorylation site, resulting in an opposite effect of phosphorylation on the enzyme activity in each tissue. The effect of PKA-induced phosphorylation on muscle isoform is the activation of the PFK-2 activity and in the inhibition of the F2,6BPase activity, leading to the synthesis of fructose-2,6-bisphosphate and the consequent PFK-1 activation, increasing the metabolic flux through glycolysis (Fig. 10.21). Furthermore, the muscle isoform of PFK-1 is itself a substrate for PKA. Phosphorylated muscle PFK-1 binds to actin filaments, resulting in further activation of the enzyme, which also becomes insensitive to the inhibitory effects of ATP, citrate, or lactate. Thus, as a consequence of adrenaline action, glycolysis is strongly stimulated in muscle cells, while in the liver this hormone activates of gluconeogenesis.

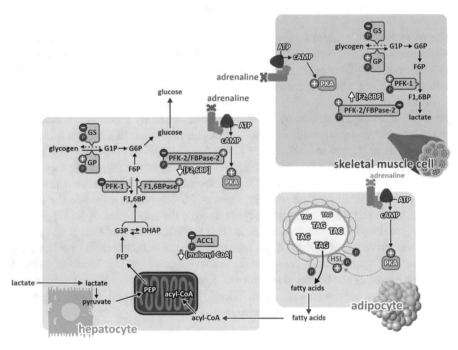

**Fig. 10.21** Effects of adrenaline on energy metabolism. In the liver, muscle, and adipose tissue cells, adrenaline binds to β-adrenergic receptors (*orange*) leading to G protein (*green*)-mediated adenylate cyclase (*red*) activation. The intracellular concentration of cAMP increases leading to the activation of PKA (*pink*), which phosphorylates different enzymes. In the adipocytes, PKA targets are perilipin (*lilac*) and HSL, resulting in the mobilization of TAGs. In the hepatocytes, the targets are (a) GP and GS, leading to glycogenolysis; (b) the bifunctional enzyme, leading to a decrease in the concentration of fructose-2,6-bisphosphate (F2,6BP) and activation of gluconeogenesis; and (c) ACC1, leading to the decrease in malonyl-CoA concentration and thus to the inhibition of fatty acid synthesis and the activation of the acyl-CoA transport into the mitochondrial matrix where they undergo β-oxidation. In the muscle cells, PKA targets are GP and GS, leading to glycogenolysis, and the muscle isoform of the bifunctional enzyme, in this case leading to an increase in F2,6BP concentration and activation of glycolysis

# Selected Bibliography

Atkinson DE (1968) The energy charge of the adenylate pool as a regulatory parameter. Interaction with feedback modifiers. Biochemistry 7:4030–4034

Harris DA (1998) Getting to grips with contraction: the interplay of structure and biochemistry. Trends Biochem Sci 23:84–87

Huxley HE (2005) Memories of early work on muscle contraction and regulation in the 1950's and 1960's. Biochem Biophys Res Commun 369:34–42

Kiens B (2006) Skeletal muscle lipid metabolism in exercise and insulin resistance. Physiol Rev 86:205–243

Oakhill JS, Scott JW, Kemp BE (2012) AMPK functions as an adenylate charge-regulated protein kinase. Trends Endocrinol Metab 23:125–132

Pedersen BK, Febbraio MA (2012) Muscles, exercise and obesity: skeletal muscle as a secretory organ. Nat Rev. Endocrinol 8:457–465

Read JA, Winter VJ, Eszes CM, Sessions RB, Brady RL (2001) Structural basis for altered activity of M- and H-isozyme forms of human lactate dehydrogenase. Proteins 43:175–185

Rose AJ, Richter EA (2005) Skeletal muscle glucose uptake during exercise: how is it regulated? Physiology 20:260–270

Szent-Gyorgyi AG (2004) The early history of the biochemistry of muscle contraction. J Gen Physiol 123:631–641

Witczak CA, Sharoff CG, Goodyear LJ (2008) AMP-activated protein kinase in skeletal muscle: from structure and localization to its role as a master regulator of cellular metabolism. Cell Mol Life Sci 65:3737–3755

## Challenging Case 10.1: Iron Muscles—Men of Steel

### Source

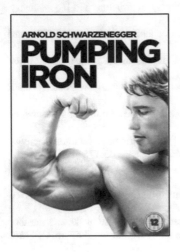

"Pumping Iron" is a documentary film about bodybuilding athletes during the preparation for the most important competition of this sport, the "Mr. Olympia," released in 1977. It was directed by Robert Fiore and George Butler, inspired in a book written by Butler and Charles Gaines in 1974: *Pumping Iron: The Art and Sport of Bodybuilding*. The documentary was a milestone for the bodybuilding sport, still unpopular in the 1970s, and was responsible for introducing to the world the young Arnold Schwarzenegger and his main rival Lou Ferrigno. Years later, in 2013, Vlad Yudin and Edwin Mejia produced "Generation Iron," which shows the life and preparation of leading bodybuilders of a new generation seeking the title of Mr. Olympia.

"Pumping Iron" DVD cover, reproduced with permission of Oak Productions, Inc

### Case Description

In the beginning of the 1890s, the circus athlete Eugen Sandow (1867–1925) became famous for his performances in strongmen competitions, a sport in which the principal task for the competitors is to lift heavy objects, being scored by time, repetitions, or maximum weight lifted.[1]

During the performances, Sandow developed a very particular style, entertaining the audience with unusual poses that highlighted his athletic and heavily muscled body (see next figure and the video in https://www.youtube.com/watch?v=7zt0E6Yh7zA). In 1894, Sandow's manager realized that most people were much more interested in appreciating his client sculptured body than his ability to lift weights.

---

[1] Chapman DL (1994) Sandow the magnificent: Eugen Sandow and the beginnings of bodybuilding. University of Illinois Press, Champaign, IL.

Eugen Sandow, referred to as the "Father of Modern Bodybuilding," in different poses: (*left*) picture by Falk, Benjamin J, 1894; (*middle*) "The Dying Gaul," illustrating his Grecian Ideal, picture by Falk, Benjamin J, 1894; (*right*) "A New Sandow Pose (VIII)," by D. Bernard & Co, Melbourne, from Sandow's Magazine of Physical Culture, 1902. (Images reproduced from Wikimedia Commons; public domain)

That was the birth of bodybuilding, a sport in which, using muscular resistance exercises, the athletes develop their musculature for esthetic purposes. In the competitions, the bodybuilders perform specified poses to be ranked by different criteria, such as symmetry, muscularity, and conditioning. The interest in the sport grew during the 1900s, leading to the appearance of bodybuilding associations. Interestingly, in 1938 the American comic book saga Superman (known as *the man of steel*) started, reinforcing the association of well-sculptured muscled bodies to heroic virtues. In 1965, Mr. Olympia, the major professional bodybuilding world contest took place, being promoted until today by the International Federation of Bodybuilders (IFBB).

The "bodybuilding world" has always attracted very devoted participants and enthusiasts, but the documentary "Pumping Iron" brought this universe to a much larger audience, popularizing the sport and significantly increasing the number of fitness addicts in the 1980s. Moreover, "Pumping Iron" brought to light one of the most famous bodybuilders of all times, Arnold Schwarzenegger. Later famous for his movie star career, and recently, also as a politician, Schwarzenegger has been the champion of Mr. Olympia for 6 consecutive years (1970–1975), which made him the third greatest bodybuilder in history. He won his first bodybuilding title at age 18 and at the age of 20 became Mr. Universe, being able to compete for Mr. Olympia after this.

The documentary also shows the great difference between Arnold Schwarzenegger and his main rival to the title in 1974, Louis Jude Ferrigno. Schwarzenegger was extremely confident and extrovert, training at the famous Gold's Gym, in Venice Beach, California, referred to as "the Mecca of bodybuilding," where he was always surrounded by women and other bodybuilders. Lou Ferrigno was reclusive, being trained by his father in a gym setup at his own house. Ferrigno failed to beat Schwarzenegger in three consecutive competitions.

Lou Ferrigno and Arnold Schwarzenegger, in 1974's Mr. Olympia contest

In the year of the release of "Pumping Iron," Ferrigno was invited to characterize the Hulk in the series "The Incredible Hulk," playing the role from 1977 to 1982. Starring in three TV movies of the "The Incredible Hulk" series, he was eternalized as the character. He also voiced Hulk until 2015.

The current bodybuilding contest has five male categories, five female categories, and one child category.[2] The difference between them is mainly in the level of muscle development and body line presented by the athlete. Athletes perform for a panel of judges who evaluate the presentation, esthetic composition, and proportion of muscle groups according to category. From the 1960s to nowadays, the sport has grown enormously and today has a large number of athletes. However, some enthusiasts and even practitioners have been criticizing the excessive body development in some categories, turning the sport into a "freak show." They think that the athletes are losing the class and the esthetics of the body line they used to have in the past (see comparison in the next figure).

---

[2] https://ifbb.com/our-disciplines/.

The evolution of bodybuilders' profile. (*left*) Eugen Sandow "A New Sandow Pose (VII)," by D. Bernard & Co, Melbourne, from Sandow's Magazine of Physical Culture, 1902; (*middle*) Arnold Schwarzenegger, when champion of the tenth Mr. Olympia contest, in 1974; (*right*) Phil Heath and Kai Greene at the Mr. Olympia 2012. (Images reproduced from Wikimedia Commons; public domain)

Dealing with the lives and the training set of the top today's bodybuilders, the first of three documentaries named "Generation Iron" was launched in 2013. This documentary highlights the dispute between the 2012 Mr. Olympia champion, Phil Heath, and his main opponent, Kai Greene. It shows how today's athletes are assisted by a team of distinct professionals, such as experienced coaches who perform incessant pose training sessions, taking up to 2 h daily dedication, and physicians, physiologists, and biochemists using the latest techniques to improve athlete performance and development.

Despite the great difference between the athletes' profile, the two documentaries address the intense bodybuilding training performed by the athletes, especially in the pre-contest period, when the extreme force used to lift the loads makes the athlete to experience the limit of pain and effort. It is also shown that resistance training using high loads plus a protein-rich energy diet seems to be a limiting factor for hypertrophy to occur.

Additionally, a taboo theme has also been addressed in the documentary: the use of "performance-enhancing drugs," the anabolic steroids. Although not openly assumed, the professional bodybuilders attest that achieving a body fitness in the level of a Mr. Olympia would be impossible without the use of these drugs. However, they warn of all the side effects caused by their use, including liver damage and cardiomegaly.

Independently whether performance-enhancing drugs are used or not, the participants of the series Generation Iron make clear how hard is the preparation for muscle development. The Dominican professional bodybuilder Víctor Martinez, who already competed ten times in Mr. Olympia, reports that his diet consists of more than 9000 calories per day, also including several supplements. Additionally, he performs two or three bodybuilding training sessions each day. Moreover, emotional control, high water intake, and exact hours of sleep and rest are required not only from the athletes but also from their coaches.

All this dedication can be explained by the commercial projection that bodybuilding has gained in recent times. Today Mr. Olympia receives US\$ 400,000 in prize money, which represents a 400-fold increase in the value payed in the first editions of the 1960s, with the winner name forever printed in the history of bodybuilding.

## Questions

1. In this chapter, we discussed the role of AMPK as a cellular energy sensor (see Sect. 10.4.1.1). This enzyme contributes to the control of energy metabolism by restoring cellular ATP levels through the activation of the catabolic pathways and the inhibition of the anabolic pathways when the ratio ADP/AMP/ATP increases (i.e., when the cellular energy charge is low).

    However, bodybuilding activities imply that energy spending activities, which decrease cellular energy charge, occur simultaneously with anabolic pathways required for the desirable muscular hypertrophy. This is possible through the activation of an important signaling pathway involved mainly in the control of protein translation: the mTOR pathway, which is essential not only for muscle hypertrophy but also for any process involving cellular proliferation.

    Search in the scientific literature the mechanism by which the mTOR pathway operates and discusses which bodybuilding training strategies described throughout this Challenging Case would activate this pathway, contributing to the muscular hypertrophy desired by athletes.

2. In June 2019, an article titled "Hardest Muscles to Build" was published on Generation Iron website.[3] The text highlights the difficulties found by the bodybuilders to build mass in specific muscles: *"Everyone has that specific set of muscles that infuriates them. That one spot where no matter how hard you train – no matter how hard you push yourself, you just don't see the results you want. But if you ask around the gym you'd probably find that there are some muscles most people agree to have a problem with. Those universal complaints that most bodybuilders have even after years of hitting the gym. (…) Calves are often either the most complained about muscle to build mass or the most overlooked."*

    Taking into account the type of fibers that compose the calves, explain why it is so hard to develop these muscles.

3. Dietary supplements associated with physical exercise are widely used both for fat reduction and muscular hypertrophy. The "fitness" market is always launching different types of supplements, often as miraculous promises of increased performance. Although the effectiveness of some of these supplements is controversial, others have proven to be safe and efficient to increase exercise perfor-

---

[3] GI Team The hardest muscles to isolate and build mass. https://generationiron.com/hardest-muscles/.

mance when wisely used under expert surveillance. This is the case of creatine, which is widely by most high-resistance exercise practitioners.[4]

Explain the relative benefits of using creatine and how it can contribute to muscle mass gain.

## Biochemical Insights

1. The mTOR (from "mammalian target of rapamycin") pathway integrates intra-cellular and extracellular signals, regulating cellular metabolism, growth, prolif-eration, and survival.[5] The central player in this signaling pathway is the protein mTOR, a protein kinase that works as the catalytic subunit of two distinct protein complexes known as mTOR complex 1 (mTORC1) and mTOR complex 2 (mTORC2). These two complexes mediate cellular responses to different sig-nals, as summarized in the next figure.

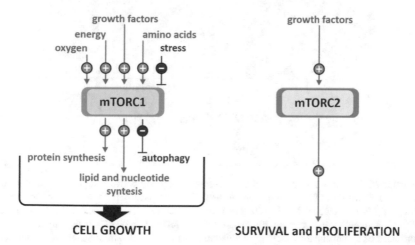

Stimuli and outcomes of mTORC1- and mTORC2-mediated signaling pathways

Muscle hypertrophy is mediated by mTORC1 through different stimuli.[6] This involves mainly the activation of proteins synthesis through the phosphorylation

---

[4] Kreider RB et al (2017) International society of sports nutrition position stand: safety and efficacy of creatine supplementation in exercise, sport, and medicine. J Int Soc Sports Nutr 14:18.

[5] Saxton RA, Sabatini DM (2017) mTOR signaling in growth, metabolism, and disease. Cell 168:960–976.

[6] Yoon MS (2017) mTOR as a key regulator in maintaining skeletal muscle mass. Front Physiol 8:788.

of two key effectors of mRNA translation, p70S6 kinase 1 (S6K1), and eIF4E binding protein (4EBP).

To be active, the mTORC1 should be recruited to the lisosomal membrane, which occurs through the interaction between raptor ("regulatory protein associated with mTOR"), a component of mTORC1, and Rheb (from "Ras homologue enriched in brain"), a protein localized at the lisosomal membrane (see next figure).

Additionally, mTORC1 is negatively regulated by TSC ("tuberous sclerosis complex"), a protein complex that maintain Rheb in its inactive form. When phosphorylated, TSC dissociates from Rheb, which becomes active, and then activates mTOR.

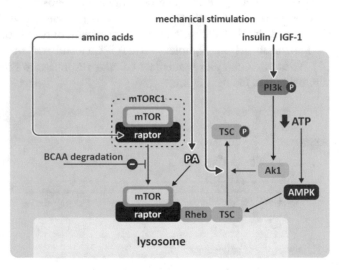

Regulation of mTORC1 activity in skeletal muscles. Interaction between raptor ("regulatory protein associated with mTOR," a component of the mTORC1) and Rheb ("Ras homologue enriched in brain," a protein located at the lisosomal membrane) recruits mTOR to the lisosomal membrane, where it is active. Different stimuli are involved in this process. mTORC1 is directly activated by amino acids or by phosphatidic acid (PA), this latter being generated by the cleavage of phosphatidylcholine by phospholipase D, which is in turn activated by mechanical stretch. Additionally, mechanical stimulus and insulin or insulin growth factor 1 (IGF-1) signaling pathway induce the phosphorylation of TSC ("tumor suppressor tuberous sclerosis complex"), an inhibitor of Rheb-mediated mTORC1 activation. Low levels of ATP activate AMPK, which also phosphorylates TSC (Figure adapted from the article cited in the footnote 6).

The mechanism by which the training strategies pointed out in the case description promote muscular hypertrophy desired by bodybuilding athletes are the following:

– "resistance training using high loads"
  Resistance exercise is an important activator of mTOR pathway because mechanical stretch caused by repeated contractions damages the sarcolemma (muscle cell plasma membrane), by activating the enzyme phospholipase D,

which hydrolyzes the phospholipids. This increases the intracellular concentration of phosphatidic acid (PA), which directly promotes mTOR activation. Additionally, the mechanical stimulus increases the phosphorylation of TSC, which dissociates from Rheb, allowing this protein to activate mTOR.

- "a protein-rich energy diet"
  Amino acids seem to be essential for muscle hypertrophy. Besides being directly used in protein synthesis, they also play an important role in activating mTORC1 as they stimulate mTOR interaction with raptor, favoring the complex to migrate to the lysosome.

- "diet (…) of more than 9,000 calories per day"
  The nutrient availability and the cellular energy status are among the most important factors for mTORC1 activation. As discussed in Chap. 8, insulin is one of the main hormones involved in the fed state. Insulin binding to its receptor causes receptor autophosphorylation leading to the recruitment of IRS-1, which, in turn, recruits and activates PI3K, which phosphorylates membrane phosphatidylinositol bisphosphate (see Sect. 8.4.2). The resulting phosphatidylinositol triphosphate recruits PDK1, which phosphorylates and activates Akt. This protein acts as a negative regulator of the TSC complex by phosphorylating it. Thus, with inactive TSC, mTORC1 activation occurs via Rheb. On the other hand, energy deficiency activates AMPK, which, when activated, enhances the inhibitory effect of TSC on mTORC1 activation.

- "performance enhancing drugs, the anabolic steroids"
  Steroid hormones, such as testosterone and its derivatives, significantly increase the amount of insulin growth factor (IGF-1), which binds to insulin receptors promoting mTORC1 activation via PI3K/Akt/mTOR.

2. Muscles are composed of two major types of fibers: the slow-twitch, also known as type 1 or red fibers, and the fast-twitch, or type 2 or white fibers (see Sect. 10.2). Slow-twitch fibers are adapted to prolonged, but low force, work and are very resistant to fatigue. These fibers are highly dependent on oxidative phosphorylation, requiring a high supply of $O_2$. Fast-twitch fibers work at low levels of $O_2$, using muscle stocks of glycogen as the major metabolic substrate. This type of fiber contracts quickly and powerfully but undergoes fatigue rapidly since anaerobic metabolism cannot sustain the ATP demand for long periods of contraction.

   Resistance training is one of the most important factors responsible for muscle hypertrophy. Resistance exercises consist of intense and prolonged muscle contractions that can lead to muscle fatigue, causing local points of muscular fiber damage, which triggers hypertrophy (see question 1). There are several hypotheses to explain how exercise results in muscle fatigue. One of the most accepted is the reduction of intracellular pH due to an excessive lactate production during the repeated contractions. Low pH impairs myosin activity as well as the glycolytic pathway reactions and the consequent production of ATP through the mechanism of substrate-level phosphorylation (see Sect. 6.1). Thus, without the necessary input of ATP, the contraction does not occur. Low pH also inhibits $Ca^{2+}$ release from sarcoplasmic reticulum, affecting muscle contraction.

The calves are formed by two main muscles: the gastrocnemius and the soleus (see below figure).

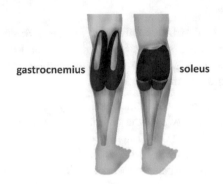

gastrocnemius    soleus

Localization of the gastrocnemius and soleus muscles

The gastrocnemius contains about 50% of each type of fiber, while the soleus is mostly composed of type 1 fibers. Intense exercise causes the gastrocnemius muscle to reach fatigue faster, limiting movement. On the other hand, the soleus, which is responsible for calf volume as it is a long muscle localized below the gastrocnemius, rarely reaches fatigue, making difficult the development of calf hypertrophy.

Recent studies highlight that the best way to hypertrophy the calf is to use a low load and a higher repetition number, so that the soleus muscle can perform as many contractions as possible. In addition, calves should be the first group to be stimulated in a resistance training session, thus preventing their phosphocreatine and glycogen stores from being depleted (see the role phosphocreatine and glycogen as an energy source for the muscle fibers in Sect. 10.3).

3. Creatine is a dietary supplement widely used by both bodybuilding athletes for the purpose of strength gain and by athletes of other sports as an ergogenic aid. It is an important substrate in providing muscle energy. Inside the muscle cells, the enzyme creatine kinase (CK) phosphorylates creatine to phosphocreatine (PCr).

When ATP is degraded during muscle contraction, generating ADP and Pi, CK catalyzes the reverse reaction, the transfer of phosphate group of phosphocreatine to ADP, regenerating the ATP (see Sect. 10.3). Thus, PCr serves as a local phosphate source for resynthesizing ATP through the process of substrate-level phosphorylation. Therefore, keeping the muscle stores of PCr high ensures a higher muscular activity as well as its readiness.

Almost 95% of creatine is stored in skeletal muscle, 2/3 of the total in the form of PCr. In general, muscle creatine stores are around 120 mmol/kg muscle mass, and about 1–3 g are degraded and excreted daily. Creatine replacement occurs through liver synthesis, dietary intake, or supplementation. When supplemented, creatine stocks can reach 160 mmol/kg of muscle mass. The International Society of Sports Nutrition's recommendation for creatine use is to consume 0.3 g/kg/day for 5–7 days, followed by 3–5 g/day to keep stocks high[4].

Creatine has also an immediate effect on increasing lean mass as it induces water retention in muscle fiber due to its osmotic characteristic, increasing the volume to the tissue. However, its function in providing an immediate source of energy for muscle contraction enables a greater contraction capacity, especially for intense and short-duration exercises such as resistance exercise. Since the more intense the contractions during a resistance exercise, the more effective the induction of protein synthesis, keeping skeletal muscle saturated with creatine is a positive factor for hypertrophy.

## *Final Discussion*

Anabolic steroids became very popular among bodybuilders and other athletes after the decade of 1970. Since then, a growing number of men and women have been using these substances without a medical prescription, in order to increase athletic performance and to help them gain a muscular and lean figure.[7] Several studies have observed that the widespread use of anabolic steroids in Western countries had a great influence on the idea of how a healthy and strong human body "should" look like. Comparing the bodies of movie stars and super heroes from the 1960s and the 2000s suggests that nowadays spectators grow up admiring body images that are impossible to be obtained without the use of anabolic steroids. Additionally, anabolic steroids can be easily bought over the Internet, with little or no control over the actual content of each vial. Unlike other drugs, their use gives the impression of a healthy lifestyle, which may pass unnoticed or even be supported by the users' friends and family members… at least while nothing goes wrong.

The problems of using anabolic steroids go beyond the adverse effects, which are dangerous enough—from increased risk of heart and liver diseases to stroke, cancer, anxiety, depression, suicide, aggressiveness, baldness, gynecomastia (growing of breasts in men), hirsutism (growing of face and body hair in women), shrinking of testicles and hypertrophy of clitoris, to name a few.[8] Even if a user manages to prevent these effects (usually by taking many other drugs at the same time), the prolonged use of anabolic steroids may lead to a chemical and psychological syndrome of dependence. In this case, the user finds himself or herself unable to cease the use of anabolic steroids, usually leading them to concentrate all their time and resources

---

[7] Kanayama G, Pope HG (2017) History and epidemiology of anabolic androgens in athletes and non-athletes. Mol Cell Endocrinol 464:4–13.

[8] Kanayama G et al (2018) Public health impact of androgens. Curr Opin Endocrinol Diabetes Obes 25:218–223.

on activities related to muscle training and the acquisition of more drugs, despite the occurrence of deleterious effects.[9]

This is why the use of anabolic steroids is considered a public health concern in many countries. In Denmark, for instance, any person working out in a government-registered gym must provide urine samples for the screening of anabolic steroids.[10] If doping is detected, that person must pay a fine and can be banned from training and taking part in official sports activities for many years. In countries like the UK and Australia, users of anabolic steroids can go to harm reduction centers and exchange their used syringes for new ones, in an attempt to reduce the transmission of HIV and hepatitis.[11]

Therefore, those using or considering the use of anabolic steroids is strongly advised to look for natural and healthy ways to boost their physical training if they wish, as well as to keep a strong network of clinical and psychological support, addressing issues like self-esteem and the search for an overall satisfying life, beyond the limits of iron weights.

*This case had the collaboration of Dr. Lorena O. Fernandes-Siqueira (Instituto de Bioquímica Médica Leopoldo de Meis, Universidade Federal do Rio de Janeiro, Brazil) and MD Julio X. Amaral (Addictions Department, Institute of Psychiatry, Psychology and Neuroscience, King's College London, UK).*

---

[9] Amaral JMX, Cruz MS  Use of androgenic anabolic steroids by patients under treatment for substance use disorder: case series. J Bras Psiquiatr 66:120–123.

[10] Anti Doping Danmark. https://www.antidoping.dk/.

[11] Kimergard A, McVeigh J (2014) Environments, risk and health harms: a qualitative investigation into the illicit use of anabolic steroids among people using harm reduction services in the UK. BMJ Open 4:e005275.

# Challenging Case 10.2: Armstrong's Race to the Sun

## Sources

Not even the most farfetched fictional story would come close to the real-life story of Lance Armstrong. This is what one thinks when watching two films about Armstrong's extraordinary career as a cyclist and the following doping scandal that

stripped all his titles and banned him from competitions for life: "The Armstrong Lie," an 2013 American documentary film directed by Alex Gibney, and "The Program," a 2015 biographical drama film, directed by Stephen Frears, starring Ben Foster as Armstrong.

DVD covers of "The Armstrong lie," Sony Pictures Classics, and "The Program"; reproduced with permission

## Case Description

Lance Edward Armstrong became a professional cyclist in 1992, emerging as a rising star between 1993 and 1996. However, his very promising career was abruptly interrupted in the end of 1996, when, at age of 25, he was diagnosed with an advanced testicular cancer with metastases spread to brain, lungs, and abdomen. Armstrong's chances of survival were virtually none. Immediately after the diagnosis, he underwent an orchiectomy (surgery to remove one or both testicles; in the case of Lance Armstrong, it was only the affected testicle) and started an intense chemotherapy treatment. During the treatment, the affected areas in the brain were also surgically removed.

Contrary to all predictions, only 4 months after the diagnosis Armstrong was declared free of his cancer. He immediately engaged in a very intense training program, returning to professional cycling with outstanding performances, including the extraordinary seven consecutive Tour de France titles, from 1999 to 2005. Since the first Tour de France win, Armstrong faced persistent doping allegations, which he vehemently denied, claiming that he had never tested positive for any performance-enhancing drugs, although he was the most tested athlete in the world.

Armstrong retired as an idol in 2005, being the source of inspiration for millions around the globe. However, he couldn't stay away from the competitions and decided to return in 2009. This inspired the director Alex Gibney to film a documentary, to be named "The Road Back," about Armstrong's comeback to the 2009's Tour de France. However, just before the film launch, Armstrong's doping scandal erupted, resulting in his lifetime ban from competition and the stripping of all his titles from August 1998 onward. Gibney had to shelve the project but decided to reopen it after Armstrong's public confession during the famous interview with Oprah Winfrey.[12] Gibney reedit the documentary, including a new interview with the cyclist. In this new version, named "The Armstrong Lie," the main question pursued by the director was why Armstrong decided to risk everything he had achieved and return to the competitions. A report in the Rolling Stone magazine suggests the answer: a typical case of "returning to the scene of the crime."[13]

While in "The Armstrong Lie" one has a psychological perspective of the real character, looking into his eyes when he denies and confesses his crime with the same unperturbed expression, the dramatized version of Armstrong biography, "The Program," by Stephen Frears, fills in the gaps of undocumented facts with plausible playing scripts. The film shows how the Armstrong team took advantage of a range of means to enhance performance following a doping program developed by the Italian physician Michele Ferrari, an expert in blood transfusions and other illicit means of helping cyclists to achieve the best results.

One of the main pieces of the program was EPO, an endogenous protein that stimulates red blood cell production. Recombinant human EPO (rhEPO) is used in the treatment for severe anemia and, although illicit as a performance-enhancing drug, in the 1990s could be bought in pharmacies without prescription. This, together with the fact that there was no accurate test to detect rhEPO until 2001, made its use to be an almost perfect way to enhance performance.

At those times, the use of rhEPO was inferred indirectly by testing the athlete's blood for an unnaturally high hematocrit. The Union Cycliste Internationale (UCI, International Cycling Union) adopts a hematocrit over 50 to ban the racers. A strategy commonly used by Armstrong's team to confound blood tests was to quickly administer a saline infusion to dilute the blood and decrease the hematocrit when noticing an imminent test. The US Anti-Doping Agency (USADA) report that banned Armstrong from competition for life and that stripped all his titles describes how this practice occurred at the 1998 World Championships: "*Armstrong's doctor literally smuggled past a UCI official a liter of saline concealed under his rain coat and administered it to Armstrong to lower his hematocrit right before a blood check.*"[14]

Even when accurate blood and urine tests for rhEPO became available, Armstrong and his teammates resorted to an older method that lead to the same outcome: trans-

[12] https://www.youtube.com/watch?v=N_0PSZ59Aws.

[13] Syckle KV 10 Truths From 'The Armstrong Lie'. Rolling Stone. https://www.rollingstone.com/movies/movie-news/10-truths-from-the-armstrong-lie-245369/. Accessed 7 Nov 2013.

[14] U.S. Postal Service Pro Cycling Team Investigation, USADA, 2012; http://cyclinginvestigation.usada.org/.

fusions of their own blood, even during the events. He also used growth hormone and testosterone to build muscle, as well as cortisone to control pain help in injury recovery. In its 1000-page report, USADA labeled Armstrong a "serial" cheat running *"the most sophisticated, professionalized and successful doping programme that sport has ever seen ... His goal [of winning the Tour de France] led him to depend on EPO, testosterone and blood transfusions but also, more ruthlessly, to expect and to require that his team-mates would likewise use drugs to support his goals if not their own"*.

## Questions

1. The USADA report concluded that Armstrong had used over the course of his career performance-enhancing practices, including applications of EPO. Search in the literature for the biological function of EPO and its regulation.
2. Discuss how EPO would work as a performance-enhancing drug.
3. rhEPO has been banned as a performance-enhancing drug since the early 1990s, but only after 2000, an accurate method for rhEPO detection was officially adopted by the World Anti-Doping Agency (WADA). Search the literature and describe the biochemical principle of the method used since then.
4. In the film "The Program," when Lance Armstrong asks the supervision of Michele Ferrari for his return to competitions after the cancer therapy, they have the following dialogue:

   Ferrari: "What is your $VO_{2max}$?"
   Armstrong: "84"
   Ferrari: "What does that mean?"
   Armstrong: "It is my maximal oxygen consumption?"
   Ferrari: "Literally, yes. But what it really means is that you were born to lose. When your compatriot Greg LeMond won the Tour the France, his $VO_{2max}$ was 93. So, in a straight race between you and a man like him, you will always lose."

   Define $VO_{2max}$; describe how this parameter can be measured and its importance to determine the capacity of endurance sports athletes.

5. Altitude training is recommended by some coaches as a mean to improve performance of endurance athletes in competitions at or near sea level. Discuss the rationale behind this type of training.
6. Interval training is considered a powerful tool to enhance performance in endurance training routines.[15] This type of training is characterized by intermittent periods of intense exercise separated by short periods of recovery, following two main basic protocols: (a) the high-intensity interval training (HIIT), in which near-maximal intensity (80–95% of maximal heart rate) exercise is performed, and (b) the sprint interval training (SIT), in which the athletes reach supramaximal intensities during the exercise period.

---

[15] MacInnis MJ, Gibala MJ (2017) Physiological adaptations to interval training and the role of exercise intensity. J Physiol 595:2915–2930.

Suggest some parameters that could be measured to gain insights into the mechanisms by which interval training strategies improve athletes' aerobic capacity.

## Biochemical Insights

1. EPO is erythropoietin, an endogenous glycoprotein secreted by kidney cells that acts on bone marrow stimulating red blood cell production (erythropoiesis). EPO binds to an specific receptor on the red cell progenitor and activates a signaling cascade that results in differentiation and proliferation of the erythroid cell.

   Blood EPO concentration in normal oxygen levels is around 10 mU/mL. However, during hypoxic stress, which may occur due to anemia or other situations in which oxygen delivery to cells is low, EPO production may increase up to 1000-fold, enhancing erythropoiesis.

   The mechanism by which EPO production is controlled by oxygen availability has been unraveled by Gregg L. Semenza and Peter J. Ratcliffe, who shared the 2019 Nobel Prize in Physiology or Medicine with William Kaelin Jr. for the discovery of hypoxia-inducible factor (HIF), a transcriptional factor responsible for the regulation of gene expression in response to hypoxia (see next figure). HIF is a heterodimeric protein composed by HIF1α and HIF1β subunits. Under normal oxygen levels (normoxia), HIF1α is hydroxylated by oxygen-sensitive prolyl-4-hydroxylases (PHD, see also Box 3.7). Hydroxylated HIF1α is polyubiquitinated by VHL ("von Hippel–Lindau" protein), which tags it to degradation by the proteasomes (cellular organelles responsible for protein turnover). In hypoxic conditions, HIF is stabilized and translocated into the nucleus, where it dimerizes with HIF1β, enhancing the transcription of target genes, including EPO gene.

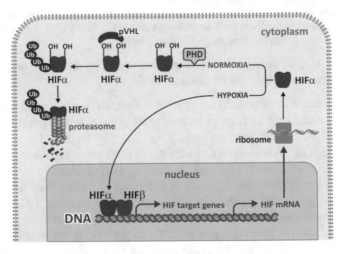

Regulation of gene expression in response to hypoxia. In normoxia, hydroxylation of HIFα by PHD favors its interaction with VHL, which catalyzes its polyubiquitination, tagging it for proteasome degradation. In hypoxia, HIFα migrates to the nucleus where it interacts with HIFβ to activate gene expression

2. Endurance exercises, such as long-distance cycling or running, depend on the cardiorespiratory system efficiency to take oxygen from atmosphere to muscle mitochondria. The easiest way to improve oxygen delivery to muscle cells is to increase blood hemoglobin (Hb) mass, since Hb is a key player in the transport of oxygen in the body (see Sect. 3.3.3). The tetrameric structure of Hb enables a cooperative binding of oxygen to its monomers. The transition interval to near saturation corresponds to oxygen partial pressure found in peripheral tissues, facilitating Hb saturation in the lungs as well as oxygen delivery to the tissues. In the muscles, oxygen binds with high affinity to the monomeric myoglobin (see Fig. 3.42), enabling oxygen storage in the muscular tissue.

The increased availability of oxygen in the muscle cells favors oxidative phosphorylation, enhancing the cellular capacity to synthesize ATP using as energy source the fatty acids, either taken up from the blood or mobilized from TAGs stored in the adipocytes associated with the muscular tissue or inside the myocyte itself (intramyocellular TAGs) (see Sect. 10.4.2.1).

EPO-induced erythropoiesis is expected to increase hematocrit and the respective Hb mass thus enhancing the oxygen transport capacity of the blood. Indeed, this has been observed in several studies.[16]

3. Endogenous and the recombinant EPOs present identical primary amino acid sequence but show differences in their glycosylation patterns [15]. The endogenous protein contains four glycosylation sites (three N-linked and one O-linked) resulting in about 40% of the protein mass corresponding to polysaccharides. The first generation rhEPOs are named "epoetins" in which Greek letters ($\alpha$, $\beta$, $\theta$, $\zeta$) identify their different glycosylation patterns. Two additional N-linked glycosylation sites were introduced to rhEPO in order to increase its plasma half-life from about 24 h to 48 h when administered subcutaneously, being this new generation EPO named "darbepoetin $\alpha$" or "Aranesp." To further increase plasma half-life (to about 140 h), a 30 kDa methoxy polyethylene glycol polymer was covalently link to rhEPO, being this form named "CERA," from "continuous erythropoietin receptor activator."

The variations in the glycosylation patterns alter the protein charge distribution resulting in distinct isoelectric points (pI), i.e., the pH in which the protein net charge in neutral. This allows separating the proteins using the technique of isoelectrical focusing (IEF; see the following figure). The pI value of the endogenous EPO ranges from 3.7 to 4.7, while hyposulfated polysaccharide chains of epoetins $\alpha$ and $\beta$ result in a more basic pI (~4.4–5.1). Aranesp presents a more acidic pI range (~3.7–4).[17]

---

[16] Sgrò P et al (2018) Effects of erythropoietin abuse on exercise performance. Phys Sportsmed 46:105–115.

[17] Delanghe JR et al (2008) Testing for recombinant erythropoietin. Am J Hematol 83:237–241.

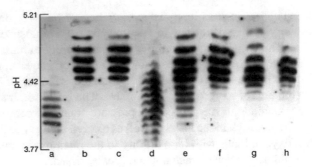

Detection of EPO in urine samples by IEF. (**a**) Commercial human endogenous EPO; (**b**) rhEPO-β (Neorecormon, France); (**c**) rhEPO-α (Eprex, France); (**d**) urine sample from a control subject; (**e**, **f**) urine from two patients treated with Neorecormon EPO for post-hemorrhagic anemia; (**g**, **h**) urine from two cyclists from Tour de France 1998. (Figure reproduced with permission from Lasne F and Ceaurriz J (2000) [18])

4.  $VO_{2max}$ is the maximum amount of oxygen one can consume during a very intense physical activity. It is expressed as the volume (usually in milliliters) of oxygen consumed per kilogram of body mass per minute. So, when lance Armstrong answers "84," this means that his $VO_{2max}$ is 84 mL $O_2$/(kg min).

    $VO_{2max}$ can be calculated by measuring oxygen uptake during an exercise of progressively increased intensity. When oxygen consumption remains constant despite an increase in exercise intensity, it means that $VO_{2max}$ was reached.

    Although this is the most accurate way to obtain the $VO_{2max}$ value, there are alternative simpler procedures that allow one to estimate $VO_{2max}$ value from physiological responses during submaximal exercise. The simplest way is to use the relationship: $VO_{2max} = 15 \times (HR_{max}/HR_{rest})$, in which HR is the heart rate. Thus, measuring the heart rate in rest and during an extenuated exercise, it is possible to estimate the $VO_{2max}$. This estimation was validated in a study with well-trained men, which showed a very good agreement between the experimental and theoretical values.[19] However, the applicability of this method to other groups still needs direct validation.

    $VO_{2max}$ is an excellent indicator of aerobic endurance capacity and cardiovascular fitness as it reflects how efficiently the cells use oxygen, which means the capacity of oxidizing nutrient generating ATP for muscle activity.

5.  As altitude increases, the atmospheric pressure falls almost linearly reaching 50% or 30% of the sea level value at 5500 m or 8900 m (the summit of Everest), respectively.[20] Thus, although regardless of the altitude, the percentage of oxygen in the air remains the same (approximately 21%), the partial pressure of inspired oxygen decreases, reducing the driving pressure for gas exchange in the lungs and consequently the oxygen availability for the exercising tissues

---

[18] Lasne F, Ceaurriz J (2000) Recombinant erythropoietin in urine. Nature 405:635.

[19] Uth N et al (2004) Estimation of $VO_{2max}$ from the ratio between $HR_{max}$ and $HR_{rest}$—the heart rate ratio method. Eur J Appl Physiol 91:111–115.

[20] Peacock AJ (1998) ABC of oxygen: oxygen at high altitude. BMJ 317:1063–1066.

mitochondria. Therefore, training in high altitudes is similar to training under hypoxic conditions.

Since it has been observed that although the immediate exposure to an oxygen shortage condition impairs endurance training, after some weeks at high altitude, athletes usually adapt and performance recovers. This made some coaches believe that when returned to sea level, the athlete would experience an enhancement in performance.

Although the literature shows many controversial studies in this regard, a meta-analysis compiling data from 51 articles evaluated systematically the potential increase in endurance performance in elite or sub-elite athletes of different exercise modalities (cyclists, runners, swimmers, triathletes, skiers, among others) after adaptation to six protocols of natural or artificial altitude.

Two natural altitude protocols were evaluated, namely, the live-high train-high (LHTH) and live-high train-low (LHTL). In the first, the athlete stays and trains at high altitude, while in the second they stay and sleep at altitude but descend regularly for training sessions. This last protocol has been shown to be the best for enhancing endurance performance in both elite and sub-elite athletes.[21] This fact encouraged the coaches to subject the athletes to artificial exposure to hypoxic conditions, using nitrogen houses, hypobaric chambers, altitude tents, or hypoxic inhalers in between the training sessions, so that they would adapt to hypoxia without the need to travel up and down a mountain.

The meta-analysis study evaluated four protocols of artificial altitude training. Three of them consisting of artificial LHTL with daily exposure to a simulated moderate altitude condition for periods: (a) long and continuous (8–18 h); (b) brief and continuous (1.5–5 h); or (c) brief (<1.5 h) and intermittent. The fourth protocol was an artificial live-low train-high (LLTH), with the athlete only training in a simulated altitude condition.

Most of the studies provided estimates for effects on power output, some on $V_{O2max}$, and fewer evaluated other potential mediators. The conclusion was that only natural LHTL provided enhancing endurance performance for elite athletes, while for sub-elite athletes besides natural LHTL, also natural LHTL, artificial brief intermittent LHTL and, with a lower probability, long continuous LHTL were also effective.

Hypoxic exposure increases several parameters, such as erythrocyte volume, maximal aerobic capacity, and capillary density, being an interesting choice of a legal ergogenic aids to improve endurance performance.

6. Training strategies for endurance athletes mainly aim to improve the aerobic (i.e., oxygen-dependent) energy metabolism. This would be reached through with different adaptations, including:

(a) The direct improvement of muscular cellular respiration capacity through:

- Increase in mitochondrial content
- Increase in the activity of mitochondrial respiration complexes, i.e., the increase their functional capacity to produce ATP

---

[21] Bonetti DL, Hopkins WG (2009) Sea-level exercise performance following adaptation to hypoxia: a meta-analysis. Sports Med 39:107–127.

(b) The increase in different factors that ultimately result in an enhancement oxygen delivery to the muscle cells to support the aerobic activity, such as:

- Muscle capillary density
- Hb mass
- $VO_{2max}$
- Maximal stroke volume (i.e., the volume of blood pumped from the heart ventricle per beat)
- Maximal cardiac output (which corresponds to the product of the heart rate and the stroke volume)
- Blood volume

A number of studies evaluated the modulation of some of these parameters in response to different training approaches, but the results were not conclusive, probably due to differences in the training protocols used as well as in participants characteristics.

One of these studies evaluated the mitochondrial function in the muscle cells of 29 cyclists performing three different training protocols consisting of two types of interval training, SIT or HIIT, and a continuous training protocol (STCT, sub-lactate threshold continuous training), used as a control.[22] The athletes performed 12 cycling sessions for 4 weeks, with the training scheme showed in the next figure.

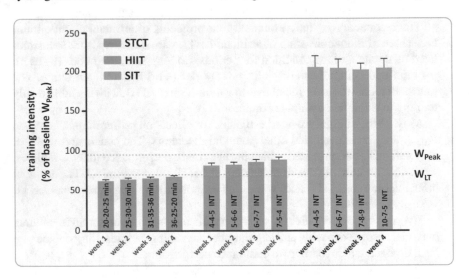

Training protocols used by three groups of cyclists (9 following the STCT protocol, 11 HIIT, and 9 SIT) matched by age, height, and $V_{O2peak}$. The numbers in each bar represent the duration in minutes (STCT) or the number of intervals (HIIT, SIT) for each of the training sessions. The STCT and HIIT groups were matched for total work. (Figure adapted from the article cited in the footnote 22, with permission)

---

[22] Granata C et al (2016) Training intensity modulates changes in PGC-1a and p53 protein content and mitochondrial respiration, but not markers of mitochondrial content in human skeletal muscle. FASEB J 30:959–970.

Resting muscle biopsy samples (*vastus lateralis*) were obtained before and 72 h after the training protocol has been completed. The results showed that SIT resulted in an increased maximum coupled and uncoupled mitochondrial respiration (see next figure), although mitochondrial content, measured by the classical method of quantifying the activity of the citrate synthase remained unchanged.

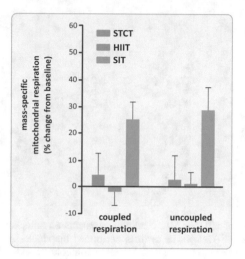

Effects of different training protocols on mitochondrial respiration. (Figure adapted from the article cited in the footnote 22, with permission)

Another study measured the same parameters, post-training citrate synthase (CS) maximal activity and mass-specific oxidative phosphorylation capacities comparing a HIIT protocol with a moderate-intensity continuous training protocol (MICT).[23] An interesting approach used in this study was that the comparisons were made using the "within-subject model," in which the effects of different training protocols are evaluated within the same individual. For this, the cyclist pedals using one leg, with the non-exercising leg resting on a stationary platform. The cycle ergometer used is adapted to eliminate the need to pull up on the pedal, so the cyclist feels as the two legs were cycling, although training with only one leg at a time.

In the study, the participants completed six sessions of HIIT (4 × 5 min at 65% $W_{peak}$) or MICT (30 min at 50% $W_{peak}$), over 2 weeks, with one leg performing HIIT

---

[23] MacInnis MJ et al (2017) Superior mitochondrial adaptations in human skeletal muscle after interval compared to continuous single-leg cycling matched for total work. J Physiol 595:2955–2968.

and the other leg performing MICT. The results showed that the post-training citrate synthase activity and mass-specific oxidative phosphorylation capacities were greater in HIIT relative to MICT (see next figure).

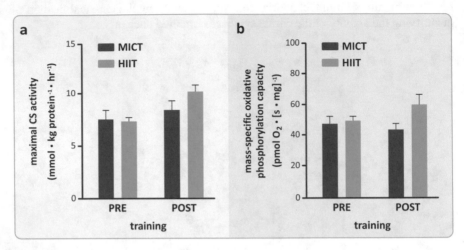

Effects of six sessions of HIIT or MICT training protocols on mitochondria content (*left*) and mitochondrial respiration (*right*). The cyclists performed the single-leg cycling protocol with each leg completing either HIIT or MICT. Mitochondria content was estimated by measuring citrate synthase (CS) activity, and mitochondrial respiration was measured through the specific oxidative phosphorylation capacity through complexes I and II. (Figure adapted from the article cited in the footnote 23, with permission)

## *Final Discussion*

Although the Armstrong case can be considered the biggest doping scandal in the history of cycling, it seems to be just the tip of an iceberg. In a press release from October 2012, the UCI announced that Armstrong stripped wins would not be allocated to other riders, acknowledging that "*a cloud of suspicion would remain hanging over this dark period, but while this might appear harsh for those who rode clean, they would understand there was little honour to be gained in reallocating places.*"[24]

---

[24] UCI Communications Service. Press release: UCI takes decisive action in wake of Lance Armstrong affair. https://www.uci.org/inside-uci/press-releases/press-release-uci-takes-decisive-action-in-wake-of-lance-armstrong-affair-153863-a66a. Accessed 26 Oct 2012.

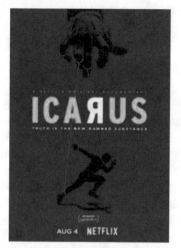

"Icarus" DVD cover; reproduced with permission

Additionally, doping practices are clearly not restricted to cycling but are spread throughout all sport competitions.

A much broader view of the dimension of doping in professional sports appears in the 2017 documentary film "Icarus," directed by Bryan Fogel and winner of the Best Documentary Feature at the Oscars in 2018. Interestingly, the Armstrong case was the starting point for the film production, which, in the end, unravels one of the biggest scandals in sporting history. As pointed out in one of the film reviews, *"what actually happened was a bit like tugging on an errant thread and having the entire clothing industry unravel right on top of you."*[25]

When Brian Fogel decided to film "Icarus," his first intention was to prove that the system to test doping on athletes was flawed. Fogel is also an amateur cyclist who was hugely disappointed and disturbed with the Lance Armstrong doping case. His idea was to make a film examining how easy it is to escape from doping tests. For this, he decided to subject himself to a doping program specially prepared to avoid positive results and compete in the Haute Route, one of the hardest amateur bike races in the world, in which he had competed before completely drug-free.

To design the doping program, he firstly contacted Don Catlin, a prestigious scientist who founded the UCLA Olympic Analytical Laboratory, the first and largest testing facility of performance-enhancing drugs in the world. However, worried that Fogel's plan would affect his reputation, Catlin quit the project. Although leaving Fogel's plan, Catlin suggested that he talk to Grigory Rodchenkov, the former director of Russia's national anti-doping laboratory. This suddenly transformed Fogel's personal experiment into a geopolitical thriller.

Although working officially in anti-doping control, Rodchenkov helped the development of the doping program of Russian athletes from 2005 to 2015, when the World Anti-Doping Agency (WADA) connected him to a state-sponsored doping program in Russia, which culminates with Russia partially banned from the 2016 Rio de Janeiro Olympics and totally banned from the 2018 Pyeongchang Winter Olympics. In 2019, the WADA banned Russia from all major sporting events for 4 years.

---

[25] Icarus GS A doping house of cards tumbles down. The Atlantic. https://www.theatlantic.com/entertainment/archive/2017/08/icarus-review-netflix/535962/. Accessed 6 Aug 2017.

## Challenging Case 10.3: K. V. Switzer's Run for Rights

### Source

For those who follow the Olympics, it is difficult to believe that only in 1984 women's marathon was added to the Olympics program, almost a century after the first international Olympic Games held in modern history (the 1896 Athens Olympics). A key event that changed the rules that prohibited women from competing in long-distance racing was due to Kathrine V. Switzer, who, using her initials (K. V. Switzer) in the entry form for the 1967's Boston Marathon, was assigned the number 261 and became the first woman to officially run that race. As she comments in the video on which this case is based on,[26] completing the race was not easy. However, the main difficulty had nothing to do with women being too weak or fragile to run a long-distance race, but the fierceness with which a race official tried to remove her from the race. Thanks to her teammates, she got rid of the official and finished the complete marathon. This experience encouraged her to fight for women's rights in sports.

Kathrine Switzer at the 2011 Berlin Marathon Expo; reproduced from Wikimedia Commons, under a CC-BY license

### Case Description

Kathrine Switzer was a journalism student at Syracuse University. She loved to run, but there was no women's running team at the university, so she started to train unofficially with the men's cross-country team. There, she met Arnie Briggs, the university mailman who was also a veteran of 15 Boston Marathons. One day, during an evening training session, when he was telling her many stories of Boston Marathons, they had the decisive dialogue[27]:

> Kathrine: "Oh, let's quit talking about the Boston Marathon and run the damn thing!"
> Arnie: "No woman can run the Boston Marathon"
> Kathrine: "Why not? I'm running 10 miles a night!"

---

[26] Legacy of 261 & Kathrine Switzer, SC Featured, ESPN Stories, ESPN on YouTube. https://www.youtube.com/watch?v=U6CoScOIK_I&t=2s.

[27] Kathrine Switzer The real story of Kathrine Switzer's 1967 Boston Marathon; Kathrine Switzer website: https://kathrineswitzer.com/1967-boston-marathon-the-real-story/.

Arnie argued that women were too fragile to run such distance and couldn't believe that a woman, Roberta Gibb, completed the race unofficially in 1966. Instigated by this fact, Arnie challenged Kathrine: "*If any woman could do it, you could, but you would have to prove it to me. If you ran the distance in practice, I'd be the first to take you to Boston.*"

Although Kathrine had no problem to sign up for the race since she used her initials instead of her first name, a few miles after the race started a press truck approached, in which the journalists became excited to see a woman in the race, and, more importantly, a woman wearing a number! This called the attention of race officials, among them Jock Semple, who aggressively tried to remove Kathrine from the race (see next figure and the related video[28]).

Kathrine Switzer in the 1967 Boston Marathon; reproduced under CC-BY license

The situation was very stressful but, instead of making Kathrine quit the race, gave her a reason to continue, as she tells in her memoir[27]: "*I knew if I quit, nobody would ever believe that women had the capability to run 26-plus miles. If I quit, everybody would say it was a publicity stunt. If I quit, it would set women's sports back, way back, instead of forward. If I quit, I'd never run Boston. If I quit, Jock Semple and all those like him would win. My fear and humiliation turned to anger.*"

Kathrine's experience in Boston Marathon made her to realize that she should do something to change the women's inequality in sports[27]: "*The reason there are no intercollegiate sports for women at big universities, no scholarships, prize money, or any races longer than 800 meters is because women don't have the opportunities to prove they want those things. If they could just take part, they'd feel the power and accomplishment and the situation would change. After what happened today, I felt responsible to create those opportunities.*"

---

[28] "Kathrine Switzer: First Woman to Enter the Boston Marathon, by MAKERS.com; https://youtu.be/fOGXvBAmTsY.

The fact that Jock Semple attempted to rip Kathrine's number off and to remove her from the race took place just in front of the press truck made this episode highly documented. On the day after, photographs of race officials chasing after Switzer spread in the national newspapers. The timing of when the incident happened was also opportune: it was coincident with the midst of the women's liberation movement.[29]

The repercussion of the episode was an important step for the rules to be changed, although this happened very slowly. In 1972, women were allowed to compete officially in the Boston Marathon. During the 1970s, more and more women began to compete in marathons, with the first Women's International Marathon Championship occurring in 1973, in Waldniel, Germany[29].

However, including a women's marathon in the Olympic Games was not under consideration. The reasons claimed by the Olympics' organizers for the exclusion were (1) the fragile women's physical condition, albeit this could not be sustained for long with the increasing success of women performances in marathons, and (2) the requirement that the sport must be widely practiced to justify its inclusion (the rules stated that women's sport to be included must be practiced in at least 25 countries on at least two continents)[29]. Again, Kathrine Switzer had a crucial role to pave the way for the changes. She convinced the director of Avon cosmetic company to sponsor a series of women's races. In 1979, the Avon Marathon attracted over 250 world class entrants from 25 countries. Even though, the resistance to include women marathon in the Olympics still persisted for more 5 years, and only in 1984, women's marathon was finally included in the Olympics Games.

## Questions

1. In 1992, a scientific correspondence published in the Nature magazine predicted, based on the progression of the world records' in the Olympic running events (from 200 m to marathon), that in 1998 women would be running as fast as men in a marathon[30] (see below figure).

---

[29] Lovett C The fight to establish the women's race. Olympic Marathon, Chap. 25, 1997. Reproduced in the website "Marathon guide.com". http://www.marathonguide.com/history/olympicmarathons/chapter25.cfm. Accessed 28 Jan 2020.

[30] Whipp BJ, Ward SA (1992) Will women soon outrun men? Nature 335:25.

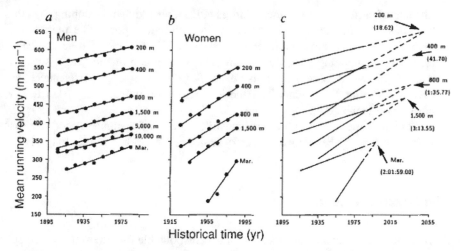

World records' progression for men and women with their best fit linear regressions (solid lines) and the extrapolation for their intersection (dashed lines). (Figure adapted from the article cited in the footnote 30, with permission)

However, this has not happened: the current female marathon record is 2:14:04, set by Brigid Kosgei, in the Chicago Marathon, on October 13, 2019, while the current male record is 2:01:39, set by Eliud Kipchoge, in the Berlin Marathon, on September 16, 2018.

Considering the main hormonal differences between men and women, explain the faster times recorded by men.

2. Performance in different types of sports is determined by a number of factors, including the characteristics of the contractile apparatus. The maximum speed in which the muscle can contract as well as the maximum force and power produced by the muscle is highly dependent on the fiber type composition. Muscle fibers are classified according to the expression profile of myosin heavy chain isoforms in type 1 and type 2 (see Sect. 10.2). The distribution of the fiber types in a particular muscle is determined genetically, but some factors, such as training, age, and sex, can also influence in the proportion of fiber types in the muscles.

A study published in 2019 compared the fiber type composition of the *vastus lateralis* muscle in men and women. Analyses of muscles from 114 males and 101 females showed a higher number of type 1 fibers in women than in men (63.4% vs. 51.8%, respectively); naturally, the opposite occurred regarding the type 2 fibers (36.6% vs. 48.2%).[31] Considering only fiber type composition as a variable, suggest the type of exercise that would be better performed by each sex, taking into account the properties of each type of muscle fiber.

---

[31] Jeon Y et al (2019) Sex- and fiber-type-related contractile properties in human single muscle fiber. J Exerc Rehabil 15:537–545.

3. The use of performance-enhancing drugs (PED), also known as doping, by athletes has become increasingly common. In endurance sports, like marathons,
   long-distance runs, and cycling, the most used PEDs are testosterone and EPO
   (see also Challenging Case 10.2). However, sympathomimetic drugs, such as
   amphetamines and cocaine, are also used illegally to enhance performance, as
   occurred in May 2013, when the American marathon runner Mary Akor tested
   positive for clenbuterol.[32]

   Search in the literature the mechanism of action of clenbuterol and discuss
   what would be the benefit for the performance of a marathon runner to use it.

## Biochemical Insights

1. Testosterone, the main male sex hormone, is responsible for important characteristics that increase men's performance. This hormone induces the expression of
   insulin-like growth factor (IGF-1) and growth hormone, increasing proteins synthesis and decreasing proteolysis, leading to muscle hypertrophy.[33] This provides
   men's a better performance in terms strength.

   Especially important for the performance in endurance sports such as the
   marathon, testosterone is known to stimulate erythropoiesis in different ways.[34]
   Directly, this hormone induces the production of blood cells by the bone marrow
   and stimulates glycolytic metabolism and incorporation of iron by erythrocytes.
   Indirectly, it induces the proliferation of renal cells and, consequently, the
   increase in erythropoietin production and secretion, inducing the production of
   erythrocytes (see Challenging Case 10.2). In addition, men's left ventricle tends
   to be 20 to 25% larger than that of women. This makes men cardiac output to be
   higher and, consequently, their $VO_2$ as well. These factors together make tissue
   oxygenation more efficient in men, providing a greater aerobic capacity than in
   women.

   Estrogen is the primary female sex hormone. Although this hormone is produced in both men and women, its levels are significantly higher in women of
   reproductive age. One of the estrogen effects is to increase fat stores in the
   body[35]; one would hypothesize that this hormone makes women more resistant
   to long-term aerobic sports than men. However, since estrogen acts on almost all
   the organs, affecting metabolism in different ways, such suggestion would be a

---

[32] Douglas S. Prolific marathoner Mary Akor gets 2-Year doping ban. Runner's News. https://www.
runnersworld.com/races-places/a20822805/prolific-marathoner-mary-akor-gets-2-year-doping-
ban/. Accessed 18 Dec 2013.

[33] Fink J et al (2018) The role of hormones in muscle hypertrophy. Phys Sportsmed 46:129–134.

[34] Shahani S et al (2009) Androgens and erythropoiesis: past and present. J Endocrinol Investig
32:704–716.

[35] Lee MJ, Fried SK (2017) Sex-dependent depot differences in adipose tissue development and
function: role of sex steroids. J Obes Metab Syndr 26:172–180.

little hasty and it indeed has not been proved. It is known that estrogen reduces bone resorption, thus increasing bone formation, but it also reduces the stiffness of tendons and ligaments, making women more susceptible to develop injuries.[36] Additionally, it is important to note that the estrogen levels in women vary ten- to 100-fold depending on the period of the menstrual cycle, while in menopause, the hormone concentrations are constantly low.

In conclusion, even though women continue to improve their performance in sports, the hormonal differences between sexes may explain why men still perform better than women in marathons.

2. Type 1 fibers are rich in mitochondria and myoglobin and are irrigated by a large number of blood vessels, conferring to these fibers an oxidative profile. They are adapted to prolonged work, being highly resistant to fatigue. Type 2 fibers are adapted to work at low levels of $O_2$. They contract quickly and powerfully but are more susceptible to fatigue due to the low ATP yield provided by anaerobic metabolism.

Thus, based only on fiber composition, one would expect that women would be favored in long-distance or endurance runs, soccer, or cycling, while men would excel as sprinters, bodybuilders, and weightlifters. Indeed, type 1 fibers are usually found in abundance in elite endurance athletes, while the amount of type 2 fibers is proportionally higher in elite strength and power athletes.[37] However, although muscle fiber type interconversion upon training is still a controversial issue, several studies support that a careful manipulation of exercise variables may shift type 1 to type 2 fibers and vice versa.

3. Sympathomimetic drugs are compounds that stimulate or cause the effect of catecholamines on the sympathetic nervous system, mimicking adrenaline action. Their action is classified as indirect, when they increase the presynaptic release of catecholamines or inhibit their reuptake, or direct, when they specifically bind to adrenergic receptors, acting as agonists of either α- or β-adrenergic receptors (see the types of adrenergic receptors in Sect. 10.5.1.1).

Clenbuterol acts as an agonist of the β2-adrenergic receptors, causing positive chronotropic, and inotropic cardiac effects, i.e., it increases the frequency and strength of cardiac contraction, enhancing cardiac output and accelerating blood circulation. In addition, it acts as a potent bronchodilator, increasing the breathing capacity. Together, these actions increase the respiratory capacity and oxygen distribution to tissues. A third and important action of clenbuterol is to induce the secretion of both insulin and glucagon, strongly increasing glycogenolysis and glucose uptake by tissues, favoring muscle activity.

---

[36] Chidi-Ogbolu N, Baar K (2019) Effect of estrogen on musculoskeletal performance and injury risk. Front Physiol 9:1834.

[37] Wilson JM (2012) The effects of endurance, strength, and power training on muscle fiber type shifting. J Strength Cond Res 26:1724–1729.

## *Final Discussion*

The study published by Whipp and Ward in 1992 titled "Will women soon outrun men?" cited in question 1, predicted that women would outperform men in marathons in 1998, which did not happen.

Actually, the interpretation given by the authors on the data they collected had been shown to be easily refutable, as discussed, for instance, in an article published in the scientific periodical *The Journal of Physiology* by Prof Michael J. Joyner, Distinguished Investigator from the Mayo Clinic, USA[38]: *"One interesting caveat about this paper was the extent to which the authors might or might not have been pulling an elaborate spoof on the scientific community in a high profile journal like Nature to illustrate the limits of linear regression and causal inferences."*

Interestingly, Whipp and Ward themselves end their article posing questions that, in the end, reveal the flaws in their analysis, namely:

"Why is the world record progression in the various events so linear over an interval of approximately a century?"
*"Why is the slope of the record progression so similar from the sprints to the 10,000 m?"*
"Why is the record progression in the marathon appreciably greater?"
"Why are the record progressions for women increasing in such a rapid rate relatively to men?"

In his article, Prof. Joyner discusses some points that disprove, at least in part, some of these questions. An interesting point raised by him was that the evolution of the records depends not only on the physiological limiting factors associated with endurance performance but also on historical and sociological factors, such as increased training intensity and volume, globalization, more competitive opportunities, and professionalization. This results in a performance progress profile that is not linear, as claimed by Whipp and Ward, but nearly hyperbolic, in which physiological and social components would contribute to the evolution of the marathon records' times, as represented in the next figure.

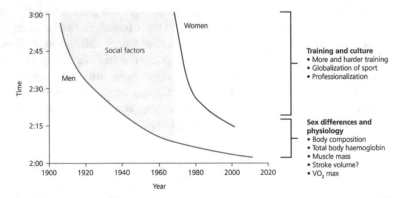

Progression of world best marathon times for men and women over the last century, pointing out the social and physiological factors that explain sex differences in human endurance exercise performance. (Figure reproduced with permission from the article cited in the footnote 38)

---

[38] Joyner MJ (2017) Physiological limits to endurance exercise performance: influence of sex. J Physiol 595:2949–2954.

Women's performance was influenced by those social factors in a much shorter time frame (about 15 years) than occurred for men, for whom these factors were gradually introduced during a century of evolution in sports. This explains the rapid initial rate of women's records progression when compared to that of men.

On the other hand, the physiological limitations are based on sex characteristics and are difficult to overcome. In this context, the main physiological difference between women and men that affects performance achievements and records in elite endurance competitions is $VO_{2max}$. Women's $VO_{2max}$ values are, on average, approximately 10% lower than those observed for men. This seems to be caused by a combination of higher body fat in women and lower red cell mass for a given body weight.

Nevertheless, even if women do not reach the records of men, both sexes are evolving in sport and it should not be underestimated to what extent the limits of the human body can be overcome. In this sense, Prof Joyner ends his article with a nice reflection about his personal experience[38]: *"On a personal note, starting in the late 1970s and into the 1980s I was lucky enough to participate in distance running as a college undergraduate and into medical school and beyond. During this time, distance running became widely open to women and I participated on successful co-ed teams. In this context, it was an incredible privilege to have a 'front row seat' as my female teammates made what was once thought to be dangerous and impossible, routine. The larger lesson is that it is a grave mistake to underestimate or limit any individual or group of people on the basis of superficial preconceived notions and biases. I am forever grateful to these women for teaching me this lesson."*

Prof. Joyner's plot on the progression of marathon records also raises an interesting reflection: if the profile of the records' curve is hyperbolic limit, there is an absolute universal mark that will never be surpassed. The asymptotic limit of the curve would be this absolute physiological limit, which, according to the figure would be about 2 h for men and impossible to determine yet for women.

*This case had the collaboration of Dr. Lorena O. Fernandes-Siqueira (Instituto de Bioquímica Médica Leopoldo de Meis, Universidade Federal do Rio de Janeiro, Brazil).*

# Chapter 11
# Control of Body Weight and the Modern Metabolic Diseases

The increase in the body weight in the world population has become one of the most important public health problems, as it is a major risk factor for several pathologies such as cardiovascular diseases and diabetes.

A systematic analysis evaluating the worldwide changes in the body mass index (BMI, defined as the body mass divided by the square of the height) was performed by the Global Burden of Metabolic Risk of Chronic Diseases Collaborating Group. They used health examination surveys and epidemiological studies from 199 countries and territories in the world, including 9.1 million participants, and showed that the mean worldwide BMI increased by $0.4$ $kg/m^2$ per decade for men and $0.5$ $kg/m^2$ per decade for women. Figure 11.1 shows the difference over 28 years of the percentage of the obese (BMI $\geq 30$ $kg/m^2$) or overweight (BMI $\geq 25$ $kg/m^2$) people, in distinct areas of the globe. It is clear that although there are some differences among the regions, the increase in body weight seems to be a global phenomenon.

Body weight is a result of the balance between food intake and energy expenditure. Several peripheral mediators secreted by different cells in the body act on the central nervous system, influencing the feeding behavior by controlling appetite or satiety, as well as regulating body energy expenditure by changing the metabolic rate and controlling thermogenesis. In this chapter, we will discuss the mechanisms by which the body weight is controlled as well as the main proposals on how the impairment of this control may cause the modern metabolic diseases.

© Springer Nature Switzerland AG 2021
A. T. Da Poian, M. A. R. B., *Integrative Human Biochemistry*,
https://doi.org/10.1007/978-3-030-48740-9_11

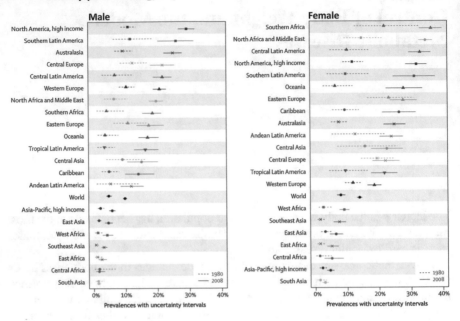

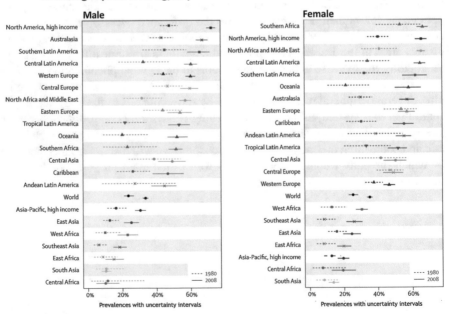

**Fig. 11.1** Prevalence of obesity (**a**) or overweight (**b**) in 1980 and 2008 among males (*left*) and females (*right*) in different areas of the world. (Figure reproduced from Finucane et al. Lancet 377:557–567, 2011, with permission from Elsevier)

## 11.1   Humoral Control of Food Ingestion

Although eating is a complex behavior that involves distinct areas of the brain, including the sensory areas that process the information of food taste, smell, and appearance, and the cortex, which is responsible for the psychological component of the appetite and satiety, the hypothalamus may be seen as a central player in the control of feeding.

The hypothalamus is located below the thalamus, just above the brain stem, in the ventral part of the diencephalon (Fig. 11.2). Anatomically, it is divided in several regions; the following ones are directly involved in the control of energy balance: arcuate nucleus (ARC), ventromedial hypothalamus (VMH), dorsomedial hypothalamus (DMH), paraventricular nucleus (PVN), and lateral hypothalamus (LH) (Fig. 11.2).

The hypothalamus induces anorexigenic (appetite-suppressing) or orexigenic (appetite-stimulating) behaviors by sensing several substances that are secreted by different cells in the body as a response to food intake or to the increase in adiposity. For instance, leptin, insulin, cholecystokinin (CCK), and peptide YY (PYY) provide the anorexigenic signals, while ghrelin is the main orexigenic mediator, as it will be discussed in the next sections.

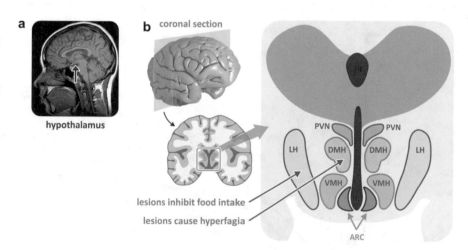

**Fig. 11.2** (**a**) Magnetic resonance image of the human brain showing the localization of the hypothalamus. (Figure reproduced from the free media repository Wikimedia Commons). (**b**) Human brain section showing the anatomical localization of the hypothalamic regions: *III* third ventricle, *ARC* arcuate nucleus, *VMH* ventromedial hypothalamus, *DMH* dorsomedial hypothalamus, *PVN* paraventricular nucleus, *LH* lateral hypothalamus. The *blue arrow* indicates the area whose lesion caused hyperphagia, and the *orange arrow* indicates the area whose lesion caused the inhibition of food intake

### 11.1.1   A Historical Perspective of the Role of Hypothalamus in Food Intake

The first evidence showing that the hypothalamus controls the body energy balance came from a number of studies conducted in the 1940s. These experiments showed that lesions in areas of the medial basal hypothalamus caused either hyperphagia and obesity or anorexia, depending on the specific region that was lesioned (Fig. 11.2).

More than just revealing its role in the control of food ingestion, these experiments led to the hypothesis that the hypothalamus would act as a sensor of the feeding status of the organism, stimulating or inhibiting food intake depending on the amount of the energy stored in the body. A question that immediately raised was how this sensing mechanism would operate. An attractive hypothesis was that it occurred through the circulating metabolites detected by the hypothalamus.

A common approach to investigate the participation of a circulating mediator in a given phenomenon was through parabiosis experiments. In these experiments, the blood vessels of two animals were surgically connected to allow the exchange of circulating factors between them. In a classical experiment performed in 1959, parabiotic pairs of rats were subjected to a lesion in the hypothalamus of one of the animals. While the lesioned rats became obese, as expected, the unlesioned partners stopped eating and lost weight, indicating that a circulating factor produced by the obese animals, for which the lesion turned the animal unresponsive, inhibited food intake in the normal ones (Fig. 11.3).

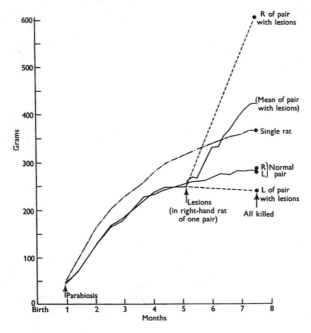

**Fig. 11.3** Growth curves of parabiotic rats. When the animals were 5-month-old, lesions were made in the hypothalamus of the right-hand member of the parabiotic pair. The last point in the curve corresponds to the individual body weights after death and separation of the parabiotic pairs. The *dashed lines* indicate the presumed body weight of each member of the pair based on the weight at the end of the experiment. The growth curve of a single rat is also shown for comparison. (Figure reproduced with permission from Hervey, J. Physiol. 145:336–352, 1959)

Other parabiosis experiments were performed in the 1950s taking advantage of the description of two spontaneous recessive mutations in mice that result in a phenotype of extreme obesity caused by hyperphagia. The two mutated genes were called *ob* (from the obese phenotype; Fig. 11.4) and *db* (from diabetic, and also obese, phenotype).

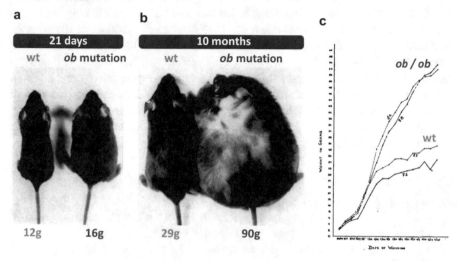

**Fig. 11.4** Photographs and growth curves reproduced from the original article that described the phenotype of the animal with the spontaneous mutation in the *ob* gene. (**a**) Normal and *ob/ob* mice at 21 days of age. (**b**) Normal and *ob/ob* mice at 10 months of age. (**c**) Growth curves of two normal and two *ob/ob* mice from birth to 4 months of age. (Figures reproduced with permission from Ingalls et al. J. Hered. 41:317–318, 1950)

When a homozygotic mouse for a mutation in the *ob* gene (*ob/ob* mouse) was joined through parabiosis to a normal mouse, the obese animal stopped eating in excess and tended to normalize its weight. This suggested that the product of the *ob* gene, which is defective in the *ob/ob* mouse, would be the circulating factor that signalizes to the hypothalamus the excess of body weight, leading to the inhibition of food intake. In contrast, when a *db/db* mouse was connected to a normal mouse, only the normal animal lost weight, suggesting that the mutation caused a failure in the response to the "obesity" factor, which acts in the normal mouse inhibiting food ingestion (Fig. 11.5).

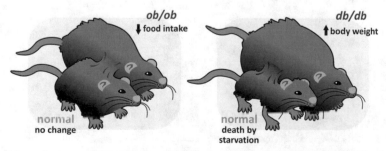

**Fig. 11.5** Schematic representation of the parabiosis experiments with *ob/ob* and *db/db* mice. When an *ob/ob* mouse was connected to a normal mouse, the obese animal stopped eating in excess and lost weight, while no changes occurred in the normal mouse. When a *db/db* mouse was connected to a normal one, only the normal mouse lost weight

## 11.1.2　Leptin: A Hormone Indicative of Adiposity

The parabiosis experiments with the *ob/ob* and *db/db* mice described in the previous section indicated that the amount of adipose tissue accumulated in the body is regulated by the endocrine system. However, the hormone (the circulating factor) responsible for the transmission of the information of the increase in body weight to the hypothalamus remained unknown until 1994, when the product of the *ob* gene was finally discovered. More interestingly, it was found that this protein is expressed almost exclusively in the adipose tissue. This protein was named leptin (from the Greek word *leptos*, which means thin), and its discovery represented a revolution in the study of obesity both due to the expectations it raised on the pharmaceutical industry and because it revealed a complete new function for the adipose tissue, whose vision changed dramatically from an inert tissue of energy storage to an endocrine organ (Box 11.1).

---

**Box 11.1　The Adipose Tissue as an Endocrine Organ**

The discovery of leptin opened the way to the identification of a series of proteins secreted by the adipose tissue, which were named adipokines (see below table). These proteins act as hormones or cytokines and are mainly involved in the regulation of energy metabolism or in inflammation. This made the white adipose tissue recognized as a dynamic endocrine organ instead of only a lipid storage tissue.

**Active proteins secreted by the adipose tissue**

| Name | Molecular nature | Main features and functions |
|---|---|---|
| Leptin | 16 kDa protein | Expressed mainly in adipocytes. Controls energy balance (see main text) |
| Adiponectin | 30 kDa protein that forms multimeric complexes | Expressed exclusively in adipocytes. Blood levels inversely correlate to adiposity. Increases lipid catabolism and insulin sensitivity |
| Resistin | 75 kDa protein | Expressed mainly in adipocytes. Secreted during adipogenesis. Increases insulin resistance |
| Visfatin | 52 kDa protein | Extracellular isoform of the enzyme nicotinamide phosphoribosyltransferase of the NAD biosynthesis pathway. Stimulates insulin secretion |
| Apelin | 55 amino acid precursor that generates several active fragments | Expressed in many cell types including adipocytes. Increases cardiac contractility and lowers blood pressure |
| Retinol binding protein-4 (RBP-4) | 20 kDa protein | Expressed mainly in hepatocytes and adipocytes. Transports retinol in the blood but also decreases insulin sensitivity |

(continued)

**Box 11.1**  (continued)

| Name | Molecular nature | Main features and functions |
|---|---|---|
| Tumor necrosis factor α (TNF-α) | 17 kDa protein that forms soluble trimers | Pro-inflammatory cytokine secreted mainly by macrophages but also by many other cells including adipocytes. Triggers the inflammatory response and regulates cell death but also increases insulin resistance (see Sect. 11.3.1) |
| Interleucine 6 (IL-6) | 24 kDa protein | Produced by many cell types. Acts as a pro-inflammatory cytokine |
| Monocyte chemotactic protein-1 (MCP-1) | 13 kDa protein | Produced by many cell types. Acts on the inflammatory process but also impairs insulin signaling in skeletal muscle cells |
| Plasminogen activator inhibitor-1 (PAI-1) | 47 kDa protein | Expressed in endothelium but also in other cell types such as adipocytes. Inhibits the proteases involved in the degradation of blood clots |

Immediately after the discovery of leptin, massive investments were made in the development of new treatments that envisaged the elimination of obesity by simply administrating leptin to obese people. However, as high as the expectations were the disappointments when it became clear that in humans the mutations in the leptin gene accounted for a very small number of obesity cases and the simple administration of leptin in the remaining obesity cases did not lead to weight loss. However, despite the failure of using leptin to treat obesity, it soon became clear that the plasma levels of leptin directly correlated with the BMI and to the percentage of fat in the body (Fig. 11.6).

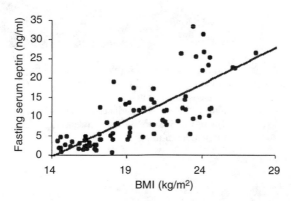

**Fig. 11.6**  Correlation of plasma leptin concentrations with BMI in a sample of 41 obese children (aged 6–9 years old) and the same number of non-obese children (control group), matched by age and sex. (Figure reproduced with permission from Valle et al. Int. J. Obes. Relat. Metab. Disord. 27:13–18, 2003)

Leptin is a small protein of 167 amino acid residues forming an α-helical structure that resembles the structure of some cytokines (Fig. 11.7). Leptin effects are
mediated by its binding to specific receptors expressed by neurons in the central
nervous system, the inhibition of appetite being mainly due to its binding to receptors of the hypothalamic neurons (see also Sect. 11.1.5).

The leptin receptor is a homodimeric transmembrane protein that is constitutively associated to the enzyme Janus kinase 2 (JAK2, Box 11.2). Leptin binding
promotes conformational changes in the receptor that mediate an increase in the
tyrosine kinase activity of JAK2, resulting in its autophosphorylation and in the
phosphorylation of different intracellular targets, leading to the activation of different signaling pathways, including the JAK/STAT3, the phosphatidylinositol-3-
kinase (PI3K), and mitogen-activated protein kinase (MAPK) pathways
(Fig. 11.7).

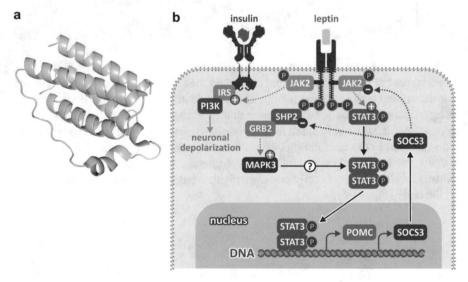

**Fig. 11.7** (**a**) Leptin structure. (**b**) Leptin signaling pathway. Leptin binding to its receptors in the
hypothalamus leads to the activation of JAK2 and to the recruitment of STAT3, IRS, and SHP,
which become active. STAT3 migrates to the nucleus, inducing the expression of POMC and
SOCS3, the feedback inhibitor of the pathway. IRS mediates the activation of PI3K, which induces
neuronal depolarization involved in the rapid responses independent on gene expression. Activation
of MAPK pathway is mediated by SHP recruitment

**Box 11.2  The Origin of the Janus Kinase Name**

In the Roman mythology, Janus (in Latin, *Ianus*) is the god of the gateways, beginnings, and transitions, who guarded the door of Heaven. He is represented as having two faces looking in opposite directions (see figure with  Janus representation in an old Roman coin), meaning that he can see the future and the past simultaneously. The first month of the year is named after him, as in January we look back at the last year and forward to the next. The reference to Janus in JAK's name is related to the fact that the members of this family of kinases possess two near-identical phosphate-transferring domains, one with kinase activity and the other that negatively regulates the kinase activity of the first.

Janus head represented in a Roman coin dated of 225–214 BC. (Figure reproduced from the free media repository Wikimedia Commons)

JAK/STAT3 is the predominant pathway that mediates leptin effects. When STAT3 is phosphorylated by JAK2, it migrates to the nucleus, where it regulates the expression of several genes. Among them are those encoding some important neuropeptides involved in the appetite control, such as neuropeptide Y (NPY), which has its expression inhibited, and the pro-opiomelanocortin (POMC), corticotropin-releasing hormone (CRH), and the cocaine- and amphetamine-regulated transcript (CART), which have their expression induced (see Sect. 11.1.5 for details on the role of these neuropeptides in the appetite control). STAT3 also induces the expression of feedback inhibitors of the signaling pathway, such as SOCS3 (suppressor of cytokine signaling 3) (Fig. 11.7).

The active JAK2 also recruits proteins from the family of the insulin receptor substrates (IRS), which mediate the activation of the PI3K pathway (for more details, see Sect. 8.4.2). Thus, there is a cross talk between the signaling pathways triggered by leptin and insulin at the hypothalamic level, so that the leptin action on the hypothalamus is positively modulated by insulin and vice versa. PI3K regulates the electrophysiological properties of the hypothalamic neurons, modulating the release of neurotransmitters in the synapses, which also contribute to the reduction of feeding. It is important to mention that the blood concentration of both insulin and leptin are increased in response to food ingestion. Since the hypothalamic neurons are also rich in insulin receptors, both leptin and insulin contribute to the triggering of the anorexigenic responses after feeding.

In summary, the action of leptin on the hypothalamic neurons alters the gene expression pattern in these cells and causes cell depolarization, resulting in the activation or inhibition of the secretion of different neuropeptides (see Sect. 11.1.5). It is important to mention that leptin effects can be seen as part of the long-term effects on the control of the body weight. Leptin does not promote an abrupt interruption in food ingestion but influences the amount of food intake and its relationship with the body energy expenditure over time. There are other hormones and mediators that directly promote the postprandial satiety or trigger appetite. These short-term mediators are mainly secreted by the gastrointestinal tract, as discussed in the next sections.

### 11.1.3   Intestinal Peptides: Triggers of Postprandial Satiety

The satiety feeling just after a meal is a result of several signals, mainly arising from the gastrointestinal tract, that act on the brain, providing information regarding the quality and the quantity of food ingested. These signals are involved in the short-term regulation of food ingestion, especially the control of the meal size.

The first experiments suggesting the role of the intestine in promoting satiety were performed using a surgical procedure previously developed by Ivan P. Pavlov in his classic experiments of the decade of the 1890s, which revealed the role of the central nervous system in the regulation of gastric secretion in dogs (see Box 11.3). This procedure consists of establishing an esophagic fistula in animals, so that when they eat, the food is tasted and swallowed but does not accumulate in the stomach and does not pass into the small intestine. This is usually referred as sham feeding (Box 11.3).

---

**Box 11.3  Pavlov and the Sham Feeding Experiments**

Ivan Petrovich Pavlov was a Russian scientist whose research on the physiology of digestion led him to be awarded the Nobel Prize in Physiology or Medicine in 1904. Pavlov developed a surgical method that establishes fistulas in various organs, enabling the continuous observation of their functions. This opened a new era in the development of physiology, whose procedures until then were based essentially on vivisection methods, which allowed only an instantaneous picture of the analyzed process. The classical Pavlov's experiments were performed on sham-fed dogs, to which a gastric fistula was applied (see below figure). Within a few minutes after the beginning of sham feeding, gastric juice begins to flow without ceasing for hours.

Pavlov's esophagostomy associated to the application of a gastric fistula used in the sham-feeding experiments

Pavlov demonstrated that when the vagus nerves were lesioned, secretion of gastric juice during sham feeding was absent, showing the reflex nature of the first phase of gastric juice secretion.

A number of experiments performed in the 1970s showed that when the animals were sham-fed, satiety did not occur, revealing that the satiety feeling originates from the passage of food through the gastrointestinal tract. These experiments allowed the subsequent identification of several intestinal peptides involved in triggering satiety and whose plasma levels rapidly increase after a meal (Fig. 11.8). In this section we will comment on the secretion profile and the action of three of these peptides: cholecystokinin (CCK), glucagon-like peptide-1 (GLP-1), and peptide YY (PYY).

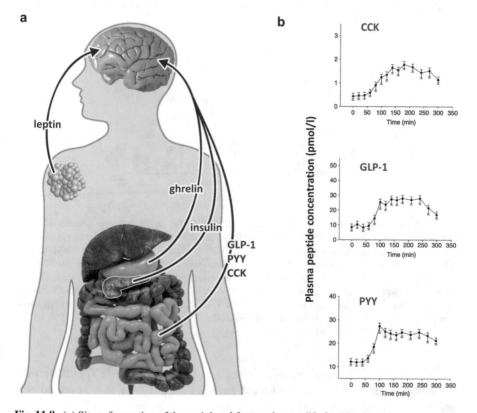

**Fig. 11.8** (a) Sites of secretion of the peripheral factors that modify food intake and energy expenditure through direct effects on the brain: ghrelin is secreted by the stomach prior to the meals, inducing food intake (Sect. 11.1.4); CCK, GLP-1, and PYY are secreted by the intestine just after the meals, interrupting eating; leptin is secreted by the adipose tissue as a signal of adiposity (Sect. 11.1.2); and insulin is secreted by the pancreas in response to nutrient ingestion (Sect. 8.4). (b) Plasma concentrations of intestinal peptides in response to breakfast ingestion, which occurred 60 min after the beginning of the measurements. Data are from 12 subjects, presented as mean ± SEM. *CCK* cholecystokinin, *GLP-1* glucagon-like peptide-1, *PYY* peptide tyrosine tyrosine. (Figure reproduced with permission from Vidarsdottir et al. Eur. J. Endocrinol. 162:75–83, 2010)

CCK is secreted by the I cells, mainly located in the proximal duodenum, suppressing food intake and decreasing meal size. CCK was one of the first gastrointestinal peptides to be demonstrated as an important trigger of the postprandial satiety. Experiments in the 1970s using sham-fed rats, which ate continuously due to the absence of intestinal stimulation, showed that the intraperitoneal injection of CCK suppressed feeding in a dose-dependent manner (Fig. 11.9).

Long-chain fatty acids and proteins are particularly effective in inducing CCK secretion, although carbohydrate-rich meals also promote an increase in CCK plasma levels. CCK seems to act on the brain through a paracrine stimulation of the vagal afferent nerve fibers, whose terminals are proximal of the intestinal I cells, suggesting that the CCK plasma levels are not the best indicator of its potential action.

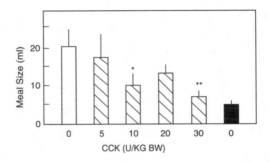

**Fig. 11.9** Result of an experiment in which sham intake of a liquid diet was quantified in rats ($n = 8$) upon intraperitoneal injection of CCK in the concentrations indicated in the figure (*hatched bars*), with the "0" (*white bar*) corresponding to saline injection. Animals in which the gastric fistula was maintained closed were used as control (*black bar*). The animals were submitted to food deprivation for a period of 3 h followed by food ingestion until satiety. The total volume of food intake was quantified for each animal. *Vertical lines* are the standard errors with * corresponding to $p = 0.05$ and ** corresponding to $p < 0.01$ for CCK vs. saline. (Figure reproduced with permission from Kraly et al. J. Comp. Physiol. Psychol. 92:697–707, 1978. Copyright © 1978 by the American Psychological Association)

The L cells of the distal intestine secrete both GLP-1 and PYY, although the regulation of the release of each of these peptides differs. PYY is a 34 amino acid residue peptide with structural homology to the neuropeptide Y (NPY) and the pancreatic polypeptide (PP). PYY is named after its tyrosine content: five residues, including one at both the N- and C-terminals. Its secretion is triggered by the direct contact of the L cells with the nutrients, as well as in response to duodenal lipids. This latter stimulus is probably mediated by CCK. PYY binds to Y receptors in ARC neurons inhibiting NPY release and thus causing anorexigenic effects (see Sect. 11.1.5). GLP-1 is a peptide derived from the pro-glucagon precursor, and its secretion is stimulated especially after a carbohydrate-rich meal, although fats are also potent secretagogues (i.e., a substance that promotes secretion). A body of evidence suggests that, as CCK, GLP-1 reduces food intake and promotes satiety by

acting on vagal afferent nerve terminals close to the L cells. The satiety promoted
by the increase in secretion of both GLP-1 and PYY seems to be the main reason for
the success of bariatric surgeries, currently considered the most effective surgical
treatment for morbid obesity (see Box 11.4).

**Box 11.4  Bariatric Surgery**

Bariatric surgery is a weight loss procedure based on the removal of a portion
of the stomach (gastrectomy), the reduction of its size through the introduction of a medical device (gastric banding), or the creation of a small stomach
pouch which is connected to the intestine skipping the duodenum (gastric
bypass). It has become the treatment of choice for individuals with severe
obesity, being recommended for people with BMI > 40 or BMI > 35 when
serious coexisting medical conditions occur. Bariatric surgery may be classified as purely restrictive (as in the case of the adjustable gastric band) to
mostly malabsorptive (as occurs in the biliopancreatic diversion), which usually results in clinical complications related to malabsorption of macronutri-
ents. The surgical approaches predominantly used today are the Roux-en-Y
gastric bypass and the adjustable gastric band (see below figure). The Roux-
en-Y gastric bypass produces better results in terms of weight loss, probably
because it is associated to an increase of the plasma concentrations of the
satiety peptides PPY and GLP-1 (see below figure). More recently, a variation
of this procedure was proposed, in which a gastric fundus resection is combined to the bypass. This promotes a more consistent decrease in ghrelin
secretion together with an increase in PPY and GLP-1 release, which makes
the response even more effective.

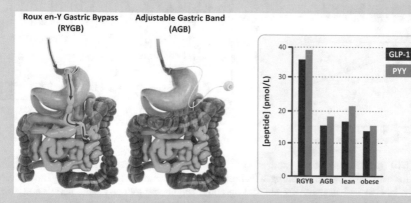

Schematic representation of the predominantly used bariatric surgery: the Roux-en-Y gastric bypass (*left*) and the adjustable gastric band (*right*) and changes in plasma concentrations of GLP-1 and PYY 1 h after breakfast in lean (*n* = 15) or obese (*n* = 12) subjects or
patients submitted to RYGB (*n* = 6) or GB (*n* = 6). (Based on data from the study described
in le-Roux et al. Ann. Surg. 243:108–114, 2006)

## 11.1.4    Ghrelin: The Main Orexigenic Hormone

Ghrelin was discovered in 1999 as an endogenous ligand of the growth hormone secretagogue receptor, from which its name is derived. Indeed, one of the actions of ghrelin is to stimulate growth hormone release, but its role in the control of appetite seems to be independent of this previously identified function.

The active form of ghrelin is an acylated peptide of 28 amino acid residues, secreted mainly by the stomach (Fig. 11.10a). Ghrelin concentration in blood increases during fasting and reaches its lowest level just after a meal, indicating a physiological role for this hormone in meal initiation (Fig. 11.10b). Measurements of plasma ghrelin levels in gastrectomized patients confirmed that the stomach is a major source of circulating ghrelin in humans (Fig. 11.10b). Indeed, the fact that ghrelin is an important signal to start feeding may explain the decrease in feeding behavior after gastrectomy (see Box 11.4).

Many studies have shown that ghrelin administration into the brain ventricle in rodents induces food intake through the increase of the expression of hypothalamic NPY, an opposite effect of that observed for leptin. In agreement, the relationship of blood ghrelin concentrations and BMI is the opposite of that observed for leptin: plasma ghrelin concentrations inversely correlate with BMI (Fig. 11.10c).

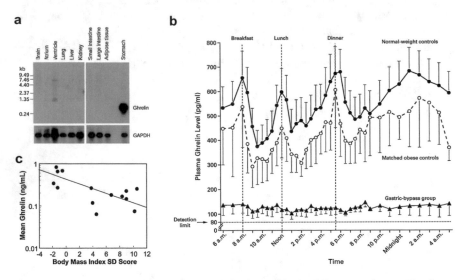

**Fig. 11.10** (**a**) Analysis of the expression of ghrelin mRNA in different rat tissues. The figure shows the result of a northern blotting, a technique in which the mRNA for a specific gene is radioactively labeled and detected by autoradiography. (Figure reproduced with permission from Kojima et al. Nature 402:656–660, 1999) (**b**) Mean 24-h plasma ghrelin concentration during a day in ten normal-weight human subjects, five obese subjects who underwent a proximal Roux-en-Y gastric bypass and five obese subjects who had recently lost weight by dieting and were matched to the subjects in the gastric bypass group according to final body mass index, age, and sex. The *dashed lines* indicate breakfast, lunch, and dinner. (Figure reproduced with permission from Cummings et al. N. Engl. J. Med. 346:1623–1630, 2002) (**c**) Relationship between the mean 24-h plasma ghrelin concentration and BMI in five lean and nine obese girls. (Figure reproduced with permission from Foster et al. Pediatr. Res. 62:731–734, 2007)

Ghrelin receptor, the growth hormone secretagogue receptor (GHSR), belongs to the class of G protein-coupled receptors (for more details on G protein-coupled receptors, see Sect. 10.5.1). GHSR may be coupled to different types of G protein, including type $G_q$, which mediates changes in ion currents in the neurons through the activation of phosphatidylinositol-specific phospholipase C. Phospholipase C hydrolyzes the phosphatidylinositol 4,5-bisphosphate ($PIP_2$) in inositol 1,4,5-triphosphate ($IP_3$) and diacylglycerol (DAG), which induce $Ca^{2+}$ release from ER and the activation of protein kinase C (PKC), leading to the inhibition of $K^+$ channels and causing neuronal depolarization (Fig. 11.11). Additionally, GHSR, through G proteins type $G_q$ or $G_i$, mediates the activation of the MAPK or PI3K pathways, leading to changes in neuronal gene expression, which includes the induction of expression of the orexigenic NPY (see next section) (Fig. 11.11).

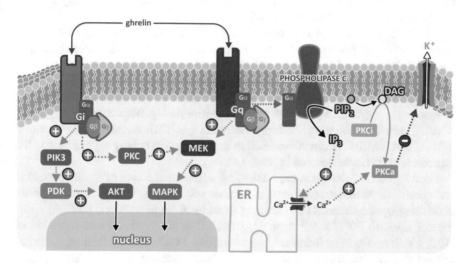

**Fig. 11.11** The binding of ghrelin to GHSR results in the release of GDP and binding of GTP to the G protein α-subunit coupled to the receptor. Neuronal depolarization is triggered by ghrelin binding to GHSR coupled to G protein type $G_q$, which activates PLC, generating $IP_3$ and DAG, resulting in an increase in the intracellular $Ca^{2+}$ concentration and the activation of PKC, which inhibits the $K^+$ channels. Changes in the neuronal gene expression are achieved through ghrelin binding to GHSR coupled to G protein type $G_q$ or $G_i$, which activates MAPK or PI3K pathways

## 11.1.5 The Arcuate Nucleus and the Melanocortin System

The main target of the peripheral mediators in the hypothalamus is the arcuate nucleus (ARC). ARC is a region of the medial basal hypothalamus adjacent to the third ventricle (see Fig. 11.2). The blood–brain barrier in this region seems to be more permeable, facilitating the contact of the ARC neurons with humoral and metabolic factors coming from the peripheral circulation, including the anorexigenic and the orexigenic factors discussed in the previous sections.

Two subsets of ARC neurons play a central role in the regulation of energy balance: the NPY/AgRP neurons and the POMC neurons (Fig. 11.12). It is important to mention that the name of these neuron populations is derived from the name of the neuropeptides secreted by them, which, in turn, were named as a reference to an initial function described for them, usually unrelated to their role in the energy balance control. Thus, although we will point out the origin of these peptides' nomenclature, the names themselves will not be useful for the understanding of their functions in the energy balance control.

The NPY/AgRP neurons express NPY, which stimulates food ingestion. These neurons also express the agouti-related protein (AgRP), a protein related to fur pigmentation in rodents that when mutated causes obesity in mice.

The POMC neurons express pro-opiomelanocortin (POMC), whose cleavage originates several endocrine- and neuroendocrine-active peptides, including the anorexigenic peptides $\alpha$- and $\beta$-melanocyte-stimulating hormones ($\alpha$-MSH and $\beta$-MSH) and the adrenocorticotropic hormone (ACTH). Among these peptides, $\alpha$-MSH and $\beta$-MSH are directly related to the increase in energy expenditure in the organism, as it will be detailed in Sect. 11.2.

Food ingestion behavior is established by a concerted action of the anorexigenic and orexigenic hormones on these two neuron populations. During fasting, ghrelin is secreted by the stomach cells leading to an increase in its blood level. In ARC, ghrelin binds to NPY/AgRP neurons, stimulating the synthesis and secretion of NPY, which results in an increase in appetite (Fig. 11.12). During feeding, gastrointestinal peptides, such as PYY, are released from the intestine cells. PYY also acts on NPY/AgRP neurons, but its binding to these neurons inhibits NPY release, diminishing appetite. As a response to the increase of adiposity, the adipocytes secrete leptin, which binds to both NPY/AgRP and POMC neurons. The effects of leptin binding to each of these neurons are antagonic: it inhibits the NPY/AgRP neurons and stimulates POMC neurons, stopping food ingestion by a simultaneous decrease in the NPY-induced appetite and increase in satiety promoted by anorexigenic POMC-derived peptides (Fig. 11.12).

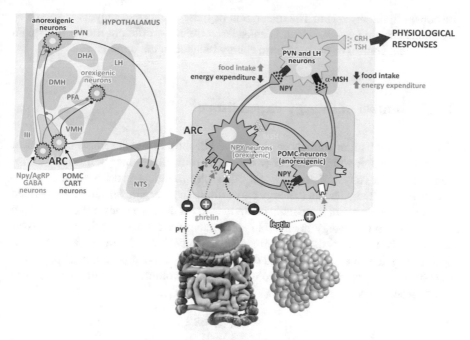

**Fig. 11.12** Schematic representation of the subpopulation of the ARC neurons: the NPY/AgRP and the POMC neurons. PYY and leptin inhibit NPY/AgRP neurons, decreasing appetite by the reduction of NPY release. Leptin activates POMC neurons, leading to the release of POMC-derived peptides, such as α-MSH, which bind to melanocortin receptors (such as MC4R) in other hypothalamic areas (mainly in the PVN and LH), promoting satiety. Ghrelin stimulates NPY/AgRP neurons, inducing the expression and secretion of NPY and AgRP, which are antagonists of melanocortin receptors, triggering hunger and decreasing energy expenditure

The next question to be posed is how the ghrelin and leptin signals are deciphered by the hypothalamus producing their respective physiological responses: the changes in the feeding behavior and the energy expenditure rate in the body. The answer resides in the fact that both NPY/AgRP and POMC neurons are also connected to neurons in other hypothalamic areas (mainly in the PVN and LH; Fig. 11.12, see also Fig. 11.2), which mediate the recognition of the feeding status to a systemic response in the organism. The PVN and LH neurons express the melanocortin receptors (MCR1–MCR5), whose agonists are the POMC-derived peptides and the antagonist is NPY. Upon stimulation by POMC-derived peptides, PVN and LH neurons secrete the corticotropin-releasing hormone (CRH) and thyrotropin-releasing hormone (TRH), which mediate anorexigenic, thermogenic responses, and even physical activity (behavioral) responses.

In summary, the orexigenic and anorexigenic signals (such as ghrelin and leptin) bind to the ARC neurons leading to the secretion of POMC-derived peptides or NPY. These peptides act on the PVN and LH neurons, stimulating or inhibiting the release of CRH and TRH, which in turn regulate the anorexigenic and thermogenic responses in the organism. The integration of these events and players is known as the melanocortin system, which is considered the most efficient model to explain the neuronal control of the long-term energy balance in the organism.

## 11.2 Control of Energy Expenditure

Energy metabolism in animals can be defined in short as the ensemble of chemical processes through which the energy obtained from food—energy intake—is converted to work (physical activity and other physiological processes) and heat—energy expenditure. The total energy expenditure in an organism can be subdivided for didactic purposes in three main components: (a) the basal metabolic rate, which corresponds to the energy involved in all metabolic reactions required for the cellular functions; (b) the adaptive thermogenesis, which consists in the production of heat in response to cold or diet; and (c) the physical activity, both the spontaneous, such as that necessary to maintain posture, and the voluntary, including exercise in sports, leisure, and other activities (Fig. 11.13).

**Fig. 11.13** The three components of energy expenditure: the basal metabolic rate (BMR), represented in *blue*, corresponds to energy involved in the metabolic reactions necessary to maintain cellular functions; the adaptive thermogenesis, represented in *green*, corresponds to the regulated energy release in the form of heat as a response to changes in ambient temperature or diet; and physical activity, represented in *red*, corresponds to the energy expenditure during voluntary or involuntary physical activities

### 11.2.1 Adaptive Thermogenesis

The adaptive thermogenesis is one of the components of energy expenditure and can be defined as the production of heat due to a regulated increase in the metabolic rate. It is controlled by the brain, both through the stimulation of the sympathetic system and through the hypothalamus–pituitary–thyroid axis, as a response to triggering signals including cold exposure and food intake (Fig. 11.14).

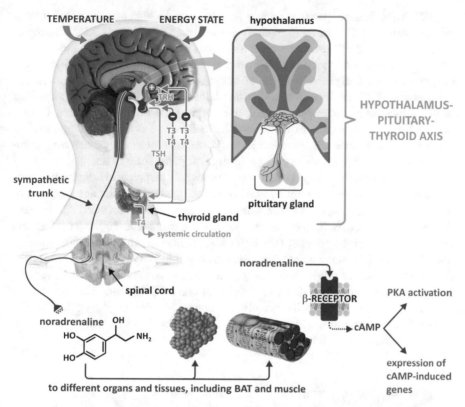

**Fig. 11.14** Changes in the ambient temperature and in diet are sensed by the brain resulting in noradrenaline release by the sympathetic nerves and in the activation of hypothalamus–pituitary–thyroid axis. Noradrenaline acts on brown adipose tissue (BAT) and muscles through its binding to adrenergic receptors, which leads to an increase in intracellular AMPc concentration and the consequent modulation of the activities of some PKA-regulated enzymes, as well as the induction of the expression of the AMPc-induced genes. Thyroid hormones (T3 and T4) play a major role in controlling energy expenditure (see Sect. 11.2.2). They are released by the thyroid gland as a response to the thyroid-stimulating hormone (TSH) produced by pituitary, whose secretion is stimulated by the hypothalamic thyrotropin-releasing hormone (TRH). T3 and T4 exert a negative feedback control over the hypothalamus and pituitary

### 11.2.1.1   Cold-Induced Thermogenesis

Shivering is one of the most primitive responses induced by exposure to cold, which is especially important to the adaptive thermogenesis in adult humans and large mammals. However, in human newborns and other small mammals, the most important cold-induced adaptive thermogenesis involves the release of noradrenaline by the sympathetic nerve terminals, particularly within the brown adipose tissue (BAT).

BAT thermogenic capacity is mainly sustained by the high activity of the uncoupling protein-1 (UCP1 or thermogenin) in this tissue (see Sect. 6.2.5). Binding of noradrenaline to BAT adrenergic receptors leads to an increase in the intracellular cAMP concentration, which in turn mediates the modulation of enzymatic activities, as well as the expression of the cAMP-induced genes (Fig. 11.14), including the gene for UCP1 and those that codify the deiodinase 2 (D2), the enzyme involved in the intracellular production of the active form of the thyroid hormone, T3 (see Sect. 11.2.2). A synergic effect between T3-dependent and cAMP-dependent actions, which includes the increase in fatty acid oxidation associated with mitochondria uncoupling, results in energy dissipation in the form of heat.

For many years, the role of BAT in the thermogenic response in humans has been considered to be relevant only in newborns due to the insignificant amounts of brown fat found in adults. However, a number of new findings revealed that in addition to the classical BAT, there is an inducible thermogenic adipose tissue, also referred to as beige adipocytes, which are interspersed in white fat depots in adult humans. The origin of the beige adipocytes is not completely clear, but it seems that they have a different origin than that of the classical brown fat cells. Conversely, strong evidence supports a common origin for classical BAT and skeletal muscle cells. Interestingly, muscle cells recently also appear as potential players in the cold- and diet-induced thermogenic responses, with a mechanism involving a controlled cycling of calcium (see Box 11.5).

---

**Box 11.5 Thermogenic Control by the Skeletal Muscle Cells**

Exposure to cold induces norepinephrine (NE) release by the sympathetic nerve terminals within the skeletal muscle. NE binding to β-adrenergic receptors in muscle cells mediates the activation of the sarcoplasmic reticulum calcium ATPase (SERCA). Two proteins are known to associate to SERCA in muscle, phospholamban and a recently identified protein named sarcolipin. Sarcolipin uncouples SERCA-mediated ATP hydrolysis from $Ca^{2+}$ pumping, resulting in the dissipation of energy in the form of heat. Simultaneously, NE induces the "browning" of the adipose tissue in a combined action with at least two other muscle-derived peptides: irisin and natriuretic peptide, which are secreted by skeletal and cardiac muscles, respectively. One of the main responses that lead to the "browning" of the adipocytes is the expression of UCP1 in the mitochondria, which induces heat production (see Sect. 6.2.5).

(continued)

**Box 11.5**  (continued)

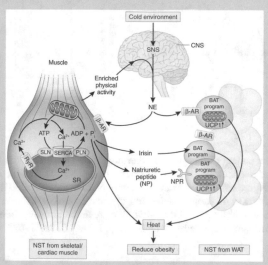

Schematic representation of the thermogenic responses in the muscle and in the adipose tissue. *CSN* central nervous system, *SNS* sympathetic nervous system, *NE* norepinephrine, *β-AR* β-adrenergic receptor, *SERCA* sarcoplasmic reticulum Ca²⁺-ATPase, *PLN* phospholamban, *SLN* sarcolipin, *RyR* ryanodine receptor, *SR* sarcoplasmic reticulum, *BAT* brown adipose tissue, *NST* non-shivering. (Figure reproduced by permission from Macmillan Publishers Ltd., from Kozac and Young, Nat. Med. 18:1458–1459, 2012)

### 11.2.1.2   Diet-Induced Thermogenesis

Adaptive thermogenesis also occurs in response to diet. The diet-induced thermogenesis can be defined as the increase in energy expenditure above the basal fasting level as a response to food intake. This phenomenon is illustrated in Fig. 11.15a, which represents the mean pattern of the diet-induced thermogenesis throughout a day. Indeed, feeding increases the metabolic rate by 25–40%, while starvation reduces it to up to 40%, and these compensatory changes in metabolism may be the explanation for the low efficacy of treatments for obesity.

This can be exemplified by a study in which the energy expenditure was measured in obese and non-obese volunteers submitted to overfeeding or underfeeding diets that resulted in controlled gain or loss of weight, respectively. The results of this study, shown in Fig. 11.15b, revealed that a 10% increase or 10% decrease in body weight was associated to compensatory changes in energy expenditure in both the obese and the non-obese groups of individuals.

The mechanisms through which food intake affects the metabolic rate are not completely clear, but they seem to involve the "browning" of the adipose tissue (see previous topic) and the melanocortin system described in Sect. 11.1.5.

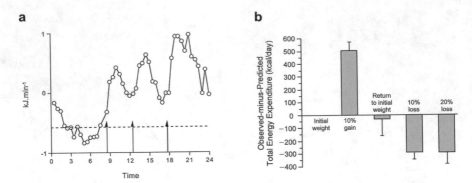

**Fig. 11.15** (a) Mean pattern of diet-induced thermogenesis measured throughout the day in 37 subjects. The *dashed line* indicates the level of basal metabolic rate, and the *arrows* indicate the time of the meals. (Figure reproduced from Westerterp Nutr. Metab. 1:5, 2004) (b) Changes in the 24-h energy expenditure of 41 subjects subjected to controlled diets that resulted in 10% weight gain or 10% or 20% weight loss, as well as when they returned to their original weights. (Figure reproduced with permission from Leibel et al. N. Engl. J. Med. 332:621–628, 1995)

## 11.2.2   Role of Thyroid Hormones

The thyroid hormones thyroxine (T4) and triiodothyronine (T3) are iodinated hormones produced in the thyroid gland that play a major role in the control of energy expenditure by mediating the increase in the metabolic rate. This can be clearly illustrated by the fact that patients with hyperthyroidism show an up to 50% increase in the total body energy expenditure, while in severe hypothyroidism, it can fall by as much as 50%.

The thyroid gland is a butterfly-shaped endocrine organ composed of two lobes, located on the anterior side of the neck (Fig. 11.16a). The name thyroid comes from the Greek word *thyreos*, which means shield, in a reference of its position around the larynx and trachea. Histological observation of the thyroid shows spherical follicles formed by a single layer of polarized epithelial follicular cells (Fig. 11.16b). The basolateral membrane of the follicular cells is in contact with the bloodstream from where iodide is taken up through a sodium/iodide co-transporter (NIS, from $Na^+/I^-$ symporter) (Fig. 11.16c). This mechanism makes the iodide concentration inside the thyroid 20- to 50-fold higher than in the blood. The follicular cells surround a region called follicular lumen, which is filled with a colloidal substance composed mainly by a large glycoprotein named thyroglobulin. Around the follicles, the parafollicular cells form the thyroid parenchyma, which produce the hormone calcitonin, involved in calcium homeostasis.

The first step of thyroid hormone synthesis is called the "organification" of iodine, which consists in the iodide oxidation followed by its incorporation to tyrosine residues of the thyroglobulin (Fig. 11.16c). Thyroperoxidase (TPO), an enzyme located on the apical membrane of the follicular cells, catalyzes iodide oxidation using hydrogen peroxide as the oxidizing agent, and the subsequent iodination of thyroglobulin, generating either a mono-iodinated tyrosine (MIT) or di-iodinated tyrosine (DIT). The next step in the synthesis of thyroid hormones is also catalyzed by TPO and consists of the coupling of two neighboring iodotyrosyl residues through the formation of an ether bond between the iodophenol part of a donor iodotyrosyl and the hydroxyl group of the acceptor iodotyrosyl residue (Fig. 11.16c). The cleavage of the iodophenol group of the tyrosyl donor forms an alanine side chain that remains in the thyroglobulin polypeptide chain as dehydroalanine. The coupling of two DITs generates T4, whereas when a DIT and a MIT are coupled, the product is T3.

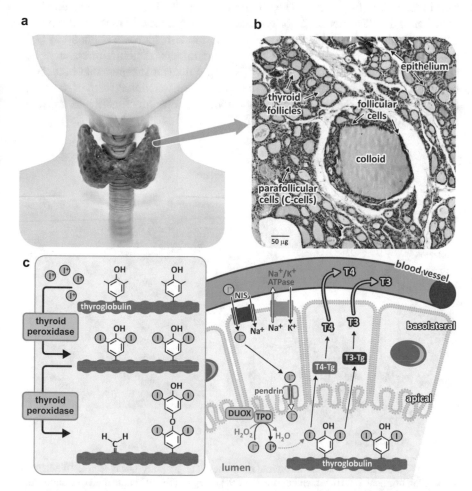

**Fig. 11.16** (**a**) Anatomy of the thyroid gland. (**b**) Histological image of thyroid follicles showing the follicular cells, the colloid, and the parafollicular cells. (**c**) Iodine organification and the synthesis of thyroid hormones. The follicular cells take up iodide from the blood through the transporter NIS (Na⁺/I⁻ symporter) located on their basolateral membranes. The activity of the Na⁺/K⁺-ATPase on the follicular cell membrane maintains a low intracellular concentration of Na⁺. At the apical membrane, the enzyme thyroperoxidase (TPO) oxidizes the iodide and incorporates the iodine to tyrosine residues of thyroglobulin in the lumen of the follicles. Iodinated tyrosines are conjugated forming T3 or T4 (*inset*)

The secretion of the thyroid hormones depends on the endocytosis of the iodin-ated thyroglobulin from the colloid, which is stimulated by the thyroid-stimulating hormone (TSH) released from the pituitary gland. Thyroglobulin is then digested by lysosomal proteases, and the thyroid hormones are released in the bloodstream, where they circulate bound to specific binding proteins, mainly the thyroxine-binding globulin (TBG) and transthyretin (formerly known as thyroxine-binding prealbumin) but also by albumin. After transport into the cells, the thyroid hor-mones bind to nuclear receptors (THR) that belong to the large superfamily of the steroid receptors (Fig. 11.17). The affinity of T3 for THR is about 100-fold higher than that of T4. THR acts as a hormone-activated transcription factor that regulates gene expression through binding to specific hormone-responsive elements in DNA (Fig. 11.17). THR can bind to DNA either as monomers, as homodimers, or as het-erodimers with the retinoid X receptor (RXR), a nuclear receptor that binds 9-cis retinoic acid. The heterodimer seems to be the major functional form of the recep-tor. Additionally, in the absence of the hormone, the receptors function as transcrip-tion inhibitors, repressing the expression of some genes.

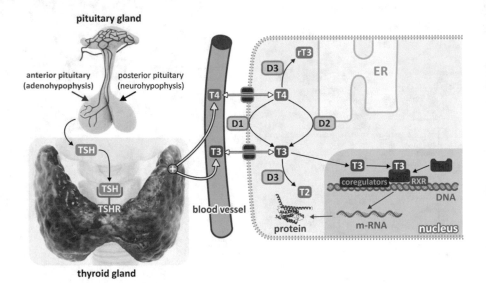

**Fig. 11.17** Mechanism of action of thyroid hormones. The pituitary gland secretes the thyroid-stimulating hormone (TSH), which binds to the TSH receptor (TSHR) in the thyroid gland, stimu-lating the release of T4 (in higher amounts) and T3. Circulating T4 can be converted to T3 on the surface of some cells, especially liver and kidney cells, in a reaction catalyzed by the type 1 deio-dinase (D1). T3 and T4 are transported into the cell. At the membrane of ER, T4 is converted to T3 through the action of the type 2 deiodinase (D2). Both T3 and T4 can also be inactivated by type 3 deiodinase (D3), which converts T4 to the inactive reverse T3 (rT3) or T3 to the inactive diiodo-thyronine (T2). T3 binds to the thyroid hormone receptors (THR), and the complex regulates the expression of specific genes by binding to DNA either as monomers, as homodimers, or as het-erodimers with the retinoid X receptor (RXR)

It is important to bear in mind that although T3 is more biologically active than T4 due to its higher affinity to THR, the major coupling reaction occurring inside the thyroid is the one that generates T4 (it is estimated that the thyroid gland secretes about 80 μg T4 and only 5 μg T3 each day). Thus, a critical step for the thyroid hormones action is the conversion of T4 into T3, which occurs mainly in the peripheral tissues. This reaction is catalyzed by the deiodinases, enzymes that remove one iodine atom from the outer ring of T4. There are three isoforms of deiodinases (Fig. 11.17). Type 1 deiodinase (D1) is expressed on the outer face of the plasma membrane of liver and kidney cells, being responsible for the production of T3 in the serum. Type 2 deiodinase (D2) is an intracellular isoform of the enzyme, expressed on the membrane of the endoplasmic reticulum, mainly in cardiac and skeletal muscles, adipose tissue, the central nervous system, and thyroid and pituitary glands. D2 is responsible for the control of cytoplasmic levels of T3. Type 3 deiodinase (D3) inactivates the thyroid hormones by catalyzing the removal of an iodine atom from the inner ring, which converts T4 to the inactive reverse T3 or T3 to inactive T2.

Thyroid hormones regulate a broad range of physiological processes including growth, development, and energy metabolism. The effects of thyroid hormones on energy metabolism can be clearly recognized in cases of hypo- or hyperthyroidism. Almost all patients with hyperthyroidism exhibit weight loss even under a high food intake diet, whereas weight gain is a very usual condition of hypothyroidism patients. This happens because the general effect of the thyroid hormones is to increase the metabolic rate by simultaneously activating, directly or indirectly, the metabolic pathways that consume and produce ATP. In this context, thyroid hormones are also critical for thermogenesis since the heat that is automatically released from the metabolic reactions is consequently increased.

The regulation of the metabolic rate by thyroid hormones occurs through different mechanisms depending on the tissue (Fig. 11.18). In muscle, their action includes the upregulation of the expression and the activity of $Na^+/K^+$-ATPase and sarcoplasmic reticulum $Ca^{2+}$-ATPase, increasing ATP consumption by these enzymes (see Sect. 10.3). Consequently, the cellular respiratory activity is increased to maintain the ATP levels. In BAT, thyroid hormones enhance noradrenaline signaling pathway as well as the expression of UCP1, sustaining the thermogenic effect mediated by the sympathetic nervous system (see Sect. 11.2.1).

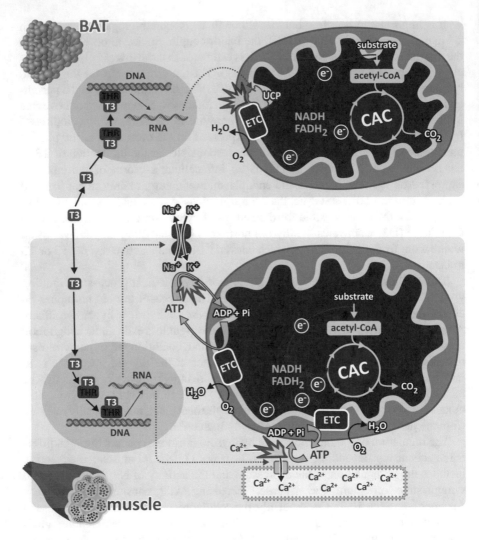

**Fig. 11.18** Main metabolic effects of thyroid hormones in the peripheral tissues. Thyroid hormones induce the expression of UCP1 in BAT, increasing cellular respiratory activity, and of Na$^+$/K$^+$-ATPase and Ca$^{2+}$-ATPase in muscle, increasing ATP consumption by these enzymes. These effects accelerate the metabolic rate and lead to thermogenesis. The *dashed arrows* indicate the enzymes whose expression is induced by the thyroid hormones, and the *fire* represents the thermogenic effect

In addition to these actions on the peripheral tissues, it is now evident that the metabolic effects of the thyroid hormones are also mediated by their action on the central nervous system, specifically on hypothalamic areas such as ARC, VMH, and PVH (Fig. 11.19). In ARC, T3 upregulates the expression of the neuropeptides AgRP and NPY and downregulates the expression of POMC, leading to hyperphagia (see Sect. 11.1.5). Simultaneously, T3 inhibits AMPK in VMH, which activates the sympathetic nervous system leading to the enhancement of BAT thermogenic program (see Sect. 11.2.1). Thus, these combined effects of thyroid hormones on hypothalamic regions promote an increase both in food intake and energy expenditure. Additionally, it is also evident that the central action of thyroid hormones also controls glucose metabolism in different tissues, although the mechanism through which this occurs are still unknown.

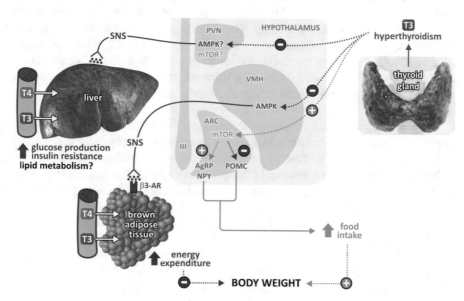

**Fig. 11.19** Central actions of thyroid hormones. In ARC, T3 regulates feeding through mammalian target of rapamycin (mTOR), which upregulates the expression of AgRP and NPY and downregulates the expression of POMC, increasing food intake. In VMH, T3 inhibits AMPK, activating the sympathetic nervous system and the thermogenic program in BAT. Additionally, T3 action on PVH is probably involved in the control of hepatic glucose homeostasis

## 11.3   Obesity and the Metabolic Syndrome

Obesity is caused by a massive accumulation of adipose tissue, which is primarily a result of an imbalance between energy intake and energy expenditure. The increase in adipose mass is strongly associated with the development of insulin resistance and type 2 diabetes, which together with hyperlipidemia, glucose intolerance, and hypertension, characterize a clinical entity sometimes referred to as metabolic syndrome.

A crucial clue for the understanding of the metabolic syndrome was the discovery that obesity itself causes an inflammatory state in the metabolic tissues, which in turn is tightly linked to the development of insulin resistance. In this section, we will discuss this metabolic-triggered inflammation, which has been termed metaflammation, a chronic, low-grade inflammatory response that seems to be initiated by the excess of nutrients.

## 11.3.1    Chronic Inflammation and Insulin Resistance in Obesity

Adipose tissue occupies a central position in the development of the obesity-induced inflammation. The first observation that clearly linked obesity and chronic inflammation was that the pro-inflammatory cytokine tumor necrosis factor-α (TNF) is overexpressed and secreted in higher levels by adipocytes of obese individuals (Fig. 11.20a). After this first discovery, it became clear that in addition to TNF, several other cytokines and inflammatory mediators were produced in high levels by adipose and also by other metabolic tissues from obese individuals, generating a chronic inflammatory state. Additionally, immune cells such as macrophages are recruited to and infiltrate these inflamed tissues, amplifying the inflammatory response.

The connection between the chronic inflammatory state in obesity and the development of insulin resistance became apparent from the discovery that the signaling pathway triggered by TNF (and by other overexpressed cytokines) blocks insulin action downstream of the activation of its receptor (Fig. 11.20b). In summary, what occurs is the following. In response to the inflammatory signals, intracellular kinases, mainly JNK (c-jun N-terminal kinase) and IKK-β (inhibitor of nuclear factor-κB kinase-β), are activated. These kinases have as one of their substrates IRS (the insulin receptor substrate), which is phosphorylated by them in Ser residues instead of the Tyr residues that are phosphorylated in response to the insulin signaling pathway (see details of the insulin signaling pathway in Sect. 8.4.2). IRS phosphorylation in Ser residues impairs its phosphorylation in Tyr by the insulin receptor, leading to the inhibition of insulin-mediated cellular responses.

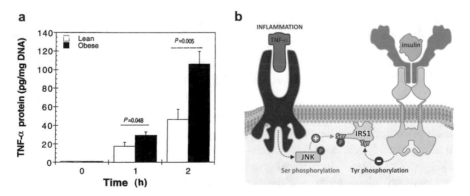

**Fig. 11.20** (**a**) Comparison of TNF secretion during 2 h by the same amount of adipocytes explanted from the adipose tissue from 18 lean and 19 obese female subjects. (Figure reproduced with permission from Hotamisligil et al. J Clin Invest 95:2409–2415, 1995) (**b**) Integration of inflammation and insulin signaling pathways. Binding of TNF to its cellular receptor triggers the activation of the kinases JNK and IKK-β, which catalyze the Ser phosphorylation of IRS. IRS phosphorylation in Ser residues blocks its phosphorylation in Tyr residues by the insulin receptor, inhibiting insulin action

At a first glance, it is intriguing that the control of metabolism would be so tightly linked to the inflammatory process. Under an evolutionary perspective, one can understand the development of the immune and the metabolic responses as basic requirements for survival. The efficient use of the available nutrients and the ability to successfully eliminate the pathogens have been essential features since the days of primitive organisms. Clues sustaining the coevolution of these processes can be found in nature, for example, in the case of the insects, in which the control of metabolic homeostasis and the immune responses are carried out by the same organ, the fat body (Box 11.6).

**Box 11.6   Coevolution of Immune and Metabolic Responses**
In insects, a single organ, the fat body, accumulates the functions that are carried out in mammals by the liver, the adipose tissue, and the immune cells (see below figure). The fat body is the largest organ in the insect body cavity, and it is the major site of intermediate metabolism in these organisms. It metabolizes and stores lipids, carbohydrates, and proteins; it is the target for the majority of the insect hormones and also synthesizes the hemolymph proteins. Additionally, the fat body cells express in high levels the innate immune receptors, proteins that recognize molecular patterns specific of pathogens. Through the activation of these receptors, the fat body produces and secretes a number of antimicrobial agents and regulators of cellular immune response. It is interesting to note that the metabolic organs of the more complex organisms contain resident immune cells, such as the Kupffer cells in the liver and the macrophages in the adipose tissue.

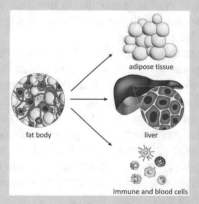

Organization of the metabolic and immune cells in insect and mammals. (Figure reproduced with permission from Hotamisligil, Nature 444:860–867, 2006)

Another remarkable evidence of the common evolutionary origin of the metabolic and immune responses is the fact that macrophages and adipocytes

(continued)

**Box 11.6** (continued)

show an extensive genetic and functional overlap, with several genes expressed equally by pre-adipocytes, adipocytes, and macrophages, including many inflammatory genes (see the following figure).

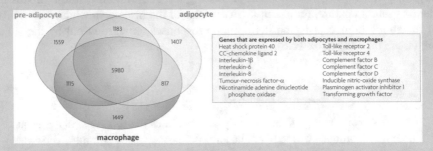

Transcriptional profile overlapping in pre-adipocytes, adipocytes and macrophages (*left*), and the list of the inflammatory genes expressed in both adipocytes and macrophages (*right*). (Figure reproduced with permission from Hotamisligil and Erbay, Nat Rev Immunol 8:923–934, 2008)

## 11.3.2   Origin of Inflammation in Obesity

The discovery that obesity is linked to a chronic inflammatory state raises the question of which are the triggers of inflammation in this case. This issue is now under intense investigation, but the obesity-related factors that initiate the inflammatory process have not been identified so far.

The classical inflammatory response is driven by the contact of the host cells with molecular components specifically present in pathogens (known as pathogen-associated molecular patterns, PAMPs) or exposed during tissue injury (the damage-associated molecular patterns, DAMPs). These molecules are recognized by cellular receptors, known as pattern recognition receptors (PRR), which have a central role in the innate immune response. PRR can be membrane-bound receptors, such as the Toll-like receptors (TLRs), which are located on the plasma or endosomal membranes, or cytoplasmic receptors, such as the NOD-like receptors (NLRs) and the RIG-I-like receptors (RLRs). The binding of PAMPs or DAMPs to the PRR triggers a chain of signaling events that result in the activation of a number of transcription factors, which induce the expression of several pro-inflammatory cytokines, including TNF-α, and antimicrobial molecules, such as interferons (Fig. 11.21).

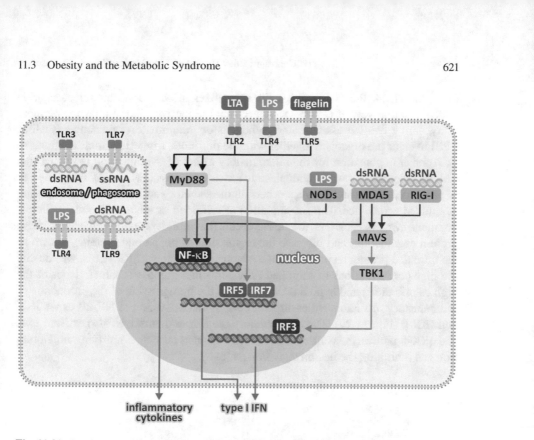

**Fig. 11.21** Sensing microbial molecular patterns by host innate immune system. Toll-like receptors (TLRs), NOD-like receptors (NLRs), RIG-like receptors (RLRs), and C-type lectin receptors (CLRs) are pattern recognition receptors (PRR) expressed on cellular membranes or in the cytoplasm of a variety of cells in host tissues. PPRs bind to PAMPs or DAMPs, including lipoteichoic acid (LTA), lipopolysaccharide (LPS), flagelin, double-stranded RNA (dsRNA), and single-stranded RNA (ssRNA). Ligand binding to PRR triggers different signaling pathways that culminate with the activation of transcription factors, such as the IRFs, AP-1, and NF-κB, inducing the expression of interferons α and β (IFNα, β) and several pro-inflammatory cytokines. They recognize different microbial components, such as cell walls, microbial (and modified host) nucleic acids, bacterial motor flagellin, or stress (danger)-induced molecules. The information of ligand binding to PRR is transmitted to the transcription factors through some adaptor proteins, including MyD88, TRAM, TRIF, or MAVS. Additionally, some NLRs can activate caspase-1, which cleaves the pro-IL-1β and pro-IL-18 in the active IL-1β and IL-18 that are released from the cell

The starting signal for the inflammatory response caused by overfeeding is still unclear. One hypothesis is that nutrients themselves would be recognized by innate immune system, although probably this occurs only when nutrient are in excess. This would happen if nutrients were not the preferential ligands of PRR, binding to these receptors with low affinity. During normal feeding the blood levels of nutrients are maintained low due to their rapid metabolization. Thus, very few nutrient molecules would bind PRR and trigger inflammation. Conversely, as nutrient concentration is maintained high enough to allow PRR occupation, the innate immune response is triggered, resulting in the establishment of the inflammatory state. Indeed, there is evidence that saturated fatty acids are recognized by one of the

PRR, the TLR4, the PRR that usually recognizes the bacterial lipopolysaccharide (LPS). Alternatively, it is also speculated that the increase in the intestine permeability during feeding allows some pathogens or inflammatory molecules (such as LPS) to enter the organism together with the nutrients, triggering inflammation.

Another explanation for the inflammatory effect of overnutrition is related to a cellular response known as endoplasmic reticulum stress (ER stress). ER is the site of protein synthesis and the place where all the secretory and membrane proteins are assembled and folded. The accumulation of unfolded or misfolded proteins in this organelle triggers ER stress, a series of events that inhibit protein synthesis, increase protein degradation, and increase the expression of chaperone proteins, known as the unfolded protein response (UPR). Among the triggers of ER stress, we can cite the high concentrations of saturated fatty acids and glucose (characteristics of the high caloric diets) and hypoxia. UPR leads to the upregulation of the production of inflammatory mediators either directly through the activation of NF-κB or via JNK and IKK-β (Fig. 11.22). The inflammatory mediators themselves also activate JNK and IKK-β pathways, as discussed in the previous sections, amplifying inflammation and inhibiting the insulin signaling pathway.

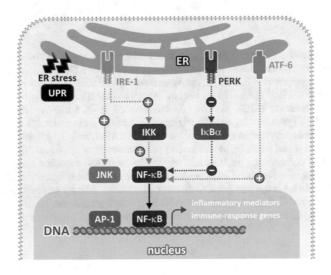

**Fig. 11.22** ER stress pathways leading to inflammation. Three components of ER membrane are implicated in the inflammatory response, IRE-1, PERK, and ATF-6. IRE-1 acts either through its association with TRAF2, which activates JNK and IKK, and consequently AP-1 and NF-κB, or through the splicing of XBP1 mRNA. PERK inhibits the translation of IκBα, an inhibitor of NF-κB, thereby increasing NF-κB transcriptional activity. ATF-6 also increases NF-κB transcriptional activity

# Selected Bibliography

Coleman DL (2010) A historical perspective on leptin. Nat Mcd 16:1097–1099

Dayan CM, Panicker V (2009) Novel insights into thyroid hormones from the study of common genetic variation. Nat Rev Endocrinol 5:211–218

Gao Q, Horvath TL (2007) Neurobiology of feeding and energy expenditure. Annu Rev Neurosci 30:367–398

Gregor MF, Hotamisligil GS (2011) Inflammatory mechanisms in obesity. Annu Rev Immunol 29:415–445

Hotamisligil GS, Erbay E (2008) Nutrient sensing and inflammation in metabolic diseases. Nat Rev Immunol 8:923–934

Ishii KJ, Koyama S, Nakagawa A, Coban C, Akira S (2008) Host innate immune receptors and beyond: making sense of microbial infections. Cell Host Microbe 3:352–363

Lago F, Dieguez C, Gómez-Reino J, Gualillo O (2007) Adipokines as emerging mediators of immune response and inflammation. Nat Clin Pract Rheumatol 3:716–724

López M, Alvarez CV, Nogueiras R, Diéguez C (2013) Energy balance regulation by thyroid hormones at central level. Trends Mol Med 19:418–427

Lowell BB, Spiegelman BM (2000) Towards a molecular understanding of adaptive thermogenesis. Nature 404:652–660

Moran TH, Dailey MJ (2011) Intestinal feedback signaling and satiety. Physiol Behav 105:77–81

Williams KW, Elmquist JK (2012) From neuroanatomy to behavior: central integration of peripheral signals regulating feeding behavior. Nat Neurosci 15:1350–1355

# Challenging Case 11.1: The Capsules of (H)eating Beauty

## Source

Public Health England campaign on the risks of DNP use. (Figure reproduced with permission of Public Health England)

In 2013, a case of 2,4-dinitrophenol (DNP) poisoning shocked the UK: the medical student Sarah Houston died at age 23 after being hospitalized due to excessive heat and discomfort. Sarah, who suffered of anorexia and bulimia at age 14, consumed, for 18 months, the slimming-aid compound DNP, which she had bought online. This case seemed to be just the beginning of a sad wave of DNP-induced deaths in the UK in the same year, including the 18-year-old rugby player Chris Mapletoft and the 28-year-old bodybuilder Sean Cleathero. In 2015, one more case was reported: 21-year-old student Eloise Parry died after taking eight DNP pills at one time. The major difficulty regarding the tragic consequences of DNP use is that its production cannot be prohibited as the compound is an efficient pesticide and is also used in the manufacture of other products. Therefore, the control of its misuse must occur through awareness programs. This led health professionals and researchers to warn: no dose is safe; DNP is a poison.[1]

## Case Description

Data from the World Health Organization (WHO) show that obesity is a growing problem worldwide, and today more than half of the world's population is overweight.[2] This is a very important public health problem since obesity is a major cause of metabolic disorders and cardiovascular disease. Many anti-obesity medications act as "anorectics," inhibiting appetite directly in the central nervous system (see details on the hypothalamic control of appetite in Sect. 11.1.5). However, the severity of the side effects of these medications made them banned in many countries. Today, the development of anti-obesity medications is focused on drugs called "thermogenics," whose mechanism of action is to make cells to consume their ener-

[1] Thomas S Deadly DNP "Public health matters", official blog of Public Health England. https://publichealthmatters.blog.gov.uk/2018/08/13/deadly-dnp/. Accessed 13 Aug 2018.

[2] https://www.who.int/gho/ncd/risk_factors/overweight_obesity/.

getic reserves not for ATP synthesis but for heat production instead. In contrast to modern interest in these drugs, the use of thermogenics for weight loss is quite old.

In the decade of 1910s, during the World War I, 2,4-dinitrophenol (DNP), an explosive compound used to produce munition, called attention for causing weight loss in people who had been exposed to it. This caused DNP to be widely prescribed as an anti-obesity drug, being used by over 100,000 people in the USA despite the adverse effects. At that time, munition factory workers died of DNP poisoning, but this was not enough to inhibit DNP use as a weight loss compound.

The onset of the Great Depression after the war marked a major transition in the beauty standards: being overweight was no longer admired and weight loss became a concern.[3] For this reason, a number of anti-obesity drugs were developed, such as Enjola, Dilex-Redusols, Dinitriso, Nitromet, and Dinitrenal, and in the 1930s many advertisements encouraged the use of these drugs as a strategy to lose weight (see the following figure).

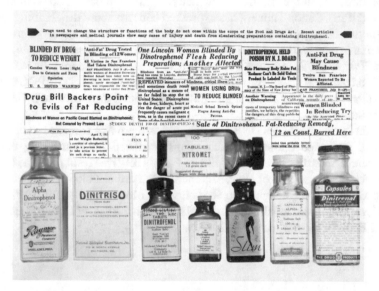

Advertisement of DNP-containing weight loss drugs in 1930s' newspapers. (Image from the US Food and Drug Administration's photostream, shared in The Commons)

In 1933, when researchers from the Stanford University proved DNP action as a weight loss drug and reported how this compound enhanced the metabolic rate,[4] its use increased substantially, including as a treatment for obese patients. However, in addition to weight loss, there were several reports of strong side effects, such as

---

[3] Scutts J The depression era's magic bullet for weight loss. The New Republic. https.//newrepublic.com/article/133751/depression-eras-magic-bullet-weight-loss. Accessed 27 May 2016.

[4] Tainter ML, Stockton AB, Cutting WC (1933) Use of dinitrophenol in obesity and related conditions. A progress report. JAMA 101:1472–1475.

vomiting, headaches, flushed skin, sweating, tachycardia, cataracts, and especially hyperthermia.[5] Not only that, deaths due to DNP consumption also increased during this period. This led DNP prescription for human use to be banned by US Food and Drug Administration (FDA) in 1938.

More recently, DNP has become popular among body-building practitioners since it promotes a huge fat loss while preserving muscle mass. This was the reason that led Lewis Brown, 25 years old, to take large numbers of DNP pills in 2015. Addicted to the gym, Brown attempted to lose fat, despite being aware of the danger of using DNP.[6] *"My body was just warming up and cooking from the inside out, I'm just lucky that I'm still alive"* he said. He went to the Addenbrooke's Hospital in Cambridge, where the doctors put him in coma for 3 days, covered in ice. Brown spent a week in intensive care, before doctors were forced to perform three surgeries to remove the entire muscle tissue at the front of his left leg. Brown was one of the few survivors of DNP poisoning.

The director of the Public Health England's National Poisons Information Service (NPIS), Simon Thomas, has expressed concern about the alarming increase in the notifications of DNP use in UK[7]: *"Until 2011, DNP was a very rare poison. To put that in perspective, we only saw three cases of DNP poisoning in the five-year period 2007 to 2011, but since 2012 … we've now heard of around a total of 115 cases."* Indeed, in 2015, the growing number of cases led the INTERPOL to issue a global alert for DNP as an illicit and potentially lethal drug used as a dieting and body-building aid.

The fact that in websites DNP was not sold as a product for human use, but as a pesticide, prevented the sellers from being penalized for distributing the drug. Nevertheless, in 2018, the efforts of Eloise Parry's relatives resulted in the first conviction for the illegal sale of DNP, sentencing Bernard Rebelo to manslaughter for selling DNP to their daughter their daughter.[6] Most recently, in 2019, Andrey Shepelev was arrested in Ukraine for involvement in the sale of DNP to the 21-year-old girl Beth Shipsey, who died in 2017.[8]

---

[5] Grundlingh J et al (2011) 2,4-dinitrophenol (DNP): a weight loss agent with significant acute toxicity and risk of death. J Med Toxicol 7:205–212.

[6] Father JA 25, who was desperate for a 'beach body' was given just hours to live after overdosing on diet pills DNP that 'cooked' his body from the inside. MailOnline. https://www.dailymail.co.uk/news/article-3381099/Father-nearly-died-overdosing-diet-pills-gain-beach-body.html. Accessed 1 Jan 2016.

[7] Tatum M DNP: the dangerous diet pill pharmacists should know about. Pharmaceut J. https://www.pharmaceutical-journal.com/news-and-analysis/features/dnp-the-dangerous-diet-pill-pharmacists-should-know-about/20206616.article?firstPass=false. Accessed 13 Jun 2019.

[8] Duffin C Man suspected of selling fatal diet pills to British woman who died after buying a £150 batch is held in Ukraine. The Daily Mail. https://www.dailymail.co.uk/news/article-6641063/Man-suspected-selling-fatal-diet-pills-British-woman-arrested-Ukraine.html. Accessed 28 Jan 2019.

## Questions

1. Considering the fact that DNP acts as an uncoupler of the electron transport system (see Sect. 6.2.5) explains why hyperthermia is the main side effect of this compound.
2. Adaptive thermogenesis is defined as the production of heat due to a regulated increase in the metabolic rate (see Sect. 11.2.1). The brown adipose tissue (BAT) plays an important role in adaptive thermogenesis, especially in newborns and animals that are naturally exposed to cold. Explain how BAT is involved in thermogenesis.
3. The "browning" of white adipose tissue (WAT), i.e., the appearance of brown-like adipocytes (white adipocytes in which UCP1 expression is induced) interspersed in WAT, has been recently observed in adult humans. These adipocytes, also referred to as beige adipocytes, emerge after certain stimuli, such as cold exposure, and would be seen as inducible thermogenic adipocytes (see Sect. 11.2.1). Intriguingly, in 2013, a group of researchers identified creatine metabolism as a signature of beige adipocytes' mitochondria.[9] Based on a series of findings described in the study, they hypothesized that when UCP1-dependent thermogenesis is reduced, beige adipocytes switch their thermogenic mechanisms to a creatine-driven substrate cycle, as depicted in the next figure.

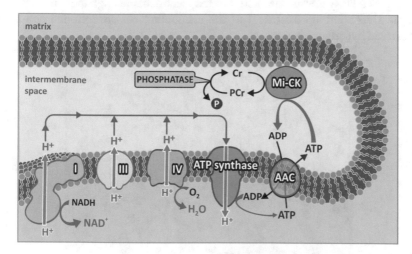

Schematic representation of the creatine-driven substrate cycle. Mitochondrial creatine kinase (Mi–CK) uses the ATP synthesized through oxidative phosphorylation to phosphorylate creatine (Cr), forming phosphocreatine (PCr) and ADP. Simultaneously, the PCr is hydrolyzed through the activity of a mitochondrial phosphatase expressed in beige adipocyte, regenerating Cr. ACC, ATP/ADP carrier. (Figure adapted from the article cited in the footnote 9)

---

[9] Kazak L (2015) A creatine-driven substrate cycle enhances energy expenditure and thermogenesis in beige fat. Cell 163:643–655.

Explain why this publication has drawn much attention from pharmacologists interested in developing anti-obesity drugs.

4. The use of the so-called fat burns or thermogenics has become very popular among practitioners of physical activity. These supplements are sold as a blend that combines a series of compounds that act by accelerating metabolism. The most commonly used compounds are caffeine, L-carnitine and yohimbine. Explain how these compounds may contribute to the loss of body fat and discuss whether they are appropriated classified as "thermogenics."

## Biochemical Insights

1. The effect of an uncoupler on oxidative phosphorylation is shown in Fig. 6.24. As the drug exemplified in this figure (carbonyl cyanide-p-trifluoromethoxyphenyl hydrazine, FCCP), DNP dissipates the proton gradient and the mitochondrial membrane potential by allowing the protons to pass through the inner mitochondrial membrane. As a consequence, the electron transport through the respiratory complexes in the inner mitochondrial membrane accelerates, increasing oxygen consumption, but the proton gradient cannot be maintained as the uncoupler moves the protons back into the matrix (see next figure). Thus, instead of being used for ATP synthesis, the energy of the electrochemical gradient is released as heat. Depending on the extent of the uncoupling, an uncontrolled and hardly reversed hyperthermia may occur.

Schematic representation of DNP uncoupling action. The phenol is a weak acid, thus capable of protonation and deprotonation depending on $H^+$ concentration. The hydrophobicity of the rest of the molecular structure allows the anion to be weakly lipid-soluble despite the charged group. Therefore, both the protonated and deprotonated forms are mobile in membranes, enabling DNP to disrupt the proton gradient through the inner mitochondrial membrane. This decreases ATP synthesis by ATP synthase and accelerates electron transport in the respiratory chain

2. The thermogenesis promoted by BAT is associated with the expression of the uncoupling protein 1 (UCP1, see Sect. 6.2.5.1), which, as DNP and the other artificial uncouplers, dissipates the mitochondrial proton gradient, accelerating respiration, which results in the conversion of stored energy (essentially lipids) into heat. Conversely to the action of the artificial uncouplers, UCP-induced heat production is tightly controlled through the regulation of its expression, in a process known as adaptive thermogenesis.

   Adaptive thermogenesis is triggered by different signals, including cold (temperature) or diet and physical activity (energy state), which activate the sympathetic system or hypothalamus-pituitary-thyroid axis, resulting in noradrenaline release (see Fig. 11.14). Binding of noradrenaline to BAT adrenergic receptors increases intracellular cAMP concentration, modulating the activity of PKA-regulated enzymes, as well as enhancing the expression of the AMPc-induced genes (see below figure). PKA phosphorylates the cAMP response element-binding protein (CREB), increasing the transcription of CREB-dependent genes, which includes UCP1 gene. Additionally, PKA-regulated enzymes include the hormone sensitive lipase (see Sect. 7.4.1), resulting in the release of free fatty acids from LDs. Since fatty acids are known to activate UCP1, lipolysis contributes to thermogenesis both providing the fuel as well as favoring uncoupling respiration.

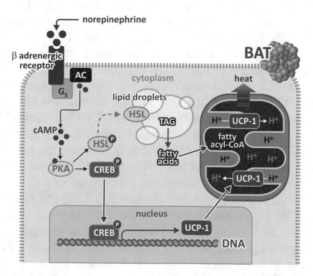

BAT-mediated adaptive thermogenesis. Noradrenaline binding to BAT adrenergic receptors increases intracellular cAMP concentration, leading to PKA activation. PKA phosphorylates CREB, which migrates to the nucleus, inducing UCP1 expression. PKA also phosphorylates hormone sensitive lipase (HSL), leading to the mobilization of TAGs from the lipid droplets. Fatty acids are oxidized in mitochondria, feeding the respiratory chain. The presence of UCP1 in the inner mitochondrial membrane dissipates the resulting H⁺ gradient, accelerating respiration and increasing heat production

3. The enzyme creatine kinase (CK) plays an important role during intense muscle contraction as a mechanism to generate ATP transiently and rapidly by transferring the phosphate group of phosphocreatine to ADP (see Sect. 10.3). CK can also catalyze the reversal reaction, the phosphorylation of creatine using ATP as the phosphate donor, forming phosphocreatine and ADP (see Fig. 10.12). If this latter reaction is coupled to mitochondrial respiration, it would work as futile substrate cycling, in which the rapid regeneration of ADP from the ATP produced in oxidative phosphorylation increases the respiratory rate, driving the oxidation of the stored lipids. The maintenance of this cycle requires the resultant phosphocreatine (PCr) to be hydrolyzed to creatine (Cr). Indeed, the study mentioned in the question correlated a mitochondrial phosphatase highly expressed in beige adipocytes with the creatine-driven substrate cycle. Thus, a compound that induces either mitochondrial phosphatase expression or creatine content in adipocytes would be efficient in increasing lipid oxidation, being an interesting candidate to be used as an anti-obesity drug.

4. "Fat burns" are widely used for losing weight. Usually, the recommendation is to consume these supplements just before physical activity since they would give an extra boost of energy to the physical activity.

Among the three most used "fat burns," the only one that actually acts as a thermogenic compound is caffeine. Caffeine is an antagonist of adenosine receptor, causing vasoconstriction and increasing heart rate, and also binds to other targets, such as the phosphodiesterases, GABA receptors, and ryanodine receptors (see also Challenging Case 4.1). In the context of thermogenesis, a recent study has shown that caffeine induces adipose tissue browning both in vitro and in vivo.[10] Using a model of adipocytes, the authors showed that caffeine treatment induced UCP1 expression and increased mitochondria biogenesis. Additionally, the mitochondria appear in tight contact with lipid droplets, suggesting an increase in lipid oxidation through the UCP1-rich mitochondria respiration. These in vitro findings correlate to the in vivo observations as the study also showed that drinking coffee significantly increased the temperature of the supraclavicular region, a region that co-localizes with brown adipose tissue in humans. Thus, caffeine can be considered a fat burn as well as a thermogenic.

L-carnitine is an amino acid derivate (see Fig. 7.10) that is essential for the transport of the long-chain fatty acid to the mitochondria, where β-oxidation takes place (see Sect. 7.4.3). Carnitine supplementation, especially in individuals whose endogenous production is low, increases lipid oxidation capacity, leading to weight loss. In this sense, carnitine classification as "fat burn" is appropriate, although it would not be considered a thermogenic compound.

Yohimbine is a compound extracted from the African tree *Pausinystalia johimbe* and the South American tree *Aspidosperma quebracho-blanco*, which has been used in the treatment of erectile dysfunction. It acts as a $\alpha_2$-adrenergic

---

[10]Velickovic K et al (2019) Caffeine exposure induces browning features in adipose tissue in vitro and in vivo. Sci Rep 9:9104.

antagonist, stimulating adrenaline release, thus enhancing the overall metabolic rate. Yohimbine has been used since the 1990s in the composition of fat burns, but its effectiveness as a fat burn compound was not proven when used separately. But even potentiating the oxidation of lipids when combined with other substances, one cannot classify its action as a thermogenic effect.

## Final Discussion

### Anti-obesity Medications: Pharmacological Strategies for Obesity Control

DNP can be considered one of the first drugs to be prescribed for weight loss. Before DNP introduction into the market, in the 1930s (see Case Description), the empirical observation that the ingestion of thyroid gland extract caused weight loss stimulated its prescription as a tool to lose weight. This occurred in the end of the nineteenth century, when a pioneer neurosurgeon, Victor Horsley, and his pupil, George Murray, showed that hypothyroidism could be treated with thyroid extracts.[11] That was the time of the so-called organotherapy, a medical practice inaugurated in 1889 by the Mauritian physiologist Brown-Séquard, who proposed that testicular extracts had a rejuvenating effect in man, testing this idea on himself by injecting extracts of monkey testis when he was 72.

After the first uses of thyroid extracts, many studies showed that thyroid hormone treatment was more efficient in promoting weight loss than dieting alone. However, there were significant negative consequences, such as the loss of mineral bone density as well as muscle protein without a comparable loss of body fat.[12]

Between the 1950s and 1970s, sympathomimetic drugs (i.e., drugs that mimic the effects of the catecholamines, the endogenous agonists of the sympathetic nervous system), such as amphetamines, amphetamine-like, and their derivatives, gained popularity to treat obesity.[13] These drugs act directly on the central nervous system by inhibiting appetite, being known as "anorectic drugs." Due to their direct action on hypothalamic receptors, they reduce food intake, but they also cause quite adverse side effects, such as increased heart rate and blood pressure, among others. To mitigate this set of unwanted effects, a combination of amphetamines and benzodiazepines, diuretics, and thyroid hormones, called the "rainbow diet pills," started to be prescribed. However, due to the risks associated with its use, this combination was soon banned.

---

[11] Henderson J (2005) Ernest Starling and 'Hormones': an historical commentary. J Endocrinol 184:5–10.

[12] Bashir A, Weaver JU (2018) Historical drug therapies in obesity. In: Practical guide to obesity medicine, pp 265–269.

[13] Rodgers RJ et al (2012) Anti-obesity drugs: past, present and future. Dis Model Mech. 5:621–626.

In the 1990s, sympathomimetics drugs that work by blocking the re-uptake of serotonin, such as sibutramine, were developed with the promise of acting as slimming pills without so many side effects.[14] However, this was another unsuccessful attempt. Within a few years, this class of anti-obesity drug was banned in most countries.

Still in the class of anorectic drugs, in the late 1990s, drugs acting as antagonists of the cannabinoid CB1 receptor, such as rimonabant, were developed. The compound was shown to inhibit appetite efficiently, causing weight loss, but psychological side effects, such as depression and anxiety, were highly associated with its use, interrupting its prescription after only 2 years of use in Europe.

Currently, there are five anti-obesity drugs approved for the long-term use: (1) orlistat, a lipase inhibitor; (2) lorcaserin, a serotoninergic drug; (3) phentermine–topiramate, an association of an amphetamine derivative and an anti-seizure; (4) naltrexone–bupropion, an association of an opioid antagonist and a noradrenaline and dopamine reuptake inhibitor; and (5) liraglutide, an analogue of GLP-1.[15] Among them, orlistat is the one that has been approved longest, showing the mildest side effects. This drug, unlike sympathomimetics, does not act on the central nervous system but on the gastrointestinal system. It inhibits the pancreatic lipase, reducing the absorption of diet lipids by 30%. Interestingly, this strategy resembles the first known intervention attempting weight loss in the history of medicine: the prescription of laxatives and purgatives by Greek physicians in the second century.

Despite orlistat has been approved since 1998, its prolonged use has been recently associated with liver damage, showing that the challenge of developing an efficient anti-obesity drug still persists. A recent attempt to minimize the unwanted effects was to combine two or more drugs in smaller doses, but the chances to create a new version of "rainbow diet pills" are slim.[16]

Currently, the most promising agents are single-peptide molecules that combine different mechanisms of action and are secreted by adipose tissue, such as leptin and adiponectin; by the pancreas, as insulin and amylin; and by the intestine, as ghrelin, CCK, PYY, and GLP-1. An example is exenatide and liraglutide, analogues of glucagon-like peptide-1 (GLP-1). GLP-1 is secreted by the intestine and increases insulin secretion. Recent clinical studies show that liraglutide, which is also used to treat type 2 diabetes, is capable of promoting weight loss, motivating its use to treat obesity. The following figure summarizes the main anti-obesity drugs used in the last decades, highlighting their site of action.

---

[14] Kang JG, Park CY (2012) Anti-obesity drugs: a review about their effects and safety. Diabetes Metab J 36:13–25.

[15] Prescription medications to treat overweight and obesity. National Institute of Diabetes and Digestive and Kidney Diseases, NIH, US Department of Health and Human Services; https://www.niddk.nih.gov/health-information/weight-management/prescription-medications-treat-overweight-obesity.

[16] Cohen PA et al (2012) The return of rainbow diet pills. Am J Publ Health 102(9):1676–1686.

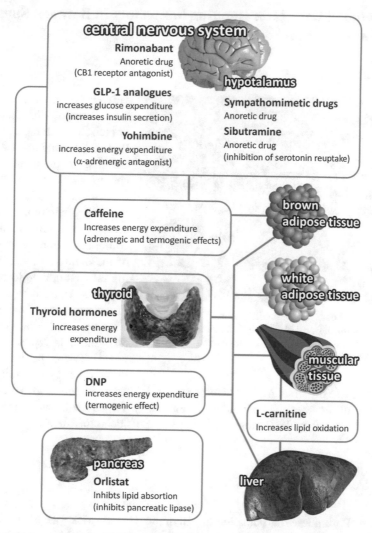

Main anti-obesity drugs with their sites of action. Sympathomimetics and serotoninergic drugs, such amphetamines and sibutramine/lorcaserin, respectively, act directly on CNS, inhibiting appetite. Other drugs also act on the CNS, such as rimonabant, a CB1 receptor antagonist that inhibits appetite; GLP-1 analogs, which increase insulin secretion and glucose utilization; and α-adrenergic antagonists, as yohimbine, which increases energy expenditure. Caffeine and thyroid hormones act both on the CNS and other organs, increasing energy expenditure. Caffeine also increases energy expenditure by increasing UCP1 expression in BAT. DNP strongly increases energy expenditure acting as an uncoupler of the oxidative phosphorylation in several organs, especially skeletal muscles. L-carnitine acts mainly on the muscle and liver, increasing fatty acid oxidation. Orlistat inhibits pancreatic lipase, reducing the absorption of fat from the intestine

*This case had the collaboration of Dr. Lorena O. Fernandes-Siqueira (Instituto de Bioquímica Médica Leopoldo de Meis, Universidade Federal do Rio de Janeiro, Brazil).*

# Challenging Case 11.2: Running to Eternity in Between Suet and Sweat

## Sources

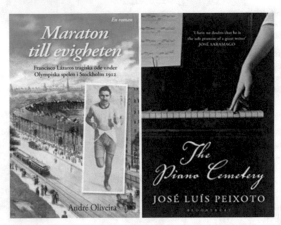

"Either I win or I die" had been supposedly said by Francisco Lázaro, the Portugal's standard-bearer, in the first participation of the country at the Olympic Games, in 1912. Indeed, Lázaro's prediction came true on the most dramatic side: he collapsed at the 30-km mark of the marathon race, becoming the first athlete to die during a modern Olympic event.

Covers of the books "Maraton till evigheten," reproduced with permission of Bertil Ekerlid; and *The Piano Cemetery*, reproduced with permission of the author and Bloomsbury Publishing Plc

Francisco Lázaro's dream of winning the marathon in Stockholm is reported in the novel "Corro para a eternidade" ("I run for eternity"), by the Portuguese diplomat André Oliveira, also translated to Swedish, with the title of "Maraton till evigheten" ("Marathon to eternity"). Francisco Lázaro's story also inspired the novel by the Portuguese writer José Luís Peixoto, *The Piano Cemetery*.

## Case Description

The first Portuguese participation in Olympic Games took place in 1912, in Stockholm, Sweden. The Portuguese delegation was composed by six athletes: the runners Armando Cortesão, António Stromp, and Francisco Lázaro, the wrestlers Joaquim Vital and António Pereira, and the fencer Fernando Correia.

Francisco Lázaro, who actually worked as a carpenter in an automobile factory in Lisbon, was the Portuguese bet for an Olympic medal. Without a coach, he used to run daily after leaving his job (see next picture).

Francisco Lázaro running back home after a working day. (Image from the Wikimedia Commons)

The fact that Francisco Lázaro had won three national marathons and had a victory in the 1912 Lisbon Marathon with a better time than that obtained by the marathon winner of the previous Olympics (1908 London Olympic Games) made every one confident for a good result in Stockholm. However, these hopeful expectations turned into a tragic end, as described, in Swedish and Portuguese, in a plate in his honor at the Gate of the Marathon of the Stockholm Olympic Stadium[17]: *"The marathon at the 1912' Olympic Games was held on 14 July 1912 under a scorching summer heat. From the Stockholm Olympic Stadium, 69 runners left (…) Thousands of people lined up along the 40.2 km course. Only 34 runners would complete the race. The young Portuguese runner Francisco Lázaro was in 18th position at the Silverdal supply point, at 30 km. Shortly ahead, near Mellanjärva, Lázaro fell down under a relentless sun and got unconscious on the path. (…) Stockholm 1912 was the first Olympic Game in which Portugal participated as a nation. Francisco Lázaro was the first fatal victim of the modern Olympic Games"* (translated by the authors).

Francisco Lázaro's cause of death is controversial. His biography in the website Sports Reference summarizes the most popular hypothesis[18]: *"He thought he could improve his performance by not sweating, and greased up his body for the race. This had an adverse effect, as his body could not cool down. Just after the half-way point, Lázaro fell for the first time, and stopped walking straight. After 30 kilometers, he eventually collapsed, and did not get up. A doctor was quickly at the scene, and the now unconscious marathon runner was quickly carried to the Seraphim hospital.*

[17] Silva RP, Lázaro F O fim trágico do primeiro português olímpico. É desporto. https://edesporto.com/francisco-lazaro-o-fim-tragico-do-157487. Accessed 3 Feb 2020.

[18] https://www.sports-reference.com/olympics/athletes/la/francisco-lazaro-1.html.

*He was taken in with a body temperature of 41 degrees Celsius. Intensive care all through the night could not revive him, and he died around 6 AM, causing a shock to the other athletes and the general public."*

The autopsy attributed the cause of death to an irreversible hydro-electrolyte imbalance, which agrees with a heat stroke and a strong dehydration after having greased up his body with suet before the race. This would have restricted the athlete's natural transpiration, leading to a fatal body fluid electrolytic imbalance. However, according to Gustavo Pires, professor at the Faculty of Human Motricity of the University of Lisbon, Lázaro was a victim of a performance-enhancing procedure practiced at that time: a mixture of turpentine essence (anesthetic) and acetic acid.[19] Francisco Lázaro's "addiction" to strychnine, a substance supposed to be used by athletes at that time to avoid fatigue, has also been explored in the media.[20] On the other hand, reports from that time state that Armando Cortesão, Lázaro's colleague in the Olympic Games, was emphatic in pointing out the cause of death: suet and heat. Doubt rules until today, bouncing between suet and sweat.

## Questions

1. Despite the controversy surrounding the cause of Francisco Lázaro's death, many sources mention that he collapsed with a body temperature of 41 °C. The increase in body temperature is a consequence of an increase in the metabolic rate, which may occur through physical activity or through adaptive thermogenesis (see Sect. 11.2). Compare the mechanisms involved in heat production due to exercise practice (voluntary physical activity) to those occurring through cold- or diet-induced thermogenesis (adaptive thermogenesis).

2. Although it is reasonable to associate the increase in heat production during physical activity to an enhanced muscle metabolic rate, a discovery from 2012 broadened the understanding of the exercise-induced thermogenesis. A study published in the scientific periodical Nature identified a new protein released from muscle cells in response to exercise.[21] This protein was named irisin and was shown to act on white adipose cells stimulating UCP1 expression and the browning of the subcutaneous adipose tissue. Discuss how this finding can be correlated to the discovery of leptin, in 1994 (see Sect. 11.1.2).

---

[19] Pereira D, Lázaro F O maratonista que morreu na estreia de Portugal nos Jogos Olímpicos. Diário de notícias. https://www.dn.pt/desportos/francisco-lazaro-o-maratonista-que-morreu-nos-primeiros-jogos-olimpicos-para-portugal-11080474.html. Accessed 15 Jul 2019.

[20] Palma T, Lázaro F A morte ao sol do carpinteiro que se fez mito na maratona. Observador. https://observador.pt/especiais/francisco-lazaro-a-morte-ao-sol-do-carpinteiro-que-se-fez-mito-na-maratona/. Accessed 6 Aug 2016.

[21] Boström P et al (2012) A PGC1α-dependent myokine that drives browning of white fat and thermogenesis. Nature 481:463–468.

3. The most popular hypothesis for Francisco Lázaro's death assumes that by greasing his body with suet before the race, Lázaro impaired natural sweating, ultimately leading to fatal body overheating. Sweat is mainly composed of water, with only a small amount of solutes (less than 1%). Considering the thermodynamic properties of water, which underlie its action as a temperature modulator, explain how sweating contributes to the maintenance of body temperature.
4. Sweating is primarily a means of thermoregulation in humans. Exposure to hyperthermic conditions, such as the increased metabolic rate during exercise or exposure to hot environment (both factors experienced by Francisco Lázaro), increases internal and skin temperatures inducing sweating, an essential mechanism for heat dissipation, thus ensuring survival.[22] Search in the literature and prepare a concise description of the general mechanisms through which sweating is regulated.

## Biochemical Insights

1. Human survival depends on the maintenance of body temperature in a very narrow range, typically between 36.5 and 37.5 °C. Body temperatures above 40 °C characterize a life-threatening condition that requires medical care.

    Intense physical activity increases ATP consumption in muscle cells up to 100-fold due to myosin ATPase, $Na^+/K^+$-ATPase sarcoplasmic reticulum and $Ca^{2+}$-ATPase activities (see Sect. 10.3). To restore ATP levels in the muscle cells, metabolic rate is greatly enhanced, especially through an increase in oxidative phosphorylation. Since all energy conversion processes are not completely efficient (there is always some loss in the form of heat), an increase in heat production is a direct consequence of the increased metabolic rate during exercise.

    On the other hand, the production of heat may be induced in a regulated manner by external triggers such as exposure to cold or food intake, in a process known as adaptive thermogenesis (see Sect. 11.2.1). Adaptive thermogenesis is controlled by the brain, mainly through noradrenaline-mediated responses that ultimately result in the expression of UCP1 and deiodinase 2 genes.
2. The discovery of leptin and the subsequent identification of a number of proteins secreted by the adipose tissue that act as hormones or cytokines (the adipokines, see Box 11.1) revealed that the adipose tissue is also an endocrine organ. In a similar way, the discovery of irisin changed the primary view of the skeletal muscle as just a component of the locomotor system, adding to the muscles, as occurred in the case of the adipose tissue, an important role as an endocrine organ. Indeed, besides irisin, several cytokines and other peptides secreted by muscles, referred to as myokines, have been identified.[23] Myokines act not only

[22] Shibasaki M, Crandall CG (2010) Mechanisms and controllers of eccrine sweating in humans. Front Biosci 2:685–696.

[23] Arhire LI et al (2019) Irisin: a hope in understanding and managing obesity and metabolic syndrome. Front Endocrinol 10:524.

on the muscle itself, controlling myogenesis, hypertrophy, and metabolism, but also regulate different activities of other organs such as the adipose tissue, bones, pancreas, liver, and brain.

Irisin was named after the Greek goddess Iris, the messenger of good news from Gods to humans, an appropriate reference as it acts as a chemical messenger transmitting the effects of exercise to other organs. Thus, irisin, like leptin, acts as a hormone.

Irisin is produced by the cleavage of the membrane-bound precursor protein fibronectin type III domain-containing protein 5 (FNDC5) in response to exercise. Its discovery has been commented in the scientific periodical Science with the slogan "light my fire"[24] as circulating irisin activates the subcutaneous adipocyte thermogenic program leading to mitochondrial heat production with the consequent increase in energy expenditure. This, together with other irisin effects on the organism, recapitulates the benefits of exercise, opening new avenues for the development of therapeutic approaches to manage obesity and metabolic syndrome.

3. Sweating or perspiration is the secretion of fluids (sweat) by the eccrine glands in the skin. In adults sweat rate may reach up to 2–4 L per hour. Evaporation of sweat from the skin surface has an efficient cooling effect due to the high heat of vaporization of water (see some examples for comparison in the next table).

| Substance | Heat of vaporization (kJ/g) |
|---|---|
| Water | 2257 |
| Ethanol | 841 |
| Acetone | 540 |
| Chloroform | 247 |

The heat of vaporization, or the enthalpy of vaporization, is the amount of energy necessary to transform a liquid substance to a gas. This means that water absorbs a lot of energy to evaporate. Thus, overheating due to the increase in the metabolic rate during exercise is prevented by sweat production and its consequent evaporation from the surface of skin. The high heat of vaporization of water is a consequence of its high heat capacity. The specific heat capacity is the amount of energy in the form of heat necessary to increase in one unit the temperature of one unit of mass of a given substance. Both the high heat capacity and heat of vaporization of water result from the extensive hydrogen bonding between water molecules (as discussed in the introduction to Chap. 2), which creates cohesion between the molecules (see next figure). This extensive tri-dimensional network keeps water liquid in the interval from 0 to 100 °C, at atmospheric pressure; evaporation requires that energy is spent to break the extensive mesh of interconnected water molecules.

---

[24] Kelly DP (2012) Irisin, light my fire. Science 336:42–43.

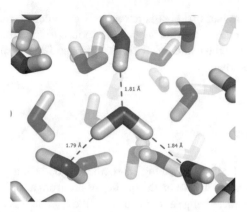

Hydrogen bonds between O (red) and H (white) atoms of neighboring water molecules. (Figure authored by Thomas Splettstoesser reproduced from Wikimedia Commons)

In liquid water, hydrogen bonds are very dynamic being formed and broken frequently. In ice, the hydrogen bonds are more static and ordered; the distance and orientation of hydrogen bonds becomes less random. This is the reason behind ice (solid water) being less dense than its liquid counterpart: the precise orientation and precise distance of hydrogen bonds between water molecules in a static arrangement determines that the average distance between molecules increase, thus increasing the volume occupied by the same mass of water. This is the reason why ice floats in liquid water. The differences in the structures of liquid and solid water are cleverly demonstrated in computer graphics simulations available at http://biomodel.uah.es/en/water/index.htm.

4. In humans, sweating is mainly controlled by the integration of internal and skin temperatures, although non-thermal factors may also regulate the sweating response. The primary thermoregulatory center is located within the hypothalamus, being connected to the eccrine sweat glands through sympathetic nerves consisting mainly of cholinergic terminals. Acetylcholine is the main neurotransmitter causing sweating in a response mediated by its binding to muscarinic receptors. Sweat secretion has been shown to be mediated by aquaporin 5, a membrane water channel protein (see Sect. 5.3.1), which is abundant in the plasma membranes of sweat secretory cells and in the excretory duct of sweat glands.[25] Exercise per se is one of the non-thermal factors that modulate human sweat rate. It has been demonstrated that sweating occurred immediately after an onset of exercise, prior to changes in the internal temperature.[26]

Other physiological responses to counteract hyperthermia include peripheral vasodilation, which consists in increasing the diameter of arterioles in the skin. The increased blood circulation near the body surface favors heat release. This is

---

[25] Nejsum LN et al (2002) Functional requirement of aquaporin-5 in plasma membranes of sweat glands. Proc Nat Acad Sci USA 99:511–516.

[26] van Beaumont W, Bullard RW (1963) Sweating: its rapid responses to muscular work. Science 141:643–646.

the reason why redness (rubor) is associated to hyperthermia. The increase in respiratory frequency is another adaptive response to hyperthermia, albeit not as important. Higher intake and exhalation of air also favor heat release to the surrounding environment.

## Final Discussion

Despite the essential role of sweating in regulating body temperature, abnormally increased sweating might impact on the quality of life from a social and psychological point of view, characterizing a clinical condition known as hyperhidrosis.[27, 28] Hyperhidrosis is not necessarily related to heat or exercise and may be caused by an overactivity of the sympathetic nervous system that disturbs of the endocrine system, some types of cancers, infections, and medications. The main regions affected are the axillae, palms, soles, or craniofacial region, but it may also be focal or generalized. Treatments for hyperhidrosis depend on the affected region and severity, including topical, injected, or oral medications as well as surgical procedures when other medical therapies fail. The main therapeutic strategies are reviewed in and summarized below.

Aluminum chloride solutions are the most effective and frequently used topical antiperspirants, especially for primary focal hyperhidrosis. The mechanism of action is basically physical, as precipitates of aluminum salts and sweat mucopolysaccharides block the eccrine sweat gland duct lumen until skin renewal.

Iontophoresis is the primary treatment for palmar and plantar hyperhidrosis. It consists in applying a direct electrical current in hands or feet submerged in tap water using specific devices. The mechanism of action is not completely understood. Some hypotheses considered are the formation of a hyperkeratotic plug that blocks sweat secretion; the interference with the sweat secretion electrochemical gradient; or the blockage of sympathetic nerve transmission.

Botulinum toxin injection is the first- or second-line treatment for axillary, palmar, plantar, or face hyperhidrosis. There are several commercially available preparations approved by the US Food and Drug Administration, being onabotulinumtoxinA (Botox®) the most commonly used. Botulinum toxin mechanism of action consists in the inhibition of acetylcholine release from the cholinergic neurons that innervate the eccrine sweat glands.

Oral anticholinergics are generally prescribed for generalized or multifocal primary hyperhidrosis, especially when other treatments fail. They act by inhibiting eccrine sweat gland secretion by competing with acetylcholine in stimulating postsynaptic muscarinic receptors.

Finally, local surgery and endoscopic thoracic sympathectomy would be considered when conservative treatments have failed.

---

[27] McConaghy JR, Fosselman D (2018) Hyperhidrosis: management options. Am Fam Physician 97:729–734.

[28] Nawrocki S, Cha J (2019) The etiology, diagnosis, and management of hyperhidrosis: a comprehensive review. J Am Acad Dermatol 81:669–680.

# Challenging Case 11.3: Malthus Meets Metabolism

## *Sources*

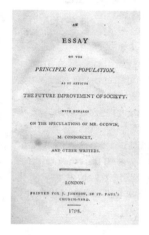

Concerns about human overpopulation are not new. Thomas Robert Malthus, an English economist and demographer, wrote in 1798 a very influential book titled *An Essay on the Principle of Population*,[29] which to this date is still very relevant and very polemic too. Malthus' text inspired many pieces of different types. Besides Malthus' book, this case is also based on the contributions by comic producers, two novelists, and a film director who offered their views of the future and proposed different solutions for the very real problem of human overpopulation.

Cover of the books *An Essay on the Principle of Population*, by Thomas Malthus, 1798 edition (image source: Leeds University Library)

## *Case Description*

The essence of the *An Essay on the Principle of Population* is the incompatibility of the growth rate of population with the availability of resources. Malthus argued that whereas population growth is exponential, there would be a moment in time in which the arithmetical linear increase in food production would not be able to sustain it. According to Malthus this situation would inevitably lead to a crisis in which food restriction would end up naturally checking population growth, but not without much suffering especially among the less privileged members of the society. Malthus was convinced that even education would not be enough to introduce self-imposed family planning because he knew that in nature whenever economic conditions improved population growth ensued. Notwithstanding, Malthus's warning about the impending disaster was accompanied by recommendations one of which was noteworthy: he suggested that social policies such as the Poor Laws should not be implemented. In his opinion, help under the guise of stipends, for example, would only encourage the growth of families, especially among the poor. Malthus's legacy is enduring.

Recently the producers of the Avengers (Marvel comics) in their 2018 film "The Infinity War," introduced the arch villain Thanos, who declares himself a Malthusian. Thanos is obviously well-read. He must also have been aware of the British author Anthony Burgess who contributed with ideas to describe and help solve the problem of overpopulation. His prescient novel, *The Wanting Seed*, written in 1962, sets the scene in a crowded London with a clearly divided society.[30] The government dealt

---

[29] Malthus TR (1798) An essay on the principle of population. J. Johnson, London.

[30] Burgess A (1962) The wanting seed. WW Norton, New York.

with the problem of limited resources by strongly promoting the recycling of materials and by encouraging homosexuality and self-sterilization and naturally, condemning birth. Euthanasia was also approved. In the book, the social mores imposed by the ruling class were reinforced by the Population Police. Later in the novel, one finds out that recycling of materials also included cannibalism, which after an insurrection becomes liberally practiced in Britain as an accepted means of survival.

Overpopulation is also the theme of Harry Harrison's novel *Make Room! Make Room!*, written in 1966.[31] Although Harrison concentrated his description on the miseries of living in a severely overpopulated and polluted New York, suffering also from the consequences of the greenhouse effect, his book inspired the making of the much more sensationalistic post-apocalyptic film "Soylent Green," directed by Richard Fleischer. "Soylent Green" actually refers to a green wafer which becomes the staple diet of the needy. This food was supposedly made of green algae and high-energy plankton. However, as the plot develops, one finds out that this consumable is made of nothing else than human flesh and that the elderly people are the providers of the raw materials. The film ends with the main character (Charlton Heston) being taken away by the police and famously shouting to the unbelieving bystanders: "Soylent Green is people!".

## Questions

1. Taking into account the knowledge you have acquired reading this book, comment whether cannibalism would be an answer to the problem of feeding the hungry in the ever-growing global human population.
2. Plan and explain an experiment to test your answer or hypothesis.
3. Human populations grow heterogeneously around the globe. Low income regions have the highest growth rates, which has puzzled social scientists for decades.[32] One would be tempted to shallowly conclude that low income countries put higher pressure on available food resources. Considering the data in Fig. 11.1, elaborate on this matter.

---

[31] Harrison H (1966) Make Room! Make Room! Doubleday, Garden City, NY.

[32] Castro J Geopolítica da Fome. Rio de Janeiro: Casa do Estudante do Brasil, 1951; Translated into English and published in the US with the title "The Geography of Hunger", in 1952. Opposite to a Malthusian vision Josué de Castro did not believe that hunger was a consequence of overpopulation but is rather a man-made phenomenon. Even more, he undertakes to quite to the contrary, he demonstrates that hunger is the cause of overpopulation. The work was disruptive and the interpretation of underdevelopment was revised.

## *Biochemical Insights*

1. Some fundamental concepts are at play here. Firstly, one would have to consider population parameters: if the global population is growing (which it is) that means that the birth rate exceeds that of the death rate and clearly the cannibalistic solution would not serve its purpose. There must be a continuous input of energy in terms of nourishment in order that people survive.

   Would this strategy work if the birth rate equalled that of the death rate? Again no. In that case, one could argue that in order to reach a grand old age, the individuals must feed throughout their lives and therefore, they too would have consumed the energy obtained from the environment. The ingested elderly would only yield the energy stored in them at the moment of the meals and not the sum total of the energy consumed during their lives. There would always be a deficit, as we could have guessed from the second law of thermodynamics. Also, you must have seen that all metabolic pathways only function when substrates for the enzymes that take part in them are available. For example, in order for the energy to be obtained from glycolysis in the form of ATP, it is necessary that glucose is consumed. In turn, glucose is obtained from the food ingested by the individual. By stopping the supply of glucose, one would eventually slow down glycolysis and as a consequence generate an energy imbalance. This is in fact the basis of dieting. Cut down on caloric intake and one would soon start losing mass. So, it is hopeless to eat your neighbour!

2. For several industrial purposes, microorganisms are grown in special tanks known as fermenters (or fermentors). To these, all the ingredients are added (the broth or growth medium) that will permit the growth (replication) of the microorganisms that will then be processed in order to extract and purify some by-product of interest. Given the right conditions, the microorganisms will grow following an exponential curve until all the substrates are consumed. When the growth medium is depleted, the microorganisms stop growing, a stage which is known as the stationary phase. Eventually, if no more broth is added to the fermenter the live microorganisms will die and get degraded (death phase). This illustrates the fact that growth requires the constant input of energy in the form of the added broth or growth medium. Without that, the colony will eventually disappear. In this example, the fermenter is a closed system.

   In nature, the open system follows the food chain established between several organisms. For example, the terrestrial food chain goes from:

$$\text{plant} \rightarrow \text{herbivores} \rightarrow \text{carnivore}$$

   Without the energy input provided by the Sun, plants would not survive and therefore, the whole chain would collapse.

3. Pressure on the availability of resources it is not only a matter of how many people are in a population but also the average consumption of food by the individuals. High income countries tend to have higher consumption of food, which

results in lifestyle-associated unhealthy conditions such as obesity. It is worth mentioning that modern discussions on the exhaustion of resources in the planet is not confined to food only. The discussion on the pressure put on the environment to generate energy for human activities (transportation, cooling/heating, manufacturing, industrial processing, etc) in the twenty-first century has replaced the concerns of the twentieth century for feeding an ever growing population.

## Final Discussion

The all-encompassing laws of thermodynamics that form a special chapter in physics are particularly present in biochemistry and students should be aware of them constantly. As far as energy is concerned, extensive and meticulous experiments carried out throughout the centuries have shown convincingly that living cells are not an exception in nature: in order to produce work, be it mechanical or chemical, cells do require the constant supply of energy.

*This case is a contribution of Prof. Franklin D. Rumjanek (Instituto de Bioquímica Médica Leopoldo de Meis, Universidade Federal do Rio de Janeiro, Brazil). In memoriam (1945–2020).*

# Credits

Illustrator (except caricatures and PDB molecular graphics): Júlio Xerfan, MD
Molecular Graphics Designer: Fabiana Carneiro, PhD
Caricaturist: Bruno Matos Vieira, PhD
Text reviewer: Franklin David Rumjanek, PhD

# Index

© Springer Nature Switzerland AG 2021
A. T. Da Poian, M. A. R. B., *Integrative Human Biochemistry*,
https://doi.org/10.1007/978-3-030-48740-9

Printed in the United States
by Baker & Taylor Publisher Services